Birkhäuser

Operator Theory: Advances and Applications

Volume 272

Founded in 1979 by Israel Gohberg

More information about this series at http://www.springer.com/series/4850

Vladimir Bolotnikov • Sanne ter Horst • André C.M. Ran
Victor Vinnikov
Editors

Interpolation and Realization Theory with Applications to Control Theory

In Honor of Joe Ball

Editors
Vladimir Bolotnikov
Department of Mathematics
The College of William and Mary
Williamsburg, VA, USA

Sanne ter Horst
Department of Mathematics
Unit for BMI, North-West University
Potchefstroom, South Africa

André C.M. Ran
Department of Mathematics
Faculty of Science, VU Amsterdam
Amsterdam, The Netherlands

Victor Vinnikov
Department of Mathematics
Ben Gurion University of the Negev
Be'er-Sheva, Israel

ISSN 0255-0156 ISSN 2296-4878 (electronic)
Operator Theory: Advances and Applications
ISBN 978-3-030-11613-2 ISBN 978-3-030-11614-9 (eBook)
https://doi.org/10.1007/978-3-030-11614-9

Mathematics Subject Classification (2010): 47-XX

This book is published under the imprint Birkhäuser, www.birkhauser-science.com by the registered company Springer Nature Switzerland AG.
The registered company address is: Gewerbestrasse 11, 6330 Cham, Switzerland

Contents

Preface

This volume in the *Operator Theory: Advances and Application* series is devoted to Joseph A. (Joe) Ball's four and a half decade long mathematical career and in celebration of his seventieth birthday on June 4, 2017. His exemplary dedication to mathematics was also recognized in a special session and dinner dedicated to Joe's Birthday at the 34th Southeastern Analysis Meeting (SEAM 2018), at Georgia Tech, organized by Quanlei Fang, and his election to the class of 2019 fellows of the American Mathematical Society.

Joe started out working on model theory and related topics for non-contractions and operators on multiply-connected domains. After he got into contact with Bill Helton from UCSD in the 1970s, more applied operator theory themes appeared in his work, involving factorization and interpolation for operator-valued functions, with extensive applications in system and control theory. This culminated in his 1990 monograph *Interpolation of rational matrix functions* (OT 45) written jointly with Israel Gohberg and Leiba Rodman. He has since worked on nonlinear control, time-varying systems and more recently on multidimensional systems and noncommutative H^∞-theory on the unit ball and polydisk, and more general domains, and these are only the main themes in his vast oeuvre. It came as a shock to many when Joe announced he was going to retire in 2016, but he quickly reassured us that it is only retirement from regular teaching. And, indeed, his productivity has not suffered under this, to the contrary. In total Joe published more than 200 research papers, over 40 proceedings papers and supervised fourteen Ph.D. students. Further details can be found in Joe's curriculum vitae, publication list and list of Ph.D. students included elsewhere in this volume. There is also a chapter that includes personal reminiscences by some of his collaborators, colleagues and friends.

However, the main bulk of this volume is reserved for fourteen research papers on topics in operator theory and it's application, by mathematicians, many of whom collaborated with Joe or were in some other way influenced by his work. We thank all the authors who contributed to this volume for their efforts as well as the referees who in many cases helped to significantly improve the manuscripts.

We dedicate this volume to Joe, with gratitude for the many things we learned from him, and we wish him many fruitful years to go.

The editors — Vladimir Bolotnikov, Sanne ter Horst,
André Ran, Victor Vinnikov

Joseph A. Ball

Operator Theory:
Advances and Applications, Vol. 272, ix–xii

Curriculum Vitae of Joseph A. Ball

Personal data

Name: Joseph (Joe) A. Ball
Date of birth: June 4, 1947
Place of birth: Washington, D.C., USA

Research field and research interests

Operator theory, engineering systems theory, robust control theory, complex analysis.

Education

1969 B.S. in Mathematics, Georgetown University, Washington, D.C.
1970 M.S. in Mathematics, University of Virginia, Charlottesville, VA.
1973 Ph.D. in Mathematics, University of Virginia, Charlottesville, VA.
Dissertation: "Unitary Perturbations of Contraction Operators"
Advisor: Marvin Rosenblum

Academic appointments

- Assistant Professor, Department of Mathematics, Virginia Tech, Blacksburg, VA, September 1973–August 1978.
- Associate Professor, Department of Mathematics, Virginia Tech, Blacksburg, VA, September 1978–August 1982.
- Professor, Department of Mathematics, Virginia Tech, Blacksburg, VA, September 1982–June 2016.
- Professor Emeritus, Department of Mathematics, Virginia Tech, Blacksburg, VA, September 2016–present.

Visiting professorships

- Mathematician, Dahlgren U.S. Navy Weapons Research Lab, Summer 1975.
- Visiting Assistant Professor, Department of Mathematics, University of California at San Diego, La Jolla, CA, January–June 1978 and September 1979–June 1980.
- Visiting Professor, Weizmann Institute of Science, Rehovot, Israel, January–June 1983.

- Visiting Professor, University of California at San Diego, La Jolla, CA, March–June 1987 and March–June 1991.
- Full Member of the Mathematical Sciences Research Institute "Holomorphic Spaces" program, Berkeley, California, September–December 1995.

Research outputs

- Over 140 papers in refereed professional journals, 60 refereed book chapters and 40 conference proceedings papers.
- two research monographs and two AMS memoirs.
- Co-editor of eight conference proceedings and special book volumes.
- Fourteen Ph.D. student dissertations directed.
- Seven M.S. students supervised.

Lectures at professional meetings

Over 165 lectures at national and international workshops and conferences, including more than

- 25 at American Mathematical Society Meetings;
- 25 at International Workshops on Operator Theory and it's Applications (IWOTA);
- 19 at International Symposia on Mathematical Theory of Network and Systems (MTNS);
- 12 at South East Analysis Meetings (SEAM);
- and 10 at conferences of the Society for Industrial and Applied Mathematics (SIAM).

Plenary and semi-plenary lectures:

- IWOTA 2005, Storrs, CT, USA. Title: Multidimensional system theory, Lax–Phillips scattering and multivariable operator theory: the polydisk setting.
- IWOTA 2006, Seoul, South Korea. Title: Transfer function realization and Nevanlinna–Pick interpolation for general classes of nonselfadjoint operator algebras.
- IWOTA 2007, Potchefstroom, South Africa. Title: Multivariable generalizations of the Schur class, completely positive kernels and multidimensionla linear systems.
- IWOTA 2009, Guanajuato, Mexico. Title Discrete-time overdetermined (Livšic) linear systems, algebraic curves, and meromorphic bundle maps: examples and applications.
- IWOTA 2010 Berlin, Germany. Title: Canonical model theory for Hilbert space row contractions.
- MTNS 2010, Budapest, Hungary. Title: Robust Control, Multidimensional Systems and Multivariable Function Theory: Commutative and Noncommutative Settings.
- IWOTA 2011, Seville, Spain. Title: The intertwining of function theory and systems engineering.

- IWOTA 2012, Sydney Australia. Title: Input/state/output linear systems and function theory: the Bergman space setting.
- IWOTA 2013, Bangalore, India. Title: Transfer function realization and zero/pole structure for multivariable rational matrix functions: the direct analysis.
- IWOTA 2015, Tbilisi, Georgia. Title: Multivariable Nevanlinna–Pick interpolation: the free noncommutative setting.

Editorial work

Member of the Editorial Board of

- Integral Equations and Operator Theory, 1984–present.
- Systems & Control Letters, 1987–1992.
- Journal of the Mathematics of Systems, Estimation, and Control, 1990–1996.
- Journal of Mathematical Analysis and Applications, 1994–present.
- Proceedings of the American Mathematical Society, 1999–2007.
- Complex Analysis and Operator Theory, 2006–present.
- The Birkhäuser book series Operator Theory: Advances and Applications, 2009–present.
- Banach Journal of Mathematical Analysis, 2010–present.
- Multidimensional Signals & Systems, 2011–present.

Refereeing and reviewing

- Refereed research papers for over 70 journals.
- Reviewer for *Zentralblatt für Mathematik und ihre Grenzgebiete* (over 110 reviews) and *Mathematical Reviews* (over 350 reviews).
- Served on numerous graduate student advisor committees.
- Served on several undergraduate honor student oral exam committees.
- Reviewer of grant proposals and prospective research monographs.

Conference organization

- Member of Steering Committee for the International Symposia on Mathematical Theory of Networks and Systems (MTNS), 2002–present.
- Member of Steering Committee for the International Workshop on Operator Theory and Applications (IWOTA), 2002–present.
- Member of IWOTA Presidium, 2009–present.
- Member of local organizing committee for IWOTA 2002, Virginia Tech, Blacksburg, VA, USA, August 6–9, 2002.
- Member of scientific committee for IWOTA 2003, Cagliari, Sardinia, Italy, June 24–27, 2003.
- Member of local organizing committee for IWOTA 2008, College of William & Mary, Williamsburg, VA, USA, July 22–26, 2008.
- Member of local organizing committee for MTNS 2008, Virginia Tech, Blacksburg, VA, USA, July 28–August 1, 2008.
- Member of local organizing committee for Southeast Analysis Meeting (SEAM) 2013, Virginia Tech, Blacksburg, VA, USA, March 15–16, 2013.

- Co-organizer of special sessions at numerous AMS, IWOTA and MTNS conferences.

Research grants and awards

- Supported participant at the NSF Operator Theory Institute at the University of New Hampshire, 1976.
- US National Science Foundation research grant, 1977–1987 (with R.F. Olin and J.E. Thomson), 1978–1998, 2000–2003.
- Co-principal investigator (with P. Kachroo of Center for Transportation Research, Virginia Tech) for Federal Highway Administration Grant DTFG61-93-X-00017-002, 1996–1997.
- Alumni Award for Research Excellence, Virginia Tech, 1997.
- US-Israeli Binational Science Foundation grant, 2000–2007 (with D. Alpay, C. Sadosky and V. Vinnikov), 2011–2014 (with D. Kaliuzhnyi-Verbovetskyi and Victor Vinnikov).
- Co-Principal investigator (with M. Klaus, L. Rodman and J.W. Helton) for NSF Grant #DMS-0126746 "Thirteenth International Workshop on Operator Theory and Applications", 2002–2003.
- Co-Principal investigator (with M. Klaus) for NSF Grant DMS-1266053 "Conference/Workshop: Southeastern Analysis Meeting SEAM 2013", 2012–2013.
- Fellow of the American Mathematical Society, class of 2019.

Memberships of professional organizations

- American Mathematical Society.
- Society of Industrial and Applied Mathematicians.

Operator Theory:
Advances and Applications, Vol. 272, xiii–xxx

Publication List of J.A. Ball

Dissertation

[1] J.A. Ball, *Unitary perturbations of contractions*, ProQuest LLC, Ann Arbor, MI, 1973. Thesis (Ph.D.) – University of Virginia.

Research monographs

[1] J.W. Helton with the assistance of J.A. Ball, C.R. Johnson and J.N. Palmer, *Operator theory, analytic functions, matrices, and electrical engineering*, CBMS Regional Conference Series in Mathematics, vol. 68, Published for the Conference Board of the Mathematical Sciences, Washington, DC; by the American Mathematical Society, Providence, RI, 1987.

[2] J.A. Ball, I. Gohberg, and L. Rodman, *Interpolation of rational matrix functions*, Operator Theory: Advances and Applications, vol. 45, Birkhäuser Verlag, Basel, 1990.

Papers in professional journals

[1] J.A. Ball, *Factorization and invariant subspaces for noncontractions*, Bull. Amer. Math. Soc. **80** (1974), 896–900.

[2] J.A. Ball, *Hardy space expectation operators and reducing subspaces*, Proc. Amer. Math. Soc. **47** (1975), 351–357.

[3] J.A. Ball, *Models for noncontractions*, J. Math. Anal. Appl. **52** (1975), no. 2, 235–254.

[4] J.A. Ball, *A norm estimate for an integral operator*, Transport Theory Statist. Phys. **4** (1975), no. 2, 67–69.

[5] J.A. Ball and W. Greenberg, *A Pontrjagin space analysis of the supercritical transport equation*, Transport Theory Statist. Phys. **4** (1975), no. 4, 143–154.

[6] J.A. Ball and A. Lubin, *On a class of contractive perturbations of restricted shifts*, Pacific J. Math. **63** (1976), no. 2, 309–323.

[7] M.B. Abrahamse and J.A. Ball, *Analytic Toeplitz operators with automorphic symbol. II*, Proc. Amer. Math. Soc. **59** (1976), no. 2, 323–328.

[8] J.A. Ball, *Rota's theorem for general functional Hilbert spaces*, Proc. Amer. Math. Soc. **64** (1977), no. 1, 55–61.

[9] J.A. Ball, R.F. Olin, and J.E. Thomson, *Weakly closed algebras of subnormal operators*, Illinois J. Math. **22** (1978), no. 2, 315–326.

[10] J.A. Ball, *Factorization and model theory for contraction operators with unitary part*, Mem. Amer. Math. Soc. **13** (1978), no. 198, iv+68.

[11] J.A. Ball, *Operators of class C_{00} over multiply-connected domains*, Michigan Math. J. **25** (1978), no. 2, 183–196.

[12] J.A. Ball, *A lifting theorem for operator models of finite rank on multiply-connected domains*, J. Operator Theory **1** (1979), no. 1, 3–25.

[13] J.A. Ball, *Operator extremal problems, expectation operators and applications to operators on multiply connected domains*, J. Operator Theory **1** (1979), no. 2, 153–175.

[14] J.A. Ball and J.W. Helton, *Nonnormal dilations, disconjugacy and constrained spectral factorization*, Integral Equations Operator Theory **3** (1980), no. 2, 216–309.

[15] J.A. Ball, *Interpolation problems and Toeplitz operators on multiply connected domains*, Integral Equations Operator Theory **4** (1981), no. 2, 172–184.

[16] J.A. Ball and J.W. Helton, *Subinvariants for analytic mappings on matrix balls*, Analysis **1** (1981), no. 3, 217–226.

[17] J.W. Helton and J.A. Ball, *The cascade decompositions of a given system vs. the linear fractional decompositions of its transfer function*, Integral Equations Operator Theory **5** (1982), no. 3, 341–385.

[18] J.A. Ball and J.W. Helton, *Lie groups over the field of rational functions, signed spectral factorization, signed interpolation, and amplifier design*, J. Operator Theory **8** (1982), no. 1, 19–64.

[19] J.A. Ball and J.W. Helton, *Factorization results related to shifts in an indefinite metric*, Integral Equations Operator Theory **5** (1982), no. 5, 632–658.

[20] J.A. Ball and J.W. Helton, *A Beurling–Lax theorem for the Lie group* $\mathrm{U}(m,n)$ *which contains most classical interpolation theory*, J. Operator Theory **9** (1983), no. 1, 107–142.

[21] J.A. Ball, *Interpolation problems of Pick–Nevanlinna and Loewner types for meromorphic matrix functions*, Integral Equations Operator Theory **6** (1983), no. 6, 804–840.

[22] J.A. Ball and J.W. Helton, *Beurling–Lax representations using classical Lie groups with many applications. II.* $\mathrm{GL}(n,\mathbf{C})$ *and Wiener–Hopf factorization*, Integral Equations Operator Theory **7** (1984), no. 3, 291–309.

[23] J.A. Ball and I. Gohberg, *A commutant lifting theorem for triangular matrices with diverse applications*, Integral Equations Operator Theory **8** (1985), no. 2, 205–267.

[24] J.A. Ball and J.W. Helton, *Beurling–Lax representations using classical Lie groups with many applications. III. Groups preserving two bilinear forms*, Amer. J. Math. **108** (1986), no. 1, 95–174 (1986).

[25] J.A. Ball and I. Gohberg, *Shift invariant subspaces, factorization, and interpolation for matrices. I. The canonical case*, Linear Algebra Appl. **74** (1986), 87–150.

[26] J.A. Ball and J.W. Helton, *Interpolation problems of Pick–Nevanlinna and Loewner types for meromorphic matrix functions: parametrization of the set of all solutions*, Integral Equations Operator Theory **9** (1986), no. 2, 155–203.

[27] J.A. Ball and J.W. Helton, *Beurling–Lax representations using classical Lie groups with many applications. IV.* $\mathrm{GL}(n, \mathbf{R})$, $\mathrm{U}^*(2n)$, $\mathrm{SL}(n, \mathbf{C})$, *and a solvable group*, J. Funct. Anal. **69** (1986), no. 2, 178–206.

[28] J.A. Ball and I. Gohberg, *Pairs of shift invariant subspaces of matrices and noncanonical factorization*, Linear and Multilinear Algebra **20** (1986), no. 1, 27–61.

[29] J.A. Ball and T.L. Kriete III, *Operator-valued Nevanlinna–Pick kernels and the functional models for contraction operators*, Integral Equations Operator Theory **10** (1987), no. 1, 17–61.

[30] J.A. Ball and A.C.M. Ran, *Global inverse spectral problems for rational matrix functions*, Linear Algebra Appl. **86** (1987), 237–282.

[31] J.A. Ball and A.C.M. Ran, *Optimal Hankel norm model reductions and Wiener–Hopf factorization. I. The canonical case*, SIAM J. Control Optim. **25** (1987), no. 2, 362–382.

[32] J.A. Ball, I. Gohberg, and L. Rodman, *Minimal factorization of meromorphic matrix functions in terms of local data*, Integral Equations Operator Theory **10** (1987), no. 3, 309–348.

[33] J.A. Ball and A.C.M. Ran, *Local inverse spectral problems for rational matrix functions*, Integral Equations Operator Theory **10** (1987), no. 3, 349–415.

[34] J.A. Ball and A.C.M. Ran, *Optimal Hankel norm model reductions and Wiener–Hopf factorization. II. The noncanonical case*, Integral Equations Operator Theory **10** (1987), no. 3, 416–436.

[35] J.A. Ball, C. Foiaş, J.W. Helton, and A. Tannenbaum, *On a local nonlinear commutant lifting theorem*, Indiana Univ. Math. J. **36** (1987), no. 3, 693–709.

[36] J.A. Ball and A.C.M. Ran, *Hankel norm approximation of real symmetric rational matrix functions*, Systems Control Lett. **9** (1987), no. 2, 105–115.

[37] J.A. Ball and N. Cohen, *Sensitivity minimization in an* H^∞ *norm: parametrization of all suboptimal solutions*, Internat. J. Control **46** (1987), no. 3, 785–816.

[38] J.A. Ball, J.W. Helton, and C.H. Sung, *Nonlinear solutions of Nevanlinna–Pick interpolation problems*, Michigan Math. J. **34** (1987), no. 3, 375–389.

[39] J.A. Ball and J.W. Helton, *Shift invariant manifolds and nonlinear analytic function theory*, Integral Equations Operator Theory **11** (1988), no. 5, 615–725.

[40] J.A. Ball, C. Foiaş, J.W. Helton, and A. Tannenbaum, *A Poincaré–Dulac approach to a nonlinear Beurling–Lax–Halmos theorem*, J. Math. Anal. Appl. **139** (1989), no. 2, 496–514.

[41] D.W. Luse and J.A. Ball, *Frequency-scale decomposition of* H^∞*-disk problems*, SIAM J. Control Optim. **27** (1989), no. 4, 814–835.

[42] J.A. Ball and J.W. Helton, *Interconnection of nonlinear causal systems*, IEEE Trans. Automat. Control **34** (1989), no. 11, 1132–1140.

[43] J.A. Ball and K.F. Clancey, *An elementary description of partial indices of rational matrix functions*, Integral Equations Operator Theory **13** (1990), no. 3, 316–322.

[44] J.A. Ball, I. Gohberg, L. Rodman, and T. Shalom, *On the eigenvalues of matrices with given upper triangular part*, Integral Equations Operator Theory **13** (1990), no. 4, 488–497.

[45] J.A. Ball, I. Gohberg, and L. Rodman, *Simultaneous residue interpolation problems for rational matrix functions*, Integral Equations Operator Theory **13** (1990), no. 5, 611–637.

[46] J.A. Ball and M. Rakowski, *Minimal McMillan degree rational matrix functions with prescribed local zero-pole structure*, Linear Algebra Appl. **137/138** (1990), 325–349.

[47] A.C. Antoulas, J.A. Ball, J. Kang, and J.C. Willems, *On the solution of the minimal rational interpolation problem*, Linear Algebra Appl. **137/138** (1990), 511–573.

[48] J.A. Ball, I. Gohberg, and L. Rodman, *Common minimal multiples and divisors for rational matrix functions*, Linear Algebra Appl. **137/138** (1990), 621–662.

[49] J.A. Ball and J. Kang, *Matrix polynomial solutions of tangential Lagrange–Sylvester interpolation conditions of low McMillan degree*, Linear Algebra Appl. **137/138** (1990), 699–746.

[50] J.A. Ball and B. Taub, *Factoring spectral matrices in linear-quadratic models*, Econom. Lett. **35** (1991), no. 1, 39–44.

[51] J.A. Ball, N. Cohen, and L. Rodman, *Zero data and interpolation problems for rectangular matrix polynomials*, Linear and Multilinear Algebra **29** (1991), no. 1, 53–78.

[52] J.A. Ball, J.W. Helton, and M. Verma, *A factorization principle for stabilization of linear control systems*, International Journal of Robust and Nonlinear Control **1** (1991), 229–294.

[53] J.A. Ball and M. Rakowski, *Null-pole subspaces of nonregular rational matrix functions*, Linear Algebra Appl. **159** (1991), 81–120.

[54] J.A. Ball, I. Gohberg, and L. Rodman, *Boundary Nevanlinna–Pick interpolation for rational matrix functions*, J. Math. Systems Estim. Control **1** (1991), no. 2, 131–164.

[55] J.A. Ball and T.R. Fanney, *Uniform limits of sequences of polynomials and their derivatives*, Proc. Amer. Math. Soc. **114** (1992), no. 3, 749–755.

[56] J.A. Ball and J.W. Helton, *Inner-outer factorization of nonlinear operators*, J. Funct. Anal. **104** (1992), no. 2, 363–413.

[57] D. Alpay, J.A. Ball, I. Gohberg, and L. Rodman, *State space theory of automorphisms of rational matrix functions*, Integral Equations Operator Theory **15** (1992), no. 3, 349–377.

[58] J.A. Ball and J.W. Helton, *Nonlinear H^∞ control theory for stable plants*, Math. Control Signals Systems **5** (1992), no. 3, 233–261.

[59] J.A. Ball and E.A. Jonckheere, *The four-block Adamjan–Arov–Kreĭn problem*, J. Math. Anal. Appl. **170** (1992), no. 2, 322–342.

[60] J.A. Ball, J. Kim, L. Rodman, and M. Verma, *Minimal-degree coprime factorizations of rational matrix functions*, Linear Algebra Appl. **186** (1993), 117–164.

[61] J.A. Ball, J.W. Helton, and M.L. Walker, *H^∞ control for nonlinear systems with output feedback*, IEEE Trans. Automat. Control **38** (1993), no. 4, 546–559.

[62] J.A. Ball, I. Gohberg, and L. Rodman, *The structure of flat gain rational matrices that satisfy two-sided interpolation requirements*, Systems Control Lett. **20** (1993), no. 6, 401–412.

[63] J.A. Ball and K.F. Clancey, *Interpolation with meromorphic matrix functions*, Proc. Amer. Math. Soc. **121** (1994), no. 2, 491–496.

[64] J.A. Ball and J. Kim, *Stability and McMillan degree for rational matrix interpolants*, Linear Algebra Appl. **196** (1994), 207–232.

[65] J.A. Ball, M.A. Kaashoek, G. Groenewald, and J. Kim, *Column reduced rational matrix functions with given null-pole data in the complex plane*, Linear Algebra Appl. **203/204** (1994), 67–110.

[66] J.A. Ball, M. Rakowski, and B.F. Wyman, *Coupling operators, Wedderburn–Forney spaces, and generalized inverses*, Linear Algebra Appl. **203/204** (1994), 111–138.

[67] D. Alpay, J.A. Ball, I. Gohberg, and L. Rodman, *J-unitary preserving automorphisms of rational matrix functions: state space theory, interpolation, and factorization*, Linear Algebra Appl. **197/198** (1994), 531–566. Second Conference of the International Linear Algebra Society (ILAS) (Lisbon, 1992).

[68] D. Alpay, J.A. Ball, I. Gohberg, and L. Rodman, *The two-sided residue interpolation in the Stieltjes class for matrix functions*, Linear Algebra Appl. **208/209** (1994), 485–521.

[69] J.A. Ball, I. Gohberg, and M.A. Kaashoek, *Bitangential interpolation for input-output maps of time-varying systems: the continuous time case*, Integral Equations Operator Theory **20** (1994), no. 1, 1–43.

[70] J.A. Ball and M. Verma, *Factorization and feedback stabilization for nonlinear systems*, Systems Control Lett. **23** (1994), no. 3, 187–196.

[71] J.A. Ball and M. Rakowski, *Interpolation by rational matrix functions and stability of feedback systems: the 2-block case*, J. Math. Systems Estim. Control **4** (1994), no. 3, 261–318.

[72] J.A. Ball, I. Gohberg, and M. Rakowski, *Reconstruction of a rational nonsquare matrix function from local data*, Integral Equations Operator Theory **20** (1994), no. 3, 249–305.

[73] J.A. Ball, I. Gohberg, and M.A. Kaashoek, *Two-sided Nudel′man interpolation for input-output operators of discrete time-varying systems*, Integral Equations Operator Theory **21** (1995), no. 2, 174–211.

[74] J.A. Ball and T.R. Fanney, *Pure sub-Jordan operators and simultaneous approximation by a polynomial and its derivative*, J. Operator Theory **33** (1995), no. 1, 43–78.

[75] J.A. Ball, I. Gohberg, and M.A. Kaashoek, *A frequency response function for linear, time-varying systems*, Math. Control Signals Systems **8** (1995), no. 4, 334–351.

[76] J.A. Ball and A.J. van der Schaft, *J-inner-outer factorization, J-spectral factorization, and robust control for nonlinear systems*, IEEE Trans. Automat. Control **41** (1996), no. 3, 379–392.

[77] J.A. Ball and K.F. Clancey, *Reproducing kernels for Hardy spaces on multiply connected domains*, Integral Equations Operator Theory **25** (1996), no. 1, 35–57.

[78] J.A. Ball and J. Kim, *Bitangential interpolation problems for symmetric rational matrix functions*, Linear Algebra Appl. **241/243** (1996), 35–57.

[79] J.A. Ball, *Commutant lifting and interpolation: the time-varying case*, Integral Equations Operator Theory **25** (1996), no. 4, 377–405.

[80] J.A. Ball and V. Vinnikov, *Zero-pole interpolation for meromorphic matrix functions on an algebraic curve and transfer functions of 2D systems*, Acta Appl. Math. **45** (1996), no. 3, 239–316.

[81] J.A. Ball and J.W. Helton, *Viscosity solutions of Hamilton–Jacobi equations arising in nonlinear H_∞-control*, J. Math. Systems Estim. Control **6** (1996), no. 1, 22.

[82] J.A. Ball, I. Gohberg, and M.A. Kaashoek, *Nudelman interpolation and the band method*, Integral Equations Operator Theory **27** (1997), no. 3, 253–284.

[83] J.A. Ball and T.T. Trent, *Unitary colligations, reproducing kernel Hilbert spaces, and Nevanlinna–Pick interpolation in several variables*, J. Funct. Anal. **157** (1998), no. 1, 1–61.

[84] J.A. Ball, Yu.I. Karlovich, L. Rodman, and I.M. Spitkovsky, *Sarason interpolation and Toeplitz corona theorem for almost periodic matrix functions*, Integral Equations Operator Theory **32** (1998), no. 3, 243–281.

[85] J.A. Ball, P. Kachroo, and A.J. Krener, *H_∞ tracking control for a class of nonlinear systems*, IEEE Trans. Automat. Control **44** (1999), no. 6, 1202–1206.

[86] J.A. Ball and V. Vinnikov, *Zero-pole interpolation for matrix meromorphic functions on a compact Riemann surface and a matrix Fay trisecant identity*, Amer. J. Math. **121** (1999), no. 4, 841–888.

[87] J.A. Ball, W.S. Li, D. Timotin, and T.T. Trent, *A commutant lifting theorem on the polydisc*, Indiana Univ. Math. J. **48** (1999), no. 2, 653–675.

[88] J.A. Ball, M.V. Day, and P. Kachroo, *Robust feedback control of a single server queueing system*, Math. Control Signals Systems **12** (1999), no. 4, 307–345.

[89] J.A. Ball, M.V. Day, T. Yu, and P. Kachroo, *Robust L_2-gain control for nonlinear systems with projection dynamics and input constraints: an example from traffic control*, Automatica J. IFAC **35** (1999), no. 3, 429–444.

[90] J.A. Ball and J. Chudoung, *Comparison theorems for viscosity solutions of a system of quasivariational inequalities with application to optimal control with switching costs*, J. Math. Anal. Appl. **251** (2000), no. 1, 40–64.

[91] D. Alpay, J.A. Ball, and V. Bolotnikov, *On the bitangential interpolation problem for contractive valued functions in the polydisk*, J. Operator Theory **44** (2000), no. 2, 277–301.

[92] D. Alpay, J.A. Ball, and Y. Peretz, *System theory, operator models and scattering: the time-varying case*, J. Operator Theory **47** (2002), no. 2, 245–286.

[93] J.A. Ball and V. Bolotnikov, *On a bitangential interpolation problem for contractive-valued functions on the unit ball*, Linear Algebra Appl. **353** (2002), 107–147.

[94] J.A. Ball, J. Chudoung, and M.V. Day, *Robust optimal stopping-time control for nonlinear systems*, Appl. Math. Optim. **46** (2002), no. 1, 1–29.

[95] J.A. Ball and V. Bolotnikov, *A tangential interpolation problem on the distinguished boundary of the polydisk for the Schur–Agler class*, J. Math. Anal. Appl. **273** (2002), no. 2, 328–348.

[96] J.A. Ball, J. Chudoung, and M.V. Day, *Robust optimal switching control for nonlinear systems*, SIAM J. Control Optim. **41** (2002), no. 3, 900–931.

[97] J.A. Ball and V. Bolotnikov, *A bitangential interpolation problem on the closed unit ball for multipliers of the Arveson space*, Integral Equations Operator Theory **46** (2003), no. 2, 125–164.

[98] J.A. Ball and M.A. Petersen, *Non-linear minimal square spectral factorization*, Internat. J. Control **76** (2003), no. 12, 1233–1247.

[99] J.A. Ball and T. Malakorn, *Multidimensional linear feedback control systems and interpolation problems for multivariable holomorphic functions*, Multidimens. Systems Signal Process. **15** (2004), no. 1, 7–36.

[100] J.A. Ball, M.A. Petersen, and A. van der Schaft, *Inner-outer factorization for nonlinear noninvertible systems*, IEEE Trans. Automat. Control **49** (2004), no. 4, 483–492.

[101] J.A. Ball and V. Bolotnikov, *Realization and interpolation for Schur–Agler-class functions on domains with matrix polynomial defining function in* $\mathbb{C}^n$, J. Funct. Anal. **213** (2004), no. 1, 45–87.

[102] J.A. Ball, C. Sadosky, and V. Vinnikov, *Conservative linear systems, unitary colligations and Lax–Phillips scattering: multidimensional generalizations*, Internat. J. Control **77** (2004), no. 9, 802–811.

[103] J.A. Ball, C. Sadosky, and V. Vinnikov, *Conservative input-state-output systems with evolution on a multidimensional integer lattice*, Multidimens. Syst. Signal Process. **16** (2005), no. 2, 133–198.

[104] J.A. Ball and V. Bolotnikov, *Nevanlinna–Pick interpolation for Schur–Agler class functions on domains with matrix polynomial defining function in* $\mathbb{C}^n$, New York J. Math. **11** (2005), 247–290.

[105] J.A. Ball and V. Vinnikov, *Lax–Phillips scattering and conservative linear systems: a Cuntz-algebra multidimensional setting*, Mem. Amer. Math. Soc. **178** (2005), no. 837, iv+101.

[106] J.A. Ball, K.M. Mikkola, and A.J. Sasane, *State-space formulas for the Nehari–Takagi problem for nonexponentially stable infinite-dimensional systems*, SIAM J. Control Optim. **44** (2005), no. 2, 531–563.

[107] J.A. Ball, G. Groenewald, and T. Malakorn, *Structured noncommutative multidimensional linear systems*, SIAM J. Control Optim. **44** (2005), no. 4, 1474–1528.

[108] J.A. Ball, C. Sadosky, and V. Vinnikov, *Scattering systems with several evolutions and multidimensional input/state/output systems*, Integral Equations Operator Theory **52** (2005), no. 3, 323–393.

[109] J.A. Ball and A.J. Sasane, *Equivalence of a behavioral distance and the gap metric*, Systems Control Lett. **55** (2006), no. 3, 214–222.

[110] J.A. Ball and O.J. Staffans, *Conservative state-space realizations of dissipative system behaviors*, Integral Equations Operator Theory **54** (2006), no. 2, 151–213.

[111] J.A. Ball, G. Groenewald, and T. Malakorn, *Bounded real lemma for structured noncommutative multidimensional linear systems and robust control*, Multidimens. Syst. Signal Process. **17** (2006), no. 2-3, 119–150.

[112] J.A. Ball and M.W. Raney, *Discrete-time dichotomous well-posed linear systems and generalized Schur–Nevanlinna–Pick interpolation*, Complex Anal. Oper. Theory **1** (2007), no. 1, 1–54.

[113] J.A. Ball, V. Bolotnikov, and Q. Fang, *Transfer-function realization for multipliers of the Arveson space*, J. Math. Anal. Appl. **333** (2007), no. 1, 68–92.

[114] J.A. Ball and V. Bolotnikov, *Interpolation in the noncommutative Schur–Agler class*, J. Operator Theory **58** (2007), no. 1, 83–126.

[115] J.A. Ball, V. Bolotnikov, and Q. Fang, *Multivariable backward-shift-invariant subspaces and observability operators*, Multidimens. Syst. Signal Process. **18** (2007), no. 4, 191–248.

[116] J.A. Ball, P.T. Carroll, and Y. Uetake, *Lax–Phillips scattering theory and well-posed linear systems: a coordinate-free approach*, Math. Control Signals Systems **20** (2008), no. 1, 37–79.

[117] J.A. Ball, V. Bolotnikov, and Q. Fang, *Schur-class multipliers on the Arveson space: de Branges–Rovnyak reproducing kernel spaces and commutative transfer-function realizations*, J. Math. Anal. Appl. **341** (2008), no. 1, 519–539.

[118] J.A. Ball and D.S. Kaliuzhnyi-Verbovetskyi, *Conservative dilations of dissipative multidimensional systems: the commutative and non-commutative settings*, Multidimens. Syst. Signal Process. **19** (2008), no. 1, 79–122.

[119] J.A. Ball and V. Bolotnikov, *Interpolation problems for Schur multipliers on the Drury–Arveson space: from Nevanlinna–Pick to abstract interpolation problem*, Integral Equations Operator Theory **62** (2008), no. 3, 301–349.

[120] J.A. Ball, Q. Fang, G.J. Groenewald, and S. ter Horst, *Equivalence of robust stabilization and robust performance via feedback*, Math. Control Signals Systems **21** (2009), no. 1, 51–68.

[121] J.A. Ball, V. Bolotnikov, and S. ter Horst, *A constrained Nevanlinna–Pick interpolation problem for matrix-valued functions*, Indiana Univ. Math. J. **59** (2010), no. 1, 15–51.

[122] J.A. Ball, V. Bolotnikov, and S. ter Horst, *Interpolation in de Branges–Rovnyak spaces*, Proc. Amer. Math. Soc. **139** (2011), no. 2, 609–618.

[123] J.A. Ball, *Multidimensional circuit synthesis and multivariable dilation theory*, Multidimens. Syst. Signal Process. **22** (2011), no. 1-3, 27–44.

[124] J.A. Ball and A. Kheifets, *The inverse commutant lifting problem. I: Coordinate-free formalism*, Integral Equations Operator Theory **70** (2011), no. 1, 17–62.

[125] J.A. Ball, V. Bolotnikov, and S. ter Horst, *Abstract interpolation in vector-valued de Branges–Rovnyak spaces*, Integral Equations Operator Theory **70** (2011), no. 2, 227–263.

[126] J.A. Ball, G.M. Boquet, and V. Vinnikov, *A behavioral interpretation of Livšic systems*, Multidimens. Syst. Signal Process. **23** (2012), no. 1-2, 17–48.

[127] J.A. Ball and A.J. Sasane, *Extension of the ν-metric*, Complex Anal. Oper. Theory **6** (2012), no. 1, 65–89.

[128] J.A. Ball and V. Bolotnikov, *Canonical transfer-function realization for Schur multipliers on the Drury–Arveson space and models for commuting row contractions*, Indiana Univ. Math. J. **61** (2012), no. 2, 665–716.

[129] J.A. Ball and M.D. Guerra Huamán, *Test functions, Schur–Agler classes and transfer-function realizations: the matrix-valued setting*, Complex Anal. Oper. Theory **7** (2013), no. 3, 529–575.

[130] J.A. Ball and V. Bolotnikov, *Weighted Bergman spaces: shift-invariant subspaces and input/state/output linear systems*, Integral Equations Operator Theory **76** (2013), no. 3, 301–356.

[131] J.A. Ball and A. Kheifets, *The inverse commutant lifting problem. II: Hellinger functional-model spaces*, Complex Anal. Oper. Theory **7** (2013), no. 4, 873–907.

[132] J.A. Ball and V. Bolotnikov, *A Beurling type theorem in weighted Bergman spaces*, C. R. Math. Acad. Sci. Paris **351** (2013), no. 11-12, 433–436.

[133] J. Agler, J.A. Ball, and J.E. McCarthy, *The Takagi problem on the disk and bidisk*, Acta Sci. Math. (Szeged) **79** (2013), no. 1-2, 63–78.

[134] J.A. Ball and V. Bolotnikov, *Weighted Hardy spaces: shift invariant and coinvariant subspaces, linear systems and operator model theory*, Acta Sci. Math. (Szeged) **79** (2013), no. 3-4, 623–686.

[135] J.A. Ball and M.D. Guerra Huamán, *Convexity analysis and the matrix-valued Schur class over finitely connected planar domains*, J. Operator Theory **70** (2013), no. 2, 531–571.

[136] J.A. Ball and D.S. Kaliuzhnyi-Verbovetskyi, *Rational Cayley inner Herglotz–Agler functions: positive-kernel decompositions and transfer-function realizations*, Linear Algebra Appl. **456** (2014), 138–156.

[137] J.A. Ball, M. Kurula, O.J. Staffans, and H. Zwart, *De Branges–Rovnyak realizations of operator-valued Schur functions on the complex right half-plane*, Complex Anal. Oper. Theory **9** (2015), no. 4, 723–792.

[138] J.A. Ball, D.S. Kaliuzhnyi-Verbovetskyi, C. Sadosky, and V. Vinnikov, *Scattering systems with several evolutions and formal reproducing kernel Hilbert spaces*, Complex Anal. Oper. Theory **9** (2015), no. 4, 827–931.

[139] J.A. Ball and D.S. Kaliuzhnyi-Verbovetskyi, *Schur–Agler and Herglotz–Agler classes of functions: positive-kernel decompositions and transfer-function realizations*, Adv. Math. **280** (2015), 121–187.

[140] J.A. Ball, G. Groenewald, and S. ter Horst, *Bounded real lemma and structured singular value versus diagonal scaling: the free noncommutative setting*, Multidimens. Syst. Signal Process. **27** (2016), no. 1, 217–254.

[141] J.A. Ball, G. Marx, and V. Vinnikov, *Noncommutative reproducing kernel Hilbert spaces*, J. Funct. Anal. **271** (2016), no. 7, 1844–1920.

[142] J.A. Ball, K.F. Clancey, and V. Vinnikov, *Meromorphic matrix trivializations of factors of automorphy over a Riemann surface*, Oper. Matrices **10** (2016), no. 4, 785–828.

[143] J.A. Ball and V. Bolotnikov, *Contractive multipliers from Hardy space to weighted Hardy space*, Proc. Amer. Math. Soc. **145** (2017), no. 6, 2411–2425.

[144] J.A. Ball, M. Kurula, and O.J. Staffans, *A conservative de Branges–Rovnyak functional model for operator Schur functions on* $\mathbb{C}^+$, Complex Anal. Oper. Theory **12** (2018), no. 4, 877–915.

[145] J.A. Ball, G.J. Groenewald, and S. ter Horst, *Standard versus strict bounded real lemma with infinite-dimensional state space. I. The state-space-similarity approach*, J. Operator Theory **80** (2018), no. 1, 225–253.

Papers published as book chapters

[1] J.A. Ball, *A non-Euclidean Lax–Beurling theorem with applications to matricial Nevanlinna–Pick interpolation*, Toeplitz centennial (Tel Aviv, 1981), Operator Theory: Adv. Appl., vol. 4, Birkhäuser, Basel-Boston, Mass., 1982, pp. 67–84.

[2] J.A. Ball, *Invariant subspace representations, unitary interpolants and factorization indices*, Topics in operator theory systems and networks (Rehovot, 1983), Oper. Theory Adv. Appl., vol. 12, Birkhäuser, Basel, 1984, pp. 11–38.

[3] J.A. Ball and I. Gohberg, *Classification of shift invariant subspaces of matrices with Hermitian form and completion of matrices*, Operator theory and systems (Amsterdam, 1985), Oper. Theory Adv. Appl., vol. 19, Birkhäuser, Basel, 1986, pp. 23–85.

[4] J.A. Ball and A.C.M. Ran, *Left versus right canonical Wiener–Hopf factorization*, Constructive methods of Wiener–Hopf factorization, Oper. Theory Adv. Appl., vol. 21, Birkhäuser, Basel, 1986, pp. 9–38.

[5] J.A. Ball, C. Foiaş, J.W. Helton, and A. Tannenbaum, *Nonlinear interpolation theory in H^∞*, Modelling, robustness and sensitivity reduction in control systems (Groningen, 1986), NATO Adv. Sci. Inst. Ser. F Comput. Systems Sci., vol. 34, Springer, Berlin, 1987, pp. 31–46.

[6] J.A. Ball and D.W. Luse, *Sensitivity minimization as a Nevanlinna–Pick interpolation problem*, Modelling, robustness and sensitivity reduction in control systems (Groningen, 1986), NATO Adv. Sci. Inst. Ser. F Comput. Systems Sci., vol. 34, Springer, Berlin, 1987, pp. 451–462.

[7] J.A. Ball, *Nevanlinna–Pick interpolation: generalizations and applications*, Surveys of some recent results in operator theory, Vol. I, Pitman Res. Notes Math. Ser., vol. 171, Longman Sci. Tech., Harlow, 1988, pp. 51–94.

[8] J.A. Ball, I. Gohberg, and L. Rodman, *Realization and interpolation of rational matrix functions*, Topics in interpolation theory of rational matrix-valued functions, Oper. Theory Adv. Appl., vol. 33, Birkhäuser, Basel, 1988, pp. 1–72.

[9] J.A. Ball, N. Cohen, and A.C.M. Ran, *Inverse spectral problems for regular improper rational matrix functions*, Topics in interpolation theory of rational matrix-valued functions, Oper. Theory Adv. Appl., vol. 33, Birkhäuser, Basel, 1988, pp. 123–173.

[10] J.A. Ball and J.W. Helton, *Shift invariant subspaces, passivity, reproducing kernels and H^∞-optimization*, Contributions to operator theory and its applications (Mesa, AZ, 1987), Oper. Theory Adv. Appl., vol. 35, Birkhäuser, Basel, 1988, pp. 265–310.

[11] J.A. Ball and J.W. Helton, *Factorization and general properties of nonlinear Toeplitz operators*, The Gohberg anniversary collection, Vol. II (Calgary, AB, 1988), Oper. Theory Adv. Appl., vol. 41, Birkhäuser, Basel, 1989, pp. 25–41.

[12] J.A. Ball, I. Gohberg, and L. Rodman, *Two-sided Nudel′man interpolation problems for rational matrix functions*, Analysis and partial differential equations, Lecture Notes in Pure and Appl. Math., vol. 122, Dekker, New York, 1990, pp. 371–416.

[13] J.A. Ball, I. Gohberg, and L. Rodman, *Sensitivity minimization and bitangential Nevanlinna–Pick interpolation in contour integral form*, Signal Processing, Part II, IMA Vol. Math. Appl., vol. 23, Springer, New York, 1990, pp. 3–35.

[14] J.A. Ball, I. Gohberg, and L. Rodman, *Tangential interpolation problems for rational matrix functions*, Matrix theory and applications (Phoenix, AZ, 1989), Proc. Sympos. Appl. Math., vol. 40, Amer. Math. Soc., Providence, RI, 1990, pp. 59–86.

[15] J.A. Ball, I. Gohberg, and L. Rodman, *Two-sided Lagrange–Sylvester interpolation problems for rational matrix functions*, Operator theory: operator algebras and applications, Part 1 (Durham, NH, 1988), Proc. Sympos. Pure Math., vol. 51, Amer. Math. Soc., Providence, RI, 1990, pp. 17–83.

[16] D. Alpay, J.A. Ball, I. Gohberg, and L. Rodman, *Realization and factorization for rational matrix functions with symmetries*, Extension and interpolation of linear operators and matrix functions, Oper. Theory Adv. Appl., vol. 47, Birkhäuser, Basel, 1990, pp. 1–60.

[17] J.A. Ball and M. Rakowski, *Zero-pole structure of nonregular rational matrix functions*, Extension and interpolation of linear operators and matrix functions, Oper. Theory Adv. Appl., vol. 47, Birkhäuser, Basel, 1990, pp. 137–193.

[18] J.A. Ball and T.R. Fanney, *Closability of differential operators and real sub-Jordan operators*, Topics in operator theory: Ernst D. Hellinger memorial volume, Oper. Theory Adv. Appl., vol. 48, Birkhäuser, Basel, 1990, pp. 93–156.

[19] J.A. Ball and N. Cohen, *de Branges–Rovnyak operator models and systems theory: a survey*, Topics in matrix and operator theory (Rotterdam, 1989), Oper. Theory Adv. Appl., vol. 50, Birkhäuser, Basel, 1991, pp. 93–136.

[20] J.A. Ball, I. Gohberg, and L. Rodman, *The state space method in the study of interpolation by rational matrix functions*, Mathematical system theory, Springer, Berlin, 1991, pp. 503–508.

[21] J.A. Ball, I. Gohberg, and M.A. Kaashoek, *Nevanlinna–Pick interpolation for time-varying input-output maps: the discrete case*, Time-variant systems and interpolation, Oper. Theory Adv. Appl., vol. 56, Birkhäuser, Basel, 1992, pp. 1–51.

[22] J.A. Ball, I. Gohberg, and M.A. Kaashoek, *Nevanlinna–Pick interpolation for time-varying input-output maps: the continuous time case*, Time-variant systems and interpolation, Oper. Theory Adv. Appl., vol. 56, Birkhäuser, Basel, 1992, pp. 52–89.

[23] J.A. Ball, I. Gohberg, and M.A. Kaashoek, *Reduction of the abstract four block problem to a Nehari problem*, Continuous and discrete Fourier transforms, extension problems and Wiener–Hopf equations, Oper. Theory Adv. Appl., vol. 58, Birkhäuser, Basel, 1992, pp. 121–141.

[24] J.A. Ball and M. Rakowski, *Interpolation by rational matrix functions and stability of feedback systems: the 4-block case*, Operator theory and complex analysis (Sapporo, 1991), Oper. Theory Adv. Appl., vol. 59, Birkhäuser, Basel, 1992, pp. 96–142.

[25] J.A. Ball, I. Gohberg, and M.A. Kaashoek, *Bitangential interpolation for input-output operators of time-varying systems: the discrete time case*, New aspects in interpolation and completion theories, Oper. Theory Adv. Appl., vol. 64, Birkhäuser, Basel, 1993, pp. 33–72.

[26] J.A. Ball, I. Gohberg, and L. Rodman, *Two-sided tangential interpolation of real rational matrix functions*, New aspects in interpolation and completion theories, Oper. Theory Adv. Appl., vol. 64, Birkhäuser, Basel, 1993, pp. 73–102.

[27] W.T. Ross and J.A. Ball, *Weak-star limits of polynomials and their derivatives*, Contributions to operator theory and its applications, Oper. Theory Adv. Appl., vol. 62, Birkhäuser, Basel, 1993, pp. 165–175.

[28] J.A. Ball and J. Rosenthal, *Pole placement, internal stabilization and interpolation conditions for rational matrix functions: a Grassmannian formulation*, Linear algebra for control theory, IMA Vol. Math. Appl., vol. 62, Springer, New York, 1994, pp. 21–29.

[29] J.A. Ball, *Conservative dynamical systems and nonlinear Livšic–Brodskiĭ nodes*, Nonselfadjoint operators and related topics (Beer Sheva, 1992), Oper. Theory Adv. Appl., vol. 73, Birkhäuser, Basel, 1994, pp. 67–95.

[30] J.A. Ball, I. Gohberg, and M.A. Kaashoek, *Input-output operators of J-unitary time-varying continuous time systems*, Operator theory in function spaces and Banach lattices, Oper. Theory Adv. Appl., vol. 75, Birkhäuser, Basel, 1995, pp. 57–94.

[31] J.A. Ball, *The nonlinear Nevanlinna–Pick interpolation problem*, Topics in operator theory, operator algebras and applications (Timişoara, 1994), Rom. Acad., Bucharest, 1995, pp. 1–27.

[32] J.A. Ball, I. Gohberg, and M.A. Kaashoek, *The band method and Grassmannian approach for completion and extension problems*, Recent developments in operator theory and its applications (Winnipeg, MB, 1994), Oper. Theory Adv. Appl., vol. 87, Birkhäuser, Basel, 1996, pp. 17–60.

[33] J.A. Ball, *Linear systems, operator model theory and scattering: multivariable generalizations*, Operator theory and its applications (Winnipeg, MB, 1998), Fields Inst. Commun., vol. 25, Amer. Math. Soc., Providence, RI, 2000, pp. 151–178.

[34] J.A. Ball and N.J. Young, *Problems on the realization of functions*, Operator theory and its applications (Winnipeg, MB, 1998), Fields Inst. Commun., vol. 25, Amer. Math. Soc., Providence, RI, 2000, pp. 179–185.

[35] J.A. Ball and T.T. Trent, *The abstract interpolation problem and commutant lifting: a coordinate-free approach*, Operator theory and interpolation (Bloomington, IN, 1996), Oper. Theory Adv. Appl., vol. 115, Birkhäuser, Basel, 2000, pp. 51–83.

[36] J.A. Ball, T.T. Trent, and V. Vinnikov, *Interpolation and commutant lifting for multipliers on reproducing kernel Hilbert spaces*, Operator theory and analysis (Amsterdam, 1997), Oper. Theory Adv. Appl., vol. 122, Birkhäuser, Basel, 2001, pp. 89–138.

[37] J.A. Ball and V. Vinnikov, *Hardy spaces on a finite bordered Riemann surface, multivariable operator model theory and Fourier analysis along a unimodular curve*, Systems, approximation, singular integral operators, and related topics (Bordeaux, 2000), Oper. Theory Adv. Appl., vol. 129, Birkhäuser, Basel, 2001, pp. 37–56.

[38] J.A. Ball, L. Rodman, and I.M. Spitkovsky, *Toeplitz corona problem for algebras of almost periodic functions*, Toeplitz matrices and singular integral equations (Pobershau, 2001), Oper. Theory Adv. Appl., vol. 135, Birkhäuser, Basel, 2002, pp. 25–37.

[39] J.A. Ball, K.F. Clancey, and V. Vinnikov, *Concrete interpolation of meromorphic matrix functions on Riemann surfaces*, Interpolation theory, systems theory and related topics (Tel Aviv/Rehovot, 1999), Oper. Theory Adv. Appl., vol. 134, Birkhäuser, Basel, 2002, pp. 137–156.

[40] J.A. Ball and V. Vinnikov, *Formal reproducing kernel Hilbert spaces: the commutative and noncommutative settings*, Reproducing kernel spaces and applications, Oper. Theory Adv. Appl., vol. 143, Birkhäuser, Basel, 2003, pp. 77–134.

[41] J.A. Ball and V. Vinnikov, *Overdetermined multidimensional systems: state space and frequency domain methods*, Mathematical systems theory in biology, communications, computation, and finance (Notre Dame, IN, 2002), IMA Vol. Math. Appl., vol. 134, Springer, New York, 2003, pp. 63–119.

[42] J.A. Ball and V. Bolotnikov, *Boundary interpolation for contractive-valued functions on circular domains in* $\mathbb{C}^n$, Current trends in operator theory and its applications, Oper. Theory Adv. Appl., vol. 149, Birkhäuser, Basel, 2004, pp. 107–132.

[43] J.A. Ball and V. Vinnikov, *Functional models for representations of the Cuntz algebra*, Operator theory, systems theory and scattering theory: multidimensional generalizations, Oper. Theory Adv. Appl., vol. 157, Birkhäuser, Basel, 2005, pp. 1–60.

[44] J.A. Ball, G. Groenewald, and T. Malakorn, *Conservative structured noncommutative multidimensional linear systems*, The state space method generalizations and applications, Oper. Theory Adv. Appl., vol. 161, Birkhäuser, Basel, 2006, pp. 179–223.

[45] J.A. Ball, V. Bolotnikov, and Q. Fang, *Schur-class multipliers on the Fock space: de Branges–Rovnyak reproducing kernel spaces and transfer-function realizations*, Operator theory, structured matrices, and dilations, Theta Ser. Adv. Math., vol. 7, Theta, Bucharest, 2007, pp. 85–114.

[46] J.A. Ball, A. Biswas, Q. Fang, and S. ter Horst, *Multivariable generalizations of the Schur class: positive kernel characterization and transfer function realization*, Recent advances in operator theory and applications, Oper. Theory Adv. Appl., vol. 187, Birkhäuser, Basel, 2009, pp. 17–79.

[47] J.A. Ball and V. Bolotnikov, *Canonical de Branges–Rovnyak model transfer-function realization for multivariable Schur-class functions*, Hilbert spaces of analytic functions, CRM Proc. Lecture Notes, vol. 51, Amer. Math. Soc., Providence, RI, 2010, pp. 1–39.

[48] J.A. Ball and S. ter Horst, *Multivariable operator-valued Nevanlinna–Pick interpolation: a survey*, Operator algebras, operator theory and applications, Oper. Theory Adv. Appl., vol. 195, Birkhäuser Verlag, Basel, 2010, pp. 1–72.

[49] J.A. Ball and S. ter Horst, *Robust control, multidimensional systems and multivariable Nevanlinna–Pick interpolation*, Topics in operator theory. Volume 2. Systems and mathematical physics, Oper. Theory Adv. Appl., vol. 203, Birkhäuser Verlag, Basel, 2010, pp. 13–88.

[50] J.A. Ball and V. Bolotnikov, *Canonical transfer-function realization for Schur–Agler-class functions of the polydisk*, A panorama of modern operator theory and related topics, Oper. Theory Adv. Appl., vol. 218, Birkhäuser/Springer Basel AG, Basel, 2012, pp. 75–122.

[51] J.A. Ball and V. Bolotnikov, *Canonical transfer-function realization for Schur–Agler-class functions on domains with matrix polynomial defining function in* $\mathbb{C}^n$, Recent progress in operator theory and its applications, Oper. Theory Adv. Appl., vol. 220, Birkhäuser/Springer Basel AG, Basel, 2012, pp. 23–55.

[52] J.A. Ball and A.J. Sasane, *Extension of the ν-metric: the H^∞ case*, Spectral theory, mathematical system theory, evolution equations, differential and difference equations, Oper. Theory Adv. Appl., vol. 221, Birkhäuser/Springer Basel AG, Basel, 2012, pp. 121–130.

[53] J.A. Ball and Q. Fang, *Nevanlinna–Pick interpolation via graph spaces and Kreĭn-space geometry: a survey*, Mathematical methods in systems, optimization, and control, Oper. Theory Adv. Appl., vol. 222, Birkhäuser/Springer Basel AG, Basel, 2012, pp. 43–71.

[54] J.A. Ball and V. Bolotnikov, *Interpolation in sub-Bergman spaces*, Advances in structured operator theory and related areas, Oper. Theory Adv. Appl., vol. 237, Birkhäuser/Springer, Basel, 2013, pp. 17–39.

[55] J.A. Ball and V. Bolotnikov, *De Branges–Rovnyak spaces: Basics and theory*, Operator Theory, Springer on-line reference book, Springer, Basel, 2015, pp. 631–680.

[56] J.A. Ball and V. Bolotnikov, *De Branges–Rovnyak spaces and norm-constrained interpolation*, Operator Theory, Springer on-line reference book, Springer, Basel, 2015, pp. 681–720.

[57] J.A. Ball and V. Bolotnikov, *On the expansive property of inner functions in weighted Hardy spaces*, Complex analysis and dynamical systems VI. Part 2, Contemp. Math., vol. 667, Amer. Math. Soc., Providence, RI, 2016, pp. 47–61.

[58] J.A. Ball and V. Bolotnikov, *The bitangential matrix Nevanlinna–Pick interpolation problem revisited*, Indefinite inner product spaces, Schur analysis, and differential equations, Oper. Theory Adv. Appl., vol. 263, Birkhäuser/Springer, Cham, 2018, pp. 107–161.

[59] J.A. Ball, G. Marx, and V. Vinnikov, *Interpolation and transfer-function realization for the noncommutative Schur–Agler class*, Operator theory in different settings and related applications, Oper. Theory Adv. Appl., vol. 262, Birkhäuser/Springer, Cham, 2018, pp. 23–116.

[60] J.A. Ball, G.J. Groenewald, and S. ter Horst, *Standard versus strict bounded real lemma with infinite-dimensional state space II: The storage function approach*, The diversity and beauty of applied operator theory, Oper. Theory Adv. Appl., vol. 268, Birkhäuser/Springer, Cham, 2018, pp. 1–50.

Conference proceedings papers

[1] J.A. Ball and J.W. Helton, *Interpolation with outer functions and gain equalization in amplifiers*, Mathematical theory of networks and systems (Delft, 1979), Western Periodicals Co., 1979, pp. 41–49.

[2] J.A. Ball, *Amplifier design, signed spectral factorization and signed interpolation*, Mathematical theory of networks and systems (Santa Monica, 1981), Western Periodicals Co., 1981, pp. 1–3.

[3] J.A. Ball and J.W. Helton, *Linear fractional parameterizations of matrix function spaces and a new proof of the Youla–Jabr–Bongiorno parameterization for stabilizing compensators*, Mathematical theory of networks and systems (Beer Sheva, 1983), Lect. Notes Control Inf. Sci., vol. 58, Springer, London, 1984, pp. 16–23.

[4] J.A. Ball and A.C.M. Ran, *Hankel norm approximation of a rational matrix function in terms of its realization*, Modelling, identification and robust control (Stockholm, 1985), North-Holland, Amsterdam, 1986, pp. 285–296.

[5] J.A. Ball, *The bitangential Nevanlinna–Pick interpolation problem: a geometric approach*, Proc. 26th IEEE Conference on Decision and Control (Los Angeles, 1987), IEEE, New York, 1987, pp. 1342–1343.

[6] J.A. Ball and J.W. Helton, *Well-posedness of nonlinear causal feedback systems*, Proc. 26th IEEE Conf. on Decision and Control (Los Angeles, 1987), IEEE, New York, 1987, pp. 152–154.

[7] J.A. Ball and J.W. Helton, *Sensitivity bandwidth optimization for nonlinear feedback systems*, Analysis and Control of Nonlinear Systems, North Holland, Amsterdam, 1988, pp. 123–129.

[8] J.A. Ball, *Interpolation problems for null and pole structure of nonlinear systems*, Proc. 27th IEEE Conf. on Decision and Control (Austin, 1988), IEEE, New York, 1988, pp. 14–19.

[9] J.A. Ball, N. Cohen, and L. Rodman, *On interpolation problems for rectangular matrix polynomials*, Proc. 27th IEEE Conf. on Decision and Control (Austin, 1988), IEEE, New York, 1988, pp. 1370–1372.

[10] J.A. Ball and J.W. Helton, *Factorization of nonlinear systems: toward a theory for nonlinear H^∞-infinity control*, Proc. 27th IEEE Conf. on Decision and Control (Austin, 1988), IEEE, New York, 1988, pp. 2376–2381.

[11] D.W. Luse and J.A. Ball, *Frequency-scale decomposition of H^∞-disk problems*, Linear circuits, systems and signal processing: theory and application (Phoenix, AZ, 1987), North-Holland, Amsterdam, 1988, pp. 573–580.

[12] J.A. Ball and I. Gohberg, *Cascade decompositions of linear systems in terms of realizations*, Proceedings of the 28th IEEE Conference on Decision and Control, Vol. 1–3 (Tampa, FL, 1989), IEEE, New York, 1989, pp. 2–10.

[13] J.A. Ball and J.W. Helton, *H^∞ control for nonlinear plants: connections with differential games*, Proceedings of the 28th IEEE Conference on Decision and Control, Vol. 1–3 (Tampa, FL, 1989), IEEE, New York, 1989, pp. 956–962.

[14] J.A. Ball and J.W. Helton, *Nonlinear H^∞ control theory: a literature survey*, Robust control of linear systems and nonlinear control (Amsterdam, 1989), Progr. Systems Control Theory, vol. 4, Birkhäuser Boston, Boston, MA, 1990, pp. 1–12.

[15] J.A. Ball and L.E. Carpenter, *Realizations of products and Wiener–Hopf factors for a class of matrix functions analytic in a strip*, Signal processing, scattering and operator theory, and numerical methods (Amsterdam, 1989), Progr. Systems Control Theory, vol. 5, Birkhäuser Boston, Boston, MA, 1990, pp. 291–300.

[16] J.A. Ball and M. Rakowski, *Transfer functions with a given local zero-pole structure*, New trends in systems theory (Genoa, 1990), Progr. Systems Control Theory, vol. 7, Birkhäuser Boston, Boston, MA, 1991, pp. 81–88.

[17] J.A. Ball, I. Gohberg, and L. Rodman, *Nehari interpolation problem for rational matrix functions: the generic case*, H_∞-control theory (Como, 1990), Lecture Notes in Math., vol. 1496, Springer, Berlin, 1991, pp. 277–308.

[18] J.A. Ball, I. Gohberg, and L. Rodman, *Interpolation problems for rational matrix functions and systems theory*, Recent advances in mathematical theory of systems, control, networks and signal processing, I (Kobe, 1991), Mita, Tokyo, 1992, pp. 3–12.

[19] J.A. Ball, J.W. Helton, and M. Verma, *A J-inner-outer factorization principle for the H^∞ control problem*, Recent advances in mathematical theory of systems, control, networks and signal processing, I (Kobe, 1991), Mita, Tokyo, 1992, pp. 31–36.

[20] J.A. Ball, I. Gohberg, and M.A. Kaashoek, *Time-varying systems: Nevanlinna–Pick interpolation and sensitivity minimization*, Recent advances in mathematical theory of systems, control, networks and signal processing, I (Kobe, 1991), Mita, Tokyo, 1992, pp. 53–58.

[21] J.A. Ball and M. Rakowski, *An application of valuation theory to the construction of rectangular matrix functions*, Directions in matrix theory (Auburn, 1990), Linear Algebra Appl., Vol. 162/164, 1992, pp. 730–735.

[22] J.A. Ball, J.W. Helton, and M.L. Walker, *Nonlinear H^∞ control and the bounded real lemma*, Proc. 31st Conf. on Decision and Control (Tuscon, 1992), pp. 1045–1049.

[23] J.A. Ball, I. Gohberg, and M.A. Kaashoek, *The time-varying two-sided Nudel′man interpolation problem and its solution*, Challenges of a generalized system theory (Amsterdam, 1992), Konink. Nederl. Akad. Wetensch. Verh. Afd. Natuurk. Eerste Reeks, vol. 40, North-Holland, Amsterdam, 1993, pp. 45–58.

[24] J.A. Ball, I. Gohberg, and M.A. Kaashoek, *H_∞-control and interpolation for time-varying systems*, Systems and networks: mathematical theory and applications, Vol. I (Regensburg, 1993), Math. Res., vol. 77, Akademie-Verlag, Berlin, 1994, pp. 33–48.

[25] J.A. Ball and M. Verma, *A factorization principle for stabilization of nonlinear, time-varying plants*, Systems and networks: mathematical theory and applications, Vol. II (Regensburg, 1993), Math. Res., vol. 79, Akademie-Verlag, Berlin, 1994, pp. 53–56.

[26] A.J. van der Schaft and J.A. Ball, *Inner-outer factorization of nonlinear state space systems*, Systems and networks: mathematical theory and applications, Vol. II (Regensburg, 1993), Math. Res., vol. 79, Akademie-Verlag, Berlin, 1994, pp. 529–532.

[27] A.J. van der Schaft and J.A. Ball, *Nonlinear inner-outer factorization*, Proc. 33rd IEEE Conference on Decision and Control (Orlando, 1994), IEEE, New York, 1994, pp. 2549–2552.

[28] J.A. Ball, *Nevanlinna–Pick interpolation and robust control for time-varying systems*, Smart Structures and Materials 1996: Mathematics and Control in Smart Structures (San Diego, 1996), Proceedings of SPIE – The International Society for Optical Engineering, vol. 2715, International Society for Optics and Photonics, 1996, pp. 76-86.

[29] J.A. Ball, M.V. Day, P. Kachroo, and T. Yu, *An isolated traffic intersection feedback control using H_∞*, Intelligent Transportation System (Boston, 1997), Proc. IEEE Conference on Intelligent Transportation Systems, IEEE, New York, 1997, pp. 942–947.

[30] N. Schlegel, P. Kachroo, J.A. Ball, and J.S. Bay, *Image processing based control for scaled automated vehicles*, Intelligent Transportation System (Boston, 1997), Proc. IEEE Conference on Intelligent Transportation Systems, IEEE, New York, 1997, pp. 1022–1027.

[31] J.A. Ball, M.V. Day, P. Kachroo, and T. Yu, *Robust control for signalized intersections*, Intelligent Transportation Systems (Pittsburgh, 1997), Proceedings of SPIE – The International Society for Optical Engineering, vol. 3207, International Society for Optics and Photonics, 1998, pp. 164–171.

[32] J.A. Ball and V. Vinnikov, *Noncommutative linear system theory, formal power series in noncommuting indeterminates and applications*, Multidimensional (nD) Systems (Wuppertal, 2005), Proc. nDS 2005, IEEE, New York, 2005, pp. 36-42.

[33] J.A. Ball, *Dissipative noncommutative multidimensional linear systems and robust control theory*, Multidimensional (nD) Systems (Aveiro, 2007), Proc. nDS 2007, IEEE, New York, 2007, pp. 123–129.

[34] J.A. Ball, Q. Fang, G.J. Groenewald, and S. ter Horst, *Reduction of robust performance via feedback to robust stabilization*, Proceedings of the 18th International Symposium on Mathematical Theory of Networks and Systems (Blacksburg, 2008).

[35] J.A. Ball and G.M. Boquet, *Livšic realization of 2D-behaviors with degree one autonomy*, Multidimensional (nD) Systems (Thessaloniki, 2009), Proc. nDS 2009, IEEE, New York, 2009.

[36] J.A. Ball and S. ter Horst, *A W^*-correspondence approach to multi-dimensional linear dissipative systems*, Multidimensional (nD) Systems (Thessaloniki, 2009), Proc. nDS 2009, IEEE, New York, 2009.

[37] J.A. Ball and G.M. Boquet, *Controllability of autonomous behaviors and Livšic overdetermined systems as 2D behaviors with pure autonomy degree one*, Proceedings of the 19th International Symposium on Mathematical Theory of Networks and Systems (Budapest, 2010).

[38] J.A. Ball and S. ter Horst, *Robust Control, Multidimensional Systems and Multivariable Function Theory: Commutative and Noncommutative Settings*, Proceedings of the 19th International Symposium on Mathematical Theory of Networks and Systems (Budapest, 2010).

[39] J.A. Ball and T. Malakorn, *Structured noncommutative multidimensional linear systems and scale-recursive modeling*, Proceedings of the 19th International Symposium on Mathematical Theory of Networks and Systems (Budapest, 2010).

[40] J.A. Ball and V. Bolotnikov, *System theory techniques for function theory on Bergman spaces*, Proceedings of the 21st International Symposium on Mathematical Theory of Networks and Systems (Groningen, 2014).

[41] J.A. Ball, G.J. Groenewald, and S. ter Horst, *Structured Singular Values versus Diagonal Scaling: the Noncommutative Setting*, Proceedings of the 21st International Symposium on Mathematical Theory of Networks and Systems (Groningen, 2014).

[42] J.A. Ball, G.J. Groenewald, and S. ter Horst, *Bounded Real Lemma: the infinite dimensional case*, Proceedings of the 22nd International Symposium on Mathematical Theory of Networks and Systems (Minneapolis, 2016).

Edited books

[1] S. Saitoh, D. Alpay, J.A. Ball, and T. Ohsawa (eds.), *Reproducing kernels and their applications*, International Society for Analysis, Applications and Computation, vol. 3, Kluwer Academic Publishers, Dordrecht, 1999.

[2] J.A. Ball, J.W. Helton, M. Klaus, and L. Rodman (eds.), *Current trends in operator theory and its applications*, Operator Theory: Advances and Applications, vol. 149, Birkhäuser Verlag, Basel, 2004.

[3] J.A. Ball, Y. Eidelman, J.W. Helton, V. Olshevsky, and J. Rovnyak (eds.), *Recent advances in matrix and operator theory*, Operator Theory: Advances and Applications, vol. 179, Birkhäuser Verlag, Basel, 2008.

[4] J.A. Ball, V. Bolotnikov, J.W. Helton, L. Rodman, and I.M. Spitkovsky (eds.), *Topics in operator theory. Volume 1. Operators, matrices and analytic functions*, Operator Theory: Advances and Applications, vol. 202, Birkhäuser Verlag, Basel, 2010. A tribute to Israel Gohberg on the occasion of his 80th birthday.

[5] J.A. Ball, V. Bolotnikov, J.W. Helton, L. Rodman, and I.M. Spitkovsky (eds.), *Topics in operator theory. Volume 2. Systems and mathematical physics*, Operator Theory: Advances and Applications, vol. 203, Birkhäuser Verlag, Basel, 2010. A tribute to Israel Gohberg on the occasion of his 80th birthday.

[6] J.A. Ball, R.E. Curto, S.M. Grudsky, J.W. Helton, R. Quiroga-Barranco, and N.L. Vasilevski (eds.), *Recent progress in operator theory and its applications*, Operator Theory: Advances and Applications, vol. 220, Birkhäuser/Springer Basel AG, Basel, 2012.

[7] W. Arendt, J.A. Ball, J. Behrndt, K.-H. Förster, V. Mehrmann, and C. Trunk (eds.), *Spectral theory, mathematical system theory, evolution equations, differential and difference equations*, Operator Theory: Advances and Applications, vol. 221, Birkhäuser/Springer Basel AG, Basel, 2012.

[8] J.A. Ball, M.A. Dritschel, A.F.M. ter Elst, P. Portal, and D. Potapov (eds.), *Operator theory in harmonic and non-commutative analysis*, Operator Theory: Advances and Applications, vol. 240, Birkhäuser/Springer, Cham, 2014.

Operator Theory:
Advances and Applications, Vol. 272, xxxi–xxxi

Ph.D. Students of J.A. Ball

1. Thomas R. Fanney, *Closability of Differential Operators and Subjordan Operators*, May 1989.
2. Marek Rakowski, *Zero-pole Interpolation of Nonregular Rational Matrix Functions*, December 1989.
3. Jeongook Kang [Kim], *Interpolation by Rational Matrix Functions with Minimal McMillan Degree*, December 1990.
4. Lonnie Carpenter, *Cascade Analysis and Synthesis of Transfer Functions of Infinite Dimensional Linear Systems*, May 1992.
5. Tusheng Yu, *On-line Traffic Signalization using Robust Feedback Control*, January 1998.
6. Jerawan Chudoung, *Robust Control for Hybrid Systems with Applications to Network Traffic Problems*, May 2000.
7. Tanit Malakorn, *Multidimensional Linear Systems and Robust Control*, May 2003.
8. Pushkin Kachroo, *Optimal and Feedback Control for Hyperbolic Conservation Laws*, June 2007.
9. Quanlei Fang, *Multivariable Interpolation Problems*, July 2008.
10. Grant M. Boquet, *Geometric Properties of Over-Determined Systems of Linear Partial Difference Equations*, February 2010.
11. Daniel Sutton, *Structure of Invariant Subspaces for Left Invertible Operators on Hilbert Space*, July 2010.
12. Moisés D. Guerra Huamán, *Schur class of finitely connected planar domains: the test-function approach*, April 2011.
13. Austin J. Amaya, *Beurling–Lax Representations of Shift-Invariant Spaces, Zero-Pole Data Interpolation, and Dichotomous Transfer Function Realizations: Half-Plane/Continuous-Time Versions*, April 2012.
14. Gregory Marx, *Noncommutative Kernels*, June 2017.

Operator Theory:
Advances and Applications, Vol. 272, xxxiii–xli

Personal Reminiscences

Quanlei Fang, J. William Helton, Sanne ter Horst,
Alexander Kheifets, André C.M. Ran and James Rovnyak

This chapter contains personal notes from some of Joe's students and colleagues on the occasion of his seventieth birthday.

Quanlei Fang

An ancient Chinese proverb says, "One day's teacher, a whole life's father." It means that even if someone is your teacher for only a day, you should regard him like your father for the rest of your life. The word for thesis advisor in German is "Doktorvater", a father figure who will guide you through the doctorate. The idea of a father for a doctorate was about right for how I felt Professor Ball has treated me. It is also an easy role to accept, as he was about the same age as my parents.

Professor Ball is truly knowledgeable and very much respected. He is always energetic and never stops working. Sometimes he seems to be strict and I am kinda afraid of failing to meet his expectations. But I have realized that he is just trying to help out. When I said something wrong, he would make sure to point it out. I even learned the correct pronunciation of many words from him. When I need help or guidance he responds promptly, even long after I graduated from Virginia Tech. He always encourages me to attend conferences to check out recent developments in the field. I still remember that the first IWOTA I attended was in 2005, when I was about to start my third year in graduate school. The conference was held in Connecticut. Professor Ball and Professor Klaus decided to drive together from Blacksburg to Storrs where the conference was held. I remember how grateful I was since I did not know how to drive at that time and the professors shared the long driving. During my last year of graduate school, my thesis progressed slowly due to my distraction with many other things. Professor Ball was really patient and supportive at that time.

In his free time Professor Ball enjoys singing and playing piano in the church. His wife has been very active too. I went to see their performances a few times and they were very nice. Occasionally Professor Ball and his wife Amelia would invite us (my mathematical siblings and some of his visitors at that time) to have dinner and watch football games at their home. I learned how to make sushi from Amelia. There were lots of good memories.

Figure 1. a) Univ. at Buffalo, 2009 b) New York City, 2015

Professor Ball has set an example of excellence as a researcher, advisor and mentor. I feel lucky to be one of his students and I believe my other mathematical siblings feel the same. On the occasion of his 70th birthday, I just want to say: Thank you! Have a happy, healthy, wonderful birthday, and many more to come!

J. William Helton

I first heard of Joe from Tom Kriete. Tom said they had a remarkable student at UVA who, among other things, took a problem and a week later came back with an 80 page manuscript. Tom was really struck with the Joe Ball phenomenon.

Possibly the biggest branch of Operator Theory at the time was operator model theory, and Joe already was a serious expert on this. It was discovered about this time that operator model theory was very closely related to engineering systems theory, and a few of us had at a conference on this in 1973 called the Operator Theory of Networks and Systems (OTNS). So when a second OTNS was scheduled in 1975, I contacted Joe and suggested he might be interested. Joe came and probably that is where we first met (or at least talked much). There was Joe, Doug Clark and me and about 75 theoretical engineers.

Joe missed the next OTNS (which by now is called the MTNS) in 1977; it was discouraging to lose such a powerful convert to the cause.

It turned out that Joe had not lost interest; he was even shyer, at the time, than I was and later told me that he felt the conference so exalted it would be inappropriate for him to attend.

One of the turning points in my career was getting Joe to visit San Diego for a year around 1980. Thus began a long and rewarding collaboration; every summer he would come to UCSD and also occasionally spend a quarter. This continued for about a decade and then we moved in different directions. Joe was an absolutely wonderful person to work with!

In 1981 with Joe visiting UCSD we set up the first International Workshop on Operator Theory and its Applications (IWOTA). The MTNS 1981 was set for LA and I was an organizer, so it was easy to put IWOTA as a satellite in the same hotel. Joe was a great asset in deciding who we might interest in the conference, because even then Joe had an encyclopedic knowledge of the literature. So though he did not know some of the mathematicians personally he had an opinion of their work. The conference was a big success, so much so that next summer we are expecting to have the 30th IWOTA, with Joe playing a pivotal vice presidential role on the IWOTA steering committee.

As we jump to the present, much of the work in multivariable Operator Theory depends on ideas which Joe introduced. This modern advance was introduced by Joe with Tavan Trent and brought to analytic function problems the incredibly powerful techniques of multivariable systems. Ten years later, with his student Malakorn and with Groenewald, he introduced similar techniques but for problems with operator variables. Since this is the area where I work, I am constantly feeling grateful to Joe for providing the techniques which have opened up this subject.

Joe, of course, is a backbone of our field. In addition to his work he keeps many things we depend on functioning, one of which is getting a lot of good papers refereed and published and another pertains to putting the umph in numerous conferences.

In conclusion, let us all raise a cup of coffee to Joe, in a toast to his continued bright future.

Sanne ter Horst

It must have been at one of my first international conferences, either MTNS-2004 in Louvain, Belgium, or IWOTA-2004 in Newcastle, where I first encountered Joe Ball. This was at the end of my first year as a Ph.D. student. I was working on a topic in commutant lifting theory and metric constrained interpolation, both of which Joe worked on extensively, and I had already read some of his papers. We didn't speak at that occasion, but Joe certainly made a lasting impression. Towards the end of my studies I had two ring binders full of his papers. At subsequent conferences there was more interaction, and I was very glad when Joe agreed to host me for a three months visit in the last year of my studies. One of my advisors, André Ran, who did a postdoctoral fellowship with Joe about twenty years earlier, had already prepared me: "Quite possibly he already signed you up for a colloquium." And indeed, it was during this visit that I gave my first colloquium lecture. The visit was very fruitful, I learned a lot about multidimensional systems as well as C^*-algebras and C^*-correspondences, and the basis for our first paper, jointly with Animikh Biswas and Quanlei Fang, was laid there. The paper was finished in the months following the visit, the last week in a 24 hour writing process, with the USA team taking one half of the day and I the other half from the Netherlands, to meet the IWOTA-2006 proceedings (OT 187) deadline. That was certainly a nice experience, and I was very happy when later that year I was awarded a two-year postdoctoral fellowship at Virginia Tech to continue working with Joe.

A few months later, everything was arranged, and I was back on a plane towards Roanoke, VA, the closest airport near Blacksburg. My new office was on the same floor and only four doors away from Joe's. As an operator theorist, Joe was a bit isolated, but he maintained good contact with some of the applied mathematicians, especially of the model reduction group, and was an active member of the department. He organized a weekly seminar on topics in operator theory and control systems, which was mostly frequented by his many graduate students, at that time Quanlei Fang, Grant Boquet, Daniel Sutton, Moisés Huamán and Austin Amaya, and there were many visitors, amongst others Animikh Biswas, Vladimir Bolotnikov, Gilbert Groenewald and Victor Vinnikov.

Working with Joe can be quite intense. My postdoc years at VT have been among the most productive in my career so far. Whenever I walked over to his office with a question, this would usually lead to a detailed discussion with many anecdotes. Joe has a phenomenally detailed memory, and he would come up with precise coordinates of various related publications, not only journal and year, but also where they were stored in his office, "Oh, I did the Math Review for that paper in 1984, so it must be in this drawer." It is maybe worth pointing out here that Joe wrote more than 350 reviews for AMS Mathematical Reviews. It must have been a few less ten years ago, but anyone who visited Joe can imagine how they were stored in his office, together with many other papers, books and notes and drafts of papers that were still in progress. I hardly ever left his office with less than four papers of reading material, some related to the question I had, some that just came up in the discussion. Lunches were usually enjoyed in downtown Blacksburg, with Joe's Diner as the preferred establishment, and the mathematics discussions did not stop during lunch. Typically, a few days after a colloquium lecture, over lunch Joe would give a complete account of the topic, who had worked on it and when, and also the operator theory involved in it, if any.

During these two years we collaborated on a variety of topics, ranging from robust control problems to the W^*-correspondence approach to transfer function realization and to interpolation problems in various function spaces, and also wrote some lengthy survey papers on multivariable interpolation and multidimensional systems. It is here were I learned to work on different projects and different topics at the same time. Many other academic activities I also did in my postdoc time at VT for the first time, I wrote my first refereeing reports and got some exposure to other editorial work (Joe was one of the editors of PAMS then), co-organized my first special session at a conference and taught my first classes. Joe was always helpful and patiently read through my reviews, invitation e-mails, the exams I had drafted, etc. Things I now do easily, were not as straightforward to me at that time. He is also very patient towards his students, and even small kids, as one can see from the photo below where he is teaching my youngest daughter, Lian, 18 months then, to play the piano.

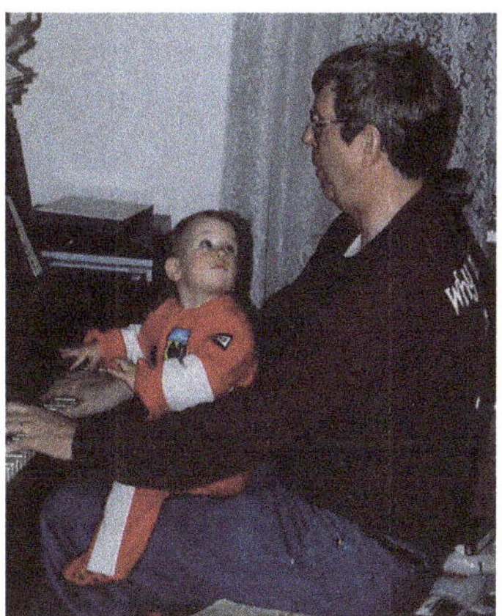

Joe learning my daughter
to play the piano, 2008

Being away from my family for long periods of time wasn't easy, but Joe made very good efforts to make me feel at home. During my two years at VT, there were countless, delicious dinners at Joe's home, all carefully prepared by his wife Amelia, usually followed by an American Football match. I should admit that despite Joe's many attempts I am still lost when it comes to the rules of the game. Although Joe works long hours, often seven days a week, he also has an interesting social life. When I was there, he played the organ in church, together with Amelia he performed with the Blacksburg Master Chorale, and they attended and sometimes participated in operas at the Roanoke opera house. Many of these performances I also attended.

After my postdoc years at VT and a hiatus of a few years, we picked up our collaboration in 2012, when I had settled at North West University in South Africa. Joe visited me several times and I went back to Blacksburg in 2016 for a three weeks visit, during which I had a chance to recall the many fond memories of my time there. Joe, I want to thank you for the many things you taught me and the important role you played in my career, and especially for the great hospitality during my postdoc years at Virginia Tech.

Alexander Kheifets

I planned to meet with Joe during my first visit to the United States over the summer of 1992. I wanted to discuss with Joe the results of my recent PhD thesis, since there was some overlap with his works on interpolation and his "Models for non-contractions" paper, especially in the part of the adaptation of the model theory of L. de Branges and J. Rovnyak. However, we did not meet then since Joe was overseas at that time. I first met Joe a couple of years later, when I was staying as a post-doc at the Weizmann Institute of Science (Israel) and Joe was frequently visiting the Institute.

In the fall of 1995 we both participated in the Holomorphic Spaces semester at MSRI, Berkeley, CA. Although we were sharing an office for the whole semester, we did not collaborate at that time: I was working on my own paper, Joe was working with Victor Vinnikov. However, we had a lot of conversations and discussions (mathematical and non mathematical). Those discussions materialized in our later joint work.

Sometimes we worked late in our office. The window of the office looked to the sunset and we had a lot of chances to watch it. Often Joe was working there with Victor Vinnikov. Then the three of us went down the hill looking for a late dinner.

Our actual collaboration started several years later in 2001, when I was staying as a visitor at The College of William and Mary. I had some ideas on extending the Abstract Interpolation Problem approach to the general Commutant Lifting problem using linear scattering systems. I tried to discuss it with several people, but only Joe got interested in the topic and we started working on it. It was a long, really exciting and fruitful collaboration. During that time I visited Blacksburg, VA several times and Joe visited me in Lowell, MA. It is Joe's mathematical taste, understanding of the subject and persistence that allowed to complete this work.

At the beginning we included the general Commutant Lifting problem into the Abstract Interpolation problem scheme. We introduced the notion of symbol of the lift (which is a measure in the general case) and we established a one-to-one correspondence between lifts and their symbols. The symbols, in turn, are solutions of the Abstract Interpolation Problem. The formula that describes the symbols of lifts of a given contraction looks rather standard (although it's precise meaning should be explained)

$$w = s_0 + s_2(I - \omega s)^{-1}\omega s_1,$$

where ω is a free parameter, and coefficients s, s_2, s_1 and s_0 are uniquely determined by the data of the problem.

However, our real objective was the inverse Commutant Lifting Problem, that consists in characterising functions s, s_2, s_1 and s_0 that appear in the above parametrization formulas for Commutant Lifting Problems. This type of inverse problems was studied extensively by V. Potapov for truncated Nevanlinna–Pick, Carathéodory–Fejér and Moment problems. For the infinite scalar Nevanlinna–Pick problem necessary conditions were obtained by R. Nevanlinna in 1929. Necessary and sufficient conditions for the Nevanlinna–Pick and some more general problems were obtained by D. Arov in the 1980s. We proved necessary and sufficient conditions for the general Commutant Lifting problem. The main difference between our "regularity" conditions and the ones obtained by D. Arov is that we stated ours in terms of the functional models (both coordinate-free and Hellinger space realization). Moreover, we considered the general (not only completely indeterminate case). For this general case we proved extremal properties of the coefficients s_2 and s_1, that are factoral minorants here (as opposed to equalities of the completely indeterminate case).

Numerous conversations with Joe played a significant role in improving my English. Joe was always willing to help and to answer my questions on how to say or write this or that, very often he volunteered to do that. I also got from Joe many insights on the traditions and culture of the country that was new to me. All this cannot be overestimated.

André Ran

In the early eighties a young man called Joe Ball spent a couple of months at the Vrije Universiteit in Amsterdam, and lectured there on work that he had recently done with Bill Helton. The lecture series was a very nice summary of a series of papers, and the participants of the seminar Analysis and Operator Theory learned a lot. This lecture series was my first encounter with Joe.

When finishing my PhD thesis became something of a certainty, and the timing of the defence became more and more certain as well, the question inevitably arose: what next? At that time, just as now, this was not something to be taken lightly. Since my first goal when starting with mathematics was to become a mathematics teacher at a high school, and since I enjoyed teaching a lot, there was certainly the option of choosing a career in that direction rather than continuing in academics. Nevertheless, after several discussions Rien Kaashoek convinced me to apply for a postdoc grant to go and visit Joe Ball for a year. And what a wonderful year it was: Joe and I wrote a series of papers, and I picked up several additional topics to work on as well. All in all, that year was decisive in my career. Joe stimulated and encouraged working on mathematics in a way that is perhaps even not deliberate. His example does the trick.

We started working on several projects: one concerning model reduction in state space terms, and one concerning inverse spectral problems in state space terms. Both projects led to multiple papers, and some of the work we did in that year formed the basis of things we did separately years later.

Joe's focus on work is exemplary, but is not appreciated by all his co-authors and friends at all times. Usually, Joe and I would have lunch at the Burger King (fries and milk to accompany our own sandwiches) and we would have a discussion on mathematics during lunch. Some time in my year there Leiba Rodman came to visit, to discuss mathematics both with Joe and with me. After one day of joining us for lunch and seeing the usual way lunch time was spent, Leiba decided he would join us for the discussion, but to have his lunch earlier: he liked to enjoy the lunch as lunch, and the discussion on mathematics for the discussion on mathematics, and never the two shall mix.

I would like to take this opportunity to thank Joe for the example he has set me, and for all the support he gave me during the early years of my career.

James Rovnyak

Joe came under the wing of Larry Shulman, a student of Louis de Branges, as an undergraduate at Georgetown University. Directly or indirectly, Larry was likely an important mentor for Joe and an influence on Joe's choice of the University of Virginia for graduate study. Larry died of leukemia, cutting a promising career tragically short.

Joe arrived at UVA in September 1969. Marvin Rosenblum's program in operator theory and classical analysis was running full steam. The faculty in operator theory around Joe's time was expanded to include Bruce Abrahamse, Jim Howland, Tom Kriete, and this writer. Graduate students were trained in a seminar that met twice a week. The seminar was one of Marvin's best ideas. A scribe was appointed and responsible to produce mimeographed notes to all participants by the next meeting. The seminar surveyed current topics in operator theory representing faculty interests and perspectives. Among the topics presented in Joe's days were canonical models, the commutant lifting theorem, and Lax–Phillips theory. Joe absorbed all and produced a masterful PhD thesis directed by Marvin Rosenblum. To quote my 1973 recommendation letter for Joe, the thesis is "a substantial treatise which explores the connections between 1) the model theories of de Branges–Rovnyak and Sz.-Nagy–Foiaş, and 2) the perturbation theories of Kato–Kuroda and de Branges–Shulman." Joe's talent was recognized early. When Joe began his teaching career, Virginia Tech was not the research powerhouse that it is today; Jimmy McShane had a high opinion of Joe and thought he should have gone to a major research university for his first job. I was present when Joe gave what was possibly his first international lecture on his work. This was at Oberwolfach, and many international mathematicians were in the audience; at the end of the lecture, Sz.-Nagy and Foiaş walked briskly up to Joe to meet this new member of the community. Joe was launched on the international scene, and the rest is history.

Sadly, I do not recall any scandalous stories relating to Joe. I remember Joe as a serious and modest student who worked hard and produced the goods. He had an enjoyable sense of humor. I was always impressed that Joe was an organist and played at a local church in his graduate student days.

Quanlei Fang
Department of Math & Computer Science
CUNY-BCC
Bronx, NY 10453, USA
e-mail: quanlei.fang@bcc.cuny.edu

J. William Helton
Department of Mathematics
University of California
San Diego, CA, USA
e-mail: helton@math.ucsd.edu

Sanne ter Horst
Department of Mathematics
Unit for BMI, North-West University
Potchefstroom 2531, South Africa
e-mail: Sanne.TerHorst@nwu.ac.za

Alexander Kheifets
Department of Mathematical Sciences
University of Massachusetts Lowell
One University Avenue
Lowell, MA 01854, USA
e-mail: Alexander_Kheifets@uml.edu

André C.M. Ran
Department of Mathematics
Faculty of Science, VU Amsterdam
De Boelelaan 1081a
1081 HV Amsterdam, The Netherlands
and
Unit for BMI, North-West University
Potchefstroom 2531, South Africa
e-mail: a.c.m.ran@vu.nl

James Rovnyak
University of Virginia
Department of Mathematics
P.O. Box 400137
Charlottesville, VA, 22904–4137, USA
e-mail: rovnyak@virginia.edu

Operator Theory:
Advances and Applications, Vol. 272, 1–42

Holomorphic Operator-valued Functions Generated by Passive Selfadjoint Systems

Yuri Arlinskiĭ and Seppo Hassi

Dedicated to Professor Joseph Ball on the occasion of his 70th birthday

Abstract. Let $\mathfrak{M}$ be a Hilbert space. In this paper we study a class $\mathcal{RS}(\mathfrak{M})$ of operator functions that are holomorphic in the domain $\mathbb{C} \setminus \{(-\infty,-1] \cup [1,+\infty)\}$ and whose values are bounded linear operators in $\mathfrak{M}$. The functions in $\mathcal{RS}(\mathfrak{M})$ are Schur functions in the open unit disk $\mathbb{D}$ and, in addition, Nevanlinna functions in $\mathbb{C}_+ \cup \mathbb{C}_-$. Such functions can be realized as transfer functions of minimal passive selfadjoint discrete-time systems. We give various characterizations for the class $\mathcal{RS}(\mathfrak{M})$ and obtain an explicit form for the inner functions from the class $\mathcal{RS}(\mathfrak{M})$ as well as an inner dilation for any function from $\mathcal{RS}(\mathfrak{M})$. We also consider various transformations of the class $\mathcal{RS}(\mathfrak{M})$, construct realizations of their images, and find corresponding fixed points.

Mathematics Subject Classification (2010). Primary 47A48, 93B28, 93C25; Secondary 47A56, 93B20.

Keywords. Passive system, transfer function, Nevanlinna function, Schur function, fixed point.

1. Introduction

Throughout this paper we consider separable Hilbert spaces over the field $\mathbb{C}$ of complex numbers and certain classes of operator-valued functions which are holomorphic on the open upper/lower half-planes $\mathbb{C}_+/\mathbb{C}_-$ and/or on the open unit disk $\mathbb{D}$. A $\mathbf{B}(\mathfrak{M})$-valued function M is called a *Nevanlinna function* if it is holomorphic outside the real axis, symmetric $M(\lambda)^* = M(\bar{\lambda})$, and satisfies the inequality

This research was partially supported by a grant from the Vilho, Yrjö and Kalle Väisälä Foundation of the Finnish Academy of Science and Letters. Yu.M. Arlinskiĭ also gratefully acknowledges financial support from the University of Vaasa.

$\operatorname{Im}\lambda \operatorname{Im} M(\lambda) \geq 0$ for all $\lambda \in \mathbb{C}\backslash\mathbb{R}$. This last condition is equivalent to the non-negativity of the kernel

$$\frac{M(\lambda) - M(\mu)^*}{\lambda - \bar{\mu}}, \quad \lambda, \mu \in \mathbb{C}_+ \cup \mathbb{C}_-.$$

On the other hand, a $\mathbf{B}(\mathfrak{M})$-valued function $\Theta(z)$ belongs to the *Schur class* if it is holomorphic on the unit disk $\mathbb{D}$ and contractive, $||\Theta(z)|| \leq 1 \ \forall z \in \mathbb{D}$ or, equivalently, the kernel

$$\frac{I - \Theta^*(w)\Theta(z)}{1 - z\bar{w}}, \quad z, w \in \mathbb{D}$$

is non-negative. Functions from the Schur class appear naturally in the study of linear discrete-time systems; we briefly recall some basic terminology here; cf. D.Z. Arov [7, 8]. Let T be a bounded operator given in the block form

$$T = \begin{bmatrix} D & C \\ B & A \end{bmatrix} : \begin{array}{c} \mathfrak{M} \\ \oplus \\ \mathcal{K} \end{array} \to \begin{array}{c} \mathfrak{N} \\ \oplus \\ \mathcal{K} \end{array} \tag{1.1}$$

with separable Hilbert spaces $\mathfrak{M}, \mathfrak{N}$, and $\mathfrak{K}$. The system of equations

$$\begin{cases} h_{k+1} = Ah_k + B\xi_k, \\ \sigma_k = Ch_k + D\xi_k, \end{cases} \quad k \geq 0, \tag{1.2}$$

describes the evolution of a *linear discrete time-invariant system*; this system is briefly denoted by $\tau = \{T, \mathfrak{M}, \mathfrak{N}, \mathfrak{K}\}$, where $\mathfrak{M}$ and $\mathfrak{N}$ are called the input and the output spaces, respectively, and $\mathfrak{K}$ is the state space. The operators A, B, C, and D are called the main operator, the control operator, the observation operator, and the feedthrough operator of τ, respectively. The subspaces

$$\mathfrak{K}^c = \overline{\operatorname{span}}\{ A^n B\mathfrak{M} : n \in \mathbb{N}_0\} \quad \text{and} \quad \mathfrak{K}^o = \overline{\operatorname{span}}\{ A^{*n} C^* \mathfrak{N} : n \in \mathbb{N}_0\} \tag{1.3}$$

are called the controllable and observable subspaces of $\tau = \{T, \mathfrak{M}, \mathfrak{N}, \mathfrak{K}\}$, respectively. If $\mathfrak{K}^c = \mathfrak{K}$ ($\mathfrak{K}^o = \mathfrak{K}$) then the system τ is said to be *controllable* (*observable*), and *minimal* if τ is both controllable and observable. If $\mathfrak{K} = \operatorname{clos}\{\mathfrak{K}^c + \mathfrak{K}^o\}$ then the system τ is said to be *simple*. Closely related to these definitions is the notion of $\mathfrak{M}$-simplicity: given a nontrivial subspace $\mathfrak{M} \subset \mathfrak{H}$ the operator T acting in $\mathfrak{H}$ is said to be $\mathfrak{M}$-*simple* if

$$\overline{\operatorname{span}}\{ T^n \mathfrak{M}, \, n \in \mathbb{N}_0\} = \mathfrak{H}.$$

Two discrete-time systems $\tau_1 = \{T_1, \mathfrak{M}, \mathfrak{N}, \mathfrak{K}_1\}$ and $\tau_2 = \{T_2, \mathfrak{M}, \mathfrak{N}, \mathfrak{K}_2\}$ are *unitarily similar* if there exists a unitary operator U from $\mathfrak{K}_1$ onto $\mathfrak{K}_2$ such that

$$A_2 = UA_1U^*, \quad B_2 = UB_1, \quad C_2 = C_1U^*, \text{ and } D_2 = D_1. \tag{1.4}$$

If the linear operator T is contractive (isometric, co-isometric, unitary), then the corresponding discrete-time system is said to be *passive* (*isometric, co-isometric, conservative*). With the passive system τ in (1.2) one associates the *transfer function* via

$$\Omega_\tau(z) := D + zC(I - zA)^{-1}B, \quad z \in \mathbb{D}. \tag{1.5}$$

It is well known that the transfer function of a passive system belongs to the *Schur class* $\mathbf{S}(\mathfrak{M},\mathfrak{N})$ and, conversely, that every operator-valued function $\Theta(\lambda)$ from the Schur class $\mathbf{S}(\mathfrak{M},\mathfrak{N})$ can be realized as the transfer function of a passive system, which can be chosen as observable co-isometric (controllable isometric, simple conservative, passive minimal). Notice that an application of the Schur–Frobenius formula (see Appendix A) for the inverse of a block operator gives with $\mathfrak{M}=\mathfrak{N}$ the relation

$$P_{\mathfrak{M}}(I-zT)^{-1}\upharpoonright\mathfrak{M}=(I_{\mathfrak{M}}-z\Omega_{\tau}(z))^{-1},\quad z\in\mathbb{D}. \tag{1.6}$$

It is known that two isometric and controllable (co-isometric and observable, simple conservative) systems with the same transfer function are unitarily similar. However, D.Z. Arov [7] has shown that two minimal passive systems τ_1 and τ_2 with the same transfer function $\Theta(\lambda)$ are only weakly similar; weak similarity neither preserves the dynamical properties of the system nor the spectral properties of its main operator A. Some necessary and sufficient conditions for minimal passive systems with the same transfer function to be (unitarily) similar have been established in [9, 10].

By introducing some further restrictions on the passive system τ it is possible to preserve unitary similarity of passive systems having the same transfer function. In particular, when the main operator A is normal such results have been obtained in [5]; see in particular Theorem 3.1 and Corollaries 3.6–3.8 therein. A stronger condition on τ where main operator is selfadjoint naturally yields to a class of systems which preserve such a unitary similarity property. A class of such systems appearing in [5] is the class of *passive quasi-selfadjoint systems*, in short *pqs-systems*, which is defined as follows: a collection

$$\tau=\{T,\mathfrak{M},\mathfrak{M},\mathcal{K}\}$$

is a *pqs*-system if the operator T determined by the block formula (1.1) with the input-output space $\mathfrak{M}=\mathfrak{N}$ is a contraction and, in addition,

$$\operatorname{ran}(T-T^*)\subseteq\mathfrak{M}.$$

Then, in particular, $F=F^*$ and $B=C^*$ so that T takes the form

$$T=\begin{bmatrix}D & C\\ C^* & F\end{bmatrix}:\begin{matrix}\mathfrak{M}\\ \oplus\\ \mathcal{K}\end{matrix}\to\begin{matrix}\mathfrak{M}\\ \oplus\\ \mathcal{K}\end{matrix},$$

i.e., T is a quasi-selfadjoint contraction in the Hilbert space $\mathfrak{H}=\mathfrak{M}\oplus\mathcal{K}$. The class of *pqs*-systems gives rise to transfer functions which belong to the subclass $\mathcal{S}^{qs}(\mathfrak{M})$ of Schur functions. The class $\mathcal{S}^{qs}(\mathfrak{M})$ admits the following intrinsic description; see [5, Definition 4.4, Proposition 5.3]: a $\mathbf{B}(\mathfrak{M})$-valued function Ω belongs to $\mathcal{S}^{qs}(\mathfrak{M})$ if it is holomorphic on $\mathbb{C}\setminus\{(-\infty,-1]\cup[1,+\infty)\}$ and has the following additional properties:

(S1) $W(z)=\Omega(z)-\Omega(0)$ is a Nevanlinna function;
(S2) the strong limit values $W(\pm 1)$ exist and $W(1)-W(-1)\le 2I$;

(S3) $\Omega(0)$ belongs to the operator ball

$$\mathcal{B}\left(-\frac{W(1)+W(-1)}{2},\; I-\frac{W(1)-W(-1)}{2}\right)$$

with the center $-\dfrac{W(1)+W(-1)}{2}$ and with the left and right radii

$$I-\frac{W(1)-W(-1)}{2}.$$

It was proved in [5, Theorem 5.1] that the class $\mathcal{S}^{qs}(\mathfrak{M})$ coincides with the class of all transfer functions of *pqs*-systems with input-output space $\mathfrak{M}$. In particular, every function from the class $\mathcal{S}^{qs}(\mathfrak{M})$ can be realized as the transfer function of a *minimal* *pqs*-system and, moreover, two minimal realization are unitarily equivalent; see [3, 5, 6]. For *pqs*-systems the controllable and observable subspaces $\mathcal{K}^c$ and $\mathcal{K}^o$ as defined in (1.3) necessarily coincide. Furthermore, the following equivalences were established in [6]:

$$T \text{ is } \mathfrak{M}\text{-simple} \iff \text{the operator } F \text{ is } \overline{\operatorname{ran}}\, C^*\text{-simple in } \mathcal{K}$$

$$\iff \text{the system} \quad \tau=\left\{\begin{bmatrix} D & C\\ C^* & F\end{bmatrix}, \mathfrak{M}, \mathfrak{M}, \mathcal{K}\right\} \text{ is minimal.}$$

We can now introduce one of the main objects to be studied in the present paper.

Definition 1.1. Let $\mathfrak{M}$ be a Hilbert space. A $\mathbf{B}(\mathfrak{M})$-valued Nevanlinna function Ω which is holomorphic on $\mathbb{C}\setminus\{(-\infty,-1]\cup[1,+\infty)\}$ is said to belong to the class $\mathcal{RS}(\mathfrak{M})$ if

$$-I\le\Omega(x)\le I,\quad x\in(-1,1).$$

The class $\mathcal{RS}(\mathfrak{M})$ will be called the combined Nevanlinna–Schur class of $\mathbf{B}(\mathfrak{M})$-valued operator functions.

If $\Omega\in\mathcal{RS}(\mathfrak{M})$, then $\Omega(x)$ is non-decreasing on the interval $(-1,1)$. Therefore, the strong limit values $\Omega(\pm1)$ exist and satisfy the following inequalities

$$-I_{\mathfrak{M}}\le\Omega(-1)\le\Omega(0)\le\Omega(1)\le I_{\mathfrak{M}}. \tag{1.7}$$

It follows from (S1)–(S3) that $\mathcal{RS}(\mathfrak{M})$ is a subclass of the class $\mathcal{S}^{qs}(\mathfrak{M})$.

In this paper we give some new characterizations of the class $\mathcal{RS}(\mathfrak{M})$, find an explicit form for inner functions from the class $\mathcal{R}(\mathfrak{M})$, and construct a bi-inner dilation for an arbitrary function from $\mathcal{RS}(\mathfrak{M})$. For instance, in Theorem 4.1 it is proven that a $\mathbf{B}(\mathfrak{M})$-valued Nevanlinna function defined on $\mathbb{C}\setminus\{(-\infty,-1]\cup[1,+\infty)\}$ belongs to the class $\mathcal{RS}(\mathfrak{M})$ if and only if

$$K(z,w):=I_{\mathfrak{M}}-\Omega^*(w)\Omega(z)-\frac{1-\bar{w}z}{z-\bar{w}}\left(\Omega(z)-\Omega^*(w)\right)$$

defines a non-negative kernel on the domains

$$\mathbb{C}\setminus\{(-\infty,-1]\cup[1,+\infty)\},\ \operatorname{Im} z>0 \ \text{ and } \ \mathbb{C}\setminus\{(-\infty,-1]\cup[1,+\infty)\},\ \operatorname{Im} z<0.$$

We also show that the transformation

$$\mathcal{RS}(\mathfrak{M}) \ni \Omega \mapsto \mathbf{\Phi}(\Omega) = \Omega_{\mathbf{\Phi}}, \quad \Omega_{\mathbf{\Phi}}(z) := (zI - \Omega(z))(I - z\Omega(z))^{-1}, \tag{1.8}$$

with $z \in \mathbb{C} \setminus \{(-\infty, -1] \cup [1, +\infty)\}$ is an automorphism of $\mathcal{RS}(\mathfrak{M})$, $\mathbf{\Phi}^{-1} = \mathbf{\Phi}$, and that $\mathbf{\Phi}$ has a unique fixed point, which will be specified in Proposition 6.6.

It turns out that the set of inner functions from the class $\mathcal{RS}(\mathfrak{M})$ can be seen as the image $\mathbf{\Phi}$ of constant functions from $\mathcal{RS}(\mathfrak{M})$: in other words, the inner functions from $\mathcal{RS}(\mathfrak{M})$ are of the form

$$\Omega_{\text{in}}(z) = (zI + A)(I + zA)^{-1}, \ A \in [-I_{\mathfrak{M}}, I_{\mathfrak{M}}].$$

In Theorem 6.3 it is proven that every function $\Omega \in \mathcal{RS}(\mathfrak{M})$ admits the representation

$$\Omega(z) = P_{\mathfrak{M}} \widetilde{\Omega}_{in}(z) {\upharpoonright} \mathfrak{M} = P_{\mathfrak{M}}(zI + \widetilde{A})(I + z\widetilde{A})^{-1} {\upharpoonright} \mathfrak{M}, \quad \widetilde{A} \in [-I_{\widetilde{\mathfrak{M}}}, I_{\widetilde{\mathfrak{M}}}], \tag{1.9}$$

where $z \in \mathbb{C} \setminus \{(-\infty, -1] \cup [1, +\infty)\}$ and $\widetilde{\mathfrak{M}}$ is a Hilbert space containing $\mathfrak{M}$ as a subspace and such that $\overline{\text{span}}\{\widetilde{A}^n \mathfrak{M} : \ n \in \mathbb{N}_0\} = \widetilde{\mathfrak{M}}$ (i.e., $\widetilde{A}$ is $\mathfrak{M}$-simple). Equality (1.9) means that an arbitrary function of the class $\mathcal{RS}(\mathfrak{M})$ admits a bi-inner dilation (in the sense of [8]) that belongs to the class $\mathcal{RS}(\widetilde{\mathfrak{M}})$.

In Section 6 we also consider the following transformations of the class $\mathcal{RS}(\mathfrak{M})$:

$$\Omega\left(\frac{z + a}{1 + za}\right) =: \Omega_a(z) \leftarrow\!\!\prec \Omega(z) \rightarrowtail \widehat{\Omega}_a(z) := (aI + \Omega(z))(I + a\Omega(z))^{-1}, \tag{1.10}$$
$$a \in (-1, 1), z \in \mathbb{C} \setminus \{(-\infty, -1] \cup [1, +\infty)\}.$$

These are analogs of the Möbius transformation

$$w_a(z) = \frac{z + a}{1 + az}, \quad z \in \mathbb{C} \setminus \{-a^{-1}\} \ (a \in (-1, 1), \ a \neq 0)$$

of the complex plane. The mapping w_a is an automorphism of $\mathbb{C} \setminus \{(-\infty, -1] \cup [1, +\infty)\}$ and it maps $\mathbb{D}$ onto $\mathbb{D}$, $[-1, 1]$ onto $[-1, 1]$, $\mathbb{T}$ onto $\mathbb{T}$, as well as $\mathbb{C}_+/\mathbb{C}_-$ onto $\mathbb{C}_+/\mathbb{C}_-$.

The mapping

$$\mathcal{RS}(\mathfrak{M}) \ni \Omega \mapsto \Omega_a(z) = \Omega\left(\frac{z + a}{1 + za}\right) \in \mathcal{RS}(\mathfrak{M})$$

can be rewritten as

$$\Omega \mapsto \Omega \circ w_a.$$

In Proposition 6.13 it is shown that the fixed points of this transformation consist only of the constant functions from $\mathcal{RS}(\mathfrak{M})$: $\Omega(z) \equiv A$ with $A \in [-I_{\mathfrak{M}}, I_{\mathfrak{M}}]$.

One of the operator analogs of w_a is the following transformation of $\mathbf{B}(\mathfrak{M})$:

$$W_a(T) = (T + aI)(I + aT)^{-1}, \quad a \in (-1, 1).$$

The inverse of W_a is given by

$$W_{-a}(T) = (T - aI)(I - aT)^{-1}.$$

The class $\mathcal{RS}(\mathfrak{M})$ is stable under the transform W_a:

$$\Omega \in \mathcal{RS}(\mathfrak{M}) \Longrightarrow W_a \circ \Omega \in \mathcal{RS}(\mathfrak{M}).$$

If T is selfadjoint and unitary (a fundamental symmetry), i.e., $T = T^* = T^{-1}$, then for every $a \in (-1,1)$ one has

$$W_a(T) = T \tag{1.11}$$

Conversely, if for a selfadjoint operator T the equality (1.11) holds for some a : $-a^{-1} \in \rho(T)$, then T is a fundamental symmetry and (1.11) is valid for all $a \neq \{\pm 1\}$.

One can interpret the mappings in (1.10) as $\Omega \circ w_a$ and $W_a \circ \Omega$, where $\Omega \in \mathcal{RS}(\mathfrak{M})$. Theorem 6.18 states that inner functions from $\mathcal{RS}(\mathfrak{M})$ are the only fixed points of the transformation

$$\mathcal{RS}(\mathfrak{M}) \ni \Omega \mapsto W_{-a} \circ \Omega \circ w_a.$$

An equivalent statement is that the equality

$$\Omega \circ w_a = W_a \circ \Omega$$

holds only for inner functions Ω from the class $\mathcal{RS}(\mathfrak{M})$. On the other hand, it is shown in Theorem 6.19 that the only solutions of the functional equation

$$\Omega(z) = \left(\Omega\left(\frac{z-a}{1-az}\right) - a\, I_{\mathfrak{M}}\right)\left(I_{\mathfrak{M}} - a\,\Omega\left(\frac{z-a}{1-az}\right)\right)^{-1}$$

in the class $\mathcal{RS}(\mathfrak{M})$, where $a \in (-1,1)$, $a \neq 0$, are constant functions Ω, which are fundamental symmetries in $\mathfrak{M}$.

To introduce still one further transform, let

$$\mathbf{K} = \begin{bmatrix} K_{11} & K_{12} \\ K_{12}^* & K_{22} \end{bmatrix} : \begin{array}{c} \mathfrak{M} \\ \oplus \\ H \end{array} \to \begin{array}{c} \mathfrak{M} \\ \oplus \\ H \end{array}$$

be a selfadjoint contraction and consider the mapping

$$\mathcal{RS}(H) \ni \Omega \;\mapsto\; \Omega_{\mathbf{K}}(z) := K_{11} + K_{12}\Omega(z)(I - K_{22}\Omega(z))^{-1}K_{12}^*,$$

where $z \in \mathbb{C} \setminus \{(-\infty,-1] \cup [1,+\infty)\}$. In Theorem 6.8 we prove that if $||K_{22}|| < 1$, then $\Omega_{\mathbf{K}} \in \mathcal{RS}(\mathfrak{M})$ and in Theorem 6.9 we construct a realization of $\Omega_{\mathbf{K}}$ by means of realization of $\Omega \in \mathcal{RS}(H)$ using the so-called *Redheffer product*; see [17, 21]. The mapping

$$\mathbf{B}(H) \ni T \mapsto K_{11} + K_{12}T(I - K_{22}T)^{-1}K_{21} \in \mathbf{B}(\mathfrak{M})$$

can be considered as one further operator analog of the Möbius transformation, cf. [18].

Finally, it is emphasized that in Section 6 we will systematically construct *explicit realizations* for each of the transforms $\mathbf{\Phi}(\Omega)$, Ω_a, and $\widehat{\Omega}_a$ as transfer functions of minimal passive selfadjoint systems using a minimal realization of the initially given function $\Omega \in \mathcal{RS}(H)$.

Basic notations. We use the symbols $\operatorname{dom} T$, $\operatorname{ran} T$, $\ker T$ for the domain, the range, and the kernel of a linear operator T. The closures of $\operatorname{dom} T$, $\operatorname{ran} T$ are denoted by $\overline{\operatorname{dom}} T$, $\overline{\operatorname{ran}} T$, respectively. The identity operator in a Hilbert space $\mathfrak{H}$ is denoted by I and sometimes by $I_{\mathfrak{H}}$. If $\mathfrak{L}$ is a subspace, i.e., a closed linear subset of $\mathfrak{H}$, the orthogonal projection in $\mathfrak{H}$ onto $\mathfrak{L}$ is denoted by $P_{\mathfrak{L}}$. The notation $T{\upharpoonright}\mathfrak{L}$ means the restriction of a linear operator T on the set $\mathfrak{L} \subset \operatorname{dom} T$. The resolvent set of T is denoted by $\rho(T)$. The linear space of bounded operators acting between Hilbert spaces $\mathfrak{H}$ and $\mathfrak{K}$ is denoted by $\mathbf{B}(\mathfrak{H}, \mathfrak{K})$ and the Banach algebra $\mathbf{B}(\mathfrak{H}, \mathfrak{H})$ by $\mathbf{B}(\mathfrak{H})$. For a contraction $T \in \mathbf{B}(\mathfrak{H}, \mathfrak{K})$ the defect operator $(I - T^*T)^{1/2}$ is denoted by D_T and $\mathfrak{D}_T := \overline{\operatorname{ran}}\, D_T$. For defect operators one has the commutation relations

$$TD_T = D_{T^*}T, \quad T^*D_{T^*} = D_TT^* \tag{1.12}$$

and, moreover,

$$\operatorname{ran} TD_T = \operatorname{ran} D_{T^*}T = \operatorname{ran} T \cap \operatorname{ran} D_{T^*}. \tag{1.13}$$

In what follows we systematically use the Schur–Frobenius formula for the resolvent of a block-operator matrix and parameterizations of contractive block operators, see Appendices A and B.

2. The combined Nevanlinna–Schur class $\mathcal{RS}(\mathfrak{M})$

In this section some basic properties of operator functions belonging to the combined Nevanlinna–Schur class $\mathcal{RS}(\mathfrak{M})$ are derived. As noted in Introduction every function $\Omega \in \mathcal{RS}(\mathfrak{M})$ admits a realization as the transfer function of a passive selfadjoint system. In particular, the function $\Omega{\upharpoonright}\mathbb{D}$ belongs to the Schur class $\mathcal{S}(\mathfrak{M})$.

It is known from [1] that, if $\Omega \in \mathcal{RS}(\mathfrak{M})$ then for every $\beta \in [0, \pi/2)$ the following implications are satisfied:

$$\begin{cases} \begin{array}{c} |z\sin\beta + i\cos\beta| \le 1 \\ z \ne \pm 1 \end{array} & \Longrightarrow \|\Omega(z)\sin\beta + i\cos\beta\, I\| \le 1 \\ \begin{array}{c} |z\sin\beta - i\cos\beta| \le 1 \\ z \ne \pm 1 \end{array} & \Longrightarrow \|\Omega(z)\sin\beta - i\cos\beta\, I\| \le 1 \end{cases}. \tag{2.1}$$

In fact, in Section 4 these implications will be we derived once again by means of some new characterizations for the class $\mathcal{RS}(\mathfrak{M})$.

To describe some further properties of the class $\mathcal{RS}(\mathfrak{M})$ consider a passive selfadjoint system given by

$$\tau = \left\{ \begin{bmatrix} D & C \\ C^* & F \end{bmatrix}; \mathfrak{M}, \mathfrak{M}, \mathcal{K} \right\}, \tag{2.2}$$

with $D = D^*$ and $F = F^*$. It is known, see Proposition B.1 and Remark B.2 in Appendix B, that the entries of the selfadjoint contraction

$$T = \begin{bmatrix} D & C \\ C^* & F \end{bmatrix} : \begin{array}{c} \mathfrak{M} \\ \oplus \\ \mathcal{K} \end{array} \to \begin{array}{c} \mathfrak{M} \\ \oplus \\ \mathcal{K} \end{array} \tag{2.3}$$

admit the parametrization

$$C = KD_F, \quad D = -KFK^* + D_{K^*}YD_{K^*}, \tag{2.4}$$

where $K \in \mathbf{B}(\mathfrak{D}_F, \mathfrak{M})$ is a contraction and $Y \in \mathbf{B}(\mathfrak{D}_{K^*})$ is a selfadjoint contraction. The minimality of the system τ means that the following equivalent equalities hold:

$$\overline{\operatorname{span}}\{F^n D_F K^*,\ n \in \mathbb{N}_0\} = \mathcal{K} \iff \bigcap_{n\in\mathbb{N}_0} \ker(KF^n D_F) = \{0\}. \tag{2.5}$$

Notice that if τ is minimal, then $\mathcal{K} = \mathfrak{D}_F$ or, equivalently, $\ker D_F = \{0\}$.

Recall from [20] the Sz.-Nagy–Foiaş characteristic function of the selfadjoint contraction F, which for every $z \in \mathbb{C} \setminus \{(-\infty, -1] \cup [1, +\infty)\}$ is given by

$$\begin{aligned}\Delta_F(z) &= \left(-F + zD_F(I - zF)^{-1}D_F\right){\restriction}\,\mathfrak{D}_F\\ &= \left(-F + z(I - F^2)(I - zF)^{-1}\right){\restriction}\,\mathfrak{D}_F\\ &= (zI - F)(I - zF)^{-1}{\restriction}\,\mathfrak{D}_F.\end{aligned}$$

Using the above parametrization one obtains the representations, cf. [5, Theorem 5.1],

$$\begin{aligned}\Omega(z) &= D + zC(I - zF)^{-1}C^* = D_{K^*}YD_{K^*} + K\Delta_F(z)K^*\\ &= D_{K^*}YD_{K^*} + K(zI - F)(I - zF)^{-1}K^*.\end{aligned} \tag{2.6}$$

Moreover, this gives the following representation for the limit values $\Omega(\pm 1)$:

$$\Omega(-1) = -KK^* + D_{K^*}YD_{K^*}, \quad \Omega(1) = KK^* + D_{K^*}YD_{K^*}. \tag{2.7}$$

The case $\Omega(\pm 1)^2 = I_{\mathfrak{M}}$ is of special interest and can be characterized as follows.

Proposition 2.1. *Let $\mathfrak{M}$ be a Hilbert space and let $\Omega \in \mathcal{RS}(\mathfrak{M})$. Then the following statements are equivalent:*

(i) $\Omega(1)^2 = \Omega(-1)^2 = I_{\mathfrak{M}}$;

(ii) *the equalities*

$$\begin{aligned}\left(\frac{\Omega(1) - \Omega(-1)}{2}\right)^2 &= \frac{\Omega(1) - \Omega(-1)}{2},\\ \left(\frac{\Omega(1) + \Omega(-1)}{2}\right)^2 &= I_{\mathfrak{M}} - \frac{\Omega(1) - \Omega(-1)}{2}\end{aligned} \tag{2.8}$$

hold;

(iii) *if $\tau = \{T; \mathfrak{M}, \mathfrak{M}, \mathcal{K}\}$ is a passive selfadjoint system* (2.2) *with the transfer function Ω and if the entries of the block operator T are parameterized by* (2.4), *then the operator $K \in \mathbf{B}(\mathfrak{D}_F, \mathfrak{M})$ is a partial isometry and $Y^2 = I_{\ker K^*}$.*

Proof. From (2.7) we get for all $f \in \mathfrak{M}$

$$\begin{aligned}||f||^2 - ||\Omega(\pm 1)f||^2 &= ||f||^2 - ||(D_{K^*}YD_{K^*} \pm KK^*)f||^2\\ &= ||(K^*(I \mp Y)D_{K^*}f||^2 + ||D_Y D_{K^*}f||^2;\end{aligned}$$

cf. [4, Lemma 3.1]. Hence

$$\Omega(1)^2 = \Omega(-1)^2 = I_{\mathfrak{M}} \Longleftrightarrow \begin{cases} K^*(I-Y)D_{K^*} = 0 \\ K^*(I+Y)D_{K^*} = 0 \\ D_Y D_{K^*} = 0 \end{cases}$$

$$\Longleftrightarrow \begin{cases} K^*D_{K^*} = D_K K^* = 0 \\ K^*YD_{K^*} = 0 \\ D_Y D_{K^*} = 0 \end{cases} \Longleftrightarrow \begin{cases} K \quad \text{is a partial isometry} \\ Y^2 = I_{\mathfrak{D}_{K^*}} = I_{\ker K^*} \end{cases}.$$

Thus (i)$\Longleftrightarrow$(iii).

Since K is a partial isometry, i.e., KK^* is an orthogonal projection, the formulas (2.7) imply that

$$K \quad \text{is a partial isometry} \Longleftrightarrow \left(\frac{\Omega(1)-\Omega(-1)}{2}\right)^2 = \frac{\Omega(1)-\Omega(-1)}{2},$$

and in this case $D_{K^*}Y = Y$, which implies that

$$Y^2 = I_{\mathfrak{D}_{K^*}} = I_{\ker K^*} \Longleftrightarrow \left(\frac{\Omega(1)+\Omega(-1)}{2}\right)^2 = I_{\mathfrak{M}} - \frac{\Omega(1)-\Omega(-1)}{2}.$$

Thus (iii)$\Longleftrightarrow$(ii). □

By interchanging the roles of the subspaces $\mathcal{K}$ and $\mathfrak{M}$ as well as the roles of the corresponding blocks of T in (2.3) leads to the passive selfadjoint system

$$\eta = \left\{ \begin{bmatrix} D & C \\ C^* & F \end{bmatrix}, \mathcal{K}, \mathcal{K}, \mathfrak{M} \right\}$$

now with the input-output space $\mathcal{K}$ and the state space $\mathfrak{M}$. The transfer function of η is given by

$$B(z) = F + zC^*(I - zD)^{-1}C, \quad z \in \mathbb{C} \setminus \{(-\infty, -1] \cup [1, +\infty)\}.$$

By applying Appendix B again one gets for (2.4) the following alternative expression to parameterize the blocks of T:

$$C = D_D N^*, \quad F = -NDN^* + D_{N^*}XD_{N^*}, \tag{2.9}$$

where $N : \mathfrak{D}_D \to \mathcal{K}$ is a contraction and X is a selfadjoint contraction in $\mathfrak{D}_{N^*}$. Now, similar to (2.7) one gets

$$B(1) = NN^* + D_{N^*}XD_{N^*}, \quad B(-1) = -NN^* + D_{N^*}XD_{N^*}.$$

For later purposes, define the selfadjoint contraction $\widehat{F}$ by

$$\widehat{F} := D_{N^*}XD_{N^*} = \frac{B(-1)+B(1)}{2}. \tag{2.10}$$

The statement in the next lemma can be checked with a straightforward calculation.

Lemma 2.2. *Let the entries of the selfadjoint contraction*

$$T = \begin{bmatrix} D & C \\ C^* & F \end{bmatrix} : \begin{matrix} \mathfrak{M} \\ \oplus \\ \mathcal{K} \end{matrix} \to \begin{matrix} \mathfrak{M} \\ \oplus \\ \mathcal{K} \end{matrix}$$

be parameterized by the formulas (2.9) *with a contraction* $N : \mathfrak{D}_D \to \mathcal{K}$ *and a selfadjoint contraction* X *in* $\mathfrak{D}_{N^*}$. *Then the function* $W(\cdot)$ *defined by*

$$W(z) = I + zDN^* \left(I - z\widehat{F}\right)^{-1} N, \quad z \in \mathbb{C} \setminus \{(-\infty, -1] \cup [1, +\infty)\}, \tag{2.11}$$

where $\widehat{F}$ *is given by* (2.10), *is invertible and*

$$W(z)^{-1} = I - zDN^*(I - zF)^{-1}N, \quad z \in \mathbb{C} \setminus \{(-\infty, -1] \cup [1, +\infty)\}. \tag{2.12}$$

The function $W(\cdot)$ is helpful for proving the next result.

Proposition 2.3. *Let* $\Omega \in \mathcal{RS}(\mathfrak{M})$. *Then for all* $z \in \mathbb{C} \setminus \{(-\infty, -1] \cup [1, +\infty)\}$ *the function* $\Omega(z)$ *can be represented in the form*

$$\Omega(z) = \Omega(0) + D_{\Omega(0)}\Lambda(z) \left(I + \Omega(0)\Lambda(z)\right)^{-1} D_{\Omega(0)} \tag{2.13}$$

with a function $\Lambda \in \mathcal{RS}(\mathfrak{D}_{\Omega(0)})$ *for which* $\Lambda(z) = z\Gamma(z)$, *where* Γ *is a holomorphic* $\mathbf{B}(\mathfrak{D}_{\Omega(0)})$*-valued function such that* $\|\Gamma(z)\| \le 1$ *for* $z \in \mathbb{D}$. *In particular,* $\|\Lambda(z)\| \le |z|$ *when* $z \in \mathbb{D}$.

Proof. To prove the statement, let the function Ω be realized as the transfer function of a passive selfadjoint system $\tau = \{T; \mathfrak{M}, \mathfrak{M}, \mathcal{K}\}$ as in (2.2), i.e., $\Omega(z) = D + zC(I - zF)^{-1}C^*$. Using (2.9) rewrite Ω as

$$\Omega(z) = D + zD_D N^*(I - zF)^{-1} N D_D = \Omega(0) + zD_{\Omega(0)} N^*(I - zF)^{-1} N D_{\Omega(0)}.$$

The definition of $\widehat{F}$ in (2.10) implies that the block operator

$$\begin{bmatrix} 0 & N^* \\ N & \widehat{F} \end{bmatrix} : \begin{matrix} \mathfrak{D}_{\Omega(0)} \\ \oplus \\ \mathcal{K} \end{matrix} \to \begin{matrix} \mathfrak{D}_{\Omega(0)} \\ \oplus \\ \mathcal{K} \end{matrix}$$

is a selfadjoint contraction (cf. Appendix B). Consequently, the $\mathbf{B}(\mathfrak{D}_D)$-valued function

$$\Lambda(z) := zN^* \left(I_{\mathcal{K}} - z\widehat{F}\right)^{-1} N, \quad z \in \mathbb{C} \setminus \{(-\infty, -1] \cup [1, +\infty)\}, \tag{2.14}$$

is the transfer function of the passive selfadjoint system

$$\tau_0 = \left\{ \begin{bmatrix} 0 & N^* \\ N & \widehat{F} \end{bmatrix}; \mathfrak{D}_{\Omega(0)}, \mathfrak{D}_{\Omega(0)}, \mathcal{K} \right\}$$

Hence Λ belongs the class $\mathcal{RS}(\mathfrak{D}_{\Omega(0)})$. Furthermore, using (2.11) and (2.12) in Lemma 2.2 one obtains

$$I + \Omega(0)\Lambda(z) = I + zDN^* \left(I - z\widehat{F}\right)^{-1} N = W(z)$$

and

$$(I+\Omega(0)\Lambda(z))^{-1}=W(z)^{-1}=I-zDN^*(I-zF)^{-1}N$$

for all $z\in\mathbb{C}\setminus\{(-\infty,-1]\cup[1,+\infty)\}$. Besides, in view of (2.9) one has $\widehat{F}-F=NDN^*$. This leads to the following implications

$$\begin{aligned}N^*\left(I-\widehat{F}\right)^{-1}N-N^*(I-zF)^{-1}N&=zN^*\left(I-\widehat{F}\right)^{-1}NDN^*(I-zF)^{-1}N\\ \Leftrightarrow zN^*\left(I-\widehat{F}\right)^{-1}N\left(I-zDN^*(I-zF)^{-1}N\right)&=zN^*(I-zF)^{-1}N\\ \Leftrightarrow \Lambda(z)\left(I+\Omega(0)\Lambda(z)\right)^{-1}&=zN^*(I-zF)^{-1}N\\ \Rightarrow \Omega(z)&=\Omega(0)+D_{\Omega(0)}\Lambda(z)\left(I+\Omega(0)\Lambda(z)\right)^{-1}D_{\Omega(0)}.\end{aligned}$$

Since $\Lambda(0)=0$, it follows from Schwarz's lemma that $||\Lambda(z)||\le|z|$ for all z with $|z|<1$. In particular, one has a factorization $\Lambda(z)=z\Gamma(z)$, where Γ is a holomorphic $\mathbf{B}(\mathfrak{D}_{\Omega(0)})$-valued function such that $\|\Gamma(z)\|\le 1$ for $z\in\mathbb{D}$; this is also obvious from (2.14). □

One can verify that the following relation for $\Lambda(z)$ holds

$$\Lambda(z)=D^{(-1)}_{\Omega(0)}(\Omega(z)-\Omega(0))(I-\Omega(0)\Omega(z))^{-1}D_{\Omega(0)},\tag{2.15}$$

where $D^{(-1)}_{\Omega(0)}$ stands for the Moore–Penrose inverse of $D_{\Omega(0)}$.

Notice also that the formula (2.13) holds for all $z\in\mathbb{C}\setminus\{(-\infty,-1]\cup[1,+\infty)\}$. A general Schur class function $\Omega\in\mathbf{S}(\mathfrak{M},\mathfrak{N})$ can be represented in the form

$$\Omega(z)=\Omega(0)+D_{\Omega(0)^*}\Lambda(z)\left(I+\Omega(0)^*\Lambda(z)\right)^{-1}D_{\Omega(0)},\quad z\in\mathbb{D}.$$

This is called a Möbius representation of Ω and it can be found in [12, 14, 18].

3. Inner functions from the class $\mathcal{RS}(\mathfrak{M})$

An operator-valued function from the Schur class is called *inner/co-inner* (or *-*inner*) (see, e.g., [20]) if it takes isometric/co-isometric values almost everywhere on the unit circle $\mathbb{T}$, and it is said to be *bi-inner* when it is both inner and co-inner.

Observe that if $\Omega\in\mathcal{RS}(\mathfrak{M})$ then $\Omega(z)^*=\Omega(\bar{z})$. Since $\mathbb{T}\setminus\{-1,1\}\subset\mathbb{C}\setminus\{(-\infty,-1]\cup[1,+\infty)\}$, one concludes that $\Omega\in\mathcal{RS}(\mathfrak{M})$ is inner (or co-inner) precisely when it is bi-inner. Notice also that every function $\Omega\in\mathcal{RS}(\mathfrak{M})$ can be realized as the transfer function of a minimal passive selfadjoint system τ as in (2.2); cf. [5, Theorem 5.1].

The next statement contains a characteristic result for transfer functions of conservative selfadjoint systems.

Proposition 3.1. *Assume that the selfadjoint system $\tau=\{T;\mathfrak{M},\mathfrak{M},\mathcal{K}\}$ in (2.2) is conservative. Then its transfer function $\Omega(z)=D+zC(I_{\mathcal{K}}-zF)^{-1}C^*$ is bi-inner and it takes the form*

$$\Omega(z)=(zI_{\mathfrak{M}}+D)(I_{\mathfrak{M}}+zD)^{-1},\quad z\in\mathbb{C}\setminus\{(-\infty,-1]\cup[1,+\infty)\}.\tag{3.1}$$

On the other hand, if τ is a minimal passive selfadjoint system whose transfer function is inner, then τ is conservative.

Proof. Let the entries of T in (2.3) be parameterized as in (2.9). By assumption T is unitary and hence $N \in \mathbf{B}(\mathfrak{D}_D, \mathcal{K})$ is an isometry and X is selfadjoint and unitary in the subspace $\mathfrak{D}_{N^*} = \ker N^*$; see Remark B.3 in Appendix B. Thus NN^* and D_{N^*} are orthogonal projections and $NN^* + D_{N^*} = I_{\mathcal{K}}$ which combined with (2.9) leads to

$$\begin{aligned}(I_{\mathcal{K}} - zF)^{-1} &= (N(I + zD)N^* + D_{N^*}(I - zX)D_{N^*})^{-1} \\ &= N(I + zD)^{-1}N^* + D_{N^*}(I - zX)^{-1}D_{N^*},\end{aligned}$$

and, consequently,

$$\begin{aligned}\Omega(z) &= D + zC(I_{\mathcal{K}} - zF)^{-1}C^* \\ &= D + zD_DN^*\left(N(I + zD)^{-1}N^* + D_{N^*}(I - zX)^{-1}D_{N^*}\right)ND_D \\ &= D + z(I + zD)^{-1}D_D^2 = (zI_{\mathfrak{M}} + D)(I_{\mathfrak{M}} + zD)^{-1},\end{aligned}$$

for all $z \in \mathbb{C} \setminus \{(-\infty, -1] \cup [1, +\infty)\}$. This proves (3.1) and this clearly implies that $\Omega(z)$ is bi-inner.

To prove the second statement assume that the transfer function of a minimal passive selfadjoint system τ is inner. Then it is automatically bi-inner. Now, according to a general result of D.Z. Arov [8, Theorem 1] (see also [10, Theorem 1], [4, Theorem 1.1]), if τ is a passive simple discrete-time system with bi-inner transfer function, then τ is conservative and minimal. This proves the second statement. □

The formula (3.1) in Proposition 3.1 gives a one-to-one correspondence between the operators D from the operator interval $[-I_{\mathfrak{M}}, I_{\mathfrak{M}}]$ and the inner functions from the class $\mathcal{RS}(\mathfrak{M})$. Recall that for $\Omega \in \mathcal{RS}(\mathfrak{M})$ the strong limit values $\Omega(\pm 1)$ exist as selfadjoint contractions; see (1.7). The formula (3.1) shows that if $\Omega \in \mathcal{RS}(\mathfrak{M})$ is an inner function, then necessarily these limit values are also unitary:

$$\Omega(1)^2 = \Omega(-1)^2 = I_{\mathfrak{M}}. \tag{3.2}$$

However, these two conditions do not imply that $\Omega \in \mathcal{RS}(\mathfrak{M})$ is an inner function; cf. Proposition 2.1 and Remark B.3 in Appendix B.

The next two theorems offer some sufficient conditions for $\Omega \in \mathcal{RS}(\mathfrak{M})$ to be an inner function. The first one shows that by shifting $\xi \in \mathbb{T}$ ($|\xi| = 1$) away from the real line then existence of a unitary limit value $\Omega(\xi)$ at a single point implies that $\Omega \in \mathcal{RS}(\mathfrak{M})$ is actually a bi-inner function.

Theorem 3.2. *Let Ω be a nonconstant function from the class $\mathcal{RS}(\mathfrak{M})$. If $\Omega(\xi)$ is unitary for some $\xi_0 \in \mathbb{T}$, $\xi_0 \neq \pm 1$. Then Ω is a bi-inner function.*

Proof. Let $\tau = \{T; \mathfrak{M}, \mathfrak{M}, \mathcal{K}\}$ in (2.2) be a minimal passive selfadjoint system whose transfer function is Ω and let the entries of T be parameterized as in (2.4).

Using the representation (2.6) one can derive the following formula for all $\xi \in \mathbb{T} \setminus \{\pm 1\}$:

$$\left\|D_{\Omega(\xi)}h\right\|^2 = \|D_{\Delta_F(\xi)}K^*h\|^2 + \|D_Y D_{K^*}h\|^2 \\ + \left\|(D_K\Delta_F(\xi)K^* - K^*YD_{K^*})\,h\right\|^2;$$

cf. [4, Theorem 5.1], [5, Theorem 2.7]. Since $\Delta_F(\xi)$ is unitary for all $\xi \in \mathbb{T}\setminus\{\pm 1\}$ and $\Omega(\xi_0)$ is unitary, one concludes that Y is unitary on $\mathfrak{D}_{K^*}$ and

$$(D_K\Delta_F(\xi_0)K^* - K^*YD_{K^*})\,h = 0 \text{ for all } h \in \mathfrak{M}.$$

Suppose that there is $h_0 \neq 0$ such that

$$D_K\Delta_F(\xi_0)K^*h_0 \neq 0 \quad \text{and} \quad K^*YD_{K^*}h_0 \neq 0.$$

Then, due to $D_K\Delta_F(\xi_0)K^*h_0 = K^*YD_{K^*}h_0$, the equalities $D_KK^* = K^*D_{K^*}$, and

$$\operatorname{ran} D_K \cap \operatorname{ran} K^* = \operatorname{ran} D_KK^* = \operatorname{ran} K^*D_{K^*},$$

see (1.12), (1.13), one concludes that there exists $\varphi_0 \in \mathfrak{D}_{K^*}$ such that

$$\begin{cases} \Delta_F(\xi_0)K^*h_0 = K^*\varphi_0 \\ YD_{K^*}h_0 = D_{K^*}\varphi_0 \end{cases}.$$

Furthermore, the equality $D_{\Omega(\xi_0)^*} = D_{\Omega(\bar{\xi}_0)} = 0$ implies

$$\left(D_K\Delta_F(\bar{\xi}_0)K^* - K^*YD_{K^*}\right)h = 0$$

for all $h \in \mathfrak{M}$. Now $YD_{K^*}h_0 = D_{K^*}\varphi_0$ leads to $\Delta_F(\bar{\xi}_0)K^*h_0 = K^*\varphi_0$. It follows that

$$\Delta_F(\xi_0)K^*h_0 = \Delta_F(\bar{\xi}_0)K^*h_0.$$

Because $\Delta_F(\bar{\xi}_0) = \Delta_F(\xi_0)^* = \Delta_F(\xi_0)^{-1}$, one obtains $\left(I - \Delta_F(\xi_0)^2\right)K^*h_0 = 0$. From

$$\Delta_F(\xi_0) = (\xi_0 I - F)(I - \xi_0 F)^{-1}$$

it follows that

$$(1 - \xi_0^2)(I - \xi_0 F)^{-2}(I - F^2)K^*h_0 = 0.$$

Since $\ker D_F = \{0\}$ (because the system τ is minimal), we get $K^*h_0 = 0$. Therefore, $D_K\Delta_F(\xi_0)K^*h_0 = 0$ and $K^*YD_{K^*}h_0 = 0$. One concludes that

$$\begin{cases} D_K\Delta_F(\xi_0)K^*h = 0 \\ K^*YD_{K^*}h = 0 \end{cases} \quad \forall h \in \mathfrak{M}.$$

The equality $\operatorname{ran} Y = \mathfrak{D}_{K^*}$ implies $K^*D_{K^*} = D_KK^* = 0$. Therefore K is a partial isometry. The equality $D_K\Delta_F(\xi_0)K^* = 0$ implies $\operatorname{ran}(\Delta_F(\xi_0)K^*) \subseteq \operatorname{ran} K^*$. Representing $\Delta_F(\xi_0)$ as

$$\Delta_F(\xi_0) = (\xi_0 I - F)(I - \xi_0 F)^{-1}K^* = \left(\bar{\xi}_0 I + (\xi_0 - \bar{\xi}_0)(I - \xi_0 F)^{-1}\right)K^*,$$

we obtain that $F(\operatorname{ran} K^*) \subseteq \operatorname{ran} K^*$. Hence $F^n D_F(\operatorname{ran} K^*) \subseteq \operatorname{ran} K^*$ for all $n \in \mathbb{N}_0$. Because the system τ is minimal it follows that $\operatorname{ran} K^* = \mathfrak{D}_F = \mathcal{K}$, i.e., K is isometry and hence T is unitary (see Appendix B). This implies that $D_{\Omega(\xi)} = 0$ for all $\zeta \in \mathbb{T} \setminus \{-1, 1\}$, i.e., Ω is inner and, thus also bi-inner. □

Theorem 3.3. *Let $\Omega \in \mathcal{RS}(\mathfrak{M})$. If the equalities (3.2) hold and, in addition, for some $a \in (-1,1)$, $a \neq 0$, the equality*

$$(\Omega(a) - aI_{\mathfrak{M}})(I_{\mathfrak{M}} - a\Omega(a))^{-1} = \Omega(0) \tag{3.3}$$

is satisfied, then Ω is bi-inner.

Proof. Let $\tau = \{T; \mathfrak{M}, \mathfrak{M}, \mathcal{K}\}$ be a minimal passive selfadjoint system as in (2.2) with the transfer function Ω and let the entries of T in (2.3) be parameterized as in (2.4). According to Proposition 2.1 the equalities (3.2) mean that K is a partial isometry and $Y^2 = I_{\ker K^*}$.

Since D_{K^*} is the orthogonal projection, $\operatorname{ran} Y \subseteq \operatorname{ran} D_{N^*}$, from (2.6) we have

$$\Omega(z) = YD_{K^*} + K(zI - F)(I - zF)^{-1}K^*.$$

Rewrite (3.3) in the form

$$\Omega(0)(I_{\mathfrak{M}} - a\Omega(a)) = \Omega(a) - aI_{\mathfrak{M}}. \tag{3.4}$$

This leads to

$$\begin{aligned}
&(-KFK^* + YD_{K^*})\left(I_{\mathfrak{M}} - a\left(YD_{K^*} + K(aI - F)(I - aF)^{-1}K^*\right)\right) \\
&\quad = YD_{K^*} + K(aI - F)(I - aF)^{-1}K^* - aI_{\mathfrak{M}},
\end{aligned}$$

$$\begin{aligned}
&(-KFK^* + YD_{K^*})\left((I - aY)D_{K^*} + K\left(I - a\,(aI - F)\,(I - aF)^{-1}\right)K^*\right) \\
&\quad = (Y - aI)D_{K^*} + K\left((aI - F)(I - aF)^{-1} - aI\right)K^*,
\end{aligned}$$

$$\begin{aligned}
&-KFK^*K\left(I - a(aI - F)(I - aF)^{-1}\right)K^* + Y(I - aY)D_{K^*} \\
&\quad = (Y - aI)D_{K^*} + K\left((aI - F)(I - aF)^{-1} - aI\right)K^*.
\end{aligned}$$

Let P be an orthogonal projection from $\mathcal{K}$ onto $\operatorname{ran} K^*$. Since K is a partial isometry, one has $K^*K = P$. The equality $Y^2 = I_{\mathfrak{D}_{K^*}}$ implies $Y(I - aY)D_{K^*} = (Y - aI)D_{K^*}$. This leads to the following identities:

$$\begin{aligned}
&K[(-FP\left(I - a(aI - F)(I - aF)^{-1}\right) - (aI - F)(I - aF)^{-1} + aI]K^* = 0, \\
&KF(I_{\mathfrak{M}} - P)(I - aF)^{-1}K^* = 0, \\
&PF(I_{\mathfrak{M}} - P)(I - aF)^{-1}P = 0.
\end{aligned}$$

Represent the operator F in the block form

$$F = \begin{bmatrix} F_{11} & F_{12} \\ F_{12}^* & F_{22} \end{bmatrix} : \begin{matrix} \operatorname{ran} P \\ \oplus \\ \operatorname{ran}(I - P) \end{matrix} \to \begin{matrix} \operatorname{ran} P \\ \oplus \\ \operatorname{ran}(I - P) \end{matrix}.$$

Define

$$\Theta(z) = F_{11} + zF_{12}(I - zF_{22})^{-1}F_{12}^*.$$

Since F is a selfadjoint contraction, the function Θ belongs to the class $\mathcal{RS}(\operatorname{ran} P)$. From the Schur–Frobenius formula (A.1) it follows that

$$(I-P)(I-aF)^{-1}P = a(I-aF_{22})^{-1}F_{12}^*(I-a\Theta(a))^{-1}P.$$

This equality yields the equivalences

$$PF(I_{\mathfrak{M}}-P)(I-aF)^{-1}P=0 \Longleftrightarrow F_{12}(I-aF_{22})^{-1}F_{12}^*(I-a\Theta(a))^{-1}P=0$$
$$\Longleftrightarrow F_{12}(I-aF_{22})^{-1}F_{12}^*=0 \Longleftrightarrow (I-aF_{22})^{-1/2}F_{12}^*=0 \Longleftrightarrow F_{12}^*=0.$$

It follows that the subspace $\operatorname{ran} K^*$ reduces F. Hence $\operatorname{ran} K^*$ reduces D_F and, therefore $F^n D_F \operatorname{ran} K^* \subseteq \operatorname{ran} K^*$ for an arbitrary $n \in \mathbb{N}_0$. Since the system τ is minimal, we get $\operatorname{ran} K^* = \mathcal{K}$ and this implies that K is an isometry. Taking into account that $Y^2 = I_{\mathfrak{D}_{K^*}}$, we get that the block operator T is unitary. By Proposition 3.1 Ω is bi-inner. □

For completeness we recall the following result on the limit values $\Omega(\pm 1)$ of functions $\Omega \in \mathbf{S}^{qs}(\mathfrak{M})$ from [5, Theorem 5.8].

Lemma 3.4. *Let $\mathfrak{M}$ be a Hilbert space and let $\Omega \in \mathbf{S}^{qs}(\mathfrak{M})$. Then:*

1. *if $\Omega(\lambda)$ is inner then*

$$\begin{gathered}\left(\frac{\Omega(1)-\Omega(-1)}{2}\right)^2 = \frac{\Omega(1)-\Omega(-1)}{2},\\ (\Omega(1)+\Omega(-1))^*(\Omega(1)+\Omega(-1)) = 4I_{\mathfrak{M}} - 2\,(\Omega(1)-\Omega(-1));\end{gathered}\tag{3.5}$$

2. *if Ω is co-inner then*

$$\begin{gathered}\left(\frac{\Omega(1)-\Omega(-1)}{2}\right)^2 = \frac{\Omega(1)-\Omega(-1)}{2},\\ (\Omega(1)+\Omega(-1))(\Omega(1)+\Omega(-1))^* = 4I_{\mathfrak{M}} - 2\,(\Omega(1)-\Omega(-1));\end{gathered}\tag{3.6}$$

3. *if (3.5)/(3.6) holds and $\Omega(\xi)$ is isometric/co-isometric for some $\xi \in \mathbb{T}$, $\xi \neq \pm 1$, then Ω is inner/co-inner.*

Proposition 3.5. *If $\Omega \in \mathcal{RS}(\mathfrak{M})$ is an inner function, then*

$$\Omega(z_1)\Omega(z_2) = \Omega(z_2)\Omega(z_1), \quad \forall z_1, z_2 \in \mathbb{C} \setminus \{(-\infty,-1] \cup [1,+\infty)\}.$$

In particular, $\Omega(z)$ is a normal operator for each $z \in \mathbb{C} \setminus \{(-\infty,-1] \cup [1,+\infty)\}$.

Proof. The commutativity property follows from (3.1), where $D = \Omega(0)$. Normality follows from commutativity and symmetry $\Omega(z)^* = \Omega(\bar{z})$ for all z. □

4. Characterization of the class $\mathcal{RS}(\mathfrak{M})$

Theorem 4.1. *Let Ω be an operator-valued Nevanlinna function defined on $\mathbb{C} \setminus \{(-\infty,-1] \cup [1,+\infty)\}$. Then the following statements are equivalent:*

(i) *Ω belongs to the class $\mathcal{RS}(\mathfrak{M})$;*

(ii) Ω *satisfies the inequality*

$$I - \Omega^*(z)\Omega(z) - (1-|z|^2)\frac{\operatorname{Im}\Omega(z)}{\operatorname{Im} z} \geq 0, \quad \operatorname{Im} z \neq 0; \tag{4.1}$$

(iii) *the function*

$$K(z,w) := I - \Omega^*(w)\Omega(z) - \frac{1-\bar{w}z}{z-\bar{w}}\left(\Omega(z) - \Omega^*(w)\right)$$

is a non-negative kernel on the domains

$\mathbb{C}\setminus\{(-\infty,-1]\cup[1,+\infty)\}$, $\operatorname{Im} z > 0$ *and* $\mathbb{C}\setminus\{(-\infty,-1]\cup[1,+\infty)\}$, $\operatorname{Im} z < 0$;

(iv) *the function*

$$\Upsilon(z) = (zI - \Omega(z))\left(I - z\Omega(z)\right)^{-1}, \quad z \in \mathbb{C}\setminus\{(-\infty,-1]\cup[1,+\infty)\}, \tag{4.2}$$

is well defined and belongs to $\mathcal{RS}(\mathfrak{M})$.

Proof. (i)$\Longrightarrow$(ii) and (i)$\Longrightarrow$(iii). Assume that $\Omega \in \mathcal{RS}(\mathfrak{M})$ and let Ω be represented as the transfer function of a passive selfadjoint system $\tau = \{T;\mathfrak{M},\mathfrak{M},\mathcal{K}\}$ as in (2.2) with the selfadjoint contraction T as in (2.4). According to (2.6) we have

$$\Omega(z) = D_{K^*}YD_{K^*} + K\Delta_F(z)K^*, \ z \in \mathbb{C}\setminus\{(-\infty,-1]\cup[1,+\infty)\}.$$

Taking into account that, see [20, Chapter VI],

$$((I - \Delta_F^*(w)\Delta_F(z))\varphi,\psi) = (1-\bar{w}z)((I-zF)^{-1}D_F\varphi,(I-wF)^{-1}D_F\psi)$$

and

$$((\Delta_F(z) - \Delta_F^*(w))\varphi,\psi) = (z-\bar{w})((I-zF)^{-1}D_F\varphi,(I-wF)^{-1}D_F\psi),$$

we obtain

$$\begin{aligned} ||h||^2 - ||\Omega(z)h||^2 &= ||K^*h||^2 - ||\Delta_F(z)K^*h||^2 \\ &\quad + ||D_YD_{K^*}h||^2 + ||(K^*YD_{K^*} - D_K\Delta_F(z)K^*)h||^2 \\ &= (1-|z|^2)||(I-zF)^{-1}D_FK^*h||^2 + ||D_YD_{K^*}h||^2 \\ &\quad + ||(K^*YD_{K^*} - D_K\Delta_F(z)K^*)h||^2. \end{aligned}$$

Moreover,

$$\operatorname{Im}(\Omega(z)h,h) = \operatorname{Im} z||(I-zF)^{-1}D_FK^*h||^2$$

and

$$\begin{aligned} &\operatorname{Im} z(||h||^2 - ||\Omega(z)h||^2) - (1-|z|^2)\operatorname{Im}(\Omega(z)h,h) \\ &\quad = \operatorname{Im} z\left(||D_YD_{K^*}h||^2 + ||(K^*YD_{K^*} - D_K\Delta_F(z)K^*)h||^2\right). \end{aligned}$$

Similarly,

$$\begin{aligned} (K(z,w)f,g) &= ((I - \Omega^*(w)\Omega(z))f,g) - \frac{1-\bar{w}z}{z-\bar{w}}((\Omega(z) - \Omega^*(w))f,g) \\ &= (D_Y^2D_{K^*}f, D_{K^*}g) \\ &\quad + ((D_K\Delta_F(z)K^* - K^*YD_{K^*})f,(D_K\Delta_F(w)K^* - K^*YD_{K^*})g). \end{aligned} \tag{4.3}$$

It follows from (4.3) that for arbitrary complex numbers $\{z_k\}_{k=1}^m \subset \mathbb{C}\setminus\{(-\infty,-1]\cup[1,+\infty)\}$, $\operatorname{Im} z_k > 0$, $k=1,\dots,n$ or $\{z_k\}_{k=1}^m \subset \mathbb{C}\setminus\{(-\infty,-1]\cup[1,+\infty)\}$, $\operatorname{Im} z_k < 0$, $k=1,\dots,n$ and for arbitrary vectors $\{f_k\}_{k=1}^\infty \subset \mathfrak{M}$ the relation

$$\sum_{k=1}^{n}(K(z_k,z_m)f_k,f_m) = \left\|D_Y D_{K^*}\sum_{k=1}^{\infty} f_k\right\|^2 + \left\|\sum_{k=1}^{\infty}(D_K\Delta_F(z_k)K^* - K^*YD_{K^*})f_k\right\|^2$$

holds. Therefore $K(z,w)$ is a non-negative kernel.

(iii)$\Longrightarrow$(ii) is evident.

(ii)$\Longrightarrow$(iv) Because $\operatorname{Im} z>0$ $(\operatorname{Im} z<0)$ $\Longrightarrow$ $\operatorname{Im}\Omega(z)\ge 0$ $(\operatorname{Im}\Omega(z)\le 0)$, the inclusion $1/z\in\rho(\Omega(z))$ is valid for z with $\operatorname{Im} z\ne 0$. In addition $1/x\in\rho(\Omega(x))$ for $x\in(-1,1)$, $x\ne 0$, because $\Omega(x)$ is a contraction. Hence $\Upsilon(z)$ is well defined on $\mathfrak{M}$ and $\Upsilon^*(z)=\Upsilon(\bar z)$ for all $z\in\mathbb{C}\setminus\{(-\infty,-1]\cup[1,+\infty)\}$. Furthermore, with $\operatorname{Im} z\ne 0$ one has

$$\operatorname{Im}\Upsilon(z) = (I-\bar z\Omega^*(z))^{-1}\left[\operatorname{Im} z(I-\Omega^*(z)\Omega(z)) - (1-|z|^2)\operatorname{Im}\Omega(z)\right] \times (I-z\Omega(z))^{-1},$$

while for $x\in(-1,1)$

$$I-\Upsilon^2(x) = (1-x^2)\,(I-x\Omega(x))^{-1}\,(I-\Omega^2(x))\,(I-x\Omega(x))^{-1}.$$

Thus, $\Upsilon\in\mathcal{RS}(\mathfrak{M})$.

(iv)$\Longrightarrow$(i) It is easy to check that if Υ is given by (4.2), then

$$\Omega(z) = (zI-\Upsilon(z))\,(I-z\Upsilon(z))^{-1},\ z\in\mathbb{C}\setminus\{(-\infty,-1]\cup[1,+\infty)\}.$$

Hence, this implication reduces back to the proven implication (i)$\Longrightarrow$(ii). □

Remark 4.2. 1) Inequality (4.1) can be rewritten as follows

$$((I-\Omega^*(z)\Omega(z))f,f) - \frac{1-|z|^2}{|\operatorname{Im} z|}|\operatorname{Im}(\Omega(z)f,f)|\ge 0,\quad \operatorname{Im} z\ne 0,\ f\in\mathfrak{M}.$$

Let $\beta\in[0,\pi/2]$. Taking into account that

$$|z\sin\beta\pm i\cos\beta|^2=1 \Longleftrightarrow 1-|z|^2=\pm 2\cot\beta\operatorname{Im} z$$

one obtains, see (2.1),

$$\begin{cases} |z\sin\beta+i\cos\beta|=1 \\ z\ne\pm 1\end{cases} \Longrightarrow \|\Omega(z)\sin\beta+i\cos\beta\, I\|\le 1$$
$$\begin{cases} |z\sin\beta-i\cos\beta|=1 \\ z\ne\pm 1\end{cases} \Longrightarrow \|\Omega(z)\sin\beta-i\cos\beta\, I\|\le 1.$$

2) Inequality (4.1) implies

$$I-\Omega^*(x)\Omega(x)-(1-x^2)\Omega'(x)\ge 0,\quad x\in(-1,1).$$

3) Formula (3.1) implies that if $\Omega \in \mathcal{RS}(\mathfrak{M})$ is an inner function, then

$$I - \Omega^*(w)\Omega(z) - \frac{1-\bar{w}z}{z-\bar{w}}\left(\Omega(z) - \Omega^*(w)\right) = 0,\ z \neq \bar{w}.$$

In particular,

$$\frac{\Omega(z)-\Omega(0)}{z} = I - \Omega(0)\Omega(z), \quad z \in \mathbb{C}\setminus\{-\infty,-1]\cup[1,+\infty)\},\ z \neq 0,$$
$$\Omega'(0) = I - \Omega(0)^2.$$

This combined with (2.15) yields $\Lambda(z) = zI_{\mathfrak{D}_{\Omega(0)}}$ in the representation (2.13) for an inner function $\Omega \in \mathcal{RS}(\mathfrak{M})$.

5. Compressed resolvents and the class $\mathbf{N}^0_{\mathfrak{M}}[-1,1]$

Definition 5.1. Let $\mathfrak{M}$ be a Hilbert space. A $\mathbf{B}(\mathfrak{M})$-valued Nevanlinna function M is said to belong to the class $\mathbf{N}^0_{\mathfrak{M}}[-1,1]$ if it is holomorphic outside the interval $[-1,1]$ and

$$\lim_{\xi\to\infty} \xi M(\xi) = -I_{\mathfrak{M}}.$$

It follows from [3] that $M \in \mathbf{N}^0_{\mathfrak{M}}[-1,1]$ if and only if there exist a Hilbert space $\mathfrak{H}$ containing $\mathfrak{M}$ as a subspace and a selfadjoint contraction T in $\mathfrak{H}$ such that T is $\mathfrak{M}$-simple and

$$M(\xi) = P_{\mathfrak{M}}(T-\xi I)^{-1}{\upharpoonright}\mathfrak{M}, \quad \xi \in \mathbb{C}\setminus[-1,1].$$

Moreover, formula (1.6) implies the following connections between the classes $\mathbf{N}^0_{\mathfrak{M}}[-1,1]$ and $\mathcal{RS}(\mathfrak{M})$ (see also [3, 5]):

$$\begin{aligned} M(\xi) \in \mathbf{N}^0_{\mathfrak{M}}[-1,1] &\Longrightarrow \Omega(z) := M^{-1}(1/z) + 1/z \in \mathcal{RS}(\mathfrak{M}),\\ \Omega(z) \in \mathcal{RS}(\mathfrak{M}) &\Longrightarrow M(\xi) := (\Omega(1/\xi)-\xi)^{-1} \in \mathbf{N}^0_{\mathfrak{M}}[-1,1]. \end{aligned} \tag{5.1}$$

Let $\Omega(z) = (zI+D)(I+zD)^{-1}$ be an inner function from the class $\mathcal{RS}(\mathfrak{M})$, then by (5.1)

$$\Omega(z) = (zI+D)(I+zD)^{-1} \Longrightarrow M(\xi) = \frac{\xi I + D}{1-\xi^2}, \quad \xi \in \mathbb{C}\setminus[-1,1].$$

The identity $\Omega(z)^*\Omega(z) = I_{\mathfrak{M}}$ for $z \in \mathbb{T}\setminus\{\pm 1\}$ is equivalent to

$$2\mathrm{Re}\,(\xi M(\xi)) = -I_{\mathfrak{M}}, \quad \xi \in \mathbb{T}\setminus\{\pm1\}.$$

The next statement is established in [2]. Here we give another proof.

Theorem 5.2. *If $M(\xi) \in \mathbf{N}^0_{\mathfrak{M}}[-1,1]$, then the function*

$$\frac{M^{-1}(\xi)}{\xi^2-1}, \quad \xi \in \mathbb{C}\setminus[-1,1],$$

belongs to $\mathbf{N}^0_{\mathfrak{M}}[-1,1]$ as well.

Proof. Let $M(\xi) \in \mathbf{N}^0_{\mathfrak{M}}[-1,1]$. Due to (5.1) $\Omega(z) = M^{-1}(1/z) + 1/z$ belongs to $\mathcal{RS}(\mathfrak{M})$. According to Theorem 4.1 the function

$$\Upsilon(z) = (zI - \Omega(z))\left(I - z\Omega(z)\right)^{-1}, \quad z \in \mathbb{C} \setminus \{(-\infty,-1] \cup [1,+\infty)\}$$

belongs to $\mathcal{RS}(\mathfrak{M})$. From the equality

$$I - z\Upsilon(z) = (1-z^2)\left(I - z\Omega(z)\right)^{-1}, \quad z \in \mathbb{C} \setminus \{(-\infty,-1] \cup [1,+\infty)\}$$

we get

$$\left(I - z\Upsilon(z)\right)^{-1} = \frac{I - z\Omega(z)}{1-z^2}.$$

Simple calculations give

$$\left(\Upsilon(1/\xi) - \xi\right)^{-1} = \frac{M^{-1}(\xi)}{\xi^2 - 1}, \quad \xi \in \mathbb{C} \setminus [-1,1].$$

Now in view of (5.1) the function $\dfrac{M^{-1}(\xi)}{\xi^2-1}$ belongs to $\mathbf{N}^0_{\mathfrak{M}}[-1,1]$. □

6. Transformations of the classes $\mathcal{RS}(\mathfrak{M})$ and $\mathbf{N}^0_{\mathfrak{M}}[-1,1]$

We start by studying transformations of the class $\mathcal{RS}(\mathfrak{M})$ given by (1.8), (1.10):

$$\mathcal{RS}(\mathfrak{M}) \ni \Omega \mapsto \mathbf{\Phi}(\Omega) = \Omega_{\mathbf{\Phi}}(z) := (zI - \Omega(z))(I - z\Omega(z))^{-1},$$

$$\mathcal{RS}(\mathfrak{M}) \ni \Omega \mapsto \Xi_a(\Omega) = \Omega_a(z) := \Omega\left(\frac{z+a}{1+za}\right), \quad a \in (-1,1),$$

and the transform

$$\mathcal{RS}(H) \ni \Omega \mapsto \mathbf{\Pi}(\Omega) = \Omega_{\mathbf{\Pi}}(z) : K_{11} + K_{12}\Omega(z)(I - K_{22}\Omega(z))^{-1}K_{12}^*, \tag{6.1}$$

which is determined by the selfadjoint contraction K of the form

$$\mathbf{K} = \begin{bmatrix} K_{11} & K_{12} \\ K_{12}^* & K_{22} \end{bmatrix} : \begin{matrix} \mathfrak{M} \\ \oplus \\ H \end{matrix} \to \begin{matrix} \mathfrak{M} \\ \oplus \\ H \end{matrix};$$

in all these transforms $z \in \mathbb{C} \setminus \{(-\infty,-1] \cup [1,+\infty)\}$.

A particular case of (6.1) is the transformation $\mathbf{\Pi}_a$ determined by the block operator

$$\mathbf{K}_a = \begin{bmatrix} aI & \sqrt{1-a^2}I \\ \sqrt{1-a^2} & -aI \end{bmatrix} : \begin{matrix} \mathfrak{M} \\ \oplus \\ \mathfrak{M} \end{matrix} \to \begin{matrix} \mathfrak{M} \\ \oplus \\ \mathfrak{M} \end{matrix}, \quad a \in (-1,1),$$

i.e., see (1.10),

$$\mathcal{RS}(\mathfrak{M}) \ni \Omega(z) \mapsto \widehat{\Omega}_a(z) := (aI + \Omega(z))(I + a\Omega(z))^{-1}.$$

By Theorem 4.1 the mapping $\mathbf{\Phi}$ given by (1.8) is an automorphism of the class $\mathcal{RS}(\mathfrak{M})$, $\mathbf{\Phi}^{-1} = \mathbf{\Phi}$. The equality (3.1) shows that the set of all inner functions of the class $\mathcal{RS}(\mathfrak{M})$ is the image of all constant functions under the transformation $\mathbf{\Phi}$. In addition, for $a,b\in(-1,1)$ the following identities hold:

$$\mathbf{\Pi}_b\circ\mathbf{\Pi}_a=\mathbf{\Pi}_a\circ\mathbf{\Pi}_b=\mathbf{\Pi}_c,\quad \Xi_b\circ\Xi_a=\Xi_a\circ\Xi_b=\Xi_c,\quad\text{where } c=\frac{a+b}{1+ab}.$$

The mapping $\mathbf{\Gamma}$ on the class $\mathbf{N}^0_{\mathfrak{M}}[-1,1]$ (see Theorem 5.2) defined by

$$\mathbf{N}^0_{\mathfrak{M}}[-1,1]\ni M(\xi)\overset{\mathbf{\Gamma}}{\mapsto} M_{\mathbf{\Gamma}}(\xi):=\frac{M^{-1}(\xi)}{\xi^2-1}\in\mathbf{N}^0_{\mathfrak{M}}[-1,1] \tag{6.2}$$

has been studied recently in [2]. It is obvious that $\mathbf{\Gamma}^{-1}=\mathbf{\Gamma}$.

Using the relations (5.1) we define the transform $\mathbf{U}$ and its inverse $\mathbf{U}^{-1}$ which connect the classes $\mathcal{RS}(\mathfrak{M})$ and $\mathbf{N}^0_{\mathfrak{M}}[-1,1]$:

$$\mathcal{RS}(\mathfrak{M})\ni\Omega(z)\overset{\mathbf{U}}{\mapsto} M(\xi):=(\Omega(1/\xi)-\xi)^{-1}\in\mathbf{N}^0_{\mathfrak{M}}[-1,1],\quad \xi\in\mathbb{C}\setminus[-1,1]. \tag{6.3}$$

$$\mathbf{N}^0_{\mathfrak{M}}[-1,1]\ni M(\xi)\overset{\mathbf{U}^{-1}}{\mapsto}\Omega(z):=M^{-1}(1/z)+1/z\in\mathcal{RS}(\mathfrak{M}), \tag{6.4}$$

where $z\in\mathbb{C}\setminus\{(-\infty,-1]\cup[1,+\infty)\}$. The proof of Theorem 5.2 contains the following commutation relations

$$\mathbf{U}\mathbf{\Phi}=\mathbf{\Gamma}\mathbf{U},\quad \mathbf{\Phi}\mathbf{U}^{-1}=\mathbf{U}^{-1}\mathbf{\Gamma}. \tag{6.5}$$

One of the main aims in this section is to solve the following realization problem concerning the above transforms: given a passive selfadjoint system $\tau=\{T;\mathfrak{M},\mathfrak{M},\mathcal{K}\}$ with the transfer function Ω, construct a passive selfadjoint systems whose transfer function coincides with $\Phi(\Omega)$, $\Xi_a(\Omega)$, $\mathbf{\Pi}(\Omega)$, and $\mathbf{\Pi}_a(\Omega)$, respectively. We will also determine the fixed points of all the mappings $\mathbf{\Phi}$, $\mathbf{\Gamma}$, Ξ_a, and $\mathbf{\Pi}_a$.

6.1. The mappings $\mathbf{\Phi}$ and $\mathbf{\Gamma}$ and inner dilations of the functions from $\mathcal{RS}(\mathfrak{M})$

Theorem 6.1.

1) *Let $\tau=\{T;\mathfrak{M},\mathfrak{M},\mathcal{K}\}$ be a passive selfadjoint system and let Ω be its transfer function. Define*

$$T_{\mathbf{\Phi}}:=\begin{bmatrix}-P_{\mathfrak{M}}T{\upharpoonright}\mathfrak{M} & P_{\mathfrak{M}}D_T\\ D_T{\upharpoonright}\mathfrak{M} & T\end{bmatrix}:\begin{matrix}\mathfrak{M}\\ \oplus\\ \mathfrak{D}_T\end{matrix}\to\begin{matrix}\mathfrak{M}\\ \oplus\\ \mathfrak{D}_T\end{matrix}. \tag{6.6}$$

Then $T_{\mathbf{\Phi}}$ is a selfadjoint contraction and $\Omega_{\mathbf{\Phi}}(z)=(zI-\Omega(z))(I-z\Omega(z))^{-1}$ is the transfer function of the passive selfadjoint system of the form

$$\tau_{\mathbf{\Phi}}=\{T_{\mathbf{\Phi}};\mathfrak{M},\mathfrak{M},\mathfrak{D}_T\}.$$

Moreover, if the system τ is minimal, then the system $\tau_{\mathbf{\Phi}}$ is minimal, too.

2) *Let T be a selfadjoint contraction in $\mathfrak{H}$, let $\mathfrak{M}$ be a subspace of $\mathfrak{H}$ and let*

$$M(\xi)=P_{\mathfrak{M}}(T-\xi I)^{-1}{\upharpoonright}\mathfrak{M}.$$

Consider a Hilbert space $\widehat{\mathfrak{H}} := \mathfrak{M} \oplus \mathfrak{H}$ *and let* $\widehat{P}_{\mathfrak{M}}$ *be the orthogonal projection in* $\widehat{\mathfrak{H}}$ *onto* $\mathfrak{M}$*. Then*

$$\frac{M^{-1}(\xi)}{\xi^2 - 1} = \widehat{P}_{\mathfrak{M}}(T_{\mathbf{\Phi}} - \xi I)^{-1}\upharpoonright \mathfrak{M}, \tag{6.7}$$

where $T_{\mathbf{\Phi}}$ *is defined by* (6.6).

3) *The function*

$$\widetilde{\Omega}(z) = (zI - T_{\mathbf{\Phi}})(I - zT_{\mathbf{\Phi}})^{-1}, \quad z \in \mathbb{C} \setminus \{(-\infty, -1] \cup [1, +\infty)\}$$

satisfies

$$\Omega(z) = P_{\mathfrak{M}}\widetilde{\Omega}(z)\upharpoonright \mathfrak{M}.$$

Proof. 1) According to (1.6) one has

$$P_{\mathfrak{M}}(I - zT)^{-1}\upharpoonright \mathfrak{M} = (I_{\mathfrak{M}} - z\Omega(z))^{-1}$$

for $z \in \mathbb{C} \setminus \{(-\infty, -1] \cup [1, +\infty)\}$. Let

$$\Omega_{\mathbf{\Phi}}(z) = (zI - \Omega(z))(I - z\Omega(z))^{-1}.$$

Now simple calculations give

$$\Omega_{\mathbf{\Phi}}(z) = \left(z - \frac{1}{z}\right)(I - z\Omega(z))^{-1} + \frac{I_{\mathfrak{M}}}{z} = P_{\mathfrak{M}}(zI - T)(I - zT)^{-1}\upharpoonright \mathfrak{M}. \tag{6.8}$$

Observe that the subspace $\mathfrak{D}_T$ is invariant under T; cf. (1.12). Let $\mathfrak{H} := \mathfrak{M} \oplus \mathfrak{D}_T$ and let $T_{\mathbf{\Phi}}$ be given by (6.6). Since T is a selfadjoint contraction in $\mathfrak{M} \oplus \mathcal{K}$, the inequalities

$$\left(\begin{bmatrix}\varphi \\ f\end{bmatrix}, \begin{bmatrix}\varphi \\ f\end{bmatrix}\right) \pm \left(\begin{bmatrix}\varphi \\ f\end{bmatrix}, T_{\mathbf{\Phi}}\begin{bmatrix}\varphi \\ f\end{bmatrix}\right) = \left\|(I \mp T)^{1/2}\varphi \pm (I \pm T)^{1/2} f\right\|^2$$

hold for all $\varphi \in \mathfrak{M}$ and $f \in \mathfrak{D}_T$. Therefore $T_{\mathbf{\Phi}}$ is a selfadjoint contraction in the Hilbert space $\mathfrak{H}$ and the system

$$\tau_{\mathbf{\Phi}} = \left\{\begin{bmatrix} -P_{\mathfrak{M}}T\upharpoonright \mathfrak{M} & P_{\mathfrak{M}}D_T \\ D_T\upharpoonright \mathfrak{M} & T\end{bmatrix}; \mathfrak{M}, \mathfrak{M}, \mathfrak{D}_T\right\}$$

is passive selfadjoint. Suppose that τ is minimal, i.e.,

$$\overline{\operatorname{span}}\,\{T^n\mathfrak{M},\ n \in \mathbb{N}_0\} = \mathfrak{M} \oplus \mathcal{K} \Longleftrightarrow \bigcap_{n=0}^{\infty} \ker(P_{\mathfrak{M}}T^n) = \{0\}.$$

Since

$$\mathfrak{D}_T \ominus \{\overline{\operatorname{span}}\,\{T^n D_T\mathfrak{M},\ n \in \mathbb{N}_0\}\} = \bigcap_{n=0}^{\infty} \ker(P_{\mathfrak{M}}T^n D_T),$$

we get $\overline{\operatorname{span}}\,\{T^n D_T\mathfrak{M} :\ n \in \mathbb{N}_0\} = \mathfrak{D}_T$. This means that the system τ_Γ is minimal.

For the transfer function $\Upsilon(z)$ of τ_{Φ} we get

$$\begin{aligned}\Upsilon(z) &= (-P_{\mathfrak{M}}T + zP_{\mathfrak{M}}D_T(I-zT)^{-1}D_T)\upharpoonright\mathfrak{M}\\ &= P_{\mathfrak{M}}\left(-T + zD_T^2(I-zT)^{-1}\right)\upharpoonright\mathfrak{M}\\ &= P_{\mathfrak{M}}(zI-T)(I-zT)^{-1}\upharpoonright\mathfrak{M},\end{aligned}$$

with $z \in \mathbb{C}\setminus\{(-\infty,-1]\cup[1,+\infty)\}$. Comparison with (6.8) completes the proof.

2) Clearly $M(\xi) = P_{\mathfrak{M}}(T-\xi I)^{-1}\upharpoonright\mathfrak{M}$ belongs to the class $\mathbf{N}^0_{\mathfrak{M}}[-1,1]$. Consequently, $\Omega(z) := M^{-1}(1/z) + 1/z \in \mathcal{RS}(\mathfrak{M})$. The function Ω is the transfer function of the passive selfadjoint system

$$\tau = \{T;\mathfrak{M},\mathfrak{M},\mathcal{K}\},$$

where $\mathcal{K} = \mathfrak{H}\ominus\mathfrak{M}$. Let $\Upsilon = \mathbf{\Phi}(\Omega)$ and $\widehat{M} = \mathbf{U}(\Upsilon)$. From (6.2)–(6.5) it follows that

$$\widehat{M}(\xi) = \frac{M^{-1}(\xi)}{\xi^2-1},\quad \xi\in\mathbb{C}\setminus[-1,1].$$

As was shown above, the function Υ is the transfer function of the passive selfadjoint system

$$\tau_{\Phi} = \{T_{\Phi};\mathfrak{M},\mathfrak{M},\mathfrak{H}\},$$

where T_{Φ} is given by (6.6). Then again the Schur–Frobenius formula (1.6) gives

$$\widehat{M}(\xi) = \widehat{P}_{\mathfrak{M}}(T_{\Phi}-\xi I)^{-1}\upharpoonright\mathfrak{M},\quad \xi\in\mathbb{C}\setminus[-1,1].$$

3) For all $z\in\mathbb{C}\setminus\{(-\infty,-1]\cup[1,+\infty)\}$ one has

$$\widetilde{\Omega}(z) = \left(z-\frac{1}{z}\right)(I-zT_{\Phi})^{-1} + \frac{1}{z}I.$$

Then

$$\begin{aligned}P_{\mathfrak{M}}\widetilde{\Omega}(z)\upharpoonright\mathfrak{M} &= \left(z-\frac{1}{z}\right)(I_{\mathfrak{M}}-z\Upsilon(z))^{-1} + \frac{1}{z}I_{\mathfrak{M}}\\ &= (zI_{\mathfrak{M}}-\Upsilon(z))(I_{\mathfrak{M}}-z\Upsilon(z))^{-1} = \Omega(z).\end{aligned}$$

This completes the proof. □

Notice that if $\Omega(z)\equiv\text{const} = D$, then $\Upsilon(z) = (zI-D)(I-zD)^{-1}$, $z\in\mathbb{C}\setminus\{(-\infty,-1]\cup[1,+\infty)\}$. This is the transfer function of the conservative and selfadjoint system

$$\Sigma = \left\{\begin{bmatrix}-D & D_D\\ D_D & D\end{bmatrix},\mathfrak{M},\mathfrak{M},\mathfrak{D}_D\right\}.$$

Remark 6.2. The block operator T_{Φ} of the form (6.6) appeared in [2] and relation (6.7) is also established in [2].

Theorem 6.3.

1) *Let* $\mathfrak{M}$ *be a Hilbert space and let* $\Omega \in \mathcal{RS}(\mathfrak{M})$. *Then there exist a Hilbert space* $\widetilde{\mathfrak{M}}$ *containing* $\mathfrak{M}$ *as a subspace and a selfadjoint contraction* $\widetilde{A}$ *in* $\widetilde{\mathfrak{M}}$ *such that for all* $z \in \mathbb{C} \setminus \{(-\infty,-1] \cup [1,+\infty)\}$ *the equality*

$$\Omega(z) = P_{\mathfrak{M}}(zI_{\widetilde{\mathfrak{M}}} + \widetilde{A})(I_{\widetilde{\mathfrak{M}}} + z\widetilde{A})^{-1}{\upharpoonright}\mathfrak{M} \tag{6.9}$$

holds. Moreover, the pair $\{\widetilde{\mathfrak{M}}, \widetilde{A}\}$ *can be chosen such that* $\widetilde{A}$ *is* $\mathfrak{M}$*-simple, i.e.,*

$$\overline{\operatorname{span}}\,\{\widetilde{A}^n\mathfrak{M} :\ n \in \mathbb{N}_0\} = \widetilde{\mathfrak{M}}. \tag{6.10}$$

The function Ω *is inner if and only if* $\widetilde{\mathfrak{M}} = \mathfrak{M}$ *in the representation* (6.10).

If there are two representations of the form (6.9) *with pairs* $\{\widetilde{\mathfrak{M}}_1, \widetilde{A}_1\}$ *and* $\{\widetilde{\mathfrak{M}}_2, \widetilde{A}_2\}$ *that are* $\mathfrak{M}$*-simple, then there exists a unitary operator* $\widetilde{U} \in \mathbf{B}(\widetilde{\mathfrak{M}}_1, \widetilde{\mathfrak{M}}_2)$ *such that*

$$\widetilde{U}{\upharpoonright}\mathfrak{M} = I_{\mathfrak{M}}, \quad \widetilde{A}_2\widetilde{U} = \widetilde{U}\widetilde{A}_1. \tag{6.11}$$

2) *The formula*

$$\Omega(z) = \int_{-1}^{1} \frac{z+t}{1+zt}\,d\sigma(t), \quad z \in \mathbb{C} \setminus \{(-\infty,-1] \cup [1,+\infty)\}, \tag{6.12}$$

gives a one-one correspondence between functions Ω *from the class* $\mathcal{RS}(\mathfrak{M})$ *and nondecreasing left-continuous* $\mathbf{B}(\mathfrak{M})$*-valued functions* σ *on* $[-1,1]$ *with* $\sigma(-1) = 0$, $\sigma(1) = I_{\mathfrak{M}}$.

Proof. 1) Realize Ω as the transfer function of a minimal passive selfadjoint system $\tau = \{T; \mathfrak{M}, \mathfrak{M}, \mathcal{K}\}$. Let the selfadjoint contraction $T_{\mathbf{\Phi}}$ be given by (6.6) and let $\widetilde{\mathfrak{M}} := \mathfrak{M} \oplus \mathfrak{D}_T$ and $\widetilde{A} := -T_{\mathbf{\Phi}}$. Then the relations (6.9) and (6.10) are obtained from Theorem 6.1. Using Proposition 3.1 one concludes that Ω is inner precisely when $\widetilde{\mathfrak{M}} = \mathfrak{M}$ in the right-hand side of (6.10). Since

$$\begin{aligned} &P_{\mathfrak{M}}(zI_{\widetilde{\mathfrak{M}}_1} + \widetilde{A}_1)(I_{\widetilde{\mathfrak{M}}_1} + z\widetilde{A}_1)^{-1}{\upharpoonright}\mathfrak{M} = P_{\mathfrak{M}}(zI_{\widetilde{\mathfrak{M}}_2} + \widetilde{A}_2)(I_{\widetilde{\mathfrak{M}}_2} + z\widetilde{A}_2)^{-1}{\upharpoonright}\mathfrak{M} \\ &\iff P_{\mathfrak{M}}(I_{\widetilde{\mathfrak{M}}_1} + z\widetilde{A}_1)^{-1}{\upharpoonright}\mathfrak{M} = P_{\mathfrak{M}}(I_{\widetilde{\mathfrak{M}}_2} + z\widetilde{A}_2)^{-1}{\upharpoonright}\mathfrak{M}, \end{aligned}$$

the $\mathfrak{M}$-simplicity with standard arguments (see, e.g., [3, 6]) yields the existence of unitary $\widetilde{U} \in \mathbf{B}(\widetilde{\mathfrak{M}}_1, \widetilde{\mathfrak{M}}_2)$ satisfying (6.11).

2) Let (6.9) be satisfied and let $\sigma(t) = P_{\mathfrak{M}}\widetilde{E}(t){\upharpoonright}\mathfrak{M}$, $t \in [-1,1]$, where $E(t)$ is the spectral family of the selfadjoint contraction $\widetilde{A}$ in $\widetilde{\mathfrak{M}}$. Then clearly (6.12) holds.

Conversely, let σ be a nondecreasing left-continuous $\mathbf{B}(\mathfrak{M})$-valued function $[-1,1]$ with $\sigma(-1) = 0$, $\sigma(1) = I_{\mathfrak{M}}$. Define Ω by the right-hand side of (6.12). Then, the function Ω in (6.12) belongs to the class $\mathcal{RS}(\mathfrak{M})$. □

Remark 6.4. If Ω is represented in the form (6.9), then the proof of Theorem 6.1 shows that the transfer function of the passive selfadjoint system

$$\widetilde{\sigma}_{\mathbf{\Phi}} = \{(-\widetilde{A})_{\mathbf{\Phi}}; \mathfrak{M}, \mathfrak{M}, \mathfrak{D}_{\widetilde{A}}\}$$

coincides with Ω. Moreover, if $\widetilde{A}$ is $\mathfrak{M}$-simple, then $\widetilde{\sigma}_{\mathbf{\Phi}}$ is minimal.

Remark 6.5. The functions from the class $\mathcal{S}^{qs}(\mathfrak{M})$ admits the following integral representations, see [5]:

$$\Theta(z) = \Theta(0) + z\int_{-1}^{1} \frac{1-t^2}{1-tz}\, dG(t),$$

where $G(t)$ is a nondecreasing $\mathbf{B}(\mathfrak{M})$-valued function with bounded variation, $G(-1) = 0$, $G(1) \le I_{\mathfrak{M}}$, and

$$\left|\left(\left(\Theta(0) + \int_{-1}^{1} t\, dG(t)\right) f, g\right)\right|^2 \le ((I - G(1))\, f, f)\, ((I - G(1))\, g, g)$$

for all $f, g \in \mathfrak{M}$.

Proposition 6.6 (cf. [2]).

1) *The mapping* $\mathbf{\Phi}$ *of* $\mathcal{RS}(\mathfrak{M})$ *has a unique fixed point*

$$\Omega_0(z) = \frac{zI_{\mathfrak{M}}}{1+\sqrt{1-z^2}}, \quad \textit{with} \quad \Omega_0(i) = \frac{iI_{\mathfrak{M}}}{1+\sqrt{2}}. \tag{6.13}$$

2) *The mapping* $\mathbf{\Gamma}$ *has a unique fixed point*

$$M_0(\xi) = -\frac{I_{\mathfrak{M}}}{\sqrt{\xi^2-1}} \quad \textit{with} \quad M_0(i) = \frac{iI_{\mathfrak{M}}}{\sqrt{2}}. \tag{6.14}$$

3) *Define the weight function* $\rho(t)$ *and the weighted Hilbert space* $\mathfrak{H}_0$ *as follows*

$$\begin{aligned}
&\rho_0(t) = \frac{1}{\pi}\frac{1}{\sqrt{1-t^2}}, \quad t \in (-1,1),\\
&\mathfrak{H}_0 := L_2([-1,1], \mathfrak{M}, \rho_0(t)) = L_2([-1,1],\ \rho_0(t)) \bigotimes \mathfrak{M}\\
&= \left\{ f(t) : \int\limits_{-1}^{1} \frac{||f(t)||^2_{\mathfrak{M}}}{\sqrt{1-t^2}}\, dt < \infty \right\}.
\end{aligned} \tag{6.15}$$

Then $\mathfrak{H}_0$ *is the Hilbert space with the inner product*

$$(f(t), g(t))_{\mathfrak{H}_0} = \frac{1}{\pi}\int\limits_{-1}^{1} (f(t), g(t))_{\mathfrak{M}}\, \rho_0(t)\, dt = \frac{1}{\pi}\int\limits_{-1}^{1} \frac{(f(t), g(t))_{\mathfrak{M}}}{\sqrt{1-t^2}}\, dt.$$

Identify $\mathfrak{M}$ *with a subspace of* $\mathfrak{H}_0$ *of constant vector functions* $\{f(t) \equiv f,\ f \in \mathfrak{M}\}$. *Let*

$$\mathcal{K}_0 := \mathfrak{H}_0 \ominus \mathfrak{M} = \left\{ f(t) \in \mathfrak{H}_0 : \int\limits_{-1}^{1} \frac{(f(t), h)_{\mathfrak{M}}}{\sqrt{1-t^2}}\, dt = 0\ \forall h \in \mathfrak{M} \right\}$$

and define in $\mathfrak{H}_0$ *the multiplication operator by*

$$(T_0 f)(t) = tf(t), \ f \in \mathfrak{H}_0. \tag{6.16}$$

Then $\Omega_0(z)$ *is the transfer function of the simple passive selfadjoint system*

$$\tau_0 = \{T_0; \mathfrak{M}, \mathfrak{M}, \mathcal{K}_0\},$$

while

$$M_0(\xi) = P_{\mathfrak{M}}(T_0 - \xi I)^{-1} \upharpoonright \mathfrak{M}.$$

Proof. 1)–2) Let $\Omega_0(z)$ be a fixed point of the mapping $\mathbf{\Phi}$ of $\mathcal{RS}(\mathfrak{M})$, i.e.,

$$\Omega_0(z) = (zI - \Omega_0(z))\,(I - z\Omega_0(z))^{-1}, \quad z \in \mathbb{C} \setminus \{(-\infty, -1] \cup [1, +\infty)\}.$$

Then

$$(I - z\Omega_0(z))^2 = (1 - z^2) I_{\mathfrak{M}}.$$

Using $\Omega_0 \in \mathcal{RS}(\mathfrak{M})$ and the Taylor expansion $\Omega_0(z) = \sum_{n=0}^{\infty} C_k z^k$ in the unit disk, it is seen that Ω_0 is of the form (6.13).

It follows that the transform $M_0 = \mathbf{U}(\Omega_0)$ defined in (6.3) is of the form (6.14) and it is the unique fixed point of the mapping $\mathbf{\Gamma}$ in (6.2); cf. (6.5).

3) For each $h \in \mathfrak{M}$ straightforward calculations, see [13, pages 545–546], lead to the equality

$$-\frac{h}{\sqrt{\xi^2 - 1}} = \frac{1}{\pi} \int_{-1}^{1} \frac{h}{t - \xi} \frac{1}{\sqrt{1 - t^2}} dt.$$

Therefore, if T_0 is the operator of the form (6.16), then

$$M_0(\xi) = P_{\mathfrak{M}}(T_0 - \xi I)^{-1} \upharpoonright \mathfrak{M}.$$

It follows that Ω_0 is the transfer function of the system $\tau_0 = \{T_0; \mathfrak{M}, \mathfrak{M}, \mathcal{K}_0\}$. □

As is well known, the Chebyshev polynomials of the first kind given by

$$\widehat{T}_0(t) = 1, \ \widehat{T}_n(t) := \sqrt{2} \cos(n \arccos t), \ n \geq 1$$

form an orthonormal basis of the space $L_2([-1, 1], \rho_0(t))$, where $\rho_0(t)$ is given by (6.15). These polynomials satisfy the recurrence relations

$$t\widehat{T}_0(t) = \frac{1}{\sqrt{2}} \widehat{T}_1(t), \quad t\widehat{T}_1(t) = \frac{1}{\sqrt{2}} \widehat{T}_0(t) + \frac{1}{2} \widehat{T}_2(t),$$

$$t\widehat{T}_n(t) = \frac{1}{2} \widehat{T}_{n-1}(t) + \frac{1}{2} \widehat{T}_{n+1}(t), \quad n \neq 2.$$

Hence the matrix of the operator multiplication by the independent variable in the Hilbert space $L_2([-1, 1], \rho_0(t))$ w.r.t. the basis $\{\widehat{T}_n(t)\}_{n=0}^{\infty}$ (the Jacobi matrix) takes the form

$$J=\begin{bmatrix} 0 & \frac{1}{\sqrt{2}} & 0 & 0 & 0 & \cdot & \cdot & \cdot \\ \frac{1}{\sqrt{2}} & 0 & \frac{1}{2} & 0 & 0 & \cdot & \cdot & \cdot \\ 0 & \frac{1}{2} & 0 & \frac{1}{2} & 0 & \cdot & \cdot & \cdot \\ 0 & 0 & \frac{1}{2} & 0 & \frac{1}{2} & 0 & \cdot & \cdot \\ \vdots & \vdots & \vdots & \vdots & \vdots & \vdots & \vdots & \vdots \end{bmatrix}.$$

In the case of vector-valued weighted Hilbert space $\mathfrak{H}_0 = L_2([-1,1],\mathfrak{M},\rho_0(t))$ the operator (6.16) is unitary equivalent to the block operator Jacobi matrix $\mathbf{J}_0 = J\bigotimes I_{\mathfrak{M}}$. It follows that the function Ω_0 is the transfer function of the passive selfadjoint system with the operator T_0 given by the selfadjoint contractive block operator Jacobi matrix

$$T_0=\left[\begin{array}{c|c} 0 & \frac{1}{\sqrt{2}}I_{\mathfrak{M}} \quad 0 \quad 0 \quad \ldots \\ \hline \begin{array}{c}\frac{1}{\sqrt{2}}I_{\mathfrak{M}} \\ 0 \\ \vdots\end{array} & \mathbf{J}_0 \end{array}\right],$$

$$\mathbf{J}_0=\begin{bmatrix} 0 & \frac{1}{2}I_{\mathfrak{M}} & 0 & 0 & 0 & \cdot & \cdot & \cdot \\ \frac{1}{2}I_{\mathfrak{M}} & 0 & \frac{1}{2}I_{\mathfrak{M}} & 0 & 0 & \cdot & \cdot & \cdot \\ 0 & \frac{1}{2}I_{\mathfrak{M}} & 0 & \frac{1}{2}I_{\mathfrak{M}} & 0 & \cdot & \cdot & \cdot \\ 0 & 0 & \frac{1}{2}I_{\mathfrak{M}} & 0 & \frac{1}{2}I_{\mathfrak{M}} & 0 & \cdot & \cdot \\ \vdots & \vdots & \vdots & \vdots & \vdots & \vdots & \vdots & \vdots \end{bmatrix}.$$

6.2. The mapping Π and Redheffer product

Lemma 6.7. *Let H be a Hilbert space, let K be a selfadjoint contraction in H and let $\Omega \in \mathcal{RS}(H)$. If $||K|| < 1$, then $(I - K\Omega(z))^{-1}$ is defined on H and it is bounded for all $z \in \mathbb{C} \setminus \{(-\infty,-1] \cup [1,+\infty)\}$.*

Proof. If $|z| \le 1$, $z \ne \pm 1$, then $||K|| < 1$ and $||\Omega(z)|| \le 1$ imply that $||K\Omega(z)|| < 1$. Hence $(I - K\Omega(z))^{-1}$ exists as bounded everywhere defined operator on H.

Now let $|z| > 1$ and $z \in \mathbb{C} \setminus \{(-\infty,-1] \cup [1,+\infty)\}$. Then there exists $\beta \in (0,\pi/2)$ such that either $|z\sin\beta - i\cos\beta| = 1$ or $|z\sin\beta + i\cos\beta| = 1$. Suppose that, for instance, $|z\sin\beta - i\cos\beta| = 1$. Then $||\Omega(z)\sin\beta - i\cos\beta I_H|| \le 1$ by (2.1).

Hence, the operator $S := \Omega(z)\sin\beta - i\cos\beta I_H$ satisfies $||S|| \le 1$ and one has

$$\Omega(z) = \frac{S + i\cos\beta\, I_H}{\sin\beta}.$$

Furthermore,

$$\begin{aligned} I - K\Omega(z) &= I - \frac{KS + i\cos\beta\, K}{\sin\beta} = \frac{1}{\sin\beta}\left((\sin\beta\, I - i\cos\beta\, K) - KS\right) \\ &= \frac{1}{\sin\beta}(\sin\beta\, I - i\cos\beta\, K)\left(I - (\sin\beta\, I - i\cos\beta\, K)^{-1}KS\right). \end{aligned}$$

Clearly

$$||(\sin\beta\, I - i\cos\beta\, K)^{-1}K||^2 \le \frac{||K||^2}{\sin^2\beta + ||K||^2\cos^2\beta} < 1,$$

which shows that $||(\sin\beta\, I - i\cos\beta\, K)^{-1}KS|| < 1$. Therefore, the following inverse operator $\left(I - (\sin\beta\, I - i\cos\beta\, K)^{-1}KS\right)^{-1}$ exists and is everywhere defined on H. This implies that

$$(I - K\Omega(z))^{-1} = \sin\beta(I - (\sin\beta I - i\cos\beta\, K)^{-1}KS)^{-1}(\sin\beta I - i\cos\beta K)^{-1}.$$

□

Theorem 6.8. *Let*

$$\mathbf{K} = \begin{bmatrix} K_{11} & K_{12} \\ K_{12}^* & K_{22} \end{bmatrix} : \begin{matrix} \mathfrak{M} \\ \oplus \\ H \end{matrix} \to \begin{matrix} \mathfrak{M} \\ \oplus \\ H \end{matrix}$$

be a selfadjoint contraction. Then the following two assertions hold:

1) *If $||K_{22}|| < 1$, then for every $\Omega \in \mathcal{RS}(H)$ the transform*

$$\begin{aligned} \Theta(z) &:= K_{11} + K_{12}\Omega(z)(I - K_{22}\Omega(z))^{-1}K_{12}^*, \\ & z \in \mathbb{C} \setminus \{(-\infty, -1] \cup [1, +\infty)\}, \end{aligned} \tag{6.17}$$

also belongs to $\mathcal{RS}(\mathfrak{M})$.

2) *If $\Omega \in \mathcal{RS}(H)$ and $\Omega(0) = 0$, then again the transform Θ defined in* (6.17) *belongs to $\mathcal{RS}(\mathfrak{M})$.*

Proof. 1) It follows from Lemma 6.7 that $(I - K_{22}\Omega(z))^{-1}$ exists as a bounded operator on H for all $z \in \mathbb{C} \setminus \{(-\infty, -1] \cup [1, +\infty)\}$. Furthermore,

$$\begin{aligned} \Theta(z) - \Theta(z)^* &= K_{12}\Omega(z)(I - K_{22}\Omega(z))^{-1}K_{12}^* - K_{12}(I - \Omega(z)^*K_{22})^{-1}\Omega(z)^*K_{12}^* \\ &= K_{12}(I - \Omega(z)^*K_{22})^{-1}\left((I - \Omega(z)^*K_{22})\Omega(z) - \Omega(z)^*(I - K_{22}\Omega(z))\right) \\ &\quad \times (I - K_{22}\Omega(z))^{-1}K_{12}^* \\ &= K_{12}(I - \Omega(z)^*K_{22})^{-1}\left(\Omega(z) - \Omega(z)^*\right)(I - K_{22}\Omega(z))^{-1}K_{12}^*. \end{aligned}$$

Thus, Θ is a Nevanlinna function on the domain $\mathbb{C} \setminus \{(-\infty, -1] \cup [1, +\infty)\}$.

Since $\mathbf{K}$ is a selfadjoint contraction, its entries are of the form (again see Proposition B.1 and Remark B.2):

$$K_{12} = ND_{K_{22}},\ K_{12}^* = D_{K_{22}}N^*,\ K_{11} = -NK_{22}N^* + D_{N^*}LD_{N^*},$$

where $N : \mathfrak{D}_{K_{22}} \to \mathfrak{M}$ is a contraction and $L : \mathfrak{D}_{N^*} \to \mathfrak{D}_{N^*}$ is a selfadjoint contraction. This gives

$$\Theta(z) = N\left(-K_{22} + D_{K_{22}}\Omega(z)(I - K_{22}\Omega(z))^{-1}D_{K_{22}}\right)N^* + D_{N^*}LD_{N^*}.$$

Denote

$$\widetilde{\Theta}(z) := -K_{22} + D_{K_{22}}\Omega(z)(I - K_{22}\Omega(z))^{-1}D_{K_{22}}.$$

Then

$$\begin{aligned}\widetilde{\Theta}(z) &= D_{K_{22}}^{-1}(\Omega(z) - K_{22})(I - K_{22}\Omega(z))^{-1}D_{K_{22}}\\ &= D_{K_{22}}(I - \Omega(z)K_{22})^{-1}(\Omega(z) - K_{22})D_{K_{22}}^{-1}\end{aligned}$$

and

$$\Theta(z) = N\widetilde{\Theta}(z)N^* + D_{N^*}LD_{N^*}.$$

Again straightforward calculations (cf. [4, 18]) show that for all $f \in \mathfrak{D}_{K_{22}}$,

$$||f||^2 - ||\widetilde{\Theta}(z)f||^2 = ||(I - K_{22}\Omega(z))^{-1}D_{K_{22}}f||^2 - ||\Omega(z)(I - K_{22}\Omega(z))^{-1}D_{K_{22}}f||^2,$$

and for all $h \in \mathfrak{M}$,

$$\begin{aligned}||h||^2 - ||\Theta(z)h||^2 = ||N^*h||^2 - ||\widetilde{\Theta}(z)N^*h||^2 + ||D_LD_{N^*}h||^2\\ + ||(D_N\widetilde{\Theta}(z)N^* - N^*LD_{N^*})h||^2.\end{aligned}$$

Since $\Omega(z)$ is a contraction for all $|z| \le 1$, $z \ne \pm 1$, one concludes that $\widetilde{\Theta}(z)$ and, thus, also $\Theta(z)$ is a contraction. In addition, the operators $\Theta(x)$ are selfadjoint for $x \in (-1, 1)$. Therefore $\Theta \in \mathcal{RS}(\mathfrak{M})$.

2) Suppose that $\Omega(0) = 0$. To see that the operator $(I - K_{22}\Omega(z))^{-1}$ exists as a bounded operator on H for all $z \in \mathbb{C} \setminus \{(-\infty, -1] \cup [1, +\infty)\}$, realize Ω as the transfer function of a passive selfadjoint system

$$\sigma = \left\{\begin{bmatrix} 0 & N \\ N^* & S\end{bmatrix}; H, H, \mathcal{K}\right\},$$

i.e., $\Omega(z) = zN(I - zS)^{-1}N^*$. Since

$$T = \begin{bmatrix} 0 & N \\ N^* & S\end{bmatrix} : \begin{matrix} H \\ \oplus \\ \mathcal{K}\end{matrix} \to \begin{matrix} H \\ \oplus \\ \mathcal{K}\end{matrix}$$

is a selfadjoint contraction, the operator $N \in \mathbf{B}(\mathcal{K}, H)$ is a contraction and S is of the form $S = D_{N^*}LD_{N^*}$, where $L \in \mathbf{B}(\mathfrak{D}_{N^*})$ is a selfadjoint contraction. It follows that the operator $N^*K_{22}N + S$ is a selfadjoint contraction for an arbitrary selfadjoint contraction K_{22} in H. Therefore, $(I - z(N^*K_{22}N + S))^{-1}$ exists on $\mathcal{K}$ and is bounded for all $z \in \mathbb{C} \setminus \{(-\infty, -1] \cup [1, +\infty)\}$. It is easily checked that the equality $\left(I - zK_{22}N(I - zS)^{-1}N^*\right)^{-1} = I + zK_{22}N\left(I - z(N^*K_{22}N + S)\right)^{-1}N^*$

holds for all $z \in \mathbb{C} \setminus \{(-\infty, -1] \cup [1, +\infty)\}$. Now arguing again as in item 1) one completes the proof. $\square$

Theorem 6.9. *Let*

$$\mathbf{S} = \begin{bmatrix} A & B \\ B^* & G \end{bmatrix} : \begin{matrix} H \\ \oplus \\ \mathcal{K} \end{matrix} \to \begin{matrix} H \\ \oplus \\ \mathcal{K} \end{matrix}, \quad \mathbf{K} = \begin{bmatrix} K_{11} & K_{12} \\ K_{12}^* & K_{22} \end{bmatrix} : \begin{matrix} \mathfrak{M} \\ \oplus \\ H \end{matrix} \to \begin{matrix} \mathfrak{M} \\ \oplus \\ H \end{matrix}$$

be selfadjoint contractions. Also let $\sigma = \{\mathbf{S}, H, H, \mathcal{K}\}$ be a passive selfadjoint system with the transfer function $\Omega(z)$. Then the following two assertions hold:

1) *Assume that $||K_{22}|| < 1$. Then $\Theta(z)$ given by (6.17) is the transfer function of the passive selfadjoint system*

$$\tau = \{\mathbf{T}, \mathfrak{M}, \mathfrak{M}, \mathcal{K}\},$$

where $\mathbf{T} = \mathbf{K} \bullet \mathbf{S}$ is the Redheffer product (see [17, 21]):

$$\mathbf{T} = \begin{bmatrix} K_{11} + K_{12}A(I - K_{22}A)^{-1}K_{12}^* & K_{12}(I - AK_{22})^{-1}B \\ B^*(I - K_{22}A)^{-1}K_{12}^* & G + B^*K_{22}(I - AK_{22})^{-1}B \end{bmatrix}, \tag{6.18}$$

where the block decomposition is w.r.t. to $\mathfrak{M} \oplus \mathcal{K}$.

2) *Assume that $A = 0$. Then the Redheffer product $\mathbf{T} = \mathbf{K} \bullet \mathbf{S}$ is given by*

$$\mathbf{T} = \begin{bmatrix} K_{11} & K_{12}B \\ B^*K_{12}^* & G + B^*K_{22}B \end{bmatrix} : \begin{matrix} \mathfrak{M} \\ \oplus \\ \mathcal{K} \end{matrix} \to \begin{matrix} \mathfrak{M} \\ \oplus \\ \mathcal{K} \end{matrix}$$

and the transfer function of the passive selfadjoint system $\tau = \{\mathbf{T}, \mathfrak{M}, \mathfrak{M}, \mathcal{K}\}$ is equal to the function Θ defined in (6.17).

Proof. By definition

$$\Omega(z) = A + zB(I - zG)^{-1}B^*, \quad z \in \mathbb{C} \setminus \{(-\infty, -1] \cup [1, +\infty)\}.$$

1) Suppose that $||K_{22}|| < 1$. Since

$$\begin{aligned} \Theta(z) &= K_{11} + K_{12}\Omega(z)(I - K_{22}\Omega(z))^{-1}K_{12}^* \\ &= K_{11} + K_{12}(I - \Omega(z)K_{22})^{-1}\Omega(z)K_{12}^*, \end{aligned}$$

one obtains

$$\begin{aligned} \Theta(z) - \Theta(0) &= K_{12}(I - \Omega(z)K_{22})^{-1}\left(\Omega(z) - \Omega(0)\right)(I - K_{22}\Omega(0))^{-1}K_{12}^* \\ &= zK_{12}\left(I - AK_{22} - zB(I - zG)^{-1}B^*K_{22}\right)^{-1} \\ &\quad \times B(I - zG)^{-1}B^*(I - K_{22}A)^{-1}K_{12}^*. \end{aligned}$$

Furthermore,

$$\begin{aligned} &\left(I - AK_{22} - zB(I - zG)^{-1}B^*K_{22}\right)^{-1}B(I - zG)^{-1} \\ &\quad = (I - AK_{22})^{-1}\left(I - zB(I - zG)^{-1}B^*K_{22}(I - AK_{22})^{-1}\right)^{-1}B(I - zG)^{-1} \end{aligned}$$

$$= (I - AK_{22})^{-1}B\left(I - z(I - zG)^{-1}B^*K_{22}(I - AK_{22})^{-1}B\right)^{-1}(I - zG)^{-1}$$
$$= (I - AK_{22})^{-1}B\left(I - z\left(G + zB^*K_{22}(I - AK_{22})^{-1}B\right)\right)^{-1}$$

and one has

$$\Theta(z) = K_{11} + K_{12}A(I - K_{22}A)^{-1}K_{12}^* + zK_{12}(I - AK_{22})^{-1}B$$
$$\times \left(I - z\left(G + zB^*K_{22}(I - AK_{22})^{-1}B\right)\right)^{-1}B^*(I - K_{22}A)^{-1}K_{12}^*.$$

Now it follows from (6.18) that $\Theta(z)$ is the transfer function of the system τ.

Next it is shown that the selfadjoint operator $\mathbf{T}$ given by (6.18) is a contraction. Let the entries of $\mathbf{S}$ and $\mathbf{K}$ be parameterized by

$$\begin{cases} B^* = UD_A, B = D_AU^* \\ G = -UAU^* + D_{U^*}ZD_{U^*} \end{cases}, \quad \begin{cases} K_{12} = VD_{K_{22}}, K_{12}^* = D_{K_{22}}V^* \\ K_{11} = -VK_{22}V^* + D_{V^*}YD_{V^*} \end{cases},$$

where V, U, Y, Z are contractions acting between the corresponding subspaces. Also define the operators

$$\Phi_{K_{22}}(A) = -K_{22} + D_{K_{22}}A(I - K_{22}A)^{-1}D_{K_{22}},$$
$$\Phi_A(K_{22}) = -A + D_AK_{22}(I - AK_{22})^{-1}D_A.$$

This leads to the formula

$$\mathbf{T} = \begin{bmatrix} V & 0 \\ 0 & U \end{bmatrix}\begin{bmatrix} \Phi_{K_{22}}(A) & D_{K_{22}}(I - AK_{22})^{-1}D_A \\ D_A(I - K_{22}A)^{-1}D_{K_{22}} & \Phi_A(K_{22}) \end{bmatrix}\begin{bmatrix} V^* & 0 \\ 0 & U^* \end{bmatrix}$$
$$+ \begin{bmatrix} D_{V^*}YD_{V^*} & 0 \\ 0 & D_{U^*}ZD_{U^*} \end{bmatrix}.$$

The block operator

$$\mathbb{J} = \begin{bmatrix} \Phi_{K_{22}}(A) & D_{K_{22}}(I - AK_{22})^{-1}D_A \\ D_A(I - K_{22}A)^{-1}D_{K_{22}} & \Phi_A(K_{22}) \end{bmatrix}$$

is unitary and selfadjoint. Actually, the selfadjointness follows from selfadjointness of the operators A, K_{22} and $\Phi_{K_{22}}(A), \Phi_A(K_{22})$. Furthermore, one has the equalities

$$||f||^2 - ||\Phi_{K_{22}}(A)f||^2 = ||D_A(I - K_{22}A)^{-1}D_{K_{22}}f||^2,$$
$$||g||^2 - ||\Phi_A(K_{22})g||^2 = ||D_{K_{22}}(I - AK_{22})^{-1}D_Ag||^2,$$
$$\left(\Phi_{K_{22}}(A)f, D_{K_{22}}(I - AK_{22})^{-1}D_Ag\right)$$
$$= \left(D_A(I - K_{22}A)^{-1}(A - K_{22})(I - K_{22}A)^{-1}D_{K_{22}}f, g\right),$$
$$\left(\Phi_A(K_{22})g, D_A(I - K_{22}A)^{-1}D_{K_{22}}f\right)$$
$$= \left(D_{K_{22}}(I - AK_{22})^{-1}(K_{22} - A)(I - AK_{22})^{-1}D_Ag, f\right).$$

These equalities imply that $\mathbb{J}$ is unitary.

Denote

$$\mathbb{W} = \begin{bmatrix} V & 0 \\ 0 & U \end{bmatrix}, \quad \mathbb{X} = \begin{bmatrix} Y & 0 \\ 0 & Z \end{bmatrix}.$$

Then

$$\mathbf{T} = \mathbb{W}\mathbb{J}\mathbb{W}^* + D_{\mathbb{W}^*}\mathbb{X}D_{\mathbb{W}^*},$$

and one obtains the equality

$$||h||^2 - ||\mathbf{T}h||^2 = ||D_{\mathbb{X}}D_{\mathbb{W}^*}h||^2 + ||(\mathbb{W}^*\mathbb{X} - D_{\mathbb{W}}\mathbb{J}\mathbb{W}^*)h||^2.$$

Thus, $\mathbf{T}$ is a selfadjoint contraction.

The proof of the statement 2) is similar to the proof of statement 1) and is omitted. □

6.3. The mapping $\Omega(z) \mapsto (a\,I + \Omega(z))\,(I + a\,\Omega(z)\,)^{-1}$

Proposition 6.10. *Let*

$$\tau = \left\{ \begin{bmatrix} A & B \\ B^* & G \end{bmatrix}; \mathfrak{M}, \mathfrak{M}, \mathcal{K} \right\}$$

be a passive selfadjoint system with transfer function Ω. Let $a \in (-1, 1)$. Then the passive selfadjoint system

$$\sigma_a = \left\{ \begin{bmatrix} (aI + A)(I + aA)^{-1} & \sqrt{1-a^2}(I + aA)^{-1}B \\ \sqrt{1-a^2}B^*(I + aA)^{-1} & G - aB^*(I + aA)^{-1}B \end{bmatrix}; \mathfrak{M}, \mathfrak{M}, \mathcal{K} \right\}$$

has transfer function

$$\widehat{\Omega}_a(z) = (a\,I + \Omega(z))(I + a\,\Omega(z))^{-1}, \quad z \in \mathbb{C} \setminus \{(-\infty, -1] \cup [1, +\infty)\}.$$

Proof. Let

$$\mathbf{K}_a = \begin{bmatrix} aI & \sqrt{1-a^2}I \\ \sqrt{1-a^2} & -aI \end{bmatrix} : \begin{matrix} \mathfrak{M} \\ \oplus \\ \mathfrak{M} \end{matrix} \to \begin{matrix} \mathfrak{M} \\ \oplus \\ \mathfrak{M} \end{matrix}, \quad \mathbf{S} = \begin{bmatrix} A & B \\ B^* & G \end{bmatrix} : \begin{matrix} \mathfrak{M} \\ \oplus \\ \mathcal{K} \end{matrix} \to \begin{matrix} \mathfrak{M} \\ \oplus \\ \mathcal{K} \end{matrix}.$$

Then the Redheffer product $\mathbf{K}_a \bullet \mathbf{S}$ (cf. (6.18)) takes the form

$$\mathbf{T} = \begin{bmatrix} (aI + A)(I + aA)^{-1} & \sqrt{1-a^2}(I + aA)^{-1}B \\ \sqrt{1-a^2}B^*(I + aA)^{-1} & G - aB^*(I + aA)^{-1}B \end{bmatrix} : \begin{matrix} \mathfrak{M} \\ \oplus \\ \mathcal{K} \end{matrix} \to \begin{matrix} \mathfrak{M} \\ \oplus \\ \mathcal{K} \end{matrix}. \tag{6.19}$$

On the other hand, for all $z \in \mathbb{C} \setminus \{(-\infty, -1] \cup [1, +\infty)\}$ one has

$$\begin{aligned} K_{11} + K_{12}\Omega(z)(I - K_{22}\Omega(z))^{-1}K_{12}^* &= aI + (1 - a^2)\Omega(z)(I + a\Omega(z))^{-1} \\ &= (a\,I + \Omega(z))(I + a\,\Omega(z))^{-1}. \end{aligned}$$

This completes the proof. □

6.4. The mapping $\Omega(z) \mapsto \Omega\left(\frac{z+a}{1+za}\right)$ and its fixed points

For a contraction S in a Hilbert space and a complex number a, $|a| < 1$, define, see [20],

$$S_a := (S - aI)(I - \bar{a}S)^{-1}.$$

The operator S_a is a contraction, too. If S is a selfadjoint contraction and $a \in (-1,1)$, then S_a is also selfadjoint. One has $S_a = W_{-a}(S)$ (see Introduction) and, moreover,

$$\begin{aligned} D_{S_a} &= \sqrt{1-a^2}(I-aS)^{-1}D_S, \\ (I - zS_a)^{-1} &= \frac{1}{1+az}(I-aS)\left(I - \frac{z+a}{1+az}S\right)^{-1}, \\ (zI - S_a)(I - zS_a)^{-1} &= \left(\frac{z+a}{1+az}I - S\right)\left(I - \frac{z+a}{1+az}S\right)^{-1}, \end{aligned} \tag{6.20}$$

where $z \in \mathbb{C} \setminus \{(-\infty, -1] \cup [1, \infty\}$. Let the block operator

$$T = \begin{bmatrix} D & C \\ C^* & F \end{bmatrix} : \begin{matrix} \mathfrak{M} \\ \oplus \\ \mathcal{K} \end{matrix} \to \begin{matrix} \mathfrak{M} \\ \oplus \\ \mathcal{K} \end{matrix} \tag{6.21}$$

be a selfadjoint contraction and let $\Omega(z) = D + zC(I - zF)^{-1}C^*$. Then from the Schur–Frobenius formula (A.1) and from the relation

$$T_a = (T - aI)(I - aT)^{-1} = \frac{1-a^2}{a}(I - aT)^{-1} - \frac{1}{a}I \quad (a \neq 0)$$

it follows that T_a can be rewritten with $c_a := 1 - a^2$ in the block form

$$T_a = \begin{bmatrix} (\Omega(a) - aI)(I - a\Omega(a))^{-1} & c_a(I - a\Omega(a))^{-1}C(I - aF)^{-1} \\ c_a(I - aF)^{-1}C^*(I - a\Omega(a))^{-1} & F_a + ac_a(I - aF)^{-1}C^*(I - a\Omega(a))^{-1}C(I - aF)^{-1} \end{bmatrix}. \tag{6.22}$$

Theorem 6.11. *Let*

$$\tau = \left\{ \begin{bmatrix} D & C \\ C^* & F \end{bmatrix}, \mathfrak{M}, \mathfrak{M}, \mathcal{K} \right\}$$

be a passive selfadjoint system with the transfer function Ω. *Then for every* $a \in (-1,1)$ *the* $\mathbf{B}(\mathfrak{M})$*-valued function*

$$\Omega\left(\frac{z+a}{1+az}\right)$$

is the transfer function of the passive selfadjoint system

$$\tau_a = \left\{ \begin{bmatrix} \Omega(a) & \sqrt{1-a^2}C(I-aF)^{-1} \\ \sqrt{1-a^2}(I-aF)^{-1}C^* & F_a \end{bmatrix}, \mathfrak{M}, \mathfrak{M}, \mathcal{K} \right\}.$$

Furthermore, if τ *is a minimal system then* τ_a *is minimal, too.*

Proof. Let

$$C = KD_F, \ D = -KFK^* + D_{K^*}YD_{K^*},$$

be the parametrization for entries of the block operator T, cf. (2.4), where $K \in \mathbf{B}(\mathfrak{D}_F, \mathcal{K})$ is a contraction and $Y \in \mathbf{B}(\mathfrak{D}_{K^*})$ is a selfadjoint contraction. From (2.6) and (6.20) we get

$$\begin{aligned}\Omega\left(\frac{z+a}{1+az}\right) &= D_{K^*}YD_{K^*} + K\left(\frac{z+a}{1+az}I - F\right)\left(I - \frac{z+a}{1+az}F\right)^{-1}K^* \\ &= D_{K^*}YD_{K^*} + K\left(zI - F_a\right)\left(I - zF_a\right)^{-1}K^*\end{aligned}$$

with $z \in \mathbb{C} \setminus \{(-\infty, -1] \cup [1, \infty\}$. The operator

$$\begin{aligned}\widehat{T}_a &= \begin{bmatrix} -KF_aK^* + D_{K^*}YD_{K^*} & KD_{F_a} \\ D_{F_a}K^* & F_a \end{bmatrix} \\ &= \begin{bmatrix} \Omega(a) & \sqrt{1-a^2}C(I-aF)^{-1} \\ \sqrt{1-a^2}(I-aF)^{-1}C^* & F_a \end{bmatrix} : \begin{array}{c}\mathfrak{M}\\ \oplus \\ \mathcal{K}\end{array} \to \begin{array}{c}\mathfrak{M}\\ \oplus \\ \mathcal{K}\end{array}\end{aligned}$$

is a selfadjoint contraction. The formula (2.6) applied to the system τ_a gives

$$\Omega_{\tau_a}(z) = D_{K^*}YD_{K^*} + K\left(zI - F_a\right)\left(I - zF_a\right)^{-1}K^*.$$

Hence $\Omega_{\tau_a}(z) = \Omega\left(\frac{z+a}{1+az}\right)$ for all $z \in \mathbb{C} \setminus \{(-\infty, -1] \cup [1, \infty\}$.

Suppose τ is minimal. This is equivalent to the relations

$$\begin{aligned}&\overline{\operatorname{span}}\,\{F^nD_FK^*\mathfrak{M} :\ n \in \mathbb{N}_0\} = \mathcal{K} \\ \iff\quad &\bigcap_{n=0}^{\infty} \ker(KF^nD_F) = \{0\} \\ \iff\quad &\bigcap_{|z|<1} \ker K(I-zF)^{-1}D_F = \{0\}.\end{aligned}$$

Using the formulas (6.20) one obtains

$$\begin{aligned}&\bigcap_{|z|<1} \ker K(I-zF_a)^{-1}D_{F_a} \\ &\quad= \bigcap_{|z|<1} \ker K\left(I - \frac{z+a}{1+az}F\right)^{-1} D_F \\ &\quad= \bigcap_{|\mu|<1} \ker K(I-\mu F)^{-1}D_F = \{0\}\end{aligned}$$

or, equivalently,

$$\overline{\operatorname{span}}\,\{F_a^nD_{F_a}K^*\mathfrak{M},\ n \in \mathbb{N}_0\} = \mathcal{K}.$$

This shows that the system τ_a is minimal. □

Remark 6.12. 1) Let T in (6.21) be represented in the form

$$T=\begin{bmatrix}K & 0\\ 0 & I\end{bmatrix}\mathbb{J}_F\begin{bmatrix}K^* & 0\\ 0 & I\end{bmatrix}+\begin{bmatrix}D_{K^*}YD_{K^*} & 0\\ 0 & 0\end{bmatrix},$$

see Remark B.3. Then

$$\begin{aligned}&\begin{bmatrix}-KF_aK^*+D_{K^*}YD_{K^*} & KD_{F_a}\\ D_{F_a}K^* & F_a\end{bmatrix}\\ &=\begin{bmatrix}\Omega(a) & \sqrt{1-a^2}C(I-aF)^{-1}\\ \sqrt{1-a^2}(I-aF)^{-1}C^* & F_a\end{bmatrix}\\ &=\begin{bmatrix}K & 0\\ 0 & I\end{bmatrix}\mathbb{J}_{F_a}\begin{bmatrix}K^* & 0\\ 0 & I\end{bmatrix}+\begin{bmatrix}D_{K^*}YD_{K^*} & 0\\ 0 & 0\end{bmatrix}.\end{aligned}$$

2) Let the transformation $\mathbf{V}_a$ with $a\in(-1,1)$ be defined by

$$\begin{bmatrix}D & C\\ C^* & F\end{bmatrix}\overset{\mathbf{V}_a}{\mapsto}\widehat{T}_a=\begin{bmatrix}\Omega(a) & \sqrt{1-a^2}C(I-aF)^{-1}\\ \sqrt{1-a^2}(I-aF)^{-1}C^* & F_a\end{bmatrix}.$$

Then for all $a,b\in(-1,1)$ one has the identities

$$\mathbf{V}_a\circ\mathbf{V}_b=\mathbf{V}_b\circ\mathbf{V}_a=\mathbf{V}_c,\ \text{where } c=\frac{a+b}{1+ab}.$$

Proposition 6.13. *The fixed points of the mapping* $\Omega(z)\mapsto\Omega\left(\frac{z+a}{1+za}\right)$, $a\in(-1,1)$, $a\neq 0$, *consist only of constant functions.*

Proof. Suppose that for some $a\in(-1,1)$, $a\neq 0$, the equality

$$\Omega\left(\frac{z+a}{1+az}\right)=\Omega(z)$$

is satisfied for all $z\in\mathbb{C}\setminus\{(-\infty,-1]\cup[1,+\infty)\}$. Then, in particular, $\Omega(0)=\Omega(a)$. Therefore from Theorem 6.11 one gets the equality $KFK^*=KF_aK^*$. Now

$$F-F_a=aD_F^2(I-aF)^{-1}$$

leads to

$$(I-aF)^{-1/2}D_FK^*=0.$$

Taking into account that $\operatorname{ran}K^*\subseteq\mathfrak{D}_F$, we get $K^*=0$. This means that $\Omega(z)\equiv\Omega(0)$. So, the fixed points of the mapping $\Omega(z)\mapsto\Omega\left(\frac{z+a}{1+za}\right)$ are the constant functions only. □

Remark 6.14. A. Filimonov and E. Tsekanovskiĭ [16] considered J-unitary operator colligations that are automorphic invariant w.r.t. a subgroup G of the Möbius transformations of the unit disk and its representations in the channel and state spaces. The characteristic function $W(z)$ of such a colligation satisfies the condition

$$W(g(z))V_g=V_gW(z),\quad\forall z\in\mathbb{D}\quad\text{and}\quad\forall g\in G,$$

where $\{V_g\}$ is a representation of G in the channel space.

6.5. The mapping $\Omega(z) \mapsto \left(\Omega\left(\frac{z+a}{1+az}\right) - a\,I\right)\left(I - a\,\Omega\left(\frac{z+a}{1+az}\right)\right)^{-1}$

Proposition 6.15. *Let $\tau = \{T; \mathfrak{M}, \mathfrak{M}, \mathcal{K}\}$ be a passive selfadjoint system with transfer function Ω. Then the passive selfadjoint system $\eta_a = \{T_a; \mathfrak{M}, \mathfrak{M}, \mathcal{K}\}$, $a \in (-1,1)$, has the transfer function*

$$\widetilde{\Omega}_a(z) = \left(\Omega\left(\frac{z+a}{1+az}\right) - a\,I_{\mathfrak{M}}\right)\left(I_{\mathfrak{M}} - a\,\Omega\left(\frac{z+a}{1+az}\right)\right)^{-1}.$$

If τ is minimal, then η_a is minimal, too.

Proof. Let T be a selfadjoint contraction in the Hilbert space $\mathfrak{H}$ and let $a \in (-1,1)$. Due to (6.20) for all $z \in \mathbb{C} \setminus \{(-\infty,-1] \cup [1,\infty\}$ one has

$$(I - zT_a)^{-1} = \frac{1}{1+az}(I - aT)\left(I - \frac{z+a}{1+az}T\right)^{-1}.$$

Moreover,

$$\begin{aligned}
&(I - aT)\left(I - \frac{z+a}{1+az}T\right)^{-1}\\
&= \left(I - \frac{z+a}{1+az}T\right)^{-1} - aT\left(I - \frac{z+a}{1+az}T\right)^{-1}\\
&= \left(I - \frac{z+a}{1+az}T\right)^{-1} + a\,\frac{1+za}{z+a}I - a\,\frac{1+za}{z+a}\left(I - \frac{z+a}{1+az}T\right)^{-1}\\
&= a\,\frac{1+za}{z+a}I + \frac{z(1-a^2)}{z+a}\left(I - \frac{z+a}{1+az}T\right)^{-1},
\end{aligned}$$

and

$$\begin{aligned}
(I - zT_a)^{-1} &= \frac{1}{1+az}\left(a\,\frac{1+za}{z+a}I + \frac{z(1-a^2)}{z+a}\left(I - \frac{z+a}{1+az}T\right)^{-1}\right)\\
&= \frac{a}{z+a}I + \frac{z(1-a^2)}{(z+a)(1+az)}\left(I - \frac{z+a}{1+az}T\right)^{-1}.
\end{aligned}$$

Let $\mathfrak{H} = \mathfrak{M} \oplus \mathcal{K}$. Since $P_{\mathfrak{M}}(I - zT)^{-1}\restriction\mathfrak{M} = (I - z\Omega(z))^{-1}$, we get

$$\begin{aligned}
&P_{\mathfrak{M}}(I - zT_a)^{-1}\restriction\mathfrak{M}\\
&= \frac{a}{z+a}I_{\mathfrak{M}} + \frac{z(1-a^2)}{(z+a)(1+az)}\left(I_{\mathfrak{M}} - \frac{z+a}{1+az}\Omega\left(\frac{z+a}{1+az}\right)\right)^{-1}
\end{aligned}$$

$$= \frac{1}{1+az}\left(I_{\mathfrak{M}} - a\,\Omega\left(\frac{z+a}{1+az}\right)\right)\left(I_{\mathfrak{M}} - \frac{z+a}{1+az}\Omega\left(\frac{z+a}{1+az}\right)\right)^{-1}.$$

Now consider the passive selfadjoint system

$$\eta_a = \{T_a; \mathfrak{M}, \mathfrak{M}, \mathcal{K}\}, \quad T_a = (T - aI)(I - aT)^{-1},$$

and let Ω_{η_a} be the transfer function of η_a. Then from

$$P_{\mathfrak{M}}(I - zT_a)^{-1}\upharpoonright\mathfrak{M} = (I_{\mathfrak{M}} - z\Omega_{\eta_a}(z))^{-1}$$

we get

$$(I_{\mathfrak{M}} - z\Omega_{\eta_a}(z))^{-1} = \frac{1}{1+az}\left(I_{\mathfrak{M}} - a\,\Omega\left(\frac{z+a}{1+az}\right)\right)\left(I_{\mathfrak{M}} - \frac{z+a}{1+az}\Omega\left(\frac{z+a}{1+az}\right)\right)^{-1}.$$

Hence,

$$\Omega_{\eta_a}(z) = \left(\Omega\left(\frac{z+a}{1+az}\right) - a\,I_{\mathfrak{M}}\right)\left(I_{\mathfrak{M}} - a\,\Omega\left(\frac{z+a}{1+az}\right)\right)^{-1}.$$

Since

$$\begin{aligned}\bigcap_{z\in\mathbb{D}} \ker\left(P_{\mathfrak{M}}(I - zT_a)^{-1}\right) &= \bigcap_{z\in\mathbb{D}} \ker\left(P_{\mathfrak{M}}\left(I - \frac{z+a}{1+az}T\right)^{-1}(I - aT)\right)\\ &= (I - aT)^{-1}\bigcap_{\mu\in\mathbb{D}} \ker\left(P_{\mathfrak{M}}(I - \mu T)^{-1}\right),\end{aligned}$$

we conclude that if τ is minimal then also η_a is minimal. □

Corollary 6.16. *Let $\tau = \{T; \mathfrak{M}, \mathfrak{M}, \mathcal{K}\}$ be a passive selfadjoint system with transfer function Ω. Let $a \in (-1, 1)$ and suppose that $\sigma_a = \{\mathcal{T}(a); \mathfrak{M}, \mathfrak{M}, \mathcal{K}\}$ is a passive selfadjoint system with transfer function $\Omega\left(\frac{z-a}{1-az}\right)$; see Theorem 6.11. Then the passive selfadjoint system*

$$\zeta_a = \{(\mathcal{T}(a))_a; \mathfrak{M}, \mathfrak{M}, \mathcal{K}\},\ (\mathcal{T}(a))_a := (\mathcal{T}(a) - aI)(I - a\mathcal{T}(a))^{-1}$$

has the transfer function

$$\Omega_{\zeta_a}(z) = (\Omega(z) - a\,I)(I - a\,\Omega(z))^{-1},\ z \in \mathbb{C} \setminus \{(-\infty, -1] \cup [1, +\infty)\}.$$

If τ is minimal then ζ_a is minimal, too.

The next result shows that the Redheffer product $\mathbf{K}_{-a} \bullet \mathbf{V}_a(T)$ coincides with $W_{-a}(T)$.

Proposition 6.17. *Let the block operator T in (6.21) be a selfadjoint contraction, let $\Omega(z) = D + zC(I - zF)^{-1}C^*$, and denote*

$$\widehat{T}_a = \begin{bmatrix} \Omega(a) & \sqrt{1-a^2}C(I - aF)^{-1} \\ \sqrt{1-a^2}(I - aF)^{-1}C^* & F_a \end{bmatrix} : \begin{matrix}\mathfrak{M}\\ \oplus \\ \mathcal{K}\end{matrix} \to \begin{matrix}\mathfrak{M}\\ \oplus \\ \mathcal{K}\end{matrix}$$

and

$$\mathbf{K}_{-a} = \begin{bmatrix} -aI & \sqrt{1-a^2}I \\ \sqrt{1-a^2} & aI \end{bmatrix} : \begin{matrix} \mathfrak{M} \\ \oplus \\ \mathfrak{M} \end{matrix} \to \begin{matrix} \mathfrak{M} \\ \oplus \\ \mathfrak{M} \end{matrix}.$$

Then the Redheffer product $\mathbf{K}_{-a} \bullet \widehat{T}_a$ *satisfies the equality*

$$\mathbf{K}_{-a} \bullet \widehat{T}_a = T_a \left(= (T - aI)(I - aT)^{-1}\right). \tag{6.23}$$

Proof. It follows from (6.19) that the mapping $\mathbf{K}_{-a} \bullet \widehat{T}_a : \mathfrak{M} \oplus \mathcal{K} \to \mathfrak{M} \oplus \mathcal{K}$ can be rewritten with $c_a = 1 - a^2$ in the form

$$\mathbf{K}_{-a} \bullet \widehat{T}_a = \begin{bmatrix} (aI-\Omega(a))(I-a\Omega(a))^{-1} & c_a(I-a\Omega(a))^{-1}C(I-aF)^{-1} \\ c_aC^*(I-aF)^{-1}(I-a\Omega(a))^{-1} & F_a+ac_a(I-aF)^{-1}C^*(I-a\Omega(a))^{-1}C(I-aF)^{-1} \end{bmatrix}.$$

Comparing this with (6.22) leads to (6.23). □

Theorem 6.18.

1) *If the function* Ω *from* $\mathcal{RS}(\mathfrak{M})$ *is inner, then the equality*

$$\Omega(z) = \left(\Omega\left(\frac{z+a}{1+az}\right) - a\, I_{\mathfrak{M}}\right)\left(I_{\mathfrak{M}} - a\,\Omega\left(\frac{z+a}{1+az}\right)\right)^{-1} \tag{6.24}$$

holds for all $a \in (-1,1)$ *and* $z \in \mathbb{C} \setminus \{(-\infty,-1] \cup [1,+\infty)\}$.

2) *If* $\Omega \in \mathcal{RS}(\mathfrak{M})$ *and* (6.24) *holds for some* $a \in (-1,1)$, $a \neq 0$, *then* Ω *is an inner function.*

Proof. 1) If $\Omega \in \mathcal{RS}(\mathfrak{M})$ is an inner function, then it takes the form (3.1) and $D = \Omega(0)$. The equality (6.24) can be verified with a straightforward calculation.

2) Suppose that (6.24) holds for some $a \in (-1,1)$. Then the equality

$$\Omega\left(\frac{z+a}{1+az}\right) - a\,I = \Omega(z)\left(I - a\,\Omega\left(\frac{z+a}{1+az}\right)\right)$$

holds for all $z \in \mathbb{C} \setminus \{(-\infty,-1] \cup [1,+\infty)\}$. Letting $z \to \pm 1$, we get the equalities $\Omega(1)^2 = \Omega(-1)^2 = I_{\mathfrak{M}}$. Moreover, with $z = 0$ we get from (6.24) the equality

$$(\Omega(a) - aI_{\mathfrak{M}})(I_{\mathfrak{M}} - a\Omega(a))^{-1} = \Omega(0).$$

Then by applying Theorem 3.3 one finally concludes that Ω is an inner function. □

6.6. The equation $\Omega(z) = \left(\Omega\left(\frac{z-a}{1-az}\right) - aI_{\mathfrak{M}}\right)\left(I_{\mathfrak{M}} - a\Omega\left(\frac{z-a}{1-az}\right)\right)^{-1}$

Theorem 6.19. *Let* $a \in (-1,1)$, $a \neq 0$. *Then the equality*

$$\Omega(z) = \left(\Omega\left(\frac{z-a}{1-az}\right) - a\, I_{\mathfrak{M}}\right)\left(I_{\mathfrak{M}} - a\,\Omega\left(\frac{z-a}{1-az}\right)\right)^{-1} \tag{6.25}$$

holds for all $z \in \mathbb{C} \setminus \{(-\infty,-1] \cup [1,+\infty)\}$ *and for some* $\Omega \in \mathcal{RS}(\mathfrak{M})$ *if and only if* Ω *is identically equal to a fundamental symmetry in* $\mathfrak{M}$.

Proof. We will use the Möbius representation (2.13) for $\Omega \in \mathcal{RS}(\mathfrak{M})$,

$$\begin{gathered}\Omega(z) = \Omega(0) + D_{\Omega(0)}\Lambda(z)\left(I + \Omega(0)\Lambda(z)\right)^{-1} D_{\Omega(0)},\\ z \in \mathbb{C} \setminus \{(-\infty,-1] \cup [1,+\infty)\},\end{gathered} \tag{6.26}$$

with a function $\Lambda \in \mathcal{RS}(\mathfrak{D}_{\Omega(0)})$ such that $\Lambda(z) = z\Gamma(z)$, where Γ is a holomorphic $\mathbf{B}(\mathfrak{D}_{\Omega(0)})$-valued function with $\|\Gamma(z)\| \le 1$ for $z \in \mathbb{D}$; see Proposition 2.3.

Equality (6.25) is equivalent to the equality

$$(\Omega(z) - aI_{\mathfrak{M}})\left(I_{\mathfrak{M}} - a\Omega(z)\right)^{-1} = \Omega\left(\frac{z+a}{1+za}\right) \; \forall z \in \mathbb{C} \setminus \{(-\infty,-1] \cup [1,+\infty)\}.$$

Now, with $z = 0$ this gives the equality

$$(\Omega(0) - aI_{\mathfrak{M}})\left(I_{\mathfrak{M}} - a\Omega(0)\right)^{-1} = \Omega(a) \Longleftrightarrow \Omega(0) - \Omega(a) = a(I_{\mathfrak{M}} - \Omega(a)\Omega(0)).$$

Denote $\Omega(0) = D$. Assume that $\mathfrak{D}_D \ne \{0\}$ and represent $\Omega \in \mathcal{RS}(\mathfrak{M})$ in the form (6.26). Furthermore, we use that $\Lambda(z) = z\Gamma(z)$. This leads to

$$-aD_D(\Gamma(a)(I+aD\Gamma(a))^{-1}D_D = a\left(I_{\mathfrak{M}} - \left(D + aD_D(\Gamma(a)(I+aD\Gamma(a))^{-1}D_D\right)D\right).$$

It follows that

$$\begin{aligned} &-\Gamma(a)(I + aD\Gamma(a))^{-1} = I - a\Gamma(a)(I + aD\Gamma(a))^{-1}D\\ \Longleftrightarrow \quad &(I + a\Gamma(a)D)^{-1}\Gamma(a) = a\Gamma(a)D(I + a\Gamma(a)D)^{-1} - I\\ \Longleftrightarrow \quad &(I + a\Gamma(a)D)^{-1}\Gamma(a) = a\Gamma(a)D(I + a\Gamma(a)D)^{-1} - I\\ \Longleftrightarrow \quad &(I + a\Gamma(a)D)^{-1}\Gamma(a) = -(I + a\Gamma(a)D)^{-1}\\ \Longleftrightarrow \quad &\Gamma(a) = -I.\end{aligned}$$

Since $\Gamma(z)$ belongs to the Schur class in $\mathfrak{M}$, we get

$$\Gamma(z) = -I, \quad z \in \mathbb{C} \setminus \{(-\infty,-1] \cup [1,+\infty)\}.$$

Hence for all $z \in \mathbb{C} \setminus \{(-\infty,-1] \cup [1,+\infty)\}$,

$$\Omega(z) = D - zD_D(I - zD)^{-1}D_D = (D - zI)(I - zD)^{-1}.$$

However, the function $(D - zI)(I - zD)^{-1}$ belongs to the class $\mathcal{RS}(\mathfrak{M})$ if and only if it is a constant function. In other words, one must have $\mathfrak{D}_D = \{0\}$. This means that $\Omega(z) \equiv D$, in $\mathbb{C} \setminus \{(-\infty,-1] \cup [1,+\infty)\}$, and here D is a fundamental symmetry in $\mathfrak{M}$ ($D = D^* = D^{-1}$). □

Appendices

A. The Schur–Frobenius formula for the resolvent

Let

$$\mathcal{U} = \begin{bmatrix} D & C \\ B & A \end{bmatrix} : \begin{matrix} \mathfrak{M} \\ \oplus \\ \mathfrak{H} \end{matrix} \to \begin{matrix} \mathfrak{M} \\ \oplus \\ \mathfrak{H} \end{matrix}$$

be a bounded block operator. Then the resolvent $R_{\mathcal{U}}(\lambda) = (\mathcal{U} - \lambda I)^{-1}$ of $\mathcal{U}$ (the Schur–Frobenius formula) takes the following block form:

$$R_{\mathcal{U}}(\lambda) = \begin{bmatrix} -V^{-1}(\lambda) & V^{-1}(\lambda)CR_A(\lambda) \\ R_A(\lambda)BV^{-1}(\lambda) & R_A(\lambda)\left(I_{\mathcal{H}} - BV^{-1}(\lambda)CR_A(\lambda)\right) \end{bmatrix}, \qquad \lambda \in \rho(\mathcal{U}) \cap \rho(A), \tag{A.1}$$

where

$$V(\lambda) := \lambda I_{\mathfrak{M}} - D + CR_A(\lambda)B, \ \lambda \in \rho(A). \tag{A.2}$$

In particular, $\lambda \in \rho(\mathcal{U}) \cap \rho(A) \iff V^{-1}(\lambda) \in \mathbf{L}(\mathfrak{M})$ and (A.1) and (A.2) imply

$$\left(P_{\mathfrak{M}} R_U(\lambda) \upharpoonright \mathfrak{M}\right)^{-1} = D - CR_A(\lambda)B - \lambda I_{\mathfrak{M}}.$$

B. Contractive 2 × 2 block operators

The following well-known result gives the structure of a contractive block operator.

Proposition B.1 ([11, 15, 19]). *The block operator* 2×2 *matrix*

$$T = \begin{bmatrix} D & C \\ B & F \end{bmatrix} : \begin{matrix} \mathfrak{M} \\ \oplus \\ \mathcal{K} \end{matrix} \to \begin{matrix} \mathfrak{N} \\ \oplus \\ \mathcal{L} \end{matrix}.$$

is a contraction if and only if $D \in \mathbf{B}(\mathfrak{M}, \mathfrak{N})$ *is a contraction and the entries* B,C, *and* F *take the form*

$$B = ND_D, \quad C = D_{D^*}G,$$
$$F = -ND^*G + D_{N^*}LD_G,$$

where the operators $N \in \mathbf{B}(\mathfrak{D}_D, \mathcal{L})$, $G \in \mathbf{B}(\mathcal{K}, \mathfrak{D}_{D^*})$ *and* $L \in \mathbf{B}(\mathfrak{D}_G, \mathfrak{D}_{N^*})$ *are contractions. Moreover, the operators* N, G, *and* L *are uniquely determined by* T. *Furthermore, the following equality holds for all* $f \in \mathfrak{M}$, $h \in \mathcal{K}$:

$$\left\| \begin{bmatrix} f \\ h \end{bmatrix} \right\|^2 - \left\| \begin{bmatrix} D & D_{D^*}G \\ ND_D & -ND^*G + D_{N^*}LD_G \end{bmatrix} \begin{bmatrix} f \\ h \end{bmatrix} \right\|^2$$
$$= \|D_N(D_Df - D^*Gh) - N^*LD_Gh\|^2 + \|D_LD_Gh\|^2.$$

Remark B.2. If $\mathfrak{N} = \mathfrak{M}$, $\mathcal{L} = \mathcal{K}$, then $T \in \mathbf{B}(\mathfrak{M} \oplus \mathcal{K})$ is a selfadjoint contraction if and only if $D = D^*$, $B = C^*$, $G = N^*$, $L = L^*$.

Remark B.3. Let F be a selfadjoint contraction in the Hilbert space $\mathcal{K}$, then the operator given by the block operator

$$\mathbb{J}_F = \begin{bmatrix} -F & D_F \\ D_F & F \end{bmatrix} : \begin{matrix} \mathfrak{D}_F \\ \oplus \\ \mathcal{K} \end{matrix} \to \begin{matrix} \mathfrak{D}_F \\ \oplus \\ \mathcal{K} \end{matrix}$$

is selfadjoint and unitary: $\mathbb{J}_F = \mathbb{J}_F = \mathbb{J}_F^{-1}$.

Let $\mathfrak{M}$ be a Hilbert space, let $K \in \mathbf{B}(\mathfrak{D}_F, \mathfrak{M})$ be a contraction and let

$$\begin{bmatrix} K & 0 \\ 0 & I \end{bmatrix} : \begin{matrix} \mathfrak{D}_F \\ \oplus \\ \mathcal{K} \end{matrix} \to \begin{matrix} \mathfrak{M} \\ \oplus \\ \mathcal{K} \end{matrix}.$$

Then for any selfadjoint contraction $Y \in \mathbf{B}(\mathfrak{D}_{K^*})$ the block operator

$$\begin{aligned} T &= \begin{bmatrix} K & 0 \\ 0 & I \end{bmatrix} \begin{bmatrix} -F & D_F \\ D_F & F \end{bmatrix} \begin{bmatrix} K^* & 0 \\ 0 & I \end{bmatrix} + \begin{bmatrix} D_{K^*} Y D_{K^*} & 0 \\ 0 & 0 \end{bmatrix} \\ &= \begin{bmatrix} -KFK^* + D_{K^*} Y D_{K^*} & KD_F \\ D_F K^* & F \end{bmatrix} : \begin{matrix} \mathfrak{M} \\ \oplus \\ \mathcal{K} \end{matrix} \to \begin{matrix} \mathfrak{M} \\ \oplus \\ \mathcal{K} \end{matrix} \end{aligned}$$

is selfadjoint contraction. Conversely, any selfadjoint contraction

$$T = \begin{bmatrix} D & C \\ C^* & F \end{bmatrix} : \begin{matrix} \mathfrak{M} \\ \oplus \\ \mathcal{K} \end{matrix} \to \begin{matrix} \mathfrak{M} \\ \oplus \\ \mathcal{K} \end{matrix}$$

has the representation

$$T = \begin{bmatrix} K & 0 \\ 0 & I \end{bmatrix} \mathbb{J}_F \begin{bmatrix} K^* & 0 \\ 0 & I \end{bmatrix} + \begin{bmatrix} D_{K^*} Y D_{K^*} & 0 \\ 0 & 0 \end{bmatrix}$$

with some contraction $K \in \mathbf{B}(\mathfrak{D}_F, \mathfrak{M})$ and some selfadjoint contraction $Y \in \mathbf{B}(\mathfrak{D}_{K^*})$. Moreover, T is unitary if and only if K is an isometry and $Y = Y^* = Y^{-1}$ in the subspace $\mathfrak{D}_{K^*} = \ker K^*$.

References

[1] Yu.M. Arlinskiĭ, *Characteristic functions of operators of the class $C(\alpha)$*, Izv. Vyssh. Uchebn. Zaved. Mat., 1991, No. 2, p. 13–21 [in Russian]. English translation in Soviet Math. (Iz. VUZ), 35 No. 2 (1991), 13–23.

[2] Yu.M. Arlinskiĭ, *Transformations of Nevanlinna operator-functions and their fixed points*, Methods Funct. Anal. Topology, **23** (2017), No. 3, 212–230.

[3] Yu.M. Arlinskiĭ, S. Hassi, H.S.V. de Snoo, *Q-functions of quasi-selfadjoint contractions*, Operator theory and indefinite inner product spaces, Oper. Theory Adv. Appl., **163** (2006), 23–54, Birkhäuser, Basel.

[4] Yu. Arlinskiĭ, S. Hassi, and H.S.V. de Snoo, *Parametrization of contractive block operator matrices and passive discrete-time systems*, Complex Anal. Oper. Theory, **1** (2007), 211–233.

[5] Yu.M. Arlinskiĭ, S. Hassi, H.S.V. de Snoo, *Passive systems with a normal main operator and quasi-selfadjoint systems*, Complex Anal. Oper. Theory, **3** (2009), No. 1, 19–56.

[6] Yu. Arlinskiĭ and L. Klotz, *Weyl functions of bounded quasi-selfadjoint operators and block operator Jacobi matrices*, Acta Sci. Math. (Szeged), **76** (2010), No. 3–4, 585–626.

[7] D.Z. Arov, *Passive linear stationary dynamical systems*, Sibirsk. Mat. Zh., **20** (1979), No. 2, 211–228 [Russian]. English translation in Sib. Math. J., 20 (1979), 149–162.

[8] D.Z. Arov, *Stable dissipative linear stationary dynamical scattering systems*, J. Operator Theory, **1** (1979), 95–126 [in Russian].

[9] D.Z. Arov and M.A. Nudel'man, *A criterion for the unitary similarity of minimal passive systems of scattering with a given transfer function*, Ukr. Mat. Zh. **52** (2000), No. 2, 147–156 [in Russian]. English translation in Ukr. Math. J., **52**, (2000), No. 2, 161–172.

[10] D.Z. Arov and M.A. Nudel'man, *Tests for the similarity of all minimal passive realizations of a fixed transfer function (scattering and resistance matrix)*, Mat. Sb., **193** (2002), No. 6, 3–24 [in Russian]. English translation in Sb.Math., **193** (2002), No.6,791–810.

[11] Gr. Arsene and A. Gheondea, *Completing matrix contractions*, J. Operator Theory, **7** (1982), 179–189.

[12] J.A. Ball and A. Lubin, *On a class of contractive perturbations of restricted shifts*, Pacific J. Math., **63** (1976), No. 2, 309–323.

[13] Yu.M. Berezansky, *Expansion in eigenfunctions of selfadjoint operators*, Amer. Math. Soc., Providence, R.I., 1968.

[14] T. Constantinescu, *Operator Schur algorithm and associated functions*, Math. Balkanica (N.S.) **2** (1988), No. 2-3, 244–252.

[15] Ch. Davis, W.M. Kahan, and H.F. Weinberger, *Norm preserving dilations and their applications to optimal error bounds*, SIAM J. Numer. Anal., **19** (1982), 445–469.

[16] A.P. Filimonov and E.R. Tsekanovskiĭ, *Automorphic-invariant operator colligations and the factorization of their characteristic operator-functions*. Funkts. Anal. Prilozh., **21** (1987), No.4, 94–95 [in Russian]. English translation in Funct. Anal. Appl., **21** (1987), No. 4, 343–344.

[17] R.M. Redheffer, *On certain linear fractional transformation*, J. Math. Phys., **39** (1960), 269–286.

[18] Yu.L. Shmul'yan, *Generalized fractional-linear transformations of operator balls*, Sibirsk. Mat. Zh. **21** (1980), No. 5, 114–131 [in Russian]. English translation in Sib. Math. J., **21** (1980), No. 5, 728–740.

[19] Yu.L. Shmul'yan and R.N. Yanovskaya, *Blocks of a contractive operator matrix*, Izv. Vyssh. Uchebn. Zaved. Mat., **7** (1981), 72–75. [in Russian]. English translation in Sov. Math. **25** (1981), No. 7, 82–86.

[20] B. Sz.-Nagy and C. Foiaş, *Harmonic analysis of operators on Hilbert space*, North-Holland, New York, 1970.

[21] D. Timotin, *Redheffer products and characteristic functions*, J. Math. Anal. Appl., **196** (1995), No. 3, 823–840.

Yuri Arlinskiĭ
Volodymyr Dahl East Ukrainian National University
pr. Central 59-A
Severodonetsk, 93400, Ukraine
e-mail: yury.arlinskii@gmail.com

Seppo Hassi
Department of Mathematics and Statistics
University of Vaasa
P.O. Box 700
65101 Vaasa, Finland
e-mail: sha@uwasa.fi

Operator Theory:
Advances and Applications, Vol. 272, 43–60

Free Bianalytic Maps between Spectrahedra and Spectraballs in a Generic Setting

Meric Augat, J. William Helton, Igor Klep and Scott McCullough

This paper, which would not exist without techniques he pioneered, is dedicated to Joe Ball on the occasion of his 70th birthday.

Abstract. Given a tuple $E = (E_1, \dots, E_g)$ of $d \times d$ matrices, the collection $\mathcal{B}_E$ of those tuples of matrices $X = (X_1, \dots, X_g)$ (of the same size) such that $\|\sum E_j \otimes X_j\| \le 1$ is a spectraball. Likewise, given a tuple $B = (B_1, \dots, B_g)$ of $e \times e$ matrices the collection $\mathcal{D}_B$ of tuples of matrices $X = (X_1, \dots, X_g)$ (of the same size) such that $I + \sum B_j \otimes X_j + \sum B_j^* \otimes X_j^* \succeq 0$ is a free spectrahedron. Assuming E and B are irreducible, plus an additional mild hypothesis, there is a free bianalytic map $p : \mathcal{B}_E \to \mathcal{D}_B$ normalized by $p(0) = 0$ and $p'(0) = I$ if and only if $\mathcal{B}_E = \mathcal{B}_B$ and B spans an algebra. Moreover p is unique, rational and has an elegant algebraic representation.

Mathematics Subject Classification (2010). 47L25, 32H02, 13J30 (Primary); 14P10, 52A05, 46L07 (Secondary).

Keywords. Bianalytic map, birational map, linear matrix inequality (LMI), spectrahedron, convex set, Positivstellensatz, free analysis, real algebraic geometry.

1. Introduction

In this article we continue our investigation of free bianalytic mappings between matrix convex domains. The results in this article stand on the bedrock of the non-commutative state space methods introduced to the operator theory community by Joe and his collaborators and they are inseparable from the profound influence of Joe's work in function theoretic operator theory and free analysis.

Fix g a positive integer. Given a positive integer n, let $M_n(\mathbb{C})^g$ denote the g-tuples $X = (X_1, \dots, X_g)$ of $n \times n$ matrices with entries from $\mathbb{C}$. Given $A \in M_d(\mathbb{C})^g$,

---

JWH was supported by the NSF grant DMS 1500835. IK was supported by the Marsden Fund Council of the Royal Society of New Zealand, and partially supported by the Slovenian Research Agency grants J1-8132, N1-0057 and P1-0222. SM was supported by NSF grants DMS 1361501 and DMS 1764231.

the set $\mathcal{D}_A(1)$ consisting of $x \in \mathbb{C}^g$ such that

$$L_A(x) = I + \sum A_j x_j + \sum A_j^* x_j^* \succeq 0$$

is a *spectrahedron*. Here $T \succeq 0$ indicates the selfadjoint matrix T is positive semidefinite. Spectrahedra are basic objects in a number of areas of mathematics; e.g., semidefinite programming, convex optimization and in real algebraic geometry [8]. They also figure prominently in determinantal representations [10, 15, 29, 33], the solution of the Lax conjecture [23], in the solution of the Kadison–Singer paving conjecture [27], and in systems engineering [9, 32].

For $X \in M_n(\mathbb{C})^g$ and still with $A \in M_d(\mathbb{C})^g$, let

$$\Lambda_A(X) = \sum A_j \otimes X_j$$

and

$$L_A(X) = I + \Lambda_A(X) + \Lambda_A(X)^* = I + \sum A_j \otimes X_j + \sum A_j^* \otimes X_j^*.$$

The *free spectrahedron* determined by A is the sequence of sets $\mathcal{D}_A = (\mathcal{D}_A(n))$, where

$$\mathcal{D}_A(n) = \{X \in M_n(\mathbb{C})^g : L_A(X) \succeq 0\}.$$

Free spectrahedra arise naturally in applications such as systems engineering [11] and in the theories of matrix convex sets, operator algebras, systems and spaces and completely positive maps [12, 21, 30]. They also provide tractable useful relaxations for spectrahedral inclusion problems that arise in semidefinite programming and engineering applications such as the matrix cube problem [7, 22].

Given a tuple $E \in M_d(\mathbb{C})^g$, the set

$$\mathcal{B}_E = \{X : \|\Lambda_E(X)\| \leq 1\}$$

is a *spectraball* [6, 13]. Spectraballs are special cases of free spectrahedra. Indeed, it is readily seen that

$$\mathcal{B}_E = \mathcal{D}_{\left(\begin{smallmatrix} 0 & E \\ 0 & 0 \end{smallmatrix}\right)}.$$

Let $M(\mathbb{C})^g$ denote the sequence $(M_n(\mathbb{C})^g)_n$. A *subset* Γ of $M(\mathbb{C})^g$ is a sequence $(\Gamma_n)_n$ where $\Gamma_n \subset M_n(\mathbb{C})^g$. (Sometimes we will write $\Gamma(n)$ in place of Γ_n.) The subset Γ is a *free set* if it is closed under direct sums and unitary similarity; that is, if $X \in \Gamma_n$ and $Y \in \Gamma_m$, then

$$X \oplus Y = \left(\begin{pmatrix} X_1 & 0 \\ 0 & Y_1 \end{pmatrix}, \dots, \begin{pmatrix} X_g & 0 \\ 0 & Y_g \end{pmatrix} \right) \in \Gamma_{n+m};$$

and if U is an $n \times n$ unitary matrix, then

$$U^* X U = \left(U^* X_1 U, \dots, U^* X_g U \right) \in \Gamma_n.$$

We say the free set $\Gamma = (\Gamma_n)_n$ is *open* if each Γ_n is open. (Generally adjectives are applied levelwise to free sets unless noted otherwise.)

A *free function* $f : \Gamma \to M(\mathbb{C})$ is a sequence of functions $f_n : \Gamma_n \to M_n(\mathbb{C})$ that *respects intertwining*; that is, if $X \in \Gamma_n$, $Y \in \Gamma_m$, $T : \mathbb{C}^m \to \mathbb{C}^n$, and

$$XT = (X_1 T, \dots, X_g T) = (TY_1, \dots, TY_g) = TY,$$

then $f_n(X)T = Tf_m(Y)$. Assuming Γ is an open free set, a free function $f : \Gamma \to M(\mathbb{C})$ is *analytic* if each f_n is analytic. Given free sets $\Gamma \subset M(\mathbb{C})^g$ and $\Delta \subset M(\mathbb{C})^h$, a *free mapping* $f : \Gamma \to \Delta$ consists of free maps $f^i : \Gamma \to M(\mathbb{C})$ such that $f(X) = \begin{pmatrix} f^1(X) & \dots & f^h(X) \end{pmatrix}$. In this case we write $f = \begin{pmatrix} f^1 & \dots & f^h \end{pmatrix}$. We refer the reader to [25, 34] for a fuller discussion of free sets and functions.

In this note, we characterize the free bianalytic maps $p : \mathcal{B}_E \to \mathcal{D}_B$ under some mild conditions on $E \in M_d(\mathbb{C})^g$ and $B \in M_e(\mathbb{C})^g$ and on p and its inverse q. These free functions take a highly algebraic form that we call *convexotonic.* A tuple $\Xi = (\Xi_1, \dots, \Xi_g) \in M_g(\mathbb{C})^g$ satisfying

$$\Xi_k \Xi_j = \sum_{s=1}^{g} (\Xi_j)_{k,s} \Xi_s$$

for each $1 \le j, k \le g$ is *convexotonic.* Convexotonic tuples naturally arise from finite-dimensional algebras. If $\{J_1, \dots, J_g\} \subset M_d(\mathbb{C})$ is linearly independent and spans an algebra, then there exists a uniquely determined tuple $\Psi \in M_g(\mathbb{C})^g$ such that

$$J_k J_j = \sum_{s=1}^{g} (\Psi_j)_{k,s} J_s \tag{1.1}$$

and Proposition 2.1 says Ψ is convexotonic.

Given a convexotonic tuple $\Xi \in M_g(\mathbb{C})^g$, the expressions $p = \begin{pmatrix} p^1 & \cdots & p^g \end{pmatrix}$ and $q = \begin{pmatrix} q^1 & \cdots & q^g \end{pmatrix}$ whose components have the form

$$p^i(x) = \sum_j x_j \left(I - \Lambda_\Xi(x)\right)^{-1}_{j,i} \qquad \text{and} \qquad q^i(x) = \sum_j x_j \left(I + \Lambda_\Xi(x)\right)^{-1}_{j,i}, \tag{1.2}$$

that is, in row form,

$$p(x) = x(I - \Lambda_\Xi(x))^{-1} \qquad \text{and} \qquad q = x(I + \Lambda_\Xi(x))^{-1}$$

are, by definition, *convexotonic.* The components of p (resp. q) are free functions with (free) domains consisting of those X for which $I - \Lambda_\Xi(X)$ (resp. $I + \Lambda_\Xi(X)$) is invertible. Hence p and q are free functions. It turns out (see [1, Proposition 6.2]) the mappings p and q are inverses of one another.

Before continuing, we would like to point out that the component functions p^i of the convexotonic map p of equation (1.2) are in fact free rational functions regular at 0. Accordingly we refer to p and q as *birational* or free birational maps. Free rational functions are most easily described and naturally understood in terms of realization theory as developed in the series of papers [3–5] of Ball–Groenewald–Malakorn. Indeed, based on those articles and on the results of [24, Theorem 3.1] and [35, Theorem 3.5]) a *free rational function regular at* 0 can, for the purposes of this article, be defined with minimal overhead as an expression of the form

$$r(x) = c^* \left(I - \Lambda_S(x)\right)^{-1} b$$

where s is a positive integer, $S \in M_s(\mathbb{C})^g$ and $b, c \in \mathbb{C}^s$ are vectors. The expression r is known as a realization. Realizations are easy to manipulate and the theory of realizations is a powerful tool. The realization r is evaluated in the obvious fashion for a tuple $X \in M_n(\mathbb{C})^g$ as long as $I-\Lambda_S(X)$ is invertible. Free polynomials are free rational functions that are regular at 0 and free rational functions regular at 0 are stable with respect to the formal algebraic operations of addition, multiplication and inversion in the sense that if r is a free rational function regular at 0 and $r(0) \neq 0$, then its multiplicative inverse r^{-1} is also a free rational function regular at 0. Thus, expressing p^i as

$$p^i = \sum_{s=1}^{g} x_s e_s^*(I - \Lambda_\Xi(x))^{-1} e_i$$

shows it is a free rational function regular at 0.

To state our main theorem precisely we need a bit more terminology. A subset $\{u^1, \dots, u^{d+1}\}$ of $\mathbb{C}^d$ is a *hyperbasis* for $\mathbb{C}^d$ if each d element subset is a basis. The tuple $A \in M_d(\mathbb{C})^g$ is *sv-generic* if there exists $\alpha^1, \dots, \alpha^{d+1}$ and $\beta^1, \dots, \beta^d$ in $\mathbb{C}^g$ such that, for each $1 \le j \le d+1$, the matrix $I - \Lambda_A(\alpha^j)^*\Lambda_A(\alpha^j)$ is positive semidefinite, has a one-dimensional kernel spanned by u^j and the set $\{u^1, \dots, u^{d+1}\}$ is a hyperbasis for $\mathbb{C}^d$; and, for each $1 \le k \le g$, the matrix $I - \Lambda_A(\beta^k)\Lambda_A(\beta^k)^*$ is positive semidefinite, has a one-dimensional kernel spanned by v^k and the set $\{v^1, \dots, v^d\}$ is a basis for $\mathbb{C}^d$. Generic tuples A satisfy this property, see [1, Remark 7.5]. Given a matrix-valued free analytic polynomial Q, the set

$$\mathcal{G}_Q = \{X \in M(\mathbb{C})^g : \|Q(X)\| < 1\} \subset M(\mathbb{C})^g$$

is a *free pseudoconvex* set.

Theorem 1.1. *Suppose $E \in M_d(\mathbb{C})^g$ and $B \in M_e(\mathbb{C})^g$. If*

(i) *E is sv-generic and linearly independent;*
(ii) *B is sv-generic and $\mathcal{D}_B$ is bounded;*
(iii) *$p : \mathcal{B}_E \to \mathcal{D}_B$ is bianalytic with $p(0) = 0$ and $p'(0) = I$; and*
(iv) *p is defined on a pseudoconvex domain containing $\mathcal{B}_E$ and $q : \mathcal{D}_B \to \mathcal{B}_E$, the inverse of p, is defined on a pseudoconvex domain containing $\mathcal{D}_B$,*

then there exist $g \times g$ unitary matrices Z and M and a tuple $\Xi \in M_g(\mathbb{C})^g$ such that

(1) *$B = M^*ZEM$;*
(2) *for each $1 \le j, k \le g$,*

$$E_k Z E_j = \sum\nolimits_s (\Xi_j)_{k,s} E_s; \tag{1.3}$$

(3) *the tuple B spans an algebra and*

$$B_k B_j = \sum\nolimits_g (\Xi_j)_{k,s} B_s;$$

(4) *Ξ is convexotonic and p is the corresponding convexotonic map $p = x(I - \Lambda_\Xi(x))^{-1}$.*

Remark 1.2. Several remarks are in order.

(i) A free spectrahedron $\mathcal{D}$ is *sv-generic* if there exists an sv-generic tuple A such that $\mathcal{D} = \mathcal{D}_A$. The article [1] contains a version of Theorem 1.1 for bianalytic mappings between sv-generic free spectrahedra (actually a weaker, but more complicated to formulate, condition from [1] that we call eig-generic would also suffice here). The sv-generic free spectrahedra are in fact generic among free spectrahedra in the sense of algebraic geometry. However, spectraballs, within the class of free spectrahedra, are *never* sv-generic in view of Lemma 4.1. Hence, Theorem 1.1 extends Theorem [1, Theorem 1.8], to the important special case of maps from spectraballs to free spectrahedra.

(ii) Let

$$F_1 = \begin{pmatrix} 0 & 1 & 0 \\ 0 & 0 & 1 \\ 0 & 0 & 0 \end{pmatrix} = \begin{pmatrix} 0 & E_1 \\ 0 & 0 \end{pmatrix} \quad \text{and} \quad F_2 = F_1^2 = \begin{pmatrix} 0 & E_2 \\ 0 & 0 \end{pmatrix}, \tag{1.4}$$

where the tuple E is given in equation (4.1). The tuple F is nilpotent. Thus, by Lemma 4.1, it is not sv-generic and the results of [1] do not apply to bianalytic maps $r : \mathcal{D}_F \to \mathcal{D}_B$ with $r(0) = 0$ and $r'(0) = I$. Moreover, $\mathcal{D}_F$ is not a spectraball by Proposition 4.3 and thus Theorem 1.1 does not directly apply either. However, as we show, there is an sv-generic tuple E and a bianalytic map $p : \mathcal{B}_E \to \mathcal{D}_F$ with $p(0) = 0$ and $p'(0) = I$ (see Proposition 4.3). On the other hand, Theorem 1.1 does apply to bianalytic maps $f : \mathcal{B}_E \to \mathcal{D}_B$. By composing f with p^{-1}, Propositions 4.4 and 4.6 classify the choices for B and all bianalytic maps between $\mathcal{D}_F$ and $\mathcal{D}_B$. In particular, these maps are convexotonic.

(iii) It is easy to check that item (1) implies $\mathcal{B}_E = \mathcal{B}_B$.

(iv) Since E is assumed linearly independent Ξ is uniquely determined by equation (1.3). Further, by Proposition 2.1, Ξ is convexotonic.

(v) Items (2) and (3) are equivalent given (1).

(vi) Note that, while p is only assumed to be bianalytic, the conclusion is that p is birational, a phenomena encountered frequently in rigidity theory in several complex variables, cf. [14].

(vii) A key ingredient in the proof of Theorem 1.1 is a suitable Positivstellensatz. Namely, p maps $\mathcal{B}_E$ into $\mathcal{D}_B$ if and only if $L_B(p(X)) \succeq 0$ for all $X \in \mathcal{B}_E$, and this equivalence feeds naturally into Positivstellensätze, a pillar of real algebraic geometry. The one used here (from [1]) is related to that of [2], which was developed in full generality in [6].

(viii) An easy argument shows, for $A \in M_d(\mathbb{C})^g$, if $\mathcal{D}_A$ is bounded, then A (really $\{A_1, \dots, A_g\}$) is linearly independent [20, Proposition 2.6(2)]. The converse fails in general; e.g., if each A_j is positive semidefinite. On the other hand, $E \in M_d(\mathbb{C})^g$ is linearly independent if and only if $\mathcal{B}_E$ is bounded [20, Proposition 2.6(1)].

There is a natural converse to Theorem 1.1. Let $\text{int}(\mathcal{D}_A)$ and $\text{int}(\mathcal{B}_A)$ denote the interiors of $\mathcal{D}_A$ and $\mathcal{B}_A$ respectively. Recall a mapping between metric spaces is *proper* if the inverse image of compact sets are compact. Thus, for open sets $\mathcal{U} \subset M(\mathbb{C})^g$ and $\mathcal{V} \subset M(\mathbb{C})^h$, a free mapping $f : \mathcal{U} \to \mathcal{V}$ is proper if each $f_n : \mathcal{U}_n \to \mathcal{V}_n$ is proper.

Proposition 1.3. *Suppose $J \in M_d(\mathbb{C})^g$ is linearly independent, spans an algebra, Ξ is the resulting convexotonic tuple,*

$$J_k J_j = \sum_{s=1}^{g} (\Xi_j)_{k,s} J_s,$$

and q is the convexotonic (birational) map,

$$q(x) = x(I + \Lambda_\Xi(x))^{-1}.$$

Then

(1) *The domain of q contains $\mathcal{D}_J$.*
(2) *q is a bianalytic map between $\operatorname{int}(\mathcal{D}_J)$ and $\operatorname{int}(\mathcal{B}_J)$; that is p, the (convexotonic) inverse of q, maps $\operatorname{int}(\mathcal{B}_J)$ into $\operatorname{int}(\mathcal{D}_J)$. In particular, q is proper.*
(3) *q maps the boundary of $\mathcal{D}_J$ into the boundary of $\mathcal{B}_J$;*
(4) *if, in addition, $\mathcal{D}_J$ is bounded, then q is a bianalytic map between $\mathcal{D}_J$ and $\mathcal{B}_J$. In particular, the domain of p contains $\mathcal{B}_J$.*

In case J does not span an algebra, we have the following corollary of Proposition 1.3.

Corollary 1.4. *Let $A \in M_d(\mathbb{C})^g$ and assume A is linearly independent (e.g., $\mathcal{D}_A$ is bounded). Let $\mathcal{A}$ denote the algebra spanned by the tuple A. If $C_1, \dots, C_h \in M_d(\mathbb{C})$ and the tuple $J = (J_1, \dots, J_{g+h}) = (A_1, \dots, A_g, C_1, \dots, C_h)$ is linearly independent and spans $\mathcal{A}$, then there is a rational map f with $f(0) = 0$ and $f'(0) = I$ such that*

(1) *f is an injective proper map from $\operatorname{int}(\mathcal{D}_A)$ into $\operatorname{int}(\mathcal{B}_J)$; and*
(2) *f maps the boundary of $\mathcal{D}_A$ into boundary of $\mathcal{B}_J$.*

Further, the tuple $\Xi \in M_{g+h}(\mathbb{C})^{g+h}$, uniquely determined by

$$J_k J_j = \sum_{s=1}^{h} (\Xi_j)_{k,s} J_s,$$

is convexotonic and

$$f(x) = \begin{pmatrix} x_1 & \cdots & x_g & 0 & \cdots & 0 \end{pmatrix} \left(I + \sum_{j=1}^{g} \Xi_j x_j \right)^{-1}.$$

For further results, not already cited, on free bianalytic and proper free analytic maps see [16–18, 26, 28, 31] and the references therein.

The remainder of the article is organized as follows. Proposition 1.3 and Corollary 1.4 are established in Section 2. Theorem 1.1 is proved in Section 3. The article concludes with several examples; see Section 4.

2. Proof of Proposition 1.3

This section gives the proof of Proposition 1.3. Implicit in the statement of that result, and used in the proof of Theorem 1.1, is the connection between finite-dimensional algebras and convexotonic tuples described in the following proposition.

Proposition 2.1. *Suppose $G \in M_{d\times e}(\mathbb{C})^g$ and $\{G_1,\dots,G_g\}$ is linearly independent, $C \in M_{e\times d}(\mathbb{C})$ and $\Psi \in M_g(\mathbb{C})^g$. If*

$$G_\ell C G_j = \sum_{s=1}^{g} (\Psi_j)_{\ell,s} G_s, \tag{2.1}$$

then the tuple Ψ is convexotonic. In particular, if $J \in M_d(\mathbb{C})^g$ is linearly independent and spans an algebra, then the tuple Ψ uniquely determined by equation (1.1) *is convexotonic.*

Proof. For notational ease let $T = CG \in M_e(\mathbb{C})^g$. The hypothesis implies T spans an algebra (but not that T is linearly independent). Routine calculations give

$$[G_\ell T_j]T_k = \sum_{t=1}^{g} (\Psi_j)_{\ell,t} G_t\, T_k = \sum_{s,t=1} (\Psi_j)_{\ell,t}(\Psi_k)_{t,s} G_s = \sum_s (\Psi_j\, \Psi_k)_{\ell,s} G_s.$$

On the other hand

$$G_\ell[T_j T_k] = G_\ell C[G_j T_k] = \sum_t G_\ell (\Psi_k)_{j,t} T_t = \sum_{s,t} (\Psi_t)_{\ell,s} (\Psi_k)_{j,t} G_s.$$

By independence of G,

$$(\Psi_j \Psi_k)_{\ell,s} = \sum_t (\Psi_k)_{j,t} (\Psi_t)_{\ell,s}$$

and therefore

$$\Psi_j \Psi_k = \sum_t (\Psi_k)_{j,t} \Psi_t$$

and the proof is complete. □

Lemma 2.2. *Suppose $F \in M_d(\mathbb{C})^g$. If $I + \Lambda_F(X) + \Lambda_F(X)^* \succeq 0$, then $I + \Lambda_F(X)$ is invertible.*

Proof. Arguing the contrapositive, suppose $I + \Lambda_F(X)$ is not invertible. In this case there is a unit vector γ such that

$$\Lambda_F(X)\gamma = -\gamma.$$

Hence,

$$\langle (I + \Lambda_F(X) + \Lambda_F(X)^*)\gamma, \gamma\rangle = \langle \Lambda_F(X)^*\gamma, \gamma\rangle = \langle \gamma, \Lambda_F(X)\gamma\rangle = -1. \qquad \square$$

Lemma 2.3. *Let $T \in M_d(\mathbb{C})$. Then*

(a) *$I + T + T^* \succeq 0$ if and only if $I + T$ is invertible and $\|(I+T)^{-1}T\| \le 1$;*
(b) *$I + T + T^* \succ 0$ if and only if $I + T$ is invertible and $\|(I+T)^{-1}T\| < 1$.*

Similarly if $I - T$ is invertible, then $\|T\| \leq 1$ if and only if $I + R + R^* \succ 0$, where $R = T(I-T)^{-1}$.

Proof. (a) We have the following chain of equivalences:

$$\begin{aligned} \|(I+T)^{-1}T\| \leq 1 \quad &\iff \quad I - \big((I+T)^{-1}T\big)\big((I+T)^{-1}T\big)^* \succeq 0 \\ &\iff \quad I - (I+T)^{-1}TT^*(I+T)^{-*} \succeq 0 \\ &\iff \quad (I+T)(I+T)^* - TT^* \succeq 0 \\ &\iff \quad I + T + T^* \succeq 0. \end{aligned}$$

The proof of (b) is the same. □

Proposition 2.4. *For $F \in M_d(\mathbb{C})^g$ we have*

$$\mathcal{D}_F = \{X \colon \|(1 + \Lambda_F(X))^{-1}\Lambda_F(X)\| \leq 1\}.$$

Proof. Immediate from Lemma 2.3. □

Proof of Proposition 1.3. Let q denote the convexotonic map associated to the convexotonic tuple Ξ in the statement of the proposition,

$$q(x) = \begin{pmatrix} x_1 & \cdots & x_g \end{pmatrix} (I + \Lambda_\Xi(x))^{-1} = x\,(I + \Lambda_\Xi(x))^{-1}.$$

Compute

$$\begin{aligned} \Lambda_J(q(x))\,\Lambda_J(x) &= \sum_{s,k=1}^{g} q^s(x) x_k J_s J_k = \sum_{j=1}^{g}\sum_{s=1}^{g} q^s(x)\left[\sum_{k=1}^{g} x_k (\Xi_k)_{s,j}\right] J_j \\ &= \sum_{j=1}^{g}\sum_{s=1}^{g} q^s(x)(\Lambda_\Xi(x))_{s,j} J_j \\ &= \sum_{j=1}^{g}\sum_{t=1}^{g} x_t \left[\sum_{s=1}^{g} (I + \Lambda_\Xi(x))^{-1}_{t,s}(\Lambda_\Xi(x))_{s,j}\right] J_j \\ &= \sum_{j=1}^{g}\sum_{t=1}^{g} x_t [(I + \Lambda_\Xi(x))^{-1}\Lambda_\Xi(x)]_{t,j} J_j. \end{aligned}$$

Hence,

$$\Lambda_J(q(x))\,(I + \Lambda_J(x)) = \sum_{j=1}^{g}\sum_{t=1}^{g} x_t[(I + \Lambda_\Xi(x))^{-1}(I + \Lambda_\Xi(x))]_{t,j} J_j = \Lambda_J(x).$$

Thus, as free (matrix-valued) rational functions regular at 0,

$$\Lambda_J(q(x)) = (I + \Lambda_J(x))^{-1}\,\Lambda_J(x) =: F(x). \tag{2.2}$$

Since J is linearly independent, given $1 \leq k \leq g$, there is a linear functional λ such that $\lambda(J_j) = 0$ for $j \neq k$ and $\lambda(J_k) = 1$. Applying λ to equation (2.2), gives

$$q^k(x) = \lambda(F(x)). \tag{2.3}$$

Since $\lambda(F(x))$ is a free rational function whose domain contains

$$\mathscr{D} = \{X : I + \Lambda_J(X) \text{ is invertible}\},$$

the same is true for q^k. (As a technical matter, each side of equation (2.3) is a rational expression. Since they are defined and agree on a neighborhood of 0, they determine the same free rational function. It is the domain of this rational function that contains $\mathscr{D}$. See [35], and also [24], for full details.) By Lemma 2.2, $\mathscr{D}$ contains $\mathcal{D}_J$ (as $X \in \mathcal{D}_J$ implies $I+\Lambda_J(X)$ is invertible). Hence the domain of the free rational mapping q contains $\mathcal{D}_J$. By Lemma 2.3 and equation (2.2), q maps the interior of $\mathcal{D}_J$ into the interior of $\mathcal{B}_J$ and the boundary of $\mathcal{D}_J$ into the boundary of $\mathcal{B}_J$.

Similarly,

$$(I - \Lambda_J(x))^{-1}\,\Lambda_J(x) = \Lambda_J(p(x)), \tag{2.4}$$

where $p(x) = x(I - \Lambda_\Xi(x))^{-1}$. Arguing as above shows the domain of p contains the set

$$\mathscr{E} = \{X : I - \Lambda_J(X) \text{ is invertible}\},$$

which in turn contains $\operatorname{int}(\mathcal{B}_J)$ (since $\|\Lambda_J(X)\| < 1$ allows for an application of Lemma 2.3). By Lemma 2.3 and equation (2.4), p maps the interior of $\mathcal{B}_J$ into the interior of $\mathcal{D}_J$. Hence q is bianalytic between these interiors. Further, if X is in the boundary of $\mathcal{B}_J$, then for $t \in \mathbb{C}$ and $|t| < 1$, we have $p(tX) \in \operatorname{int}(\mathcal{D}_J)$ and

$$\Lambda_J(p(tX)) = (I - \Lambda_J(tX))^{-1}\,\Lambda_J(tX).$$

Assuming $\mathcal{D}_J$ is bounded, it follows that $I - \Lambda_J(X)$ is invertible and thus X is in the domain of p and $p(X)$ is in the boundary of $\mathcal{D}_J$. □

Proof of Corollary 1.4. Letting $z = (z_1, \dots, z_{g+h})$ denote a $g + h$ tuple of freely non-commuting indeterminants, and Ξ the convexotonic $g + h$ tuple as described in the corollary, by Proposition 1.3 the birational mapping

$$q(z) = z(I + \Lambda_\Xi(z))^{-1}$$

is a bianalytic (hence injective and proper) mapping between $\operatorname{int}(\mathcal{D}_J)$ and $\operatorname{int}(\mathcal{B}_J)$ that also maps boundary to boundary. The mapping $\iota : \mathcal{D}_A \to \mathcal{D}_J$ defined by $\iota(x) = (x, 0)$ is proper from $\operatorname{int}(\mathcal{D}_A)$ to $\operatorname{int}(\mathcal{D}_J)$ and maps boundary to boundary. Hence, the composition

$$r(x) = p(\iota(x)) = \begin{pmatrix} x & 0 \end{pmatrix}\,(I - \Lambda_\Xi(x, 0))^{-1}$$

is a proper map from $\operatorname{int}(\mathcal{D}_A)$ into $\operatorname{int}(\mathcal{B}_J)$ that also maps boundary to boundary. □

3. Proof of Theorem 1.1

Given $E \in M_d(\mathbb{C})^g$, let

$$A = \begin{pmatrix} 0 & E \\ 0 & 0 \end{pmatrix} \in M_{2d}(\mathbb{C})^g.$$

Thus $\mathcal{B}_E = \mathcal{D}_A$ and, among other things, by assumption, there is a bianalytic map $p : \mathcal{D}_A \to \mathcal{D}_B$. It follows by the analytic Positivstellensatz [1, Theorem 1.9] applied to the matrix-valued free analytic function

$$G(x) = \Lambda_B(p(x))$$

that there exists a Hilbert space H, an isometry $\widetilde{C}$ on the range of $I_H \otimes A$ and an isometry $\mathscr{W} : \mathbb{C}^e \to H \otimes \mathbb{C}^{2d}$ such that, with $\widetilde{R} = (\widetilde{C} - I)(I_H \otimes A)$,

$$L_B(p(x)) = I+G(x)+G(x)^* = \mathscr{W}^*(I-\Lambda_{\widetilde{R}}(x))^{-*}L_{I_H\otimes A}(x)(I-\Lambda_{\widetilde{R}}(x))^{-1}\mathscr{W}. \quad (3.1)$$

That the analytic Positivstellensätze requires $\mathcal{D}_A$ to be bounded and G to extend analytically to a pseudoconvex set containing $\mathcal{D}_A$ explains the need for the hypotheses that $\mathcal{D}_A = \mathcal{B}_E$ is bounded (equivalently E is linearly independent) and p extends analytically to a pseudoconvex set containing $\mathcal{D}_A$.

Since E is sv-generic, both $\ker(E) := \cap \ker(E_j) = \{0\}$ and $\ker(E^*) = \{0\}$. In particular,

$$\operatorname{rg}(A) := \operatorname{span}\left(\bigcup_{j=1}^{g} \operatorname{rg}(A_j)\right) = \mathbb{C}^d \oplus \{0\}.$$

In particular, $\dim(\operatorname{rg}(A)) = d$. Likewise $\dim(\operatorname{rg}(A^*)) = d$ too.

The next step involves a call to [1, Lemma 7.7]. That lemma is stated in terms of conditions referred to as eig-generic, weakly eig-generic, $*$-generic and weakly $*$-generic formally defined in [1, Definition 7.3]. It is readily seen that if a g-tuple F of $N \times N$ matrices is sv-generic, then it is both eig-generic and $*$-generic (and thus weakly eig-generic and weakly $*$-generic). In particular, $\operatorname{rg}(F) = \mathbb{C}^N = \operatorname{rg}(F^*)$. Thus, both E and B are both eig-generic and $*$-generic. That E is eig-generic implies A is weakly eig-generic; and that E is $*$-generic implies A is weakly $*$-generic.

By [1, Lemma 7.7(1)], $d = \dim(\operatorname{rg}(A^*)) \leq \dim(\operatorname{rg}(B^*)) = e$. Applying [1, Lemma 7.7(1)] to $q : \mathcal{D}_B \to \mathcal{D}_A$ (so reversing the roles of A and B), it also follows that $e \leq d$. Hence $\dim(\operatorname{rg}(A^*)) = d = e = \dim(\operatorname{rg}(B^*))$ and $\dim(\operatorname{rg}(A)) = d = e = \dim(\operatorname{rg}(B))$. Thus we may now invoke (the weakly version of) [1, Lemma 7.7(4)] that says there is a vector $\lambda \in H$ and a unitary $M : \operatorname{rg}(B^*) \to \operatorname{rg}(A^*)$ and an isometry $N : \operatorname{rg}(B^*) \cap \operatorname{rg}(B) \to \operatorname{rg}(A)$ such that $\mathscr{W}v = \lambda \otimes \iota M v$ for $v \in \operatorname{rg}(B^*)$ and $\widetilde{C}(\lambda \otimes \iota N v) = \lambda \otimes \iota M v$ for $v \in \operatorname{rg}(B^*) \cap \operatorname{rg}(B)$ (where we over use ι, letting it denote the inclusions $\operatorname{rg}(A^*) \subset \mathbb{C}^{2d}$ and $\operatorname{rg}(B^*) \subset \mathbb{C}^{2d}$). This general statement in our case specializes, because $\operatorname{rg}(B^*) = \mathbb{C}^d$, $\operatorname{rg}(B) = \mathbb{C}^d$ and $\dim(\operatorname{rg}(A)) = d$, to give

(i) $M : \mathbb{C}^d \to \operatorname{rg}(A^*)$ is unitary;
(ii) $N : \mathbb{C}^d \to \operatorname{rg}(A)$ is unitary;
(iii) $\mathscr{W}v = \lambda \otimes \iota M v$, for $v \in \mathbb{C}^d$; and
(iv) $\widetilde{C}(\lambda \otimes \iota N v) = \lambda \otimes \iota M v$ for $v \in \mathbb{C}^d$.

It follows that there is a unitary mapping $Z : \operatorname{rg}(A) \to \operatorname{rg}(A^*)$ such that, for $w \in \operatorname{rg}(A)$,

$$\widetilde{C}(\lambda \otimes w) = \lambda \otimes \iota Z w.$$

Let $[\lambda] = \mathbb{C}\lambda$, the one-dimensional subspace of H spanned by the unit vector λ.

Let

$$C = \begin{pmatrix} 0 & 0 \\ Z & 0 \end{pmatrix}.$$

In particular C is isometric on the range of A. Let $R = (C - I)A$. For $1 \le j \le d$, and $\gamma \in \mathbb{C}^{2d}$,

$$\begin{aligned} \widetilde{R}_j(\lambda \otimes \gamma) &= (\widetilde{C} - I)[I_H \otimes A_j](\lambda \otimes \gamma) \\ &= (\widetilde{C} - I)(\lambda \otimes A_j\gamma) \\ &= \lambda \otimes (\iota Z - I)A_j\gamma \\ &= \lambda \otimes (C - I)A_j\gamma \\ &= \lambda \otimes R_j\gamma. \end{aligned}$$

Thus $[\lambda] \otimes \mathbb{C}^{2d}$ is invariant for the tuple $\widetilde{R}$ and further

$$\widetilde{R}_j(\lambda \otimes I) = \lambda \otimes R_j.$$

It follows that $[\lambda] \otimes \mathbb{C}^{2d}$ is invariant for the mapping $(I - \Lambda_{\widetilde{R}}(x))^{-1}$ and moreover,

$$(I - \Lambda_{\widetilde{R}}(x))^{-1}(\lambda \otimes I) = \lambda \otimes (I - \Lambda_R(x))^{-1}.$$

Finally, since $\mathscr{W}$ maps into $[\lambda] \otimes \mathbb{C}^{2d}$ and $\mathscr{W}\gamma = \lambda \otimes \iota M\gamma$,

$$W(x) := (I - \Lambda_{\widetilde{R}}(x))^{-1}\mathscr{W} = \lambda \otimes (I - \Lambda_R(x))^{-1}\iota M.$$

Since also $[\lambda] \otimes \mathbb{C}^{2d}$ is invariant for $L_{I_H \otimes A}(x)$,

$$L_{I_H \otimes A}(x)W(x) = \lambda \otimes L_A(x)(I - \Lambda_R(x))^{-1}\iota M.$$

Returning to equation (3.1) and using $\lambda^*\lambda = 1$,

$$\begin{aligned} L_B(p(x)) &= (\lambda^* \otimes (\iota M)^*(I - \Lambda_R(x))^{-*}(\lambda \otimes L_A(x)(I - \Lambda_R(x))^{-1}\iota M \\ &= M^*\iota^*(I - \Lambda_R(x))^{-*}L_A(x)(I - \Lambda_R(x))^{-1}\iota M. \end{aligned} \tag{3.2}$$

Comparing the coefficients of the x_j terms in equation (3.2) gives

$$B = M^*\iota^*CA\iota M.$$

Since $M : \mathbb{C}^d \to \operatorname{rg}(A^*)$ is unitary,

$$\iota M = \begin{pmatrix} 0 \\ U \end{pmatrix} : \mathbb{C}^d \to \mathbb{C}^{2d} = \mathbb{C}^d \oplus \mathbb{C}^d = \operatorname{rg}(A) \oplus \operatorname{rg}(A^*) \tag{3.3}$$

for a unitary mapping $U : \mathbb{C}^d \to \mathbb{C}^d$. Thus,

$$B = U^*ZEU.$$

Since

$$R = (C - I)A = \begin{pmatrix} 0 & -E \\ 0 & ZE \end{pmatrix}.$$

it follows that

$$F(x) := (I - \Lambda_R(x))^{-1} = \begin{pmatrix} I & -\Lambda_E(x)(I - \Lambda_{ZE}(x))^{-1} \\ 0 & (I - \Lambda_{ZE}(x))^{-1} \end{pmatrix}.$$

Consequently,

$$F(x)^* L_A(x) F(x) = \begin{pmatrix} I & 0 \\ -(I - \Lambda_{ZE}(x))^{-*}\Lambda_E(x)^* & (I - \Lambda_{ZE}(x))^{-*} \end{pmatrix}$$
$$\times \begin{pmatrix} I & \Lambda_E(x) \\ \Lambda_E(x)^* & I \end{pmatrix} \begin{pmatrix} I & -\Lambda_E(x)(I - \Lambda_{ZE}(x))^{-1} \\ 0 & (I - \Lambda_{ZE}(x))^{-1} \end{pmatrix}$$
$$= \begin{pmatrix} I & 0 \\ 0 & (I - \Lambda_{ZE}(x))^{-*}(I - \Lambda_E(x)^*\Lambda_E(x))(I - \Lambda_{ZE}(x))^{-1} \end{pmatrix}.$$

Hence, from equations (3.2) and (3.3),

$$L_B(p(x)) = U^*(I - \Lambda_{ZE}(x))^{-*}(I - \Lambda_E(x)^*\Lambda_E(x))(I - \Lambda_{ZE}(x))^{-1}U.$$

Further, letting

$$\widetilde{B} = CA = \begin{pmatrix} 0 & 0 \\ 0 & ZE \end{pmatrix},$$

we have

$$L_{\widetilde{B}}(p(x)) = \mathcal{W}^* F(x)^* L_A(x) F(x) \mathcal{W} \quad \text{where} \quad \mathcal{W} = \begin{pmatrix} I & 0 \\ 0 & U \end{pmatrix}.$$

By [1, Theorem 6.7], p is a convexotonic mapping determined by the (uniquely determined) convexotonic tuple Ξ satisfying

$$A_k(C - I)A_j = \sum_{s=1}^{g} (\Xi_j)_{k,s} A_s.$$

Equivalently,

$$E_k Z E_j = \sum_{s=1}^{g} (\Xi_j)_{k,s} E_s. \tag{3.4}$$

Finally to prove item (3), multiply equation (3.4) by Z on the left and use $B = U^* Z E U$ to obtain,

$$B_k B_j = \sum_{s=1}^{g} (\Xi_j)_{k,s} B_s.$$

4. Examples

In this section we take up some examples that motivate Theorem 1.1 and Corollary 1.4. First we show that a spectraball, as a member of the class of free spectrahedra, is never sv-generic.

Lemma 4.1. *Suppose $B \in M_d(\mathbb{C})^g$.*

(a) *If B is sv-generic, then* $\ker(B) := \cap_{j=1}^{g} \ker(B_j) = \{0\}$.
(b) *If B is nilpotent, then* $\ker(B) \neq \{0\}$.
(c) *If B is nilpotent, then $\mathcal{D} = \mathcal{D}_B$ is not sv-generic.*
(d) *If $\mathcal{D}$ is a spectraball, then $\mathcal{D}$ is not sv-generic.*

Remark 4.2. In fact item (a), and thus items (c) and (d), remain true with sv-generic replaced by eig-generic [1, Definition 7.3].

Proof. If $\alpha \in \mathbb{C}^g$, $u \in \mathbb{C}^d$ and $[I - \Lambda_B(\alpha)^* \Lambda_B(\alpha)]u = 0$, then $u \in \text{rg}(B^*) = \ker(B)^\perp$. Hence, if B is sv-generic, then there exists a basis $\{u^1, \dots, u^d\}$ of $\mathbb{C}^d$ such that each $u^j \in \text{rg}(B^*)$. Thus $\mathbb{C}^d = \text{rg}(B^*) = \ker(B)^\perp$ and therefore $\ker(B) = \{0\}$.

Now suppose B is nilpotent. Thus there is an N such that if β is a word whose length exceeds N, then $B^\beta = 0$. Hence there is a word α (potentially empty) such that $B^\alpha \neq 0$, but $B_j B^\alpha = 0$ for $1 \leq j \leq g$. It follows that $\{0\} \neq \text{rg}(B^\alpha) \subset \ker(B)$, proving item (b).

To prove item (c), suppose B is nilpotent and let $\mathcal{D} = \mathcal{D}_B$. Let $M \in M_m(\mathbb{C})^g$ be a *minimal defining tuple* for $\mathcal{D}$, meaning $\mathcal{D} = \mathcal{D}_M$ and if $C \in M_s(\mathbb{C})^g$ and $\mathcal{D} = \mathcal{D}_C$, then $s \geq m$. By [13, Proposition 2.2], there is a tuple J such that B is unitarily equivalent to $M \oplus J$. Since B is nilpotent, so is M. Hence $\ker(M) \neq \{0\}$ by item (b). Now suppose C is any other tuple so that $\mathcal{D}_C = \mathcal{D}_B$. Another application of [13, Proposition 2.2] gives a tuple N such that C is unitarily equivalent to $M \oplus N$. Hence $\ker(C) \neq \{0\}$ and by item (a), C is not sv-generic. Thus $\mathcal{D} = \mathcal{D}_B$ is not sv-generic.

Finally suppose $\mathcal{D}$ is a spectraball. Hence there is a positive integer e and tuple $E \in M_e(\mathbb{C})^g$ such that $\mathcal{D} = \mathcal{B}_E$. Since $\mathcal{D} = \mathcal{B}_E = \mathcal{D}_A$, where

$$A = \begin{pmatrix} 0 & E \\ 0 & 0 \end{pmatrix} \in M_{2e},$$

and A is nilpotent, item (c) implies $\mathcal{D}$ is not sv-generic. □

4.1. A spectrahedron defined by a nilpotent tuple

A spectrahedron defined by a nilpotent tuple cannot have sv-generic coefficients according to Lemma 4.1, but we give an example here of how one can overcome this problem by mapping to a spectraball.

Let

$$E_1 = I_2 \quad \text{and} \quad E_2 = \begin{pmatrix} 0 & 1 \\ 0 & 0 \end{pmatrix} \tag{4.1}$$

and let F denote the 2-tuple of 3×3 matrices given in equation (1.4). Note that $(1,1) \in \mathbb{C}^2$ is in $\mathcal{D}_F$, but $-(1,1) \notin \mathcal{D}_F$. Thus $\mathcal{D}_F$ is not rotationally invariant and hence not a spectraball. Hence Theorem 1.1 cannot be applied to bianalytic mappings $\varphi : \mathcal{D}_F \to \mathcal{D}_B$. Since F is nilpotent, $\mathcal{D}_F$ is not sv-generic (Lemma 4.1) and therefore Theorem [1, Theorem 1.8] cannot be applied to bianalytic mappings $\varphi : \mathcal{D}_F \to \mathcal{D}_B$ (even assuming B is sv-generic). On the other hand, F does span an algebra and thus Proposition 1.3 applies. A straightforward calculation shows that the origin-preserving birational map $q : \mathcal{D}_F \to \mathcal{B}_F$ of Proposition 1.3 is given by $q(x_1, x_2) = (x_1, x_2 + x_1^2)$. Evidently $\mathcal{B}_F = \mathcal{B}_E$. The following proposition summarizes the discussion above.

Proposition 4.3. *The mapping*

$$q(x_1, x_2) = (x_1, x_2 + x_1^2)$$

is a bianalytic map from $\mathcal{D}_F$ onto $\mathcal{B}_E$. Further, E is sv-generic, but $\mathcal{D}_F$ is neither a spectraball nor sv-generic.

According to Proposition 4.3, to classify bianalytic maps $f : \mathcal{D}_F \to \mathcal{D}_B$ it suffices to determine the bianalytic maps $h : \mathcal{B}_E \to \mathcal{D}_B$. Such maps are the subject of the next subsection.

4.2. Bianalytic mappings of $\mathcal{B}_E$ to a free spectrahedron $\mathcal{D}_B$

Theorem 1.1 applies in the case that B is sv-generic or has size 2.

Proposition 4.4. *Suppose $B \in M_e(\mathbb{C})^2$ and either $e = 2$ or B is sv-generic. If $f : \mathcal{B}_E \to \mathcal{D}_B$ is bianalytic, then $e = 2$ and there is a unimodular α and 2×2 unitary M such that $B = \alpha M^* E M$ and further f is the birational map*

$$f(x) = \begin{pmatrix} x_1(1-\alpha x_1)^{-1} & (1-\alpha x_1)^{-1} x_2 (1-\alpha x_1)^{-1} \end{pmatrix}.$$

Remark 4.5. The mapping f is a variant (obtained by the linear change of variable (x_1, x_2) maps to $\alpha(x_1, x_2)$) of those appearing in $g = 2$ type IV algebra (see [1, Section 8.3] or Subsubsection 4.3.4 below).

Proof of Proposition 4.4. In this case the Z in Theorem 1.1 is a unimodular multiple of the identity. Indeed, by (3.4),

$$Z = E_1 Z E_1 = (\Xi_1)_{1,1} I + (\Xi_1)_{1,2} E_2$$

and since Z is unitary, it follows that $(\Xi_1)_{1,2} = 0$ and $Z = \alpha I$. It is now easy to verify that $\Xi = \alpha E$. Hence the corresponding convexotonic map is

$$\begin{aligned} f(x) &= x(I - \Lambda_\Xi(x))^{-1} \\ &= \begin{pmatrix} x_1 & x_2 \end{pmatrix} \begin{pmatrix} 1-\alpha x_1 & -\alpha x_2 \\ 0 & 1-\alpha x_1 \end{pmatrix}^{-1} \\ &= \begin{pmatrix} x_1(1-\alpha x_1)^{-1} & (1-\alpha x_1)^{-1} x_2 (1-\alpha x_1)^{-1} \end{pmatrix}, \end{aligned}$$

as desired. □

Composing the f from Proposition 4.4 with the original $q = (x_1, x_2 + x_1^2)$, the bianalytic map between $\mathcal{D}_F$ and $\mathcal{B}_E$, gives the mapping from the original domain $\mathcal{D}_F$ to $\mathcal{D}_B$,

$$f \circ q = \begin{pmatrix} x_1(1-\alpha x_1)^{-1} & (1-\alpha x_1)^{-1} \left[x_2 + x_1^2\right] (1-\alpha x_1)^{-1} \end{pmatrix}.$$

By [1, Theorem 1.8], if G, H and K are all sv-generic and $r : \mathcal{D}_G \to \mathcal{D}_H$ and $s : \mathcal{D}_H \to \mathcal{D}_K$ are bianalytic (and extend to be analytic on pseudoconvex domains containing $\mathcal{D}_G$ and $\mathcal{D}_H$ respectively), then r, s and $r \circ s$ are convexotonic. However, generally one does not expect an arbitrary composition of convexotonic maps to be convexotonic. (See [1, Subsection 8.4].) Thus, it is of interest to note that, even though our F is not sv-generic, the map $f \circ q$ is convexotonic.

Proposition 4.6. *The map $f \circ q$ is convexotonic corresponding to the tuple $\Xi = \big(\alpha I_2 + E_2, \alpha E_2\big)$.*

Proof. Here is an outline of the computation that proves the proposition.

$$x(I - \Lambda_\Xi(x))^{-1} = x \begin{pmatrix} 1-\alpha x_1 & -(x_1+\alpha x_2) \\ 0 & 1-\alpha x_1 \end{pmatrix}^{-1}$$
$$= \begin{pmatrix} x_1 & x_2 \end{pmatrix} \begin{pmatrix} (1-\alpha x_1)^{-1} & (1-\alpha x_1)^{-1}(x_1+\alpha x_2)(1-\alpha x_1)^{-1} \\ 0 & (1-\alpha x_1)^{-1} \end{pmatrix}$$
$$= \begin{pmatrix} (x_1(1-\alpha x_1)^{-1} & x_1(1-\alpha x_1)^{-1}(x_1+\alpha x_2)(1-\alpha x_1)^{-1} + x_2(1-\alpha x_1)^{-1} \end{pmatrix}.$$

Analyzing the second entry above gives

$$\begin{aligned} &x_1(1-\alpha x_1)^{-1}(x_1+\alpha x_2)(1-\alpha x_1)^{-1} + x_2(1-\alpha x_1)^{-1} \\ &\quad = (1-\alpha x_1)^{-1}[x_1^2 + \alpha x_1 x_2 + (1-\alpha x_1)x_2](1-\alpha x_1)^{-1} \\ &\quad = (1-\alpha x_1)^{-1}[x_2 + x_1^2](1-\alpha x_1)^{-1}, \end{aligned}$$

as desired. □

4.3. Two-dimensional algebras with $g = 2$

In this section we consider, in view of Corollary 1.4, the four indecomposable algebras $\mathcal{A}$ of dimension two. In each case we choose a tuple $\mathcal{R} = (\mathcal{R}_1, \mathcal{R}_2)$ and compute the resulting convexotonic map $G : \mathcal{D}_\mathcal{R} \to \mathcal{B}_\mathcal{R}$. We adopt the names for these algebras used in [1].

4.3.1. $g = 2$ type *I* algebra. Let $\mathcal{R} = F$, where F is given by (1.4). In this case we already saw $q(x_1, x_2) = (x_1, x_2 + x_1^2)$. In this case $\mathcal{D}_F$ and $\mathcal{B}_F$ are both bounded. While the tuple F is not sv-generic, the tuple E of equation (4.1) is and moreover $\mathcal{B}_F = \mathcal{B}_E$. Hence Theorem 1.1 does indeed apply (by replacing F by E).

4.3.2. $g = 2$ type *II* algebra. Let

$$\mathcal{R}_1 = \begin{pmatrix} 1 & 0 \\ 0 & 0 \end{pmatrix}, \quad \mathcal{R}_2 = \begin{pmatrix} 0 & 1 \\ 0 & 0 \end{pmatrix}.$$

We have

$$(I + \Lambda_\mathcal{R}(x))^{-1}\Lambda_\mathcal{R}(x) = \begin{pmatrix} (1+x_1)^{-1}x_1 & (1+x_1)^{-1}x_2 \\ 0 & 0 \end{pmatrix}.$$

Hence $q = ((1+x_1)^{-1}x_1 \quad (1+x_1)^{-1}x_2)$ is a birational map from $\operatorname{int}(\mathcal{D}_\mathcal{R})$ to the spectraball $\operatorname{int}(\mathcal{B}_\mathcal{R})$ that also maps the boundary of $\mathcal{D}_\mathcal{R}$ into the boundary of $\mathcal{B}_\mathcal{R}$. On the other hand, if X_1 is skew selfadjoint, then $(X_1, 0) \in \mathcal{D}_\mathcal{R}$, so that $\mathcal{D}_\mathcal{R}$ is not bounded and, for instance, the tuple

$$\left(\begin{pmatrix} 0 & -1 \\ 1 & 0 \end{pmatrix}, \begin{pmatrix} 0 & 0 \\ 0 & 0 \end{pmatrix} \right)$$

is in $\mathcal{B}_\mathcal{R}$ but not the range of q. In this example, $\mathcal{R}$ has a (common nontrivial) cokernel and is thus not sv-generic. Hence Theorem 1.1 does not apply.

4.3.3. $g = 2$ type *III* algebra. This case, in which

$$\mathcal{R}_1 = \begin{pmatrix} 1 & 0 \\ 0 & 0 \end{pmatrix}, \quad \mathcal{R}_2 = \begin{pmatrix} 0 & 0 \\ 1 & 0 \end{pmatrix},$$

is very similar to the $g = 2$ type II case.

4.3.4. $g = 2$ type *IV* algebra. Let $\mathcal{R} = E$, where E is defined in equation (4.1), and observe

$$(I + \Lambda_{\mathcal{R}}(x))^{-1}\,\Lambda_{\mathcal{R}}(x) = \begin{pmatrix} (1+x_1)^{-1}x_1 & (1+x_1)^{-1}x_2(1+x_1)^{-1} \\ 0 & (1+x_1)^{-1}x_1 \end{pmatrix}.$$

In this case,

$$q(x) = \begin{pmatrix} x_1(1+x_1)^{-1} & (1+x_1)^{-1}x_2(1+x_1)^{-1} \end{pmatrix}$$

is bianalytic from $\operatorname{int}(\mathcal{D}_{\mathcal{R}})$ to $\operatorname{int}(\mathcal{B}_E)$ and maps boundary into boundary, but does not map boundary onto boundary. In this case $\mathcal{B}_{\mathcal{R}}$ is bounded and sv-generic and hence Theorem 1.1 does apply (with appropriate assumptions on $\mathcal{D}_B$ and $p : \mathcal{D}_{\mathcal{R}} \to \mathcal{D}_B$).

References

[1] M. Augat, J.W. Helton, I. Klep, S. McCullough: *Bianalytic Maps Between Free Spectrahedra*, Math. Ann. 371 (2018) 883–959.

[2] J. Agler, J. McCarthy: *Global holomorphic functions in several non-commuting variables*, Canad. J. Math. 67 (2015) 241–285.

[3] J.A. Ball, G. Groenewald, T. Malakorn: *Bounded Real Lemma for Structured Non-Commutative Multidimensional Linear Systems and Robust Control*, Multidimens. Syst. Signal Process. 17 (2006) 119–150.

[4] J.A. Ball, G. Groenewald, T. Malakorn: Conservative structured noncommutative multidimensional linear systems. The state space method generalizations and applications, 179–223, Oper. Theory Adv. Appl., 161, Linear Oper. Linear Syst., Birkhäuser, Basel, 2006.

[5] J.A. Ball, G. Groenewald, T. Malakorn: Structured noncommutative multidimensional linear systems. SIAM J. Control Optim. 44 (2005), no. 4, 1474–1528.

[6] J.A. Ball, G. Marx, V. Vinnikov: *Interpolation and transfer-function realization for the noncommutative Schur–Agler class*, In: Duduchava R., Kaashoek M., Vasilevski N., Vinnikov V. (eds) Operator Theory in Different Settings and Related Applications, Operator Theory: Advances and Applications, 23–116, vol 262. Birkhäuser, Cham, 2018.

[7] A. Ben-Tal, A. Nemirovski: *On tractable approximations of uncertain linear matrix inequalities affected by interval uncertainty*, SIAM J. Optim. 12 (2002) 811–833.

[8] G. Blekherman, P.A. Parrilo, R.R. Thomas (editors): *Semidefinite optimization and convex algebraic geometry*, MOS-SIAM Series on Optimization 13, SIAM, 2013.

[9] S. Boyd, L. El Ghaoui, E. Feron, V. Balakrishnan: *Linear Matrix Inequalities in System and Control Theory*, SIAM Studies in Applied Mathematics 15, SIAM, 1994.

[10] P. Brändén: *Obstructions to determinantal representability*, Adv. Math. 226 (2011) 1202–1212.

[11] M. de Oliveira, J.W. Helton, S. McCullough, M. Putinar: *Engineering systems and free semi-algebraic geometry*, in: Emerging applications of algebraic geometry (edited by M. Putinar, S. Sullivant), 17–61, Springer-Verlag, 2009.

[12] E.G. Effros, S. Winkler: *Matrix convexity: operator analogues of the bipolar and Hahn–Banach theorems*, J. Funct. Anal. 144 (1997) 117–152.

[13] E. Evert, J.W. Helton, I. Klep, S. McCullough: *Circular Free Spectrahedra*, J. Math. Anal. Appl. 445 (2017) 1047–1070.

[14] F. Forstnerič: *Proper holomorphic mappings: a survey*, in: Several complex variables (Stockholm, 1987/1988) 297–363, Math. Notes 38, Princeton Univ. Press, 1993.

[15] A. Grinshpan, D.S. Kaliuzhnyi-Verbovetskyi, V. Vinnikov, H.J. Woerdeman: *Matrix-valued Hermitian Positivstellensatz, lurking contractions, and contractive determinantal representations of stable polynomials*, in Oper. Theory Adv. Appl 255, 123–136, Birkhäuser/Springer, 2016.

[16] J.W. Helton, I. Klep, S. McCullough, N. Slinglend: *Noncommutative ball maps*, J. Funct. Anal. 257 (2009), no. 1, 47–87.

[17] J.W. Helton, I. Klep, S. McCullough: *Analytic mappings between noncommutative pencil balls*, J. Math. Anal. Appl. 376 (2011), no. 2, 407–428.

[18] J.W. Helton, I. Klep, S. McCullough: *Proper Analytic Free Maps*, J. Funct. Anal. 260 (2011) 1476–1490.

[19] J.W. Helton, I. Klep, S. McCullough: *The convex Positivstellensatz in a free algebra*, Adv. Math. 231 (2012) 516–534. (this article succeeds [20] but appeared earlier)

[20] J.W. Helton, I. Klep, S. McCullough: *The matricial relaxation of a linear matrix inequality*, Math. Program. 138 (2013) 401–445. (this article precedes [19] but appeared later)

[21] J.W. Helton, I. Klep, S. McCullough: *The tracial Hahn–Banach theorem, polar duals, matrix convex sets, and projections of free spectrahedra*, J. Eur. Math. Soc. 19 (2017) 1845–1897.

[22] J.W. Helton, I. Klep, S. McCullough, M. Schweighofer: *Dilations, Linear Matrix Inequalities, the Matrix Cube Problem and Beta Distributions*, to appear in Mem. Amer. Math. Soc., `https://arxiv.org/abs/1412.1481`

[23] J.W. Helton, V. Vinnikov: *Linear matrix inequality representation of sets*, Comm. Pure Appl. Math. 60 (2007) 654–674.

[24] D. Kaliuzhnyi-Verbovetskyi, V. Vinnikov: *Singularities of rational functions and minimal factorizations: the noncommutative and the commutative setting*, Linear Algebra Appl. 430 (2009) 869–889.

[25] D. Kaliuzhnyi-Verbovetskyi, V. Vinnikov: *Foundations of Free Noncommutative Function Theory*, Mathematical Surveys and Monographs 199, AMS, 2014.

[26] I. Klep, Š. Špenko: *Free function theory through matrix invariants*, Canad. J. Math. 69 (2017) 408–433.

[27] A.W. Marcus, D.A. Spielman, N. Srivastava: *Interlacing families II: Mixed characteristic polynomials and the Kadison–Singer problem*, Ann. of Math. (2) 182 (2015) 327–350.

[28] P.S. Muhly, B. Solel: *Schur class functions and automorphism of Hardy algebras*, Doc. Math. 13 (2008) 365–411.

[29] T. Netzer, A. Thom: *Polynomials with and without determinantal representations*, Linear Algebra Appl. 437 (2012) 1579–1595.

[30] V. Paulsen: *Completely bounded maps and operator algebras*, Cambridge Univ. Press, 2002.

[31] G. Popescu: *Free holomorphic automorphisms of the unit ball of* $B(H)^n$, J. reine angew. Math. 638 (2010) 119–168.

[32] R.E. Skelton, T. Iwasaki, K.M. Grigoriadis: *A Unified Algebraic Approach to Linear Control Design*, Taylor and Francis, 1996.

[33] V. Vinnikov: *Self-adjoint determinantal representations of real plane curves*, Math. Ann. 296 (1993) 453–479.

[34] D.-V. Voiculescu: *Free analysis questions I: Duality transform for the coalgebra of* $\partial_{X:B}$, Int. Math. Res. Not. 16 (2004) 793–822.

[35] J. Volčič: *On domains of noncommutative rational functions*, Linear Algebr Appl. 516 (2017) 69–81.

Meric Augat and Scott McCullough
Department of Mathematics
University of Florida
Gainesville, USA
e-mail: `mlaugat@math.ufl.edu`
`sam@math.ufl.edu`

J. William Helton
Department of Mathematics
University of California
San Diego, USA
e-mail: `helton@math.ucsd.edu`

Igor Klep
Department of Mathematics
The University of Auckland, New Zealand
e-mail: `igor.klep@auckland.ac.nz`

Operator Theory:
Advances and Applications, Vol. 272, 61–70

Szegő and Widom Theorems for the Neil Algebra

Sriram Balasubramanian, Scott McCullough and
Udeni Wijesooriya

In appreciation for his profound influence on operator theory and our mathematical lives, we dedicate this article to Joe Ball.

Abstract. Versions of well-known function theoretic operator theory results of Szegő and Widom are established for the Neil algebra. The Neil algebra is the subalgebra of the algebra of bounded analytic functions on the unit disc consisting of those functions whose derivative vanishes at the origin.

Mathematics Subject Classification (2010). 47B335, 30H10 (Primary) 30H05, 46E20 (secondary).

Keywords. Toeplitz operators, Szegő's Theorem, constrained algebra, Neil algebra, distinguished variety.

1. Introduction

Let $\mathbb{C}$ denote the complex numbers, $\mathbb{D} = \{|z| < 1\} \subseteq \mathbb{C}$ denote the unit disk with its boundary $\mathbb{T} = \{|z| = 1\}$. Denote by $H^2 = H^2(\mathbb{D})$ and $H^\infty = H^\infty(\mathbb{D})$ the standard Hardy spaces of functions analytic in $\mathbb{D}$ with square summable power series coefficients and bounded analytic functions on $\mathbb{D}$ respectively. Let L^p denote the L^p spaces for the $\mathbb{T}$ (identified with the corresponding L^p spaces for $[0, 2\pi]$ with respect to the measure $\frac{dt}{2\pi}$). Let $\mathscr{P}$ denote the set of analytic polynomials that vanish at 0. Thus a $p \in \mathscr{P}$ has the form,

$$p(z) = \sum_{j=1}^{n} p_j z^j$$

The first author was supported by the New Faculty Initiative Grant (MAT/15-16/836/NFIG/SRIM) of IIT Madras. The second author was supported by the NSF grant DMS-1361501.

for some positive integer n and $p_1, \dots, p_n \in \mathbb{C}$. Given a non-negative function ρ on $\mathbb{T}$ with $\log(\rho) \in L^1$ a (special case of a) well-known result of Szegő (see for instance [14] page 219) identifies the $L^2(\rho)$ distance from the constant function 1 to $\mathscr{P}$.

Theorem 1.1 (of Szegő).

$$\inf\left\{\int_0^{2\pi} |p-1|^2 \, \rho \, \frac{dt}{2\pi} : p \in \mathscr{P}\right\} = \exp\left(\int_0^{2\pi} \log(\rho) \, \frac{dt}{2\pi}\right).$$

A theorem of Widom characterizes those unimodular functions $\phi \in L^\infty$ whose distance to H^∞ is less than one in terms of Toeplitz operators. A $\phi \in L^\infty$ induces a multiplication operator $M_\phi : L^2 \to L^2$ defined by $M_\phi f = \phi f$. Let $V : H^2 \to L^2$ denote the inclusion. The operator $T_\phi = V^* M_\phi V$ is the *Toeplitz operator* with symbol ϕ.

Theorem 1.2 (Widom's invertibility criteria [11, Theorem 7.30]). *Suppose $\phi \in L^\infty$ is unimodular. There exists an $f \in H^\infty$ such that $\|f - \phi\| < 1$ if and only if T_ϕ is left invertible.*

Sarason [18] established a version of Theorem 1.1 for the annulus and Abrahamse [1, Theorems 4.1 and 4.6] established a version of Theorem 1.2 for multiply connected domains. In this paper we establish Szegő and Widom type theorems for the *Neil algebra*. The Neil algebra $\mathfrak{A}$ is the subalgebra of $H^\infty(\mathbb{D})$ consisting of those functions whose derivative vanishes at 0. It is perhaps the simplest example of a constrained algebra. As with extending classical results from the unit disc to multiply connected domains, here it is necessary to replace H^2 with a family of Hilbert–Hardy spaces that parameterize the distinction between harmonic functions and the real parts of analytic functions in $\mathfrak{A}$ either explicitly or implicitly in the statement of the results and their proofs. In addition to the references already cited, see for instance [2, 3, 8, 16] for related results on multiply connected domains, [5–7, 10, 12, 16, 17] for results on constrained algebras, [4] for results in the context of uniform algebras and finally [13] for a Pick interpolation theorem on distinguished varieties. Let $\mathfrak{A}_0$ denote those functions in $\mathfrak{A}$ that vanish at 0. Hence $\mathfrak{A}_0 = z^2 H^\infty$.

Theorem 1.3 (Szegő Theorem for $\mathfrak{A}$). *Suppose $\rho > 0$ is a continuous function on $\mathbb{T}$ and let*

$$C_\rho = \int_0^{2\pi} \log(\rho) \, \frac{dt}{2\pi}, \quad \lambda = \int_0^{2\pi} \rho(t) \exp(-it) \, \frac{dt}{2\pi}.$$

With these notations,

$$\inf\left\{\int_0^{2\pi} |1-p|^2 \, \rho \frac{dt}{2\pi} : p \in \mathfrak{A}_0\right\} = \exp(C_\rho) + \exp(-C_\rho) \, |\lambda|^2.$$

Remark 1.4. Note that $\lambda = 0$ if and only if 1 and e^{it} are orthogonal in $L^2(\rho)$ and in this case it is evident that the distance from 1 to $\mathscr{P}$ is the same as the distance from 1 to the subspace $\mathfrak{A}_0$ of $\mathscr{P}$. □

To state the analog of Theorem 1.2 for $\mathfrak{A}$ some notations are needed. Let $\mathbb{B}^2 = \{(z,w) \in \mathbb{C}^2 : |z^2| + |w|^2 = 1\}$ denote the unit ball in $\mathbb{C}^2$. To $\alpha = (a,b) \in \mathbb{B}^2$ associate the subspace $H^2_\alpha \subseteq H^2$ consisting of those $f \in H^2$ such that

$$f(0)\,b = f'(0)\,a.$$

Let $V_\alpha : H^2_\alpha \to L^2$ denote the inclusion. Hence $P_\alpha = V_\alpha V_\alpha^* : L^2 \to H^2_\alpha$ is the projection onto H^2_α. Given $\phi \in L^\infty$, define $T^\alpha_\phi : H^2_\alpha \to H^2_\alpha$ by

$$T^\alpha_\phi = V_\alpha^* M_\phi V_\alpha.$$

It is the *Toeplitz operator with symbol ϕ with respect to α* [8]. In particular, if $\phi \in \mathfrak{A}$ and $f \in H^2_\alpha$, then $V^* T^\alpha_\phi f = \phi f = T^\alpha_\phi f$.

Remark 1.5. Given $\alpha = (a,b)$ and $\beta = (c,d)$, if $ad = bc$, then $H^2_\alpha = H^2_\beta$ and likewise $T^\alpha_\phi = T^\beta_\phi$. Thus, $\mathbb{P}$, complex projective space obtained by moding out $\mathbb{B}^2$ by the relation $(a,b) = (c,d)$, is a natural choice of parameter space. For ease of exposition we accept the redundancy inherent in the use of $\mathbb{B}^2$. □

Theorem 1.6 (Inversion for $\mathfrak{A}$). *Suppose $\phi \in L^\infty$ is unimodular. The distance from ϕ to $\mathfrak{A}$ is strictly less than one if and only if T^α_ϕ is left invertible for each $\alpha \in \mathbb{B}^2$. Likewise, the distance from ϕ to the invertible elements of $\mathfrak{A}$ is strictly less than one if and only if T^α_ϕ is invertible for each $\alpha \in \mathbb{B}^2$.*

Before turning to the proofs of Theorems 1.3 and 1.6, we pause to introduce some conventions and basic background on the spaces H^2_α. For $p = 2, \infty$, the standard identification of $H^p(\mathbb{D})$ with $H^p(\mathbb{T})$, where the latter is viewed as the subspace of $L^p(\mathbb{T})$ consisting of those f with vanishing negative Fourier coefficients, will be used routinely and without comment. Let H^2_1 denote the subspace of H^2 consisting of those $f \in H^2$ whose Fourier coefficient

$$\hat{f}(1) = \int_0^{2\pi} f\, e^{-it}\, \frac{dt}{2\pi} = 0.$$

Evidently, H^2_1 is the closure of $\mathfrak{A}$ in H^2. The following Lemma can be found in [10] for instance. The first part follows from the easily verified fact that $\{a + bz, z^n : n \geq 2\}$ is an orthonormal basis for H^2_α; and the moreover part, from a standard reproducing kernel Hilbert space argument.

Lemma 1.7. *For each $\alpha = (a,b) \in \mathbb{B}^2$, the space H^2_α has reproducing kernel,*

$$k^\alpha_w(z) = k^\alpha(z,w) = (a + bz)\overline{(a + bw)} + \frac{z^2 \overline{w^2}}{1 - z\overline{w}}, \qquad z, w \in \mathbb{D}.$$

In particular,

$$\|k^\alpha_0\|^2 = k^\alpha(0,0) = |a|^2,$$

and thus $k^\alpha_w \neq 0$ with the exception of $\alpha = (0,1)$ and $w = 0$.

Moreover, if $\psi \in \mathfrak{A}$ and $w \in \mathbb{D}$, then $(T^\alpha_\psi)^* k^\alpha_w = \overline{\psi(w)} k^\alpha_w$.

2. Proof of Theorem 1.3

As a first step, observe that it suffices to prove the theorem under the additional hypothesis that $C_\rho = 0$. Indeed, if not let $\tilde{\rho} = \exp(-C_\rho)\,\rho$, so that $\int_0^{2\pi} \log(\tilde{\rho})\,\frac{dt}{2\pi} = 0$. In particular, $C_{\tilde{\rho}} = 0$ and with

$$\tilde{\lambda} = \int_0^{2\pi} \tilde{\rho}\,\exp(-it)\,\frac{dt}{2\pi} = \exp(-C_\rho)\,\lambda,$$

if Theorem 1.3 holds for $\tilde{\rho}$, then

$$\inf\left\{\int_0^{2\pi} |p-1|^2\,\tilde{\rho}\,\frac{dt}{2\pi} : p \in \mathfrak{A}_0\right\} = 1 + |\tilde{\lambda}|^2.$$

Thus,

$$\begin{aligned}\inf\left\{\int_0^{2\pi} |p-1|^2\,\rho\,\frac{dt}{2\pi} : p \in \mathfrak{A}_0\right\} &= \exp(C_\rho)\inf\left\{\int_0^{2\pi} |p-1|^2\,\tilde{\rho}\,dtn : p \in \mathfrak{A}_0\right\}\\ &= \exp(C_\rho)(1+|\tilde{\lambda}|^2) = \exp(C_\rho) + \exp(-C_\rho)\,|\lambda|^2\end{aligned}$$

as claimed. Accordingly, for the remainder of the proof, assume $C_\rho = 0$.

Let

$$\sigma = \frac{1}{\sqrt{1+|\lambda|^2}}\,(1,\lambda) \in \mathbb{B}^2.$$

In particular,

$$\|k_0^\sigma\|^2 = \frac{1}{1+|\lambda|^2}.$$

Note that, as sets, $L^2(\rho)$ and L^2 are the same and thus we may consider H^2 as a Hilbert space with the alternate inner product,

$$\langle f, g\rangle_\rho = \int_0^{2\pi} f\overline{g}\,\rho\,\frac{dt}{2\pi}.$$

To keep the distinction clear, denote this latter space by $H^2(\rho)$. Since the closure of $\mathfrak{A}_0$ in $H^2(\rho)$ is $z^2H^2 = z^2H^2(\rho)$, the objective is to find the $H^2(\rho)$-distance from 1 to z^2H^2. That is, to show

$$\inf\left\{\int_0^{2\pi} |p-1|^2\,\tilde{\rho}\,\frac{dt}{2\pi} : f \in z^2H^2\right\} = 1 + |\tilde{\lambda}|^2.$$

Since ρ is continuous and strictly positive, $\log(\rho)$ is continuous. It has Fourier series expansion

$$\log(\rho) = \sum_{j=-\infty}^{\infty} c_j e^{ijt},$$

where, because it is real-valued, $c_{-j} = \overline{c_j}$. Moreover, $c_0 = 0$ and $c_1 = \lambda$, since $C_\rho = 0$ and by the very definitions of C_ρ and λ. Letting γ denote the H^2 function represented by the series

$$\gamma = \sum_{j=1}^{\infty} c_j e^{ijt},$$

it follows that $\log(\rho) = \gamma + \gamma^*$ as elements of L^2. Further, since

$$|\exp(\pm\gamma)|^2 = \exp(\pm(\gamma+\gamma^*)) = \rho^{\pm 1},$$

both $\exp(\pm\gamma)$ are in H^∞. The mapping $U : H^2(\rho) \to H^2$ defined by $Uf = \exp(\gamma)f$ is a unitary map with inverse $U^*f = \exp(-\gamma)f$. Moreover, $U(z^2H^2) = z^2H^2$. Thus, the aim is to find the H^2-distance from $\exp(\gamma)$ to z^2H^2.

Given $f \in z^2H^2$, let $g = \exp(\gamma) - f$ and estimate, using $g(0) = 1$ and the Cauchy–Schwarz inequality,

$$\begin{aligned} \|\exp(\gamma) - f\|^2 =& \|g\|^2 \geq \frac{|\langle g, k_0^\sigma\rangle|^2}{\|k_0^\sigma\|^2} \\ =& |g(0)|^2 (1+|\lambda|^2) = 1 + |\lambda|^2. \end{aligned} \tag{2.1}$$

Let

$$f = \exp(\gamma) - (1+|\lambda|^2)k_0^\sigma$$

and note $f(0) = 0$ and $f'(0) = \gamma'(0) - \lambda = 0$. Thus $f \in z^2H^2$ and, with this choice of f, equality holds in the Cauchy–Schwarz inequality in equation (2.1).

3. Toeplitz operators on $\mathfrak{A}$

This section contains the proof of Theorem 1.6.

Lemma 3.1. *If $\phi \in L^\infty$, then $\|T_\phi^\alpha\| = \|\phi\|$ and $(T_\phi^\alpha)^* = T_{\overline{\phi}}^\alpha$.*

Proof. Since $M_\phi^* = M_{\overline{\phi}}$, it follows that $(T_\phi^\alpha)^* = V_\alpha^* M_\phi^* V_\alpha = V_\alpha^* M_{\overline{\phi}} V_\alpha = T_{\overline{\phi}}^\alpha$. Since V_α is an isometry, it follows that $\|T_\phi^\alpha\| \leq \|M_\phi\| = \|\phi\|$. Now let $V : H^2 \to L^2$ and $W : z^2H^2 \to L^2$ denote the inclusion maps. In particular, $V^*M_\phi V$ is T_ϕ, the usual Toeplitz operator with symbol ϕ. On the other hand, $W^*M_\phi W = W^*T_\phi^\alpha W$. With $U : H^2 \to z^2H^2$ given by $Uf = z^2f$, it follows that U is unitary and, for $f, g \in H^2$,

$$\langle M_\phi WUf, WUg\rangle = \langle z^2\phi f, z^2 g\rangle = \langle \phi f, g\rangle = \langle M_\phi f, g\rangle = \langle V^*M_\phi Vf, g\rangle.$$

Hence $U^*W^*M_\phi WU = V^*M_\phi V = T_\phi$ and consequently $W^*T_\phi^\alpha W$ is unitarily equivalent to T_ϕ. Hence $\|T_\phi^\alpha\| \geq \|T_\phi\|$. Since, as is well known that $\|T_\phi\| = \|\phi\|$ ([15]), the result follows. □

Let $\mathscr{B}(L^2)$ denote the bounded linear operators on L^2.

Lemma 3.2. *Giving $\mathbb{B}^2$ its usual topology and $\mathscr{B}(L^2)$ its norm topology, the mapping $\mathbb{B}^2 \ni \alpha \to P_\alpha \in \mathscr{B}(L^2)$ is continuous.*

Proof. Since $\{a + bz, z^n : n \geq 2\}$ is an orthonormal basis for H^2_α, if $f = \sum f_n z^n \in H^2$ and $\alpha = (a, b) \in \mathbb{B}^2$, then

$$P_\alpha f = (\overline{a}f_0 + \overline{b}f_1)(a + bz) + \sum_{n=2}^\infty f_n z^n.$$

Thus, letting Q denote the projection onto z^2H^2 and $F_\alpha = (a+bz)$ (a unit vector),

$$P_\alpha = F_\alpha F_\alpha^* + Q,$$

where $F_\alpha F_\alpha^* : L^2 \to L^2$ is the rank one projection operator,

$$F_\alpha F_\alpha^* f = \langle f, F_\alpha\rangle F_\alpha = (\overline{a} f_0 + \overline{b} f_1) F_\alpha.$$

Thus, if $\beta = (c,d) \in \mathbb{B}^2$, then

$$P_\alpha - P_\beta = F_\alpha F_\alpha^* - F_\beta F_\beta^* = F_\alpha (F_\alpha - F_\beta)^* + (F_\alpha - F_\beta) F_\beta^*.$$

Since $\|F_\alpha - F_\beta\| = \|\alpha - \beta\|$, the result follows. □

Let $\mathscr{M} \subseteq L^1$ denote the subspace consisting of those L^1 functions with Fourier series of the form

$$\hat{f}(-1)\exp(-it) + \sum_{j=1}^{\infty} \hat{f}(j)\exp(ijt). \tag{3.1}$$

The following lemma is the $\mathscr{M}$ version of the well-known factorization theorem for H^1 functions.

Lemma 3.3. *If $h \in \mathscr{M}$, then there exist*

(i) $\alpha \in \mathbb{B}^2$; (ii) $f \in H^2_\alpha$; *and* (iii) $g \in L^2$

such that

(a) $\overline{g} \in (H^2_\alpha)^\perp$; (b) $h = fg$; *and* (c) $\|h\|_1 = \|f\|_2 \, \|g\|_2$.

Proof. The function $\psi = zh$ is in H^1 and therefore there exists $F, G \in H^2$ such that $zh = FG$ and $\|h\|_1 = \|\psi\|_1 = \|F\|_2 \, \|G\|_2$ [11, Corollary 6.27]. Moreover, since $\psi'(0) = 0$, it follows that $F'(0)\,G(0) + F(0)\,G'(0) = 0$. There is an $\alpha = (a,b) \in \mathbb{B}^2$ such that $F \in H^2_\alpha$. (Indeed, simply choose $\alpha \in \mathbb{B}^2$ such that $aF'(0) = bF(0)$.) Thus there is a constant c and an H^2 function F_0 such that

$$F = c(a+bz) + z^2 F_0.$$

Hence, there is a constant d and H^2 function G_0 such that

$$G = d(a - bz) + z^2 G_0.$$

Let $g = \overline{z} G$, in which case $h = Fg$ and $\|g\|_2 = \|G\|_2$. Moreover,

$$\langle a + bz, \overline{g}\rangle = d \int_0^{2\pi} (a+bz)\,(d(a\overline{z} - b) + zG_0)\, \frac{dt}{2\pi} = 0$$

and, for $n \geq 2$,

$$\langle z^n, \overline{g}\rangle = \int_0^{2\pi} z^n\,(d(a\overline{z} - b) + zG_0)\, \frac{dt}{2\pi} = 0.$$

Hence $\overline{g} \in (H^2_\alpha)^\perp$. □

Recall $(L^1)^* = L^\infty$ with the equality interpreted as the isometric isomorphism determined by the mapping that assigns to $\phi \in L^\infty$ the linear functional $\lambda_\phi : L^1 \to$

$\mathbb{C}$ given by

$$\lambda_\phi(\psi) = \int_0^{2\pi} \phi\,\psi\,\frac{dt}{2\pi}.$$

Moreover, letting

$$\mathscr{M}^\perp := \left\{\phi \in L^\infty : \int_0^{2\pi} \phi\,\psi\frac{dt}{2\pi} = 0, \text{ for all } \psi \in \mathscr{M}\right\},$$

and $\pi : L^\infty \to L^\infty/\mathscr{M}^\perp$ denote the quotient mapping, the mapping $\Lambda : L^\infty/\mathscr{M}^\perp \to \mathscr{M}^*$ given by

$$\Lambda(\pi(\lambda_\phi)) = (\lambda_\phi)|_{\mathscr{M}},$$

is an isometric isomorphism. Finally, if $\phi \in \mathscr{M}$ and $\psi \in \mathfrak{A}$, then

$$\int_0^{2\pi} \phi\,\psi\,\frac{dt}{2\pi} = 0.$$

Thus, $\mathfrak{A} \subseteq \mathscr{M}^\perp$. On the other hand, $e^{ijt} \in \mathscr{M}$ for $j = -1, 1, 2, \dots$ and therefore if $\psi \in \mathscr{M}^\perp$, then its Fourier series has the form

$$\psi = \hat{\psi}(0) + \sum_{j=2}^{\infty} \hat{\psi}(j)e^{ijt}.$$

Hence $\psi \in \mathfrak{A}$ and thus we may view Λ as having domain $L^\infty/\mathfrak{A}$. The following lemma summarizes the discussion (see [9, page 88]).

Lemma 3.4. *$\Lambda : L^\infty/\mathfrak{A} \to \mathscr{M}^*$ defined by sending $\pi(\phi)$ to the linear functional $\tilde{\lambda}_\phi : \mathscr{M} \to \mathbb{C}$ given by*

$$\tilde{\lambda}_\phi(f) = \int_0^{2\pi} \phi\, f\,\frac{dt}{2\pi}$$

is an isometric isomorphism.

Lemma 3.5. *If $\phi \in L^\infty$ and $\psi \in \mathfrak{A}$, then*

$$T^\alpha_{\overline{\psi\phi}} = T^\alpha_{\overline{\psi}}\,T^\alpha_{\phi}, \qquad T^\alpha_{\psi\overline{\phi}} = T^\alpha_{\overline{\phi}}\,T^\alpha_{\psi}.$$

Proof. Let $f, g \in H^2_\alpha$ be given. Since $\psi g \in H^2_\alpha$, it follows, using Lemma 3.1, that $\langle T^\alpha_{\overline{\psi}} T^\alpha_\phi f, g\rangle = \langle T^\alpha_\phi f, (T^\alpha_{\overline{\psi}})^* g\rangle = \langle T^\alpha_\phi f, T^\alpha_\psi g\rangle = \langle V^*_\alpha \phi f, \psi g\rangle = \langle \phi f, V_\alpha \psi g\rangle = \langle \phi f, \psi g\rangle = \langle \overline{\psi}\phi f, g\rangle = \langle \overline{\psi}\phi V_\alpha f, V_\alpha g\rangle = \langle T^\alpha_{\overline{\psi}\phi} f, g\rangle$. Thus $T^\alpha_{\overline{\psi}\phi} = T^\alpha_{\overline{\psi}} T^\alpha_\phi$. Applying Lemma 3.1 to what has already been proved, $T^\alpha_{\psi\overline{\phi}} = (T^\alpha_{\overline{\psi}\phi})^* = (T^\alpha_{\overline{\psi}} T^\alpha_\phi)^* = T^\alpha_{\overline{\phi}} T^\alpha_\psi$. □

An element $\psi \in \mathfrak{A}$ is *invertible in* $\mathfrak{A}$ if it does not vanish in $\mathbb{D}$ and $\psi^{-1} = \frac{1}{\psi} \in \mathfrak{A}$.

Lemma 3.6. *Suppose $\psi \in \mathfrak{A}$. The following are equivalent.*

(i) *ψ is invertible in $\mathfrak{A}$;*
(ii) *there is an $\alpha \in \mathbb{B}^2$ such that T^α_ψ is right invertible;*
(iii) *T^α_ψ is invertible for each $\alpha \in \mathbb{B}^2$.*

Moreover, in this case $(T^\alpha_\psi)^{-1} = T^\alpha_{\psi^{-1}}$.

Proof. Evidently item (i) implies item (iii) implies item (ii). Now suppose there is an α such that $T := T_\psi^\alpha$ is right invertible. The Hilbert space H_α^2 has a reproducing kernel $k_w^\alpha(z)$ and further $T^* k_w^\alpha = \overline{\psi(w)} k_w^\alpha$ by Lemma 1.7. Since T is right invertible, T^* is bounded below; i.e., there is a $\delta > 0$ such that $\|T^* f\| \geq \delta \|f\|$ for all $f \in H_\alpha^2$. Hence,

$$|\overline{\psi(w)}| \, \|k_w^\alpha\| = \|T^* k_w^\alpha\| \geq \delta \|k_w^\alpha\|.$$

Moreover, by Lemma 1.7 $k_w^\alpha \neq 0$ for $w \neq 0$. Thus $|\frac{1}{\psi}(w)| \leq \frac{1}{\delta}$ for $w \in \mathbb{D} \setminus \{0\}$ and therefore, as $\frac{1}{\psi}$ is otherwise analytic, $|\frac{1}{\psi}|$ is bounded by $\frac{1}{\delta}$. Since $\psi \in \mathfrak{A}$ it follows that $\frac{1}{\psi} \in \mathfrak{A}$ too; i.e., item (i) holds. □

Lemma 3.7. *Suppose $\phi \in L^\infty$ is unimodular. If there exists $\psi \in \mathfrak{A}$ such that $\|\phi - \psi\| < 1$, then $T_{\overline{\phi}}^\alpha T_\psi^\alpha$ is invertible, and therefore T_ϕ^α is left invertible, for each $\alpha \in \mathbb{B}^2$. Further, if ψ is invertible in $\mathfrak{A}$, then $T_{\overline{\psi}}^\alpha T_\phi^\alpha$ is invertible, and therefore T_ϕ^α is invertible, for each $\alpha \in \mathbb{B}^2$.*

Proof. Suppose there exists $\psi \in \mathfrak{A}$ such that $\|\phi - \psi\| < 1$. In this case $\|1 - \psi\overline{\phi}\| < 1$, since $|\phi| = 1$ (unimodular). Hence, by Lemma 3.1, for a given $\alpha \in \mathbb{B}^2$,

$$1 > \|1 - \psi\overline{\phi}\| = \|T_{1-\psi\overline{\phi}}^\alpha\| = \|1 - T_{\psi\overline{\phi}}^\alpha\|.$$

In particular, $T_{\psi\overline{\phi}}^\alpha$ is invertible. Since $\psi \in \mathfrak{A}$, Lemma 3.5 applies to give, $T_{\psi\overline{\phi}}^\alpha = T_{\overline{\phi}}^\alpha T_\psi^\alpha$. Thus $T_{\overline{\phi}}^\alpha$ is right invertible. By Lemma 3.1, $(T_{\overline{\phi}}^\alpha)^* = T_\phi^\alpha$ is left invertible.

Now, assuming ψ is invertible in $\mathfrak{A}$, by Lemma 3.6, T_ψ^α is invertible. The invertibility of $T_{\overline{\phi}}^\alpha$ follows. Thus, again using Lemma 3.1, T_ϕ^α is invertible. □

Lemma 3.8. *If $\phi \in L^\infty$ and T_ϕ^α is left invertible for each $\alpha \in \mathbb{B}^2$, then there exists an $\epsilon \in (0, 1]$, such that for each $\alpha \in \mathbb{B}^2$ and $f \in H_\alpha^2$,*

$$\|T_\phi^\alpha f\| \geq \epsilon \|f\|.$$

Proof. For $\alpha \in \mathbb{B}^2$, define $X_\alpha : L^2 \to L^2$ by $X_\alpha = P_\alpha M_\phi P_\alpha + (I - P_\alpha)$. Given $\alpha \in \mathbb{B}^2$, since T_ϕ^α is left invertible, there exists an $\epsilon_\alpha \in (0, 1]$ such that $\|V_\alpha T_\phi^\alpha f\| = \|T_\phi^\alpha f\| \geq \epsilon_\alpha \|f\|$ for $f \in H_\alpha^2$. Hence, given $F = f + g$ with $f \in H_\alpha^2$ and $g \in (H_\alpha^2)^\perp$,

$$\|X_\alpha F\|^2 = \|V_\alpha T_\phi^\alpha f\|^2 + \|g\|^2 \geq \epsilon_\alpha^2 \|F\|^2.$$

Thus, $\|X_\alpha F\| \geq \epsilon_\alpha \|F\|$ for all $F \in L^2$.

To show there is an $\epsilon > 0$ such that $\|X_\alpha F\| \geq \epsilon \|F\|$ for all $\alpha \in \mathbb{B}^2$ and $F \in L^2$, we argue by contradiction. Accordingly suppose no such $\epsilon > 0$ exists. By compactness of $\mathbb{B}^2$, there is a sequence $\alpha_n = (a_n, b_n)$ from $\mathbb{B}^2$, that, by passing to a subsequence if needed, we may assume converges to some $\beta = (a, b) \in \mathbb{B}^2$ and a unit vectors $F_n \in L^2$ such that $(\|X_{\alpha_n} F_n\|)_n$ converges to 0. But then,

$$0 < \epsilon_\beta \leq \|X_\beta F_n\| \leq \|X_{\alpha_n} F_n\| + \|(X_\beta - X_{\alpha_n}) F_n\|.$$

By norm continuity (Lemma 3.2) the last term on the right-hand side tends to 0 and by assumption the first term on the right-hand side tends to 0, a contradiction.

To complete the proof, simply observe if $f \in H^2_\alpha \subseteq L^2$, then $\|\phi\| \|f\| \geq \|T^\alpha_\phi f\| = \|X_\alpha f\| \geq \epsilon \|f\|$. □

Lemma 3.9. *Suppose $\phi \in L^\infty$ is unimodular. The distance from ϕ to $\mathfrak{A}$ is strictly less than one if and only if T^α_ϕ is left invertible for each $\alpha \in \mathbb{B}^2$.*

Proof. Suppose T^α_ϕ is left invertible for each $\alpha \in \mathbb{B}^2$. In this case, Lemma 3.8 applies and thus there is an $1 \geq \epsilon > 0$ such that for each α and $f \in H^2_\alpha$,

$$\|T^\alpha_\phi f\| \geq \epsilon \|f\|.$$

Now let $h \in \mathscr{M}$ be given. By Lemma 3.3 there is an $\alpha \in \mathbb{B}^2$ and $f \in H^2_\alpha$ and a $g \in L^2$ such that $\overline{g} \in (H^2_\alpha)^\perp$ and both $h = fg$ and $\|h\|_1 = \|f\|_2 \|g\|_2$. Thus,

$$\begin{aligned}\left| \int_0^{2\pi} \phi h \frac{dt}{2\pi} \right| = \left| \int_0^{2\pi} \phi f g \frac{dt}{2\pi} \right| &= |\langle \phi f, \overline{g} \rangle| = |\langle \phi f, (I - P_\alpha)\overline{g} \rangle| \\ &= |\langle (I - P_\alpha)\phi f, \overline{g} \rangle| \leq \|(I - P_\alpha)\phi f\| \, \|g\|.\end{aligned}$$

On the other hand, using the unimodular hypothesis,

$$\begin{aligned}\|f\|^2 = \|\phi f\|^2 &= \|P_\alpha \phi f\|^2 + \|(I - P_\alpha)\phi f\|^2 \\ &= \|T^\phi_\alpha f\|^2 + \|(I - P_\alpha)\phi f\|^2 \\ &\geq \epsilon^2 \|f\|^2 + \|(I - P_\alpha)\phi f\|^2.\end{aligned}$$

Thus, $(1 - \epsilon^2)\|f\|^2 \geq \|(I - P_\alpha)\phi f\|^2$. Therefore,

$$\left| \int_0^{2\pi} \phi h \frac{dt}{2\pi} \right| \leq \sqrt{1 - \epsilon^2} \, \|f\|_2 \, \|g\|_2 = \sqrt{1 - \epsilon^2} \, \|h\|_1.$$

By Lemma 3.4, it now follows that $\|\pi(\phi)\| < 1$, where $\pi : L^\infty \to L^\infty / \mathfrak{A}$ is the quotient map; i.e., the distance from ϕ to $\mathfrak{A}$ is less than one.

Conversely, if the distance from ϕ to $\mathfrak{A}$ is less than one, then there exists a $\psi \in \mathfrak{A}$ such that $\|\phi - \psi\| < 1$. It follows from Lemma 3.7 that T^α_ϕ is left invertible. □

Proof of Theorem 1.6. All that remains to be shown is: T^α_ϕ is invertible for each $\alpha \in \mathbb{B}^2$ if and only if the distance from ϕ to the invertible elements of $\mathfrak{A}$ is at most one. If T^α_ϕ is invertible for each $\alpha \in \mathbb{B}^2$, then there exists a $\psi \in \mathfrak{A}$ such that $\|\phi - \psi\| < 1$ by Lemma 3.9. By Lemma 3.7, $T^\alpha_{\overline{\phi}} T^\alpha_\psi$ is invertible. By Lemma 3.1, $T^\alpha_{\overline{\phi}}$ is invertible and thus T^α_ψ is invertible. By Lemma 3.6, ψ is invertible in $\mathfrak{A}$.

The converse is contained in Lemma 3.7. □

References

[1] Abrahamse, M.B. *Toeplitz operators in multiply connected regions*, Amer. J. Math. 96 (1974), 261–297.

[2] Abrahamse, M.B. *The Pick interpolation theorem for finitely connected domains*, Michigan Math. J. 26 (1979), no. 2, 195–203.

[3] Abrahamse, M.B.; Douglas, R.G. *A class of subnormal operators related to multiply-connected domains*, Advances in Math. 19 (1976), no. 1, 106–148.

[4] Ahern, P.R.; Sarason, D. *The H^p spaces of a class of function algebras*, Acta Mathematica, 117 (1967), 123–163.

[5] Ball, Joseph A.; Bolotnikov, Vladimir; ter Horst, Sanne. *A constrained Nevanlinna–Pick interpolation problem for matrix-valued functions*, Indiana Univ. Math. J. 59 (2010), no. 1, 15–51

[6] Ball, Joseph A.; Guerra Huamán, Moisés, D. *Convexity analysis and the matrix-valued Schur class over finitely connected planar domains*, J. Operator Theory 70 (2013), no. 2, 531–571.

[7] Ball, Joseph A.; Guerra Huamán, Moisés, D. *Test functions, Schur–Agler classes and transfer-function realizations: the matrix-valued setting*, Complex Anal. Oper. Theory 7 (2013), no. 3, 529–575.

[8] Broschinski, Adam. *Eigenvalues of Toeplitz operators on the annulus and Neil algebra*, Complex Anal. Oper. Theory 8 (2014), no. 5, 1037–1059.

[9] Conway, John B. *A Course in Functional Analysis*, Graduate Texts in Mathematics, 96. Springer, New York, 2007.

[10] Davidson, Kenneth R.; Paulsen, Vern I.; Raghupathi, Mrinal; Singh, Dinesh. *A constrained Nevanlinna–Pick interpolation problem*, Indiana Univ. Math. J. 58 (2009), no. 2, 709–732.

[11] Douglas, Ronald G. *Banach algebra techniques in operator theory*, Pure and Applied Mathematics, Vol. 49. Academic Press, New York-London, 1972.

[12] Dritschel, Michael A.; Pickering, James. *Test functions in constrained interpolation*, Trans. Amer. Math. Soc. 364 (2012), no. 11, 5589–5604.

[13] Jury, Michael T.; Knese, Greg; McCullough, Scott. *Nevanlinna–Pick interpolation on distinguished varieties in the bidisk*, J. Funct. Anal. 262 (2012), no. 9, 3812–3838.

[14] Koosis, Paul. *Introduction to Hp spaces. With an appendix on Wolff's proof of the corona theorem*, London Mathematical Society Lecture Note Series, 40. Cambridge University Press, Cambridge-New York, 1980.

[15] Martinez-Avendaño, Rubén A.; Rosenthal, Peter. *An introduction to operators on the Hardy–Hilbert space*, Graduate Texts in Mathematics, 237. Springer, New York, 2007.

[16] Raghupathi, Mrinal. *Abrahamse's interpolation theorem and Fuchsian groups*, J. Math. Anal. Appl. 355 (2009), no. 1, 258–276

[17] Raghupathi, Mrinal. *Nevanlinna–Pick interpolation for $\mathbb{C}+BH^\infty$*, Integral Equations Operator Theory 63 (2009), no. 1, 103–125.

[18] Sarason, Donald. *The H^p spaces of an annulus*, Mem. Amer. Math. Soc. No. 56 1965.

Sriram Balasubramanian
Department of Mathematics
IIT Madras
Chennai – 600036, India
e-mail: bsriram@iitm.ac.in

Scott McCullough and Udeni Wijesooriya
Department of Mathematics
University of Florida
Gainesville, USA
e-mail: sam@ufl.edu
wudeni.pera06@ufl.edu

Operator Theory:
Advances and Applications, Vol. 272, 71–98

Block Triangular Matrices in Banach Space: Minimal Completions and Factorability

Nir Cohen

Dedicated to Joe Ball, a dear friend and esteemed colleague

Abstract. This paper considers the extension of four issues from applied matrix theory to Banach space operators under a finite partition: block LPU factorability with L, U invertible; block LU factorability with L, U not necessarily invertible; rank factorization; and, above all, the minimal rank completion problem of block triangular type. This extension requires the replacement of rank considerations by range and kernel inclusions.

LPU factorability appears as a natural condition under which the other issues can be fully analyzed. In practice it reduces to a finite sequence of complementability conditions. When the completion problem has LPU factorable data, the minimal completions of the data are factorable operators and admit a complete description; otherwise, non-factorable minimal completions may exist due to a well-known Banach space anomaly (Embry's theorem).

Mathematics Subject Classification (2010). Primary 47A20; Secondary 47A05, 15A83, 15A23.

Keywords. Banach space, rank completion, complemented operator, block LU factorization, block LPU factorization.

1. Introduction

For simplicity of notation, an *operator* will denote a bounded linear transformation between two Banach spaces, denoted $T \in B(X, Y)$.

The spectral study of triangular models for Hilbert space operators relative to an infinite lattice of invariant subspaces has a rich history going back to Volterra. Notions of triangularity in Banach space, $B(X)$, and more generally $B(X, Y)$, are more rudimentary. Here we examine several issues connected with the simpler question of triangular structure w.r.t. a *finite* partition in $B(X, Y)$,

$$X = \oplus_{j=1}^{k} X_j, \qquad Y = \oplus_{i=1}^{k} Y_i \tag{1}$$

where X_i, Y_i are closed subspaces. In matrix theory this corresponds to the study of block triangular structure of not necessarily square matrices.

The main issue treated here is the following *minimal completion problem*: given an upper triangular operator T_+ w.r.t. (1), the set of completions $\mathbf{T}(T_+)$ is the affine set of all the operators of type $T_+ + T_-$, with T_- strict lower triangular. The main question is how to define, and characterize, the subset of *minimal* completions.

In matrix theory, minimal completions are those which minimize the rank, with recent applications in several areas (e.g., [13], [37]). A full characterization of minimal completions of triangular type can be found in the work of H. Woerdeman et al. ([30, 46–48]).

As we shall see, the Banach space formulation of this problem is relevant to several areas of operator theory; furthermore, its analysis requires the extension to Banach space of other classical matrix-theoretic tools (see Section 2). These include:

(i) *The condition for invertibility of a triangular operator.* As in the case of block-triangular matrices, the obvious answer (that the main diagonal blocks must be invertible) is incomplete, see Subsection 2.5.

(ii) *Block LPU factorizations with L, U triangular and invertible and P a generalized block diagonal.* LPU factorizations (and especially LU factorizations, with $P = I$) are of considerable importance in applied matrix theory. They may be viewed as an extension of the Bruhat decomposition for the group $GL_n(\mathbb{C})$ (see, e.g., [35], [43, Thm. 10.7.1]). The main question in extending the theory to Banach space is finding a suitable definition for the middle term P. The definition used here (Definition 1) involves a canonical form which we call the *rook form.* For operators, reduction to this form is not guaranteed and depends on the partition (1). The existence of a rook form is essentially a topological question, i.e., certain blocks of the given operator should be complemented (Section 2.2).

In particular, we prove that the LPU factorability property is stable under finite rank perturbations, extending a well-known phenomenon from Fredholm theory. Although the Fredholm index may be infinite, or ill defined, its variation under finite rank perturbation is zero in a certain weak sense defined in Subsection 2.4.

(iii) *Minimal and one-sided factorability.* Every finite matrix T enjoys a *rank factorization*, $T = AB$, through a space of dimension rank(T). In Banach space, not every operator enjoys a *minimal factorization* through a natural minimal Banach space.

In the same spirit, for matrices, conditions for T to admit T' as a left or right factor are reduced to simple range and kernel inclusions. In Banach space, there are known counter-examples in which these inclusions are satisfied but T' is not a factor (Embry's theorem, [22], see Subsection 2.3).

With these issues in mind, let us return to the main theme: extending the minimal completion problem from matrices to operators. Needless to say, in Banach space the rank has to be replaced by more geometric concepts involving range and kernel; and the question whether a minimal completion is "factorable" (i.e., admits a minimal factorization through a natural space defined by the problem) is shown to be both central and non-trivial.

Initially we analyze the case in which the specified triangular data T_+ admits (strong modular) LPU factorization (Section 3.2), obtaining a full description of minimal completions for T_+. It turns out that every minimal completion is factorable and its range and kernel spaces can be described.

Secondly, when T_+ admits LPU factorization only up to a finite perturbation, it seems that every minimal completion is still in factored form, but this needs more work (Subsection 3.5).

Finally, when the LPU condition is flatly not satisfied, non-factorable minimal completions may exist. This follows from Embry's theorem, see Subsections 3.3–3.4.

The same result holds also for the slightly more general triangular model which includes *dilation*. Here, T_+ has a *strict triangular* structure, with zero main diagonal blocks.

The next natural step beyond the triangular model would be the banded model, in which the rank completion problem is fully solved in the matrix case [49] but not in Banach space. Knowledge on other patterns, such as chordal [15] and cyclic [16],[25] is still quite limited even in the matrix case.

Another topic with direct applications in operator theory, not discussed here, would be the study of minimal completions under structural side conditions (in Hilbert space: Hermitian, positive definite, normal, isometry, contractive etc; diagonal dominant, M-matrix, Z-matrix etc).

Nor do we discuss the extension of completion problems to non-closed range operators, unbounded operators, and relations. Here, Banach space complementarity, a thematic ingredient in our approach, is violated, requiring alternative methods.

The paper concludes with two related issues (Section 4). In the first, a result about simultaneous triangular completion of a pair of inverses, discussed in [51], is extended to Banach space, under a similar LPU factorability assumption (Subsection 4.1). In the second, we consider another central issue from applied linear algebra, LU factorization, $T = LU$, but without the requirement that the triangular operators L and U (nor, for this matter, T itself) be invertible. For finite matrices under the complete partition $X = Y = \mathbb{C} \oplus \cdots \oplus \mathbb{C}$, factorability is decided by certain rank inequalities (Okunev and Johnson, [36]). This criterion is extended in Subsection 4.2 to block-LU factorization, for finite matrices and for operators. Both issues in Section 4 may be viewed as cases of *reverse completion*, in which a full operator is studied in terms of minimal completions of its various "triangular parts".

2. Triangular structure and factorability

In this section we provide definitions and basic properties connected with various notions of factorization, the rook canonical form, finite perturbations and triangular structure under a finite partition in Banach space.

2.1. LPU factorization

We define the notions of LPU factorability for general operators, and *modular* LPU factorability for triangular operators. These will be needed as technical conditions under which a complete analysis of minimal completions can be developed. We denote by $K(T)$ and $R(T)$ the kernel and range (=image). Isomorphism between Banach spaces is denoted by $X \cong Y$.

Under the partition (1) any operator $T \in B(X,Y)$ has a unique matrix representation, $T = (T_{ij})_{i,j=1}^k$, with operator-valued entries $T_{ij} \in B(X_j, Y_i)$. For simplicity of notation, we refer to T as a $k \times k$ *matrix*, adopting the usual matrix notation: blocks T_{ij} of T are *entries*, and i and j are the *row and column indices* of T. To avoid confusion, T will be called a *finite matrix* if X, Y are of finite dimension.

A support for T under the given partition is a set of index pairs (i,j) which contains all the non-zero entries. We shall use the following notation for the basic types of triangularity for an operator $T = (T_{ij})_{i,j=1}^k$ in the given block partition (1):

Upper triangular	t_+	(*support is* $\{(i,j) : i \le j\}$),
Lower triangular	t_-	(*support is* $\{(i,j) : i \ge j\}$),
Upper skew triangular,	s_+	(*support is* $\{(i,j) : i+j \le k+1\}$),
Lower skew triangular,	s_-	(*support is* $\{(i,j) : i+j \ge k+1\}$).

The corresponding *strict* classes denoted as $t^0_\pm, s^0_\pm$ are defined by strict support inequalities. We use these symbols as projections. For example, $T = s_+(T) + s^0_-(T)$ is the additive representation of T as a sum of an upper and a strict lower skew triangular operators.

Definition 1. Let T be a matrix under the partition (1). T is said to be *in weak rook form w.r.t.* (1) if there exist closed subspaces

$$X'_j, W_{ij} \subset X_j, \qquad Y'_i, Z_{ij} \subset Y_i, \qquad (1 \le i,j \le k)$$

such that

(i) $X_j = \oplus_i W_{ij} \oplus X'_j$ and $Y_i = \oplus_j Z_{ij} \oplus Y'_i$.

(ii) For each ij, when we consider the matrix form of $T_{ij} \in B(X_j, Y_i)$ under the sub-partition (i), all the entries are trivial except for a single entry $Q_{ij} \in B(W_{ij}, Z_{ij})$.

We say that T is *in strong rook form* if Q_{ij} is invertible for all i,j, implying the isomorphisms $W_{ij} \cong Z_{ij}$.

We refer to Q_{ij} as the ijth (strong or weak) *pivot* of T; to W_{ij}, Z_{ij} as *pivotal subspaces*; and to X'_j, Y'_i as *residual subspaces*.

Thus, $T_{ij} = 0$ if the pivotal space W_{ij} or Z_{ij} is trivial. In the strong version, the identification of Z_{ij} with W_{ij} for all ij can be used to reduce all the pivots to identity operators.

Definition 2. Assume that $T, R \in B(X, Y)$.

(i) We say that T and R are *LU-equivalent w.r.t.* (1) if $T = LRU$ where $L \in B(Y)$ and its inverse are t_-, $U \in B(X)$ and its inverse are t_+.

(ii) T is said to *admit (weak/strong) LPU factorization* if T is LU-equivalent to a matrix R in (weak/strong) rook form.

A finite matrix admits a (strong) LPU factorization w.r.t. any partition, and pivots occupy part of a single generalized diagonal; an entire diagonal, if T is invertible. For a Banach space operator, the existence of LPU factorization is not guaranteed and may depend on the partition.

Observe that the s_- pattern is invariant under LU-equivalence. Since the s_+ pattern is not invariant, it requires a modified definition (modular LU/LPU):

Definition 3. Given T_+, R_+ of type s_+ w.r.t. the partition (1), we say that T_+, R_+ are *modular LU-equivalent* (w.r.t. the same partition) if $T_+ = s_+(LR_+U)$ with L, U as in Definition 2.
T_+ *admits (weak/strong) modular LPU factorization* if T_+ is modular LU-equivalent to an s_+ (weak/strong) rook form R_+.

Define the set of completions of T by

$$\mathbf{T}(T_+) := \{T \in B(X, Y) : s_+(T) = T_+\}.$$

T_+ and R_+ are modularly LU-equivalent iff T_+ is LU-equivalent to an element of $\mathbf{T}(R_+)$. It is easy to see that this is an equivalence relation. Under modular equivalence the map $R \to LRU$ is a bijection from $\mathbf{T}(R_+)$ onto $\mathbf{T}(T_+)$. Furthermore, if $\mathbf{T}(R_+)$ contains a matrix R_0 in (weak/strong) rook form w.r.t. (1) then clearly $R_+ = s_+(R_0)$ itself is also in rook form, inheriting the s_+ pivots of R_0.

2.2. On the role of complementarity

Every finite matrix T admits LPU factorization, and $s_+(T)$ admits modular LPU factorization, under any partition; this is not so in Banach space, under our definitions: both conditions are non-trivial, and in practice can be replaced by a sequence of complementability conditions.

We demonstrate this fact as a natural consequence of the process of reduction to rook form, which may be performed by an operator-valued version of block Gaussian elimination, simultaneously on rows and columns of the partition (1). Unlike the scalar elimination variant, this variant cannot be considered a numerical algorithm, but still has considerable theoretical importance, due to the fact that it involves a finite sequence of algebraic operations in Banach space (operator additions, operator multiplications etc.).

We recall that subspace $X_1 \subset X$ is called (topologically) complemented if it is closed and admits a closed complement $X_2 = X_1^+$ (not unique) so that $X =$

$X_1 \oplus X_2$. An operator $T \in B(X, Y)$ is called complemented if both its kernel $K(T)$ and range $R(T)$ are complemented, hence in particular T has closed range. We denote their (non-unique) complements by $K^+(T)$ and $R^+(T)$.

Many matrix results concerning kernel, range and their complements extend to the case of complemented operators. For example, the following version of the dimension theorem is valid: if $T : X \to Y$ is complemented then $X \cong K(T) \oplus R(T)$ and, by symmetry, if $K^+(T)$ and $R^+(T)$ are complements then $Y \cong R^+(T) \oplus K^+(T)$ (this observation is useful, e.g., in the analysis of infinite-dimensional smooth manifolds in connection to the implicit/inverse function theorem, see [1] Theorem 2.5.15). Furthermore, T admits a generalized "rank decomposition", i.e., $T = GRH$ where G, H are invertible and $R = \left(\begin{smallmatrix} I & 0 \\ 0 & 0 \end{smallmatrix}\right)$.

Let us examine the Gaussian procedure for LPU reduction of T, or modular *LPU* reduction of $s_+(T)$. The triangular nature of L and U means that all the row and column operations are *descending*, and pivoting (permutation of rows and columns) is allowed only within the given row or column block.

Besides the three basic Gaussian operations, the algorithm includes necessarily a fourth element, i.e., the refinement of the partition via a new sub-division. This element is a logical consequence of the fact that the final partition $W_{ij}, Z_{ij}, X'_j, Y'_i$ in Definition 1 of the algorithm is a refinement of the initial partition X_j, Y_i in (1). Refinement in the processing of the ijth block T_{ij} occurs precisely when its pivot Q_{ij} is isolated, and pivotal spaces $W_{ij} \in X_j$ and $Z_{ij} \in Y_i$ must be allocated for it. If these spaces are not complemented, the algorithm cannot proceed.

In the case of finite matrices, where *every* subspace is complemented, reduction to strong rook form is guaranteed. The ijth condition is guaranteed also when T_{ij} is a Hilbert space closed range operator, a Banach space Fredholm operator, etc.

We conclude that algorithmic reduction to strong rook form depends on a finite succession of complementarity conditions. These conditions are sufficient only, and depend on the order of the elimination steps. The question is, of course, whether a complete set of *a priori conditions* can be given which guarantees the reduction to (weak/strong) rook form. This issue needs to be studied further.

Reduction to strong rook form is a natural stage in the modelling of various completion problems. We illustrate the role of complementarity in these models in two simple examples.

Example 4. *Dilations (or extensions) of the type*

$$T_+ = \begin{pmatrix} T_{11} & 0 \\ 0 & 0 \end{pmatrix}, \qquad T = \begin{pmatrix} T_{11} & F_{12} \\ F_{21} & F_{22} \end{pmatrix} \tag{2}$$

with arbitrary unspecified F-blocks.

In general, by dilation we understand the construction of an operator $T : X \to Y$ from its block $T_{11} = P_{Y_1} T_{|X_1}$, assuming the inclusions $X_1 \subset X$ and $Y_1 \subset Y$. To assume the form (2), X_1 and Y_1 must be complemented (reaching the partition (1)). This condition fails in several important dilation problems (e.g., when X_1 or Y_1

are dense). An additional condition is required in order to reduce T_+ to strong rook form, namely, that $T_{11} \in B(X_1, Y_1)$ be complemented. Under this condition, $K(T_{11})$ and $R(T_{11})$ are complemented and T_{11} may be reduced to the form $\begin{pmatrix} I & 0 \\ 0 & 0 \end{pmatrix}$, whence T_+ is in strong rook form. □

Example 5. *The one corner problem of type*

$$T_+ = \begin{pmatrix} T_{11} & T_{12} \\ T_{21} & 0 \end{pmatrix}, \qquad T = \begin{pmatrix} T_{11} & T_{12} \\ T_{21} & F \end{pmatrix} \tag{3}$$

with F arbitrary and unspecified.

This model often occurs when the completion $T \in B(X, Y)$ is sought within the affine set $\mathbf{T} := \{T_0 + P_1 T P_2 : \ T \in B(X, Y)\}$ with $T_0 \in B(X, Y)$ fixed and P_1, P_2 projections. When P_1, P_2 are bounded, the spaces $X_2 = R(P_2)$ and $Y_2 = R(P_1)$ are complemented and one easily reaches the form (3).

For example, in many problems of the Lagrange interpolation type we have $\mathbf{T} = \{UFV + R \in B(X, Y) : \quad X \in B(X_2, Y_2)\}$ with U, V, R fixed and F arbitrary [8],[41], where the minimum (rank) completion can be analyzed (in other interpolation problems, the set $\mathbf{T}$ is subject to an additional norm restriction, e.g., [7],[20],[40]). Here, reduction to (3) is reached under a complementarity condition on $R(U) \subset Y$ and $K(V) \subset X$.

Once model (3) is reached, modular LPU factorability imposes three complementarity conditions. The first is that T_{11} be complemented, leading to

$$T_+ = \begin{pmatrix} I & 0 & B_1 \\ 0 & 0 & B_2 \\ C_1 & C_2 & 0 \end{pmatrix}, \qquad T = \begin{pmatrix} I & 0 & B_1 \\ 0 & 0 & B_2 \\ C_1 & C_2 & F \end{pmatrix}. \tag{4}$$

Up to UL equivalence we may assume that B_1 and C_1 vanish, and then the remaining two conditions are that B_2 and C_2, (*the restrictions of T_{12} and T_{21} to the residual subspaces!*), be complemented. This permits the identification of pivots, Q_{12} of B_2 and Q_{21} of C_2, reducing T_+ to strong rook form.

2.3. One-sided factorability criteria

For finite matrices, the condition which guarantees that a matrix is a one-sided factor of another matrix is simple: inclusion between the kernels or ranges of the two matrices. In addition, every finite matrix T enjoys a rank factorization, $T = T_1 T_2$, through a space of dimension rank(T). In Banach space, conditions for one-sided factorability, as well as conditions for rank factorability, properly defined, are less clear cut, with severe implications to the general completion problem. The following two theorems, based on Embry's paper [22], summarize the basic situation. We denote by $A^* \in B(Y^*, X^*)$ the Banach space adjoint of $A \in B(X, Y)$.

Theorem 6. *Given two Banach space bounded linear operators $D, E \in B(X, Y)$, consider the following conditions:*

(a) $D = FE$ *for some* $F \in B(R(E), Y)$;
(b) $R(D^*) \subset R(E^*)$;

(c) $\exists c \geq 0$ *s.t.* $\|Dx\| \leq c\|Ex\|$ *for all* $x \in X$;
(d) $K(E) \subset K(D)$.

Then

(i) *Conditions* (a–c) *are equivalent.*
(ii) *Condition* (d) *is weaker than* (c) *even in separable Hilbert space.*
(iii) *Conditions* (a, d) *are equivalent if* D *is of finite rank or if* E *is complemented.*

Proof. (i) is Embry's theorem [22]. (ii) The implication (c) to (d) is obvious. For a counter-example for the converse implication, let E be any compact real diagonal operator on ℓ^2 with trivial kernel and set $D = I$. As D is not compact, the equation $D = FE$ is impossible. (iii) If D is of finite rank, choose a basis $\{v_1, \ldots, v_k\}$ for $K^+(D)$ and define F via $F(Ev_i) = Dv_i$. Condition (d) guarantees that $F \in B(R(E), Y)$ is well defined, and by linearity $D = FE$. Again, if E is complemented then the operator $\hat{E} : K^+(E) \to R(E)$ is invertible. Condition (d) guarantees that $F = D\hat{E}^{-1} \in B(R(E), Y)$ is well defined and $D = FE$. □

Theorem 7. *Given two Banach space bounded linear operators* $A, B \in B(X, Y)$, *consider the following conditions:*

(e) $A = BC$ *for some* $C \in B(X, X)$;
(f) $R(A) \subset R(B)$;
(g) $\exists c \geq 0$ *s.t.* $\|A^* f\| \leq c\|B^* f\|$ *for all* $f \in Y^*$;
(h) $K(B^*) \subset K(A^*)$.

Then

(i) *Each condition strictly implies the next.*
(ii) *Conditions* (e–g) *are equivalent if* X, Y *are Hilbert spaces.*
(iii) *Conditions* (f, g) *are equivalent if* X, Y *are reflexive spaces.*
(iv) *Conditions* (e, f) *are equivalent if* A *is of finite rank or if* B *is complemented.*

Proof. The implications in (i) are obvious (see discussion in [22]). Counter-examples for the non-equivalence of the pairs (e,f) and (f,g) are given in [19] and [11]. As for (g,h), the same example in the proof of Theorem 6(ii) shows that (h) does not imply (g) in separable Hilbert space. Indeed, in this example we may use the Hilbert space adjoint, and both D, E are Hermitian. Item (ii) has been established in [19]; item (iii), in [22] (see also [44]). (iv) If A is of finite rank, choose a finite basis $\{v_1, \ldots, v_k\}$ for $K^+(A)$, and for each v_i choose w_i such that $Av_i = Bw_i$. Assuming (f), the operator C which takes v_i to w_i is well defined and $A = BC$. Again, if B is complemented, the operator $\hat{B} : K^+(B) \to R(B)$ is invertible and, assuming (f), $C = \hat{B}^{-1}A \in B(X, X)$ is well defined and $A = BC$. □

See [3] for the respective situations in the classical Banach spaces and [24] for unbounded operators. Observe that the last item in each theorem applies, in particular, in the case of finite matrices. In the sequel we shall need the following immediate consequences.

Corollary 8.

(i) *Let D be of finite rank. Then* $\begin{pmatrix} E \\ D \end{pmatrix}$ *is right equivalent to some* $\begin{pmatrix} E' \\ D' \end{pmatrix}$ *for which the (finite) codimension of* $K\begin{pmatrix} E' \\ D' \end{pmatrix}$ *in* $K(E')$ *equals* $\operatorname{rank}(D')$.

(ii) *Let A be of finite rank. Then* $(B \quad A)$ *is left equivalent to some* $(B' \quad A')$ *for which* $R(A')$ *and* $R(B')$ *do not intersect.*

Both conditions guarantee "linear independence": of rows of D' w.r.t. E', of columns of A' w.r.t. B'.

2.4. Finite rank perturbations

For a Fredholm operator, the Fredholm property and the Fredholm index $i(T) = \dim K(T) - \operatorname{codim} R(T)$ are known to be stable under finite rank perturbations. For general operators, we show that the index remains stable under finite rank perturbations (under a weaker interpretation) even though $\dim K(T)$ and $\dim R(T)$ are infinite; and the same extends to complementation and factorability properties.

Definition 9. Assume that $T, T' \in B(X, Y)$ are such that

$$K(T) = U_1 \oplus U_2, \qquad K(T') = U_2 \oplus U_3$$

and/or

$$R(T) = V_1 \oplus V_2, \qquad R(T') = V_2 \oplus V_3$$

and U_1, U_3, V_1, V_3 are of finite dimension. Then we may define

$$\begin{aligned} \Delta \dim K(T,T') &= \dim(U_1) - \dim(U_3), \\ \Delta \operatorname{rank}(T,T') &= \dim(V_1) - \dim(V_3), \\ \Delta i(T,T') &= \Delta \dim K(T,T') + \Delta \operatorname{rank}(T,T'). \end{aligned}$$

These values are independent of the choice of complements V_1, V_3, U_1, U_3. We interpret Δ as increment, or differential. For example, in the case of finite matrices we get $\Delta \dim K(T,T') = \dim K(T) - \dim K(T')$.

First we treat stability of complementation and index.

Proposition 10.

(i) *If $U \subset V \subset X$, and U is complemented in X and of finite codimension in V, then V is complemented in X.*

(ii) *Let T' be a finite rank perturbation of T. Then $\Delta i(T,T') = 0$. Moreover, T is complemented iff T' is.*

Proof. (i) Let $v_1, \ldots, v_k$ be a basis for a complement of U in V and let U^+ be a complement of U in X. We may represent each v_i as $u_i + u_i^+$ ($u_i \in U, u_i^+ \in U^+$). This way, $u_1^+, \ldots, u_k^+$ is also a basis for a complement of U in V. This basis has a complement in U^+, denoted V'; and it is easy to see that $V \oplus V' = X$.

(ii) It is enough to prove the assertion about the index for a rank one perturbation, and then argue by transitivity. Up to equivalence we may assume that

$$S := T' - T = \begin{pmatrix} I_r & 0 \\ 0 & 0 \end{pmatrix}, \qquad T = \begin{pmatrix} T_1 & T_2 \\ T_3 & T_4 \end{pmatrix} \tag{5}$$

where $r = 1$, T_1 is a scalar, T_2 is a bounded functional and T_3 is a vector.

We want to show that the index of $T(a) := T + aS$ is independent of a. Namely, $f(a) := \Delta i(T(a), T(0)) = 0$ for all $a \in \mathbb{C}$. There are three (overlapping) cases. a) If $T_3 \notin R(T_4)$, removing the first row removes one dimension from the kernel; and then, if $K(T_4) \subset K(T_2)$, removing the first column does not affect the range. b) If $K(T_4) \not\subset K(T_2)$, removing the first column adds one dimension to the range; and then, if $T_3 \in R(T_4)$, the first row does not affect the kernel. Both items a, b occur independently of the value of a, hence $f(a) = 0$ for all a. c) If $T_3 \in R(T_4)$ and $K(T_4) \subset K(T_2)$, kernel and range change simultaneously exactly when $T_1 + a \neq T_2(T_4^{-1}(T_3))$, and again it follows that $f(a) = 0$ for all a.

Concerning the assertion on complementarity, we return to (5), now assuming only $r \in \mathbb{N}$. Assume that $K(T)$ and $R(T)$ are complemented.

The inclusions $R\begin{pmatrix} T_2 \\ T_4 \end{pmatrix} \subset R(T)$ and $K(T) \subset K(T_3 \quad T_4)$ are of finite codimension, hence the subspaces $R\begin{pmatrix} T_2 \\ T_4 \end{pmatrix} \subset Y$ and (using item (i)) $K(T_3 \quad T_4) \subset X$ are complemented. At the same time, the inclusions $R\begin{pmatrix} T_2 \\ T_4 \end{pmatrix} \subset R(T')$ and $K(T') \subset K(T_3 \quad T_4)$ are of finite codimension, hence $K(T') \subset X$ and (using item (i)) $R(T') \subset Y$ are complemented. □

Next we treat factorization.

Theorem 11. *Assume that T, S are of type s_+ and S has finite rank. If T admits modular (weak/strong) LPU factorization w.r.t.* (1) *then so does $T + S$.*

Proof. For $k = 1$, T and $T' := T + S$ are automatically in weak rook form (see Definition 1), and for the strong version we may invoke Proposition 10(ii).

So assume $k > 1$. Given T and S, up to LU-equivalence we may assume that T is in (weak/strong) rook form. The proof contains an inductive sweep over all entries T'_{ij}, $i + j \leq k + 1$. Initially we choose $i = j = 1$ and then increase (i, j) in lexicographic order. We make the following inductive hypothesis.

Hypothesis A(ij): $T'_{i'j'} = 0$ *for all (i', j') lexicographically smaller than (i, j).*

Hypothesis A(1,1) is trivially valid. Given the current value of (i, j), since T is in rook form, T_{ij} admits a weak/strong pivot, hence by Proposition 10(ii) so does T'_{ij}.

First we work along the ith row. Let $\hat{Z} := Z_{ij} \subset Y_i \subset Y$ be the (closed) pivotal subspace associated with T'_{ij} and consider the intersection of $\hat{Z}$ with $R(T'_{ip}) \subset Y$. For $p < j$, this intersection is zero by Hypothesis A; for $p > j$ it is of finite dimension, i.e., the projected operator $T''_{ip} := P_{\hat{Z}} T'_{ip}$ is of finite rank. We may divide T''_{ip}, by column operations, into two entries: T''_{ip1} and T''_{ip2} where

$R(T''_{ip1}) \subset R(T''_{ij}, \ldots, T''_{i,p-1})$, $R(T''_{ip2}) \cap R(T''_{ij}, \ldots, T''_{i,p-1}) = \emptyset$, and T''_{ip2} is left invertible.

Using a slight adaptation of Theorem 7(iv) we find that each T''_{ip1} is of the factored form $(T''_{ij}, \ldots, T''_{i,p-1})E$ for some E. Thus, up to LU-equivalence we may assume that $T''_{ip1} = 0$. The space $\hat{Z}_2 = \cup_{p \neq j} R(T''_{ip2})$ is of finite dimension, hence is complemented in $\hat{Z}$, say with complement $\hat{Z}_1$. Divide T''_{ij} conformably into $T''_{ijq} := P_{\hat{Z}_k} T''_{ij}$, $q = 1, 2$. We claim that $K(T''_{ij1}) \subset K(T''_{ij2})$. Indeed, if this were not the case one would have $Av \in \hat{Z}_2$, creating a contradiction to the definition of $\hat{Z}_2$. At the same time, T''_{ij2} is of finite rank, hence by an adaptation of Theorem 6(iii) $T''_{ij2} = FT''_{ij1}$ for some F, and by a row operation we may assume that $T''_{ij2} = 0$.

Now T'_{ij} has shrunk row-wise to T''_{ij} and the direct sum $\hat{Z} = \hat{Z}_1 \oplus \hat{Z}_2$ in the output space Z separates $R(T''_{ij})$ from other ranges in its entire row. By an analogous argument on the jth column, a direct sum in the input space X can be found which separates a column-wise shrunk copy of $K(T''_{ij})$ from kernels in its entire column. This shrunk copy is a true weak (or strong) pivot, unique in its (refined) row and column. These rows and columns may be effectively removed, and in the resulting shrunk partition, the ijth entry becomes zero, so that Hypothesis A(i_1, j_1) holds, where (i_1, j_1) is the lexicographic follower of (i, j). This completes the inductive proof. □

2.5. Triangular structure

Formal algebraic relations valid between triangular and skew triangular block matrices extend without change to the Banach space situation. We denote by J_X (resp. J_Y) the permutation operators which reverse the order of subspaces X_j (resp. Y_i) in (1). We have

$$s_\pm J_X = t_\pm, \qquad J_Y s_\pm = t_\mp.$$

Also, $t_\pm$ is closed under products; in the non-block case, also under inversions.

In the block case, inversion requires some care. Assume that $X \cong Y$. If $X_i \cong Y_i$ then T of type $t_\pm$ is invertible iff T_{ii} are invertible, and then T^{-1} is of the same type. However, this is not the general case. Similarly, if $X_i \cong Y_{k+1-i}$ then T of type $s_\pm$ is invertible iff $T_{i,k+1-i}$ are invertible, and then T^{-1} is of the opposite type $s_\mp$. However, this is not the general case.

Proposition 12. *Assume that $X \cong Y$ and $T_+ \in B(X, Y)$ is in an s_+ strong rook form w.r.t. the partition* (1). *Then the following are equivalent:*

(i) $(T_+ + T_-)^{-1}$ *exists and is of type s_+ for all $T_- \in B(X, Y)$ of type s_-^0;*
(ii) T_+ *is invertible.*

In particular, T_+^{-1} is necessarily of type s_+, and not s_- as one might expect! (Clearly, the condition that T_+ is in strong rook form is very strong here: for example, together with the condition $X_i \cong Y_i$, it implies that T_+ must be diagonal).

The following example illustrates Proposition 12. Consider $X = Y = \mathbb{C}^2 \oplus \mathbb{C} \oplus \mathbb{C}$, with a particular matrix T_+ of type s_+ and generic matrix T_- of type s_-^0

given by

$$T_+ = T_+^{-1} = \begin{pmatrix} 0 & 0 & 1 & 0 \\ 0 & 0 & 0 & 1 \\ 1 & 0 & 0 & 0 \\ 0 & 1 & 0 & 0 \end{pmatrix}, \qquad T_- = \begin{pmatrix} 0 & 0 & 0 & 0 \\ 0 & 0 & 0 & 0 \\ 0 & 0 & 0 & a \\ 0 & 0 & b & c \end{pmatrix}.$$

The inverse of $T_+ + T_-$ is again of type s_+; indeed,

$$(T_+ + T_-)^{-1} = \begin{pmatrix} 0 & -a & I & 0 \\ -b & -c & 0 & I \\ I & 0 & 0 & 0 \\ 0 & I & 0 & 0 \end{pmatrix}.$$

To prove Proposition 12 we first establish an elementary fact.

Lemma 13. *Let $A \in B(X,Y)$ be represented in terms of the partition (1). Assume that k of the entries A_{ij} (or "pivots") are invertible and form a single generalized diagonal ($j = \sigma(i)$ for some permutation $\sigma \in S_k$). Also assume that each of these pivots is the first non-trivial entry in its row and column. Then A, A^{-1} are invertible, and A^{-1} has also k invertible pivots, on the symmetric generalized diagonal ($i = \sigma(j)$). Moreover, each of these k pivots of A^{-1} is the last non-trivial entry in its row and column.*

Proof. Let $H \in B(Y)$ be the permutation matrix so that all the k pivots of $\tilde{A} = HA$ are on the main skew diagonal. Pivots of $\tilde{A}$ are still first in their *rows*, hence $\tilde{A} \in B(Y,X)$ is of type s_- with invertible skew-diagonal pivots. So, $\tilde{A}^{-1}$ is of type s_+ and admits invertible pivots on the main skew diagonal, which are last non-trivial in their *column*. Therefore, pivots of $A^{-1} = \tilde{A}H^{-1}$ are symmetric to those of A, and are last non-trivial in their column. A dual argument, multiplying A on the right by a permutation matrix, shows that the pivots of A^{-1} are also last non-trivial in their rows, completing the proof. □

Proof of Proposition 12. If (i) holds, the choice $T_- = 0$ shows that T_+ is invertible. Conversely, if (ii) holds and T_- is of type s_- then $A := T_+ + T_-$ satisfies the assumptions of Lemma 13. Its k pivots $A_{ij} = T_{+ij}$ satisfy $i + j \leq k + 1$, hence the pivots of A^{-1}, in symmetric positions ji, satisfy the same inequality, hence are all of type s_+. They are also last non-trivial in their rows and columns, hence A^{-1} is of type s_+, proving (i). □

By a skew diagonal permutation we easily recover the analogous result for the skewed types:

Corollary 14. *Under the decomposition (1), and $T_+ \in B(X,Y)$ of type t_+ in strong rook form, the following are equivalent:*

(i) *$(T_+ + T_-)^{-1}$ exists and is of type t_- for all $T_- \in B(X,Y)$ of type t_-^0;*
(ii) *T_+ is invertible.*

Again, if $X_i \ncong Y_{k+1-i}$ A is not necessarily skew diagonal.

3. The triangular completion problem

In this section we discuss the extension from matrices to Banach space operators of the minimal rank completion problem for triangular patterns.

3.1. The problem

Before discussing minimal completions we must define minimality. Given a set Ω in matrix space, minimal elements of Ω may be defined as elements of minimal rank. Since this definition is too restrictive in Banach space, a modified definition must be considered.

Definition 15. Let $\Omega \subset B(X,Y)$. Consider the implications

$$\begin{array}{llll} \text{(I)} & K(T) \subset K(S) & \Longrightarrow & K(T) = K(S), \\ \text{(II)} & R(S) \subset R(T) & \Longrightarrow & R(S) = R(T). \end{array}$$

$T \in \Omega$ is called *kernel-maximal, range-minimal* or just *minimal* in Ω if T satisfies (I), (II) or (I, II) w.r.t. all $S \in \Omega$.

The subsets of Ω consisting of elements with these minimality types will be respectively denoted by Ω_k, Ω_r and Ω_m. For finite matrices, the set of rank-minimal elements is denoted by Ω_ρ. By definition we have in the Banach space situation

$$\Omega_\rho \subset \Omega_m = \Omega_k \cap \Omega_r, \tag{6}$$

and the number

$$mr(\Omega) := \min\{\text{rank}(T) : \ T \in \Omega\}$$

is of interest. Of course, in Banach space this number may be infinite.

We study minimality in the specific context of completion problems. Assume a general partition

$$X = \oplus_{j \in \Gamma} X_j, \qquad Y = \oplus_{i \in \Sigma} Y_i \tag{7}$$

where Γ, Σ are (say, countable) totally ordered sets and X_i, Y_i are closed subspaces. Let E_+ (the "pattern") be any subset of $\Gamma \times \Sigma$ and let E^c be the complementary pattern. A collection of elements $P_{ij} \in B(X_j, Y_i)$ $(ij \in E)$ defines a *partial matrix* P. The set of completions of P is defined as

$$\Omega := \mathbf{T}(P) = \{T \in B(X,Y) : T_{ij} = P_{ij} : \ ij \in E\}.$$

The minimal completion problem consists of the analysis of minimal elements in this set, in the senses defined above. In matrix theory, the (necessarily finite) patterns E studied initially were triangular [30, 46–48] and banded [49], admitting closed formulae for the minimal rank mr.

It is of interest to extend the study of minimal completions to Banach space operators, as a natural extension of several constructions in dilation theory, and due to possible connections with the representation theory of Banach algebras of infinite types. In the present paper we study exclusively *finite triangular patterns.* Here, we may always fix the pattern (up to row and column permutations) to be either the full upper left triangular pattern $E = \{(i,j) : \ i+j \leq k+1\}$ or the strict pattern $E_0 = \{(i,j) : \ i+j \leq k\}$. In this context we adhere to a more concrete notation. For example, when T_+ is supported on E we may choose

$$T_+ = \begin{pmatrix} T_{11} & T_{12} & \cdots & T_{1,k-1} & T_{1,k} \\ T_{21} & T_{22} & \cdots & T_{2,k-1} & 0 \\ \cdots & \cdots & \cdots & \cdots & \cdots \\ T_{k1} & 0 & \cdots & 0 & 0 \end{pmatrix}, \qquad T_{ij} \in B(X_j, Y_i). \tag{8}$$

Completions $T \in \mathbf{T}(T_+)$ of T_+ are of the form $T = T_+ + T_-$ with T_- supported on the complement pattern E^c. In terms of Subsection 2.1 we may define the set of completions as

$$\mathbf{T}(T_+) = \{T \in B(X,Y) : \ s_+(T) = T_+\}.$$

To conclude, the problem treated here may be stated as follows.

Definition 16. Given $T_+ \in B(X,Y)$ supported on the pattern E, the *minimal completion problem* for T_+ is the study of minimal elements in the affine set $\mathbf{T}(T_+)$.

As before, in the matrix case the minimum rank may be defined as

$$mr(T_+) = \min\{\text{rank}(T) : \ T \in \mathbf{T}(T_+)\} \tag{9}$$

The literature does not provide a general formula for mr for general non-banded patterns (see, e.g., [15, 16, 25]). The numerical calculations involved may become intractable when the matrix size increases (the problem is NP hard, [26]). A challenge prize offered in 2006 by a leading media services provider (Netflix), related to a particular application to data analysis, has spurred a large number of research papers on fast numerical algorithms for approximating mr for large, mainly sparse, partial matrices of arbitrary pattern (see, e.g., [12, 13, 32, 34, 38]).

Within operator theory, triangular and banded completions appear in several contexts involving Hankel and Toeplitz operators, moment problems, Schur sequences and interpolation. For matrix Toeplitz models see, e.g., [27–29] and [50]; for the infinite problem see [21]. These issues are intimately related with Wiener–Hopf theory. Indeed, for example, the construction of a Hardy space Hankel operator H_W on the unit circle from its symbol W may be viewed as a succession of one-step completions of triangular type. As long as W is rational, mr for the finite sections stabilizes on the value $\deg(W)$ as the completion size increases (Kronecker's theorem, see, e.g., [23]). A Banach space formulation may be helpful in analyzing non-rational symbols on arbitrary curves.

Completion analysis in the large may be relevant to several extension/dilation problems, mostly in Hilbert space: contraction models [6], self-adjoint extensions of certain densely defined symmetric operators [2], normal extensions of subnormal operators [17].

The complementarity condition discussed in Section 2 is automatically satisfied not just for matrices but also for finite rank operators, e.g., integral operators with separable kernel [47], Fredholm operators, closed range operators in Hilbert space etc. In several other important dilation problems, e.g., involving dilation from a dense subspace, such as Hahn–Banach extensions, this condition is not satisfied and, hence, the techniques described here are inadequate.

3.2. Analysis of the problem

In this section, a complete analysis of minimal completions of T_+ (of type s_+) is provided, under the assumption that T_+ admits a strong LPU factorization w.r.t. the block partition (1). Initially we shall assume that T_+ is already in strong rook form and characterize the representation of the full set of its completions, $\mathbf{T}(T_+)$. Completions have the form $T = T_+ + T_-$ relative to (1), where the second term is an arbitrary operator of type s^0_-.

The rook form of T_+ imposes, via Definition 1, a sub-partition of (1) consisting of pivotal and residual subspaces. We re-arrange these spaces in three levels:

$$X = W \oplus X', \qquad Y = Z \oplus Y', \tag{10}$$

$$X' = \oplus_{j=1}^{k} X'_j, \quad W = \oplus_{j=1}^{k} W_j, \qquad Y' = \oplus_{i=1}^{k} Y'_i, \quad Z = \oplus_{i=1}^{k} Z_i, \tag{11}$$

$$W_j = \oplus_{i=0}^{k+1-j} W_{ij}, \quad Z_i = \oplus_{j=0}^{k+1-i} Z_{ij}. \tag{12}$$

This re-arrangement involves nothing more than a finite permutation of row and column indices. W, X' and Z, Y' are the respective total pivotal and residual subspaces in X and Y. This finer partition induces the following representation of operators $T \in B(X, Y)$, including all the completions T of T_+. The first-order partition (10) induces a 2×2 matrix structure:

$$T = \begin{pmatrix} A & B \\ C & D \end{pmatrix}. \tag{13}$$

Under the second-order partition (11), each of the matrices A, B, C, D assumes a $k \times k$ matrix structure (some row or column indices may vanish, if some of the spaces W_j, X'_j, Z_i, Y'_i is trivial). Each of the matrices A, B, C, D can therefore be divided into its s_+/s^0_- parts w.r.t. (11) ($A = A_+ + A_-$ etc.); and it is readily seen that

$$T_+ = \begin{pmatrix} A_+ & B_+ \\ C_+ & D_+ \end{pmatrix}, \qquad T_- = \begin{pmatrix} A_- & B_- \\ C_- & D_- \end{pmatrix}.$$

Proposition 17. *Let T_+ in (8) be in strong rook form. Then $T_+ = \begin{pmatrix} A_+ & 0 \\ 0 & 0 \end{pmatrix}$. Moreover, for every completion $T \in \mathbf{T}(T_+)$ in (13) represented in the basis (10)–(12), A is invertible and A^{-1} is of type s_+.*

Proof. (i) The vanishing of B_+, C_+, D_+ in the representation of T_+ follows from the fact that T_+ (assumed to be in rook form) consists entirely of s_+ pivots, which are all part of A. Next, given the completion T, due to the position and invertibility of the pivots Q_{ij}, $s_+(A)$ satisfies the conditions of Proposition 12, independent of $s^0_-(A)$, hence their sum $A = A_+ + A_-$ has an inverse of type s_+. □

In view of the representation of the full set of completions, $\mathbf{T}(T_+)$, elaborated in Proposition 17, we can now characterize the subset $\mathbf{T}_m(T_+)$ of minimal completions. Define the set of factored solutions, in terms of (13):

$$\mathbf{T}_f(T_+) := \left\{ T \in \mathbf{T}(T_+) : \ \exists \tilde{B}, \tilde{C} : \ T = \begin{pmatrix} A & A\tilde{B} \\ \tilde{C}A & \tilde{C}A\tilde{B} \end{pmatrix} \right\}. \tag{14}$$

According to Proposition 17, A is invertible, hence

$$\mathbf{T}_f(T_+) := \{T \in \mathbf{T}(T_+) : \ D = CA^{-1}B\}. \tag{15}$$

For every data triple

$$(A, B, C) \in B(W, Z) \times B(X', Z) \times B(W, Y')$$

let $\mathbf{T}_{A,B,C}(T_+)$ be the set of completions $T \in \mathbf{T}(T_+)$ represented by (13) with arbitrary D. It follows from (15) that $\mathbf{T}_{A,B,C}(T_+)$ contains at most a single factored solution.

A priori, for the set $\mathbf{T}_{A,B,C}(T_+)$ to be non-empty, B, C must be of the s_-^0 type and A^{-1} of type s_+ (Proposition 17). In this case, $CA^{-1}B$ is necessarily of type s_-^{00}, i.e., of type s_- missing its main *two* skew diagonals. Thus the factored minimal solution does exist, and its D part (a priori of type s_-^0) is of type s_-^{00}.

With these preliminaries, the following central result provides the complete solution of the minimal completion problem for strong rook forms.

Theorem 18. *Let T_+ in* (8) *be in strong rook form.*

(i) *For any s_-^0 choice of A_-, B, C, $\mathbf{T}_{A,B,C}(T_+)$ contains a unique minimal completion of the form* (15), *for which D is of type s_-^{00}.*

(ii) *W.r.t.* (6) *we have $T_+ \in \mathbf{T}_m(T_+) = \mathbf{T}_k(T_+) = \mathbf{T}_r(T_+) = \mathbf{T}_f(T_+)$.*

(iii) *Every $T \in \mathbf{T}_m(T_+)$ satisfies*

$$K^+(T) \cong W \cong Z \cong R(T), \qquad K(T) \cong X', \qquad R^+(T) \cong Y'. \tag{16}$$

(iv) *The minimal completion is unique iff $Z = Z_{11}$, namely, $T_{ij} = 0$ for all $i + j \leq k + 1$ with $(i, j) \neq (1, 1)$.*

Proof. (i) Every completion $T \in \mathbf{T}_{A,B,C}(T_+)$, represented by (13), is uniquely determined by the parameter $F = F(D) = D - CA^{-1}B$ via the Schur complement decomposition $T = T_F := G\tilde{T}_F H$ with

$$G = \begin{pmatrix} I & 0 \\ CA^{-1} & I \end{pmatrix}, \quad \tilde{T}_F = \begin{pmatrix} A & 0 \\ 0 & F \end{pmatrix}, \quad H = \begin{pmatrix} I & A^{-1}B \\ 0 & I \end{pmatrix}. \tag{17}$$

The parameter $F = 0$ corresponds to the unique *factored* completion in this set,

$$T_0 = \begin{pmatrix} I \\ CA^{-1} \end{pmatrix} A \begin{pmatrix} I & A^{-1}B \end{pmatrix} \in \mathbf{T}_f(T_+). \tag{18}$$

As commented earlier, this completion D is of type s_-^{00}. $\tilde{T}_0$ is a *unique minimum* among the operators $\tilde{T}_F$ in this set, in terms of both kernel and range, due to its direct sum form in (17). Since G, H are fixed and invertible, it follows that T_0 is a unique minimum in $\mathbf{T}_{A,B,C}(T_+)$. For all $T_F \in \mathbf{T}_{A,B,C}(T_+)$ $(F \neq 0)$ we have the *strict* inclusions $R(T_0) \subset R(T_F)$ and $K(T_F) \subset K(T_0)$.

(ii) Every range minimal/kernel maximal completion T defines a triple A, B, C and, moreover, must be range minimal/kernel maximal within the subset $\mathbf{T}_{A,B,C}(T_+)$. By the above analysis, T is the completion T_0 in (18). Therefore, every range minimal or kernel maximal completion is factored.

Next we show the converse: every factored completion is minimal. By the analysis in item (i), this boils down to showing that if T_1, T_2 are two factored completions, we cannot have a strict inclusion between their ranges, or kernels. In view of (18) we have

$$T_1 = \begin{pmatrix} I \\ C_1 A_1^{-1} \end{pmatrix} A_1 (I \quad A_1^{-1} B_1), \qquad T_2 = \begin{pmatrix} I \\ C_2 A_2^{-1} \end{pmatrix} A_2 (I \quad A_2^{-1} B_2). \tag{19}$$

The inclusion $R(T_1) \subset R(T_2)$ implies that for all $x \in Z$ there exists $y \in Z$ so that $\begin{pmatrix} x \\ C_1 A_1^{-1} x \end{pmatrix} = \begin{pmatrix} y \\ C_2 A_2^{-1} y \end{pmatrix}$. So $y = x$ and it follows that the two ranges are equal, hence the inclusion cannot be strict. A dual argument shows that the kernels cannot be strictly contained, completing the proof of the chain of equalities in (ii). When we specialize item (i) to the class $\mathbf{T}_{0,0,0}(T_+)$ we get $T_+ \in \mathbf{T}_f(T_+)$, completing the proof of (ii).

(iii) If $T \in \mathbf{T}_m(T_+)$ then $T = T_0$ is factored, we have $F = 0$ in (17), readily implying that the diagonal term $\tilde{T}_F = T_0$ in (17) satisfies (16); given that G, H are invertible, so does T.

(iv) The only case in which no pivot of T_+ meets an entry of T_- in its row or column is when $T_{ij} = 0$ for all $(i,j) \neq (1,1)$. In this case, the matrices B, C are trivial and $F = 0$ can only be attained with $D = 0$, obtaining $T_- = 0$, i.e., $T = T_+$ as the only minimal solution. On the other hand, if another pivot exists besides P_{11}, there is always enough freedom in choosing non-zero B or C, or both, and then a unique choice of D leading to a minimal completion with non-zero s_-^0 part. □

We now use the full power of LU-equivalence to extend the validity of Theorem 18, removing the condition that T_+ is in strong rook form.

Corollary 19. *If T_+ is of type s_+ and admits strong modular LPU factorization w.r.t.* (1) *then*

(i) $\mathbf{T}_m(T_+) = \mathbf{T}_r(T_+) = \mathbf{T}_k(T_+) = \mathbf{T}_f(T_+) \neq \emptyset$.

(ii) *Every minimal completion $T \in \mathbf{T}_m(T_+)$ satisfies*

$$K^+(T) \cong W \cong Z \cong R(T), \qquad K(T) \cong X', \qquad R^+(T) \cong Y'$$

where W, Z, X', Y' are the pivotal and residual spaces of the associated rook form.

Proof. Assume the modular LU-equivalence $T_+ = s_+(LR_+U)$ with R_+ in strong rook form. The relation $T = LRU$ defines a linear bounded bijection between completions $T \in \mathbf{T}(T_+)$ and $R \in \mathbf{T}(R_+)$, which preserves the sense of minimality. Therefore, the validity of Theorem 18 extends from R_+ to T_+. □

It should be noted that R_+ (as a minimal completion of itself) is mapped to LR_+U, as a minimal completion of $T_+ = s_+(LR_+U)$. Thus, in general, T_+ is not a minimal completion of itself, unlike the situation in Theorem 18.

We also remark that the special basis (10)–(12) used in Theorem 18 is not directly available in Corollary 19. Thus, solutions of T_+ can be represented indirectly as $\mathbf{T}_f(T_+) = \{LTU : \quad T \in \mathbf{T}_f(R_+)\}$, and are equally factored. The fact that every minimal completion in Theorem 18 and Corollary 19 is factored is significant from several points of view. In view of Eqs. (17), (19), $T \in \mathbf{T}(T_+)$ admits the parametrization

$$T = \begin{pmatrix} I \\ U_1 \end{pmatrix} (A_+ + A_-)(I \quad V_1) + \begin{pmatrix} 0 \\ I \end{pmatrix} F(0 \quad I) \tag{20}$$

which is affine in F. Minimal completions correspond to the choice $F = 0$, obtaining the factored form $T = UAV$. Note however that for an arbitrary parameter F in (20) $T = T_F$ is not guaranteed to be a completion of T_+.

We have seen that if a completion is minimal then it factors through the pivotal space Z (up to isomorphism, this is the smallest space with this property). The converse is not always true: for example, if Z is infinite-dimensional and X', Y' are finite-dimensional then, trivially, *every* completion $T \in \mathbf{T}(T_+)$ factors through Z.

Besides the *non-dilated* triangular model (8) analyzed so far,

$$\mathbf{T}(T_+) := \{T \in B(X,Y) : \ s_+(T) = T_+\} \qquad T_+ \textit{ of type } s_+, \tag{21}$$

we consider next the slightly more general *dilated* triangular model

$$\mathbf{T}(T_+) := \{T \in B(X,Y) : \ s_+^0(T) = T_+\} \qquad T_+ \textit{ of type } s_+^0. \tag{22}$$

Clearly, among the minimal completions for (22) we find completions with $T_{1k} = 0$ and $T_{k1} = 0$, and if we add these two entries to the data we get a problem of type (21); however, not all the minimal completions for (22) are of this type.

In fact, if T_+ in (22) does not admit a strong LPU factorization some of its minimal completions may be non-factored, as we show below. As a study case for this model we consider the case of pure dilations discussed in Example 4. Recall that a *factored* dilation is a dilation of the form $T = \begin{pmatrix} A & AB \\ CA & CAB \end{pmatrix}$ with $C \in B(R(A), Y_2)$ and $B \in B(X_2, \mathrm{Ker}(A)^+))$.

Applying to dilations the factorability criteria (Theorems 6 and 7), we reach the following result:

Proposition 20. *A dilation problem* (3) *may admit non-factored minimal solutions. A necessary condition for this to occur is that T_{11} is not complemented, and in particular, of infinite rank.*

Proof. Consider the special case $T_+ = (A\ 0)$ and its dilation $T = (A\ F)$. Factored solutions have the form $T = A(I\ M) = (A\ AM)$, and range minimal solutions have the form $T = (A\ B)$ with $R(B) \subset R(A)$. So clearly we have $\mathbf{T}_f(T_+) \subset \mathbf{T}_r(T_+)$, and the inclusion may be strict in some cases, in view of item (i) of Theorem 7(i). By item (iv) of the same theorem, in such cases $A = T_{11}$ cannot be complemented.

In a similar way, consider the dilation $T_+ = \begin{pmatrix} A \\ 0 \end{pmatrix}$ with $T = \begin{pmatrix} A \\ F \end{pmatrix}$. A dual argument shows that $\mathbf{T}_f(T_+) \subset \mathbf{T}_k(T_+)$, and the inclusion may be strict due to

Theorem 6(i); but not if A is complemented; and in view of item (iii) of the same theorem, in such a case T is not complemented. □

Using a similar argument, it can be seen that the non-dilated model (21) may also lead to non-factorable minimal completions. For example, the one corner problem (5) with T_{11} non-complemented and $T_{12} = 0$ or $T_{21} = 0$ (but not both). The whole topic requires further study.

3.3. Finite matrices

In the case of finite matrices, the LPU condition is automatically satisfied. Moreover, the minimal completion rank $mr(T_+)$ is finite and (at least when the pattern is triangular) admits a combinatorial formula. For the dilated problem (22) we may remove the last row and column, obtaining a smaller non-dilated model of type (21).

For the model (21) a simple formula for $mr(T_+)$ has been obtained in [46], and the uniqueness of the minimal completion has been characterized in [30]. Below we show that these results can be viewed as corollaries of the more general results obtained here (Theorem 18 and Corollary 19). The precise statement is as follows. Define for T_+ in (8)

$$\begin{aligned} M_m &= (T_{ij})_{i=1}^{k+1-m}\,{}_{j=1}^{m}, & r_m &= \operatorname{rank}(M_m) \quad (m = 1, \dots, k), \\ S_m &= (T_{ij})_{i=1}^{k-m}\,{}_{j=1}^{m}, & r'_m &= \operatorname{rank}(S_m) \quad (m = 2, \dots, k), \end{aligned} \tag{23}$$

the maximal and sub-maximal s_+ blocks of T_+ and their respective ranks.

Corollary 21. *Assume that T_+, of type s_+ w.r.t. the partition* (1), *has finite rank.*

(i) *A completion $T \in \mathbf{T}(T_+)$ is minimal iff it is rank minimal.*
(ii) $mr(T_+) = \sum_{i=1}^{k} r_i - \sum_{i=2}^{k} r'_i$.
(iii) $\mathbf{T}_m(T_+)$ *is a singleton iff* $r_1 = r'_2 = r_2 = \cdots = r'_n = r_n$.

Proof. According to Corollary 19(ii), minimal completions (in the sense of Definition 15) have isomorphic ranges, hence have the same rank, say, r_1. On the other hand, rank minimal completions are minimal under Definition 15, and have rank $mr(T_+)$. We conclude that $r_1 = mr(T_+)$, proving (i). Next, reduction from T_+ to its (strong) rook form R_+ maintains invariant the pattern, as well as the numbers $mr(T_+), r_i, r'_i$. Therefore, in proving items (ii–iii), it is enough to verify the claims for R_+ in rook form. Let Q_{ij} be the pivots of R_+ and set $q_{ij} = \operatorname{rank}(Q_{ij})$.

(ii) The left-hand side $mr(R_+)$ of the given equation is equal to $\operatorname{rank}(R)$ for any $R \in \mathbf{T}_m(R_+)$. According to Theorem 18(ii), this rank is equal to $\dim(Z) = \sum_{i+j \le k+1} q_{ij}$. In the right-hand side, the number of occurrences of each q_{ij} in $\sum_{m=1}^{k} r_m$ is easily shown to be a unit greater than in $\sum_{m=2}^{k} r'_m$. Thus, the right-hand side too is equal to $\sum_{i+j \le k+1} q_{ij}$.

(iii) It can be checked that for R_+ in rook form the identity $r_i = r'_{i+1}$ ($i = 1, \dots, k-1$) occurs exactly when row $n+1-i$ of R_+ is zero, and the identity $r_i = r'_i$ ($i = 2, \dots, k$) occurs exactly when column i is zero. These rows and columns combined cover all the specified entries of R_+ except R_{11}. Therefore, these identities are simultaneously satisfied iff $R_{ij} = 0$ for all $(i,j) \neq (1,1)$. By Theorem 18, this occurs iff $\mathbf{T}_m(T_+)$ is a singleton. □

In the special case $k = 1$ we get the trivial fact $mr(T_+) = \text{rank}(T_+)$, with a trivially unique completion. In the case $k = 2$ (the one corner problem, Example 5), we get $mr(T_+) = r_1 + r_2 - r_2'$, with uniqueness iff $r_1 = r_2' = r_2$.

3.4. Finite rank perturbations

Corollary 21 provides a combinatorial formula for mr for finite rank operators, i.e., finite rank perturbations T_+ (of type s_+ in the given partition) of the zero operator $T_+' = 0$. Needless to say, both operators T_+, T_+' admit LPU factorizations.

Here we aim at extending this combinatorial treatment to the more general situation of a finite rank perturbation $T_+' = T_+ + S$ involving an *arbitrary* pair T_+, T_+' of operators of type s_+, even without assuming LPU factorability. Although both $mr(T_+)$ and $mr(T_+')$ may be infinite, the *differential* $\Delta mr(T_+, T_+')$, rather than mr itself, is finite and can be described combinatorially, at least in some cases. This differential is defined below, in the spirit of the definitions given in Subsection 2.4.

Specifically, the map $T \to T' := T + S$ maps completions of T_+ bijectively into completions of T_+'; and, by Proposition 10(ii) we have $\Delta i(T, T') = 0$. Unfortunately, this map may fail to preserve minimality, and the validity of the identity $\Delta i(T, T') = 0$ for pairs of *minimal* completions, T of T_+ and T' of T_+', is not obvious.

As a result, at this point we can only provide a *putative* definition of the differential Δmr. Namely, it is hoped that with every minimal completion T of T_+ there exists a minimal completion T' of T_+' (and vice versa) such that

$$\begin{aligned} R(T) &= V_1 \oplus V_2, & R(T') &= V_2 \oplus V_3, \\ K(T) &= U_1 \oplus U_2, & K(T') &= U_2 \oplus U_3 \end{aligned}$$

where V_1, V_3, U_1, U_3 are of finite dimension, and so that the finite numbers $\dim(U_1) - \dim(U_3)$ and $\dim(V_3) - \dim(V_1)$ are equal (i.e., $\Delta i(T, T') = 0$ in the sense of Proposition 10(ii)). In such a situation we may define $\Delta mr(T_+, T_+') := \dim(U_1) - \dim(U_3)$, independent of the choice of complements U_1, U_3, V_1, V_3. Recall that a finite-dimensional subspace of any metric space (not necessarily closed) is complemented with a well-defined codimension [31].

There are two ways to extend the formula for mr in Corollary 21 to a formula for Δmr valid for finite rank perturbations, which we describe as vertical and horizontal. Under the vertical extension, we consider M_i, S_i (resp. M_i', S_i'), the maximal and sub-maximal blocks of T_+ (resp. T_+') as in (23). The conjectured formula is

$$\Delta mr(T_+, T_+') = \sum_{i=1}^{k} \Delta \,\text{rank}(M_i, M_i') - \sum_{i=2}^{k} \Delta \,\text{rank}(S_i, S_i').$$

In spite of the simplicity of this formula, and finiteness of its right-hand side, the analysis of the left-hand side is unclear, for the reason described earlier.

The alternative, horizontal extension of the formula is given by items (i, ii) of Proposition 22 below, specialized to a particular study case. Namely, it is assumed that T'_+ is supported on its 1,1 entry, and $T_{11} = T'_{11}$. This is already a non-trivial case, since no complementarity conditions are assumed on T_{11}. In this special case, the problem and its solution may be stated in terms of T_+ alone.

Proposition 22. *Assume that T_+ is of type s_+ w.r.t.* (1) *and the entries T_{ij} $(i+j \leq k+1,\ i+j \geq 3)$ are of finite rank. Let M_i, S_i be the maximal and sub-maximal blocks of T_+. Then*

$$\text{(i)} \qquad \Delta mr(T, M_1) = \sum_{i=2}^{k} \Delta \operatorname{rank}(M_i, S_i),$$

$$\text{(ii)} \qquad \Delta mr(T, M_k) = \sum_{i=2}^{k} \Delta \dim K(S_i, M_{i-1}).$$

Proof. First we show (i). For any completion T of T_+ denote by T_j the restriction of T to $B(X_1 \oplus \cdots \oplus X_j, Y)$. We have $T_1 = M_1$ and $T_k = T$. Also set $\rho_i := \Delta \operatorname{rank}(M_i, S_i.)$ Clearly,

$$\Delta \operatorname{rank}(T, M_1) = \sum_{j=2,\ldots,k} \Delta \operatorname{rank}(T_j, T_{j-1}) \geq \sum_{j=2,\ldots,k} \rho_j,$$

establishing an inequality in (i). Next we show that we have equality if T is a minimal completion. We use induction on m and show that it is possible to complete T_m from T_{m-1} so that $\Delta \operatorname{rank}(T_m, M_1) = \sum_{j=2}^{m} \rho_j$. Each step of this type is reduced to a one corner problem of type (3); one needs to complete the 2,2 entry, given that the 1,2 and 2,1 entries are of finite rank. Using Corollary 8 we may assume, up to an L-type left equivalence, that the ranges of the 1,1 and 1,2 entries do not intersect. The second range is a subspace of finite dimension ρ_m, and it is clearly possible to choose the 2,2 entry so that also $\Delta \operatorname{rank}(T_m, T_{m-1}) = \rho_m$: for example, the zero completion.

This argument completes the proof of (i). The proof of (ii) is similar. □

We see that the obvious condition for rank calculations (LPU factorability) may become superfluous in certain cases involving finite rank perturbations. Another possibility of weakening this condition may be to replace complementarity by quasi-complementarity (see, e.g., in [33]), requiring more study.

A natural question is whether Proposition 22 extends to the case in which T'_+ is supported on a single entry which is *not* the 1,1 entry, say T_{pq} with $3 \leq p+q \leq k+1$. We observe that the case $p = q = 1$ is unique in terms of finiteness of the right-hand side in Proposition 22. This case is also special in terms of both uniqueness and factorability. For uniqueness, see Theorem 18(iv) and Corollary

21(iii). To elucidate the role of factorability, consider the special case $k = 2$, i.e., the one corner problem (3). We examine the three possible options for T'_+ :

$$T'_{+1} = \begin{pmatrix} T_{11} & 0 \\ 0 & 0 \end{pmatrix}, \qquad T'_{+2} = \begin{pmatrix} 0 & T_{12} \\ 0 & 0 \end{pmatrix}, \qquad T'_{+3} = \begin{pmatrix} 0 & 0 \\ T_{21} & 0 \end{pmatrix}.$$

In all the three problems, we obtain T_+ by adding a finite rank perturbation to the two zero s_+ entries, and then complete T_+ by adding the $2,2$ (i.e., s_+) entry. For T_{+1}, completions have a direct sum form, and comparison with the middle term in (17) is instructive. Just as argued there, we see that the only minimal (range/kernel) dilation is T_+ itself, which is in factored form, fully recovering Corollary 19 but without the LPU assumption. On the other hand, dilating T_{+2} and T_{+3} may lead to non-factored minimal dilations of the type encountered in Proposition 20.

4. Two reverse completion problems

In this section we consider two problems in which completion analysis of triangular operators w.r.t. a fixed partition (1) appears as part of the solution and not as part of the problem. Namely, we start with a complete operator T, form for it a partial matrx $T_+ = s_+(T)$ w.r.t. some partition, and then study T in terms of minimal completions of T_+.

4.1. Reverse completion for T and T^{-1}

The following problem analyzed in [51] in the context of finite matrices is generalized here to Banach space. Assume Banach spaces $X \cong Y$, under the partition (1) (but not necessarily $X_i \cong Y_i$ for all i). Given an invertible operator $T \in B(X,Y)$ and its inverse $S = T^{-1}$, consider two completion problems:

(i) Completing $T_+ := s_+(T)$ by operators of type s_-^0;
(ii) Completing $S_- := s_-^0(S)$ by operators of type s_+.

An interesting relationship between the problems (i) and (ii) has been observed in [51]:

Lemma 23. *For $T \in GL_n(\mathbb{C})$ under the complete partition $X = Y = \mathbb{C} \oplus \cdots \oplus \mathbb{C}$, the minimal ranks for* (i) *and* (ii) *add up to n.*

In Banach space, admitting arbitrary block structure, we obtain the following generalization:

Proposition 24. *Assume that T is invertible and admits a strong LPU factorization w.r.t.* (1). *Set $S = T^{-1}$. Then $T_+ = s_+(T)$ and $S_- = s_-^0(S)$ admit modular factorizations of respective strong type LPU and UPL. Moreover, there exist minimal completions T' of T_+, S' of S_-, so that $R(T') = K(S')$ and $K(T') = R(S')$.*

Proof. First assume that T, hence also S, is in invertible strong rook form. T has all its pivots on a single generalized diagonal of the type $(i, \sigma(i))$ with $\sigma \in S_k$ a permutation. S has all its pivots on the adjoint diagonal, $(\sigma(i), i)$. Therefore pivot coordinates of T_+ are $\{(i, \sigma(i)) : \ i + \sigma(i) \le k + 1\}$. Similarly, pivots of S_- are $\{(si(i), i) : \ i + \sigma(i) > k + 1\}$. In particular, T_+, S_- are also in strong rook form. Now, from Theorem 18(i) T_+ is its own minimal completion, and so is S_-, and they satisfy the equalities in the Theorem. Indeed, in terms of pivotal spaces we have $R(T_+) = \oplus\{Z_i : \ i + \sigma(i) \le k + 1\} = K(S_-)$ and $K(T_+) = \oplus\{W_{\sigma(i)} : \ i + \sigma(i) > k + 1\} = R(S_-)$, completing the proof for this case.

More generally, if $T = LRU$ and R is in strong rook form (LPU), we have $S = U^{-1}R^{-1}L^{-1}$ (UPL). Also, by definition, T_+ admits the modular LPU factorization $T_+ = s_+(LRU)$ and S_- admits the modular UPL factorization $S_- = s_-^0(U^{-1}R^{-1}L^{-1})$. In addition, in view of part 1 of the proof, $T' := LR_+U$ and $S' := U^{-1}(R^{-1})_-L^{-1}$ form a pair of minimal completions for T_+ and S_- and satisfies the theorem. □

4.2. Block LU factorization

The LU/UL factorization of invertible matrices T has numerous applications in engineering and numerical linear algebra, and interest in extending it to Banach space goes back at least to [9]. The application we have in mind is Wiener–Hopf theory. For example, the Wiener–Hopf factorization of a symbol $W(\lambda)$ w.r.t. the circle is associated with a UL factorization of its infinite Toeplitz matrix (for similar connections see [10, 45]). LU/UL factorization, in this generality, is viewed as a special type of LPU/UPL factorization with trivial middle term, and the invertibility constraint on L and U creates a serious complication. Only a few cases have been treated so far, including the following: every diagonally dominant operator in a separable Banach space is LU factorable (w.r.t. a Borel measure; [4, 39]); certain integral operators with Lipschitz kernel are LU factorable [42].

The removal of the invertibility condition on L and U (and, *a fortiori*, on T has also been considered, removing much of the difficulties involved, but the resulting theory has fewer applications. This problem turns out to be another example of inverse completion.

In the case of $n \times n$ matrices, with the complete partition and (non-block) LU factorization, the exact necessary and sufficient condition for non-invertible LU factorability has been established by Okunev and Johnson [36]. Its extension to block-LU factorizations (Lemma 25 below) is straightforward but appears to be new, in spite of ample discussion on block LU algorithms in the numerical analysis literature (see, e.g., articles cited in [18]). The extension to Banach space operators is weaker since, once again, it requires LPU factorability (now with L, U invertible) as a necessary technical condition.

The setup is as follows. Assume Banach spaces X, V, Y under the partition

$$X = \oplus_{i=1}^k X_i, \qquad V = \oplus_{i=1}^k V_i', \qquad Y = \oplus_{i=1}^k Y_i. \tag{24}$$

Given a block matrix $T = (T_{ij})_{i,j=1}^k \in B(X, Y)$, we say that T admits an LU factorization if $T = LU$ where $L = (L_{ij})_{i,j=1}^k \in B(X, V)$ is of type t_- and $U = (U_{ij})_{i,j=1}^k \in B(V, Y)$ is of type t_+. Namely, $L_{ij} = 0$ and $U_{ji} = 0$ whenever $i < j$.

We do not impose the condition that L, U are invertible, as required earlier in the LPU case. However, if X, V, Y are isomorphic and L, U (and T!) are invertible, this means that T is LU-equivalent to an identity matrix, a special type of LPU factorization.

For the analysis of LU factorability for the given operator $T \in B(X,Y)$, we consider together with T the sequence of one corner operators T_m $(m = 1, \dots, k)$ defined by

$$(T_{m+})_{ij} = T_{ij} \quad (\min\{i,j\} \leq m), \qquad (T_{m+})_{ij} = 0 \quad (i,j > m).$$

By a completion of T_{m+} we mean completion of the missing blocks T_{ij} $(i,j > m)$, a one-corner completion.

Lemma 25. *Assume that X, V, Y are finite-dimensional and set $r_m = mr(T_m)$ (as a one-block problem) and $v_m = \dim V_m$. Then $T \in B(X,Y)$ admits a (block) non-invertible LU factorization w.r.t. (24) iff $r_m \leq v_m$ for all m.*

The criterion announced in [36] is the special case of Lemma 25 in which the partition is complete: $X = V = Y = \mathbb{C} \oplus \cdots \oplus \mathbb{C}$. Here, when T is non-singular, the factorability condition simplifies to the classical condition that the n principal minors of T are non-zero. The simplification is a simple application of Corollary 21(ii).

The original proof of Okunev and Johnson generalizes in a straightforward way to a proof of Lemma 25. In addition, the lemma is subsumed by the following Banach space result.

Theorem 26. *Assume that $T \in B(X,Y)$ admits a strong LPU factorization under (1). Given V in (24), define $V_i = V_1' \oplus \cdots \oplus V_i'$.*

- (i) *A necessary condition for T to admit a non-invertible LU factorization w.r.t. (24) is that for all m the pivotal space W_m of P_m admits an isomorphic embedding onto a subspace of V_m.*
- (ii) *A sufficient condition is the existence, for all m, of an isomorphic embedding of W_m onto a complemented subspace of V_m.*

Proof. Assume that $T = LRU$ is an LPU factorization, with R in strong rook form. It is sufficient to prove that R satisfies the theorem. Thus, from now on we assume that T itself is in rook form.

First we prove necessity, as argued in [36]. If $T = LU$ is an LU factorization, set $L_m = P_{V_m} L \in B(V_m, Y)$ (projected) and $U_m = U_{|V_m} \in B(X, V_m)$ (restricted). It is easy to see that $L_m U_m$ is a completion of P_m and factors through V_m. Since the pivotal space of P_m is W_m, this is possible only if W_m embeds as a subspace of V_m.

Next we prove sufficiency. We assume that W_m embeds as a complemented subspace of V_m, and show that T admits an LU factorization by constructing triangular matrices L_m and U_m inductively in such a way that $L_m U_m$ is a completion of P_m. Denote by W_m'' a complement of W_m in V_m. Under the direst sum $V_m = W_m \oplus W_m''$ we may write $L_m = (L_{m1} \ \ L_{m2})$ and $U_m = \begin{pmatrix} U_{m1} \\ U_{m2} \end{pmatrix}$.

We shall inductively require that $L_{m2} = 0$ and $U_{m2} = 0$. Also define operators $T'_m \in B(X, Y)$ via

$$(T'_{m+})_{ij} = T_{ij} \quad (\min\{i, j\} = m), \qquad (T_{m+})_{ij} = 0 \quad (otherwise)$$

(under the convention $T'_0 = 0$ and $T'_k = T$). T'_m inherit the pivots from T, hence are also in strong rook form. Denoting by W'_m their pivotal spaces, we have $W_m = W'_1 \oplus \cdots \oplus W'_m = W_{m-1} \oplus W'_m$. In the mth induction step, we perform a factorization $T'_m = L'_m U'_m$ through the pivotal space W'_m, where $L'_m \in B(W'_m, Y)$ (resp. $U'_m \in B(X, W'_m)$) is supported away from the first $i - 1$ block rows (resp. columns). If $m \geq 2$, L_{m-1} and U_{m-1} have been constructed in the $m - 1$ st step. In the direct sum $V_m = W_{m-1} \oplus W'_m \oplus W''_m$ we define $L_m = \begin{pmatrix} L_{m-1,1} & L'_m & 0 \end{pmatrix}$ and $U_m = \begin{pmatrix} U_{m-1,1} \\ U'_m \\ 0 \end{pmatrix}$. By induction it follows that $L_m U_m = \sum_{i=1}^{m} L'_i U'_i$ for all m. When the inductive construction is completed, we define $L = L_k$ and $U = U_k$. We have $T = \sum_{m=1}^{k} T'_m = \sum_{m=1}^{k} L'_m U'_m = LU$, obtaining the desired factorization.

The construction guarantees that L, U are triangular. For example, in the mth column of U the non-zero entries are restricted to the space W_m which embeds as a subspace of V_m. □

There is interest in further generalizations of these results. These include (block) UL (rather than LU) factorization and extensions to infinite partitions, associated with factorization of infinite (block) matrices.

(i) The case of (block) UL-factorizations with k finite is deduced from Theorem 26 by permutation of coordinates.

(ii) In the case of LU factorization of an infinite matrix, T_{ij} $(i, j \geq 1)$, Theorem 26 only provides a *necessary* condition: namely, that W_m embeds in V_m for all $m \in \mathbb{N}$. Even if these embeddings are all complemented, this condition is not guaranteed to be sufficient, since boundedness of L and U must be demonstrated; in some versions, also their invertibility. See, e.g., [4],[5]. In addition, verification of the LPU condition in a finite number of (block) steps may be impossible.

(iii) The case of UL-factorizations of the same one-sided type of infinite matrices is more complicated since the spaces V_m and W_m are typically of infinite dimension for all m.

References

[1] R. Abraham, J.E. Marsden, T. Ratiu, *Manifolds, Tensor Analysis and Applications.* 3rd. Edition, Springer, 2001.

[2] A. Alonso and B. Simon, The Birman–Kreĭn–Vishik theory of self-adjoint extensions of semibounded operators. *J. Operator Theory* **4** (1980), 251–270.

[3] A.D. Andrew, WM Patterson. Range inclusion and factorization of operators on classical Banach spaces. *J. Math. Anal. Appl.* **156** (1991), 40–43.

[4] K.T. Andrews, J.D. Ward, Factorization of diagonally dominant operators on $L_1([0,1],X)$. *Trans. Amer. Math. Soc.* **291** (1985), no. 2, 789–800.

[5] K.T. Andrews, J.D. Ward, LU- factorization of order bounded operators on Banach sequence spaces. *J. Approximation Theory* **48** (1986), 169–180.

[6] G. Arsene and A. Gheondea, Completing matrix contractions. *J. Operator Theory* **7** (1982), 179–189.

[7] J.A. Ball, I. Gohberg, L. Rodman, Interpolation of rational matrix functions, *Operator theory: advances and applications*, Vol. 45, Birkhäuser Verlag, Basel, 1990.

[8] J.A. Ball, J. Kang, Matrix polynomial solutions of tangential Lagrange–Sylvester interpolation conditions of low McMillan degree. *Linear Algebra Appl.* **137/138** (1990), 699–746.

[9] M.A. Barkar and I.C. Gohberg, On factorization of operators in Banach spaces, *AMS Translations* **90** (1970), 103–133.

[10] A. Böttcher, M. Halwass, Wiener–Hopf and spectral factorization of real polynomials by Newton's method. *Linear Algebra Appl.* **438** (2013), no. 12, 4760–4805.

[11] R. Bouldin, A counterexample in the factorization of Banach space operators. *Proc. Amer. Math. Soc.* **68** (1978), no. 3, 327.

[12] E.J. Candès, B. Recht. Exact matrix completion via convex optimization. *Found. of Comput. Math*, 9 (2008), 717–772.

[13] E.J. Candès, Y. Plan, Matrix completion with noise. *Proc. IEEE.* **98** (2010), no. 6, 925–936.

[14] D. Carlson. Generalized inverse invariance, partial orders, and rank-minimization problems for matrices. *Current trends in matrix theory* (*Auburn, Ala.*, 1986), North-Holland, New York, 1987, 81–87.

[15] N. Cohen, C.R. Johnson, L. Rodman, H.J. Woerdeman, Ranks of completions of partial matrices. *The Gohberg anniversary collection, Vol. I* (Calgary, AB, 1988), pp. 165–185, Operator Theory: Advances and Applications **40**, Birkhäuser, Basel, 1989.

[16] N. Cohen, E. Pereira. The cyclic completion rank problem with regular blocks, *Linear Multilinear Algebra*, **66** (2018), no. 5, 861–868.

[17] J.B. Conway, *The Theory of Subnormal Operators*, Mathematical Surveys and Monographs, **36**. American Mathematical Society, Providence, RI, 1991.

[18] J.W. Demmel, N.J. Higham, R.S. Schreiber. Stability of block LU factorization. *Numerical Linear Algebra with applications* **2** (1995), no. 2, 173–190.

[19] R.G. Douglas, On majorization, factorization and range inclusion of operators on Hilbert space, *Proc. Amer. Math. Soc.* **17** (1966), 413–415.

[20] H. Dym, *J contractive matrix functions, reproducing kernel Hilbert spaces and interpolation.* AMS (Regional conference series in Math, 71) 1989.

[21] R.L. Ellis, D.C. Lay, Rank preserving extensions of band matrices, *Linear Multilinear Algebra* **26** (1990), 147–179.

[22] M.R. Embry, Factorization of operators on Banach Space. *Proc. Amer. Math. Soc.* **38** (1973), 587–590.

[23] S. Feldman, G. Heinig, Parametrization of minimal rank block Hankel matrix extensions and minimal partial realization. *Integral Equations Operator Theory*, **33** (1999), 153–171.

[24] M. Forough, Majorization, range inclusion, and factorization for unbounded operators on Banach spaces. *Linear Algebra Appl.* **449** (2014), 60–67.

[25] B. Grossmann, H.J. Woerdeman. Fractional minimal rank. *Linear Multilinear Algebra*, in print.

[26] N.J. A. Harvey, D.R. Karger, S. Yekhanin. The complexity of matrix completion. *Proc SODA (17th annual ACM-SIAM symposium on Discrete algorithm)*, 2006, 1103–1111.

[27] G. Heinig, P. Jankowski, Kernel structure of block Hankel and Toeplitz matrices and partial realization. *Linear Algebra Appl.* **175** (1992), 1–30.

[28] I.S. Iohvidov, *Hankel and Toeplitz matrices and forms*, Birkhäuser, Boston, Mass., 1982.

[29] E. Jonckheere, Chingwo Ma, A simple Hankel interpretation of the Berlekamp–Massey algorithm, *Linear Algebra Appl.* **125** (1989), 65–76.

[30] M.A. Kaashoek, H.J. Woerdeman, Unique minimal rank extensions of triangular operators. *J. Math. Anal. Appl.* **131** (1988), 501–516.

[31] M.S. Moslehian, A survey of the complemented subspace problem. *Trends in Mathematics*, Information Center for Mathematical Sciences, 9.1 (2006), 91–98.

[32] Shiqian Ma, D. Goldfarb, Lifeng Chen. Fixed point and bregman iterative methods for matrix rank minimization. *J. Math. Programming* **128** (2011), no. 1-2, 321–353.

[33] F.J. Murray, Quasi-complements and closed projections in reflexive Banach spaces. *Trans. Amer. Math. Soc.* **58** (1945), 77–95.

[34] F. Nan, *Low rank matrix completion.* Master's thesis, MIT (Computation for design and optimization program), 2009. URI: http://hdl.handle.net/1721.1/55077

[35] O.H. Odeh, D.D. Olesky, P. van den Driessche, Bruhat Decomposition and Numerical Stability. *SIAM J. Matrix Analysis Appl.* **19** (1998), no. 1, 89–98.

[36] P. Okunev, C.R. Johnson, Necessary and sufficient conditions for existence of the LU factorization of an arbitrary matrix, arXiv:math.NA/0506382. William and Mary NSF-REU report, 1997.

[37] M. Le Pendu, X. Jiang, C. Guillemot, Light Field inpainting via Low Rank Matrix completion, *IEEE Trans. Image Processing*, to appear, 2018.

[38] B. Recht, M. Fazel, and P.A. Parrilo. Guaranteed minimum rank solutions of linear matrix equations via nuclear norm minimization. *SIAM Rev.* **52** (2010), no. 3, 471–501.

[39] P.W. Smith, I.D. Ward, Factorization of diagonally dominant operators on ℓ_1. *Illinois J. Math.* **29** (1985), no. 3, 370–381.

[40] Ju.L. Smul'jan, Operator balls. *Integral Equations Operator Theory* **13** (1990), no. 6, 864–882.

[41] Y. Tian, S. Cheng, The maximal and minimal ranks of $A - BXC$ with applications. *New York J. Math.* **9** (2003), 345–362.

[42] A. Townsend, L.N. Trefethen. Continuous analogues of matrix factorizations. *Proc. A.* **471** (2015), no. 2173, 20140585, 21 pp.

[43] E.E. Tyrtyshnikov, *A brief introduction to numerical analysis*, Springer/Birkhäuser 1997.

[44] R.V. Vershinin, Imbedding of the image of operators and reflexivity of Banach spaces. *Ukr. Math. J.* **51** (1999), no. 2, 293–296.

[45] V. Vigon, LU-factorization versus Wiener–Hopf Factorization for Markov chains. *Acta Appl. Math.* **128** (2013), no. 1, 1–37.

[46] H.J. Woerdeman, The lower order of lower triangular operators and minimal rank extensions, *Integral Equations Operator Theory* **10** (1987), 849–879.

[47] H.J. Woerdeman, *Matrix and operator extensions*, CWI Tracts., 68, CWI, Amsterdam, 1989.

[48] H.J. Woerdeman, Minimal rank completions for block matrices. *Linear Algebra Appl.* **121** (1989), 105–122.

[49] H.J. Woerdeman, Minimal rank completions of partial banded matrices. *Linear Multilinear Algebra* **36** (1993), 59–68.

[50] H.J. Woerdeman, Toeplitz minimal rank completions, *Linear Algebra Appl.* **202** (1994), 267–278.

[51] H.J. Woerdeman, A matrix and its inverse: revisiting minimal rank completions. *Operator Theory: Advances and Applications*, 179, Birkhäuser, Basel, 2008, 329–338.

Nir Cohen
Department of Mathematics – UFRN
Universidade Federal do Rio Grande do Norte
Natal, Brazil
e-mail: nir53cohen@gmail.com
nir@ccet.ufrn.br

Operator Theory:
Advances and Applications, Vol. 272, 99–116

Multipliers of Drury–Arveson Space: A Survey

Quanlei Fang

Dedicated to Prof. Joseph Ball on the occasion of his 70th Birthday

Abstract. The Drury–Arveson Space, as a Hilbert function space, plays an important role in multivariable operator theory. We give a brief survey of some aspects of the multipliers on the space.

Mathematics Subject Classification (2010). Primary 47B32,47A48, 47A57 secondary 32A35, 47A10.

Keywords. Drury–Arveson space, multipliers, Schur class multipliers, reproducing kernel Hilbert space.

1. Introduction

Let $\mathbb{B}^d$ denote the open unit ball $\{z : |z| < 1\}$ in $\mathbb{C}^d$. The Drury–Arveson space H_d^2 ([9, 39]) is the reproducing kernel Hilbert space associated with the kernel

$$K_w(z) = \frac{1}{1 - \langle z, w\rangle}, \quad z, w \in \mathbb{B}^d, \quad \langle z, w\rangle = z_1 w_1 + \cdots + z_d w_d$$

which is a natural multivariable analogue of the Szegő kernel of the classical Hardy space H^2 of the unit disk. Note that H_d^2 coincides with H^2 when $d = 1$.

An orthonormal basis of H_d^2 is given by $\{e_\alpha\}$ where

$$e_\alpha = \sqrt{\frac{|\alpha|!}{\alpha!}} z^\alpha.$$

Here and in what follows, we use standard multivariable notations: for multi-integers $\alpha = (\alpha_1, \dots, \alpha_d) \in \mathbb{Z}_+^d$ and points $z = (z_1, \dots, z_d) \in \mathbb{C}^d$ we set

$$|\alpha| = \alpha_1 + \alpha_2 + \cdots + \alpha_d, \qquad \alpha! = \alpha_1!\alpha_2! \cdots \alpha_d!, \qquad z^\alpha = z_1^{\alpha_1} z_2^{\alpha_2} \cdots z_d^{\alpha_d}.$$

For functions $f, g \in H_d^2$ with Taylor expansions

$$f(z) = \sum_{\alpha \in \mathbb{Z}_+^d} c_\alpha z^\alpha \quad \text{and} \quad g(z) = \sum_{\alpha \in \mathbb{Z}_+^d} d_\alpha z^\alpha,$$

their inner product is given by

$$\langle f, g \rangle = \sum_{\alpha \in \mathbf{Z}_+^d} \frac{\alpha!}{|\alpha|!} c_\alpha \overline{d_\alpha}.$$

The Drury–Arveson space H_d^2 can be viewed in many different ways: It can be identified as the symmetric Fock space over $\mathbb{C}^d$; it is a member of the family of Besov–Hardy–Sobolev spaces; it is a prototype for a complete Pick space; it is a free Hilbert module over the polynomial ring $\mathbb{C}[z_1, \ldots, z_d]$ with the identification of each variable z_i with each multiplication operator M_{z_i}. A panoramic view of most operator theoretic and function theoretic aspects of this space can be found in [67].

A holomorphic function f on $\mathbb{B}^d$ is said to be a *multiplier* of the Drury–Arveson space H_d^2 if $fg \in H_d^2$ for every $g \in H_d^2$. Every multiplier is in H_d^2 since $1 \in H_d^2$. Throughout the paper, we denote the collection of multipliers of H_d^2 by $\mathcal{M}_d$. For each $f \in \mathcal{M}_d$, the multiplication operator M_f defined by $M_f g = fg$ is necessarily bounded on H_d^2 [9], and the operator norm $\|M_f\|$ is also called the multiplier norm of f. We write $\|f\|_{\mathcal{M}_d}$ for its multiplier norm:

$$\|f\|_{\mathcal{M}_d} = \sup\{\|fg\| : g \in H_d^2, \|g\| \leq 1\}.$$

This norm gives $\mathcal{M}_d$ the structure of an operator algebra.

Multipliers are an important part of operator theory on H_d^2. For example, if $\mathcal{E}$ is a closed linear subspace of H_d^2 which is invariant under $M_{z_1}, \ldots, M_{z_d}$, then there exist $\{f_1, \ldots, f_k, \ldots\} \subset \mathcal{M}_d$ such that the operator

$$M_{f_1} M_{f_1}^* + \cdots + M_{f_k} M_{f_k}^* + \cdots$$

is the orthogonal projection from H_d^2 onto $\mathcal{E}$ (see p. 191 in [10]). It is known that $\mathcal{M}_d$ is the home for the multivariate von Neumann inequality hence plays a similar role of H^∞, the algebra of bounded holomorphic functions on the unit disk to higher dimensions. The multiplier algebra $\mathcal{M}_d$ is exactly the image of the free-semigroup algebra $\mathcal{F}_d$ (generated by d letters) after applying a point-evaluation map associated with points in the unit ball. Interpolation problems for Schur multipliers (contractive multipliers) H_d^2 related to multidimentional system theory have been intensively studied over the past few decades (see [12, 24, 41, 50, 54]). A corona theorem for the Drury–Averson space multipliers was proved by Costea, Sawyer and Wick [33]. Clouâtre and Davidson studied Henkin measures for $\mathcal{M}_d$ in [30] and the ideals of the closure of the polynomial multipliers on the Drury–Arveson space in [32].

In this paper we survey some results and methods related to multipliers of H_d^2. Results are presented below without proof but with references. This survey is not intended to be comprehensive in any way and we will have to limit the article to the aspects that are most familiar to us. The present paper is organized as follows. Following this introduction, in Section 2 we review an important class of multipliers – Schur class multipliers or the unit ball of $\mathcal{M}_d$. We revisit transfer function realization and related Nevanlinna–Pick type interpolation problems for these multipliers. In Section 3 we discuss polynomial multipliers and nonpolynomial multipliers. A corona theorem and some spectral properties for multipliers of H^2_d are included in Section 4. In Section 5 we consider commutators involving multipliers and localizations. In the last section we discuss the problem of characterization of multipliers in $\mathcal{M}_d$.

2. Schur class multipliers

We will start with a special class of multipliers, the so-called Schur class multipliers. In single variable complex analysis the Schur class is the set of holomorphic functions $S(z)$ that are bounded by one on the unit disk. Schur functions are important, not only because they arise in diverse areas of classical analysis and operator theory, but also because they have connections with linear system theory and engineering.

Recall that the Schur class plays a prominent role in classical moment and interpolation problems. One of the best known examples is the **Nevanlinna–Pick interpolation problem**:

Given points $z_1, \dots, z_n$ in the unit disk $\mathbb{D}$ and complex numbers $w_1, \dots, w_d$. Find a Schur function $S(z)$ such that $S(z_i) = w_i$ for $i = 1, \dots, n$.

A solution to this problem exists if and only if the associated Pick matrix

$$\left\{ \frac{1 - w_i \bar{w}_j}{1 - z_i \bar{z}_j} \right\}_{i,j=1}^{n}$$

is positive semidefinite. There are several approaches dealing with the problem. One operator-theoretic approach can be described as follows:

Let M_S be the operator of multiplication by a Schur function $S(z)$ on the Hardy space H^2 of the unit disk. Then M_S commutes with M_z, the operator of multiplication by the coordinate z (the shift operator) and any contraction on H^2 which commutes with M_z has this form for some Schur function. The commutation relation is preserved under compressions of the operators to any invariant subspace E of the backward shift M_z^*. The generalized interpolation theorem of Sarason [66] showed that every contraction on E which commutes with the compression of M_z to E is associated with some Schur function. Particular choice of the invariant subspace leads to solutions to the Nevanlinna–Pick problem. The celebrated commutant lifting theorem by Sz.-Nagy–Foiaş [47] extends the conclusion to arbitrary Hilbert space contraction operators. In the book [20], Ball,

Gohberg, and Rodman considered the interpolation of rational matrix functions. They emphasized the state space approach and transfer function realization.

When leaving the univariate setting there are different interesting multivariable counterparts of the classical Schur class [1]. Here we review some aspects of the class of contractive operator-valued multipliers for the Drury–Arveson space.

For a Hilbert space $\mathcal{Y}$, we use notation $H_{\mathcal{Y}}(k_d)$ for the Drury–Arveson space of $\mathcal{Y}$-valued functions. Given two Hilbert spaces $\mathcal{U}$ and $\mathcal{Y}$, we denote by $\mathcal{S}_d(\mathcal{U},\mathcal{Y})$ the class of $\mathcal{L}(\mathcal{U},\mathcal{Y})$-valued functions S on $\mathbb{B}^d$ such that the multiplication operator $M_S : f \mapsto S \cdot f$ defines a contraction from $H_{\mathcal{U}}(k_d)$ into $H_{\mathcal{Y}}(k_d)$, or equivalently, such that the de Branges–Rovnyak kernel

$$K_S(\lambda, z) = \frac{I_{\mathcal{Y}} - S(\lambda)S(z)^*}{1 - \langle \lambda, z\rangle} \tag{2.1}$$

is positive on $\mathbb{B}^d \times \mathbb{B}^d$.

It is readily seen that the class $\mathcal{S}_1(\mathcal{U},\mathcal{Y})$ is the classical Schur class. In general, it follows from $K_S \geq 0$ that S is holomorphic and takes contractive values on $\mathbb{B}^d$. However, for $d > 1$ there are holomorphic contractive-valued functions on $\mathbb{B}^d$ not in $\mathcal{S}_d$. The class $\mathcal{S}_d(\mathcal{U},\mathcal{Y})$ can be characterized in various ways similarly to the one-variable situation. Here we review the characterizations of these multipliers in terms of realizations due to Ball, Trent and Vinnikov in [24]. We review this result in the form we used in [17].

Theorem 2.1. *Let S be an $\mathcal{L}(\mathcal{U},\mathcal{Y})$-valued function defined on $\mathbb{B}^d$. The following are equivalent:*

1. *S belongs to $\mathcal{S}_d(\mathcal{U},\mathcal{Y})$.*
2. *The kernel*
$$K_S(\lambda, z) = \frac{I_{\mathcal{Y}} - S(\lambda)S(z)^*}{1 - \langle \lambda, z\rangle} \tag{2.2}$$
is positive on $\mathbb{B}^d \times \mathbb{B}^d$, i.e., there exists an operator-valued function $H\colon \mathbb{B}^d \to \mathcal{L}(\mathcal{H},\mathcal{Y})$ for some auxiliary Hilbert space $\mathcal{H}$ so that
$$K_S(\lambda, z) = H(\lambda)H(z)^*. \tag{2.3}$$
3. *There exists a Hilbert space $\mathcal{X}$ and a unitary connecting operator (or colligation) $\mathbf{U}$ of the form*
$$\mathbf{U} = \begin{bmatrix} A & B \\ C & D \end{bmatrix} = \begin{bmatrix} A_1 & B_1 \\ \vdots & \vdots \\ A_d & B_d \\ C & D \end{bmatrix} : \begin{bmatrix} \mathcal{X} \\ \mathcal{U} \end{bmatrix} \to \begin{bmatrix} \mathcal{X}^d \\ \mathcal{Y} \end{bmatrix} \tag{2.4}$$
so that $S(\lambda)$ can be realized in the form
$$\begin{aligned} S(\lambda) &= D + C\left(I_{\mathcal{X}} - \lambda_1 A_1 - \cdots - \lambda_d A_d\right)^{-1}(\lambda_1 B_1 + \cdots + \lambda_d B_d) \\ &= D + C(I - Z(\lambda)A)^{-1} Z(\lambda) B, \end{aligned} \tag{2.5}$$

where we set

$$Z(\lambda) = \begin{bmatrix} \lambda_1 I_{\mathcal{X}} & \dots & \lambda_d I_{\mathcal{X}} \end{bmatrix}, \quad A = \begin{bmatrix} A_1 \\ \vdots \\ A_d \end{bmatrix}, \quad B = \begin{bmatrix} B_1 \\ \vdots \\ B_d \end{bmatrix}. \tag{2.6}$$

4. *There exists a Hilbert space* $\mathcal{X}$ *and a contractive connecting operator* $\mathbf{U}$ *of the form* (2.4) *so that* $S(\lambda)$ *can be realized in the form* (2.5).

In analogy with the univariate case, a realization of the form (2.5) is called *coisometric, isometric, unitary* or *contractive* if the operator $\mathbf{U}$ has the said property. It turns out that a more useful analogue of "coisometric realization" in the classical univariate case is not the condition that $\mathbf{U}^*$ be isometric, but rather that $\mathbf{U}^*$ be isometric on a certain subspace of $\mathcal{X}^d \oplus \mathcal{Y}$.

Definition 2.2. A realization (2.5) of $S \in \mathcal{S}_d(\mathcal{U}, \mathcal{Y})$ is called *weakly coisometric* if the adjoint $\mathbf{U}^* : \mathcal{X}^d \oplus \mathcal{Y} \to \mathcal{X} \oplus \mathcal{U}$ of the connecting operator is contractive and isometric on the subspace $\begin{bmatrix} \mathcal{D} \\ \mathcal{Y} \end{bmatrix} \subset \begin{bmatrix} \mathcal{X}^d \\ \mathcal{Y} \end{bmatrix}$ where

$$\mathcal{D} := \overline{\operatorname{span}}\{Z(z)^*(I_{\mathcal{X}} - A^* Z(z)^*)^{-1} C^* y : \ z \in \mathbb{B}^d, \, y \in \mathcal{Y}\} \subset \mathcal{X}^d. \tag{2.7}$$

For any $S \in \mathcal{S}_d(\mathcal{U}, \mathcal{Y})$, the associated kernel K_S (2.1) is positive on $\mathbb{B}^d \times \mathbb{B}^d$ so we can associate with S the de Branges–Rovnyak reproducing kernel Hilbert space $\mathcal{H}(K_S)$. In parallel to the univariate case, $\mathcal{H}(K_S)$ is the state space of certain *canonical functional-model realization* for S (see [13]).

Definition 2.3. We say that the contractive operator-block matrix

$$\mathbf{U} = \begin{bmatrix} A & B \\ C & D \end{bmatrix} : \begin{bmatrix} \mathcal{H}(K_S) \\ \mathcal{U} \end{bmatrix} \to \begin{bmatrix} \mathcal{H}(K_S)^d \\ \mathcal{Y} \end{bmatrix}$$

is a *canonical functional-model colligation* for the given function $S \in \mathcal{S}_d(\mathcal{U}, \mathcal{Y})$ if

1. The operator $A = \begin{bmatrix} A_1 \\ \vdots \\ A_d \end{bmatrix}$ solves the Gleason problem for $\mathcal{H}(K_S)$, i.e.,

$$f(z) - f(0) = \sum_{j=1}^{d} z_j (A_j f)(z) \quad \text{for all} \quad f \in \mathcal{H}(K_S).$$

2. The operator $B = \begin{bmatrix} B_1 \\ \vdots \\ B_d \end{bmatrix}$ solves the Gleason problem for S:

$$S(z)u - S(0)u = \sum_{j=1}^{d} z_j (B_j u)(z) \quad \text{for all} \quad u \in \mathcal{U}.$$

3. The operators $C : \mathcal{H}(K_S) \to \mathcal{Y}$ and $D : \mathcal{U} \to \mathcal{Y}$ are given by

$$C \colon f \mapsto f(0), \quad D \colon u \mapsto S(0)u.$$

It was shown in [18] that any Schur-class function S with associated de Branges–Rovnyak space $H(K_S)$ finite-dimensional and not $(M_{z_1}^*, \dots, M_{z_d}^*)$-invariant does not admit a contractive commutative realization. Here a realization is said to be *commutative* if the state space operators $A_1, \dots, A_d$ commute with each other. The following result in [18] shows when a Schur-class function S admits a commutative weakly coisometric realization.

Theorem 2.4. *A Schur-class function $S \in \mathcal{S}_d(\mathcal{U}, \mathcal{Y})$ admits a commutative weakly coisometric realization if and only if the following conditions hold:*

1. *The associated de Branges–Rovnyak space $H(K_S)$ is $(M_{z_1}^*, \dots, M_{z_d}^*)$-invariant, and*
2. *the inequality*

$$\sum_{j=1}^{d} \|M_{z_j}^* f\|_{H(K_S)}^2 \leq \|f\|_{H(K_S)}^2 - \|f(0)\|_{\mathcal{Y}}^2 \tag{2.8}$$

holds for all $f \in H(K_S)$.

Furthermore, if conditions (1) *and* (2) *are satisfied, then there exists a commutative canonical functional model colligation for S. Moreover, the state-space operators tuple is equal to the Drury–Arveson backward shift restricted to $H(K_S)$: $A_j = M_{z_j}^*|_{H(K_S)}$ for $j = 1, \dots, d$.*

Note that condition (2) in Theorem 2.4 means that M_z^* is a contractive solution to the Gleason problem for $H(K_S)$ ([48]). Weakly coisometric realizations for an $S \in \mathcal{S}_d(\mathcal{U}, \mathcal{Y})$ can be constructed in certain canonical way as follows. Upon applying Aronszajn's construction to the kernel K_S, (which is positive on $\mathbb{B}^d$ by Theorem 2.1), one gets the de Branges–Rovnyak space $H(K_S)$. A weakly coisometric realization for S with the state space equal to $H(K_S)$ (and output operator C equal to evaluation at zero on $H(K_S)$) will be called a *generalized functional-model realization.*

As shown in [17], any function $S \in \mathcal{S}_d(\mathcal{U}, \mathcal{Y})$ admits a generalized functional-model realization. In the univariate case, this reverts to the well-known de Branges–Rovnyak functional-model realization [26, 27]. Another parallel to the univariate case is that *any* observable (i.e., the observability operator $\mathcal{O}_{C,\mathbf{A}}$ is injective: $C(I_{\mathcal{X}} - Z(\lambda)A)^{-1}x = 0$ implies $x = 0$.) weakly coisometric realization of a Schur-class function $S \in \mathcal{S}_d(\mathcal{U}, \mathcal{Y})$ is unitarily equivalent to some generalized functional-model realization (observability is a minimality condition that is fulfilled automatically for every generalized functional-model realization). However, in contrast to the univariate case, this realization is not unique in general (even up to unitary equivalence); moreover, a function $S \in \mathcal{S}_d(\mathcal{U}, \mathcal{Y})$ may admit generalized functional-model realizations with the same state space operators $A_1, \dots, A_d$ and different input operators B_j's.

Recently Jury and Martin studied the case when the realization is unique in [60–62]. They introduced the notion of a *quasi-extreme* multiplier of the Drury–Arveson space H_d^2 for a multiplier associated with a unique generalized functional-model realization. They gave some characterizations of these multipliers. Here are some characterizations of quasi-extremity, which imply that every quasi-extreme multiplier of H_d^2 is in fact an extreme point of the unit ball of the multiplier algebra $\mathcal{M}_d$. (The converse statement, namely whether or not every extreme point is quasi-extreme, remains an open question.)

Theorem 2.5. *Let S be a contractive multiplier of H_d^2 (to ease the notation assume $\mathcal{U} = \mathcal{Y} = \mathbb{C}$). The following are equivalent:*

(1) *S is quasi-extreme.*
(2) *the only multiplier T satisfying*

$$M_T^* M_T + M_S^* M_S \leq I \tag{2.9}$$

is $T \equiv 0$.
(3) *There is a unique contractive solution $(X_1, \dots X_d)$ to the Gleason problem in $H(K_S)$.*
(4) *There exists a contractive solution $(X_1, \dots X_d)$ such that the equality*

$$\sum_{j=1}^{d} \|X_j f\|_f^2 = \|f\|_{H(K_S)}^2 - |f(0)|^2$$

holds for every $f \in H(K_S)$.
(5) *$H(K_S)$ does not contain the constant functions.*

Let $A = (A_1, \dots, A_d)$ be a commutative d-tuple of bounded, linear operators on the Hilbert space $\mathcal{X}$. If $C \in \mathcal{L}(\mathcal{X}, \mathcal{Y})$, then the pair (C, A) is said to be *output-stable* if the associated observability operator

$$\widehat{\mathcal{O}}_{C,A} \colon x \mapsto C(I - \lambda_1 A_1 - \cdots - \lambda_d A_d)^{-1} x$$

maps $\mathcal{X}$ into $H_{\mathcal{Y}}(k_d)$, or equivalently (by the closed graph theorem), the observability operator is bounded. Just as in the single-variable case, there is a system-theoretic interpretation (in the context of multidimensional systems) for this operator (see [15] for details). The following is a theorem about the left-tangential operator-argument interpolation (**LTOA**) problem formulated for the Drury–Arveson Schur-multiplier class $\mathcal{S}_d(\mathcal{U}, \mathcal{Y})$.

Theorem 2.6. *Suppose that we are given an auxiliary Hilbert space $\mathcal{X}$ together with commutative d-tuples*

$$Z^{(1)} = (Z_1^{(1)}, \dots, Z_d^{(1)}), \dots, Z^{(N)} = (Z_1^{(N)}, \dots, Z_d^{(N)}) \in \mathcal{L}(\mathcal{X})^d,$$

i.e., $Z_k^{(i)} \in \mathcal{L}(\mathcal{X})$ for $i = 1, \dots, N$ and $k = 1, \dots, d$ and for each fixed i, the operators $Z_1^{(i)}, \dots, Z_d^{(i)}$ commute pairwise, with the property that each d-tuple $Z^{(i)}$ has joint spectrum contained in $\mathbb{B}^d$ (or each $(X_i^*, Z^{*(i)})$ is an output stable pair). Assume in addition that we are given operators $X_1, \dots, X_N$ in $\mathcal{L}(\mathcal{Y}, \mathcal{X})$ and operators

$Y_1, \ldots, Y_N$ *in* $\mathcal{L}(\mathcal{U}, \mathcal{X})$. *Then there is an* $S \in \mathcal{S}_d(\mathcal{U}, \mathcal{Y})$ *so that*

$$(X_i S)^{\wedge L}(Z^{(i)}) := \sum_{n \in \mathbb{Z}_+^d} (Z^{(i)})^n X_i \mathcal{S}_n = Y_i \quad \text{for} \quad i = 1, \ldots, N$$

if and only if the associated Pick matrix

$$\mathbb{P}_{LTOA} := \left[\sum_{n \in \mathbb{Z}_+^d} (Z^{(i)})^n (X_i X_j^* - Y_i Y_j^*)(Z^{(j)})^{n*} \right]_{i,j=1}^N$$

is positive semidefinite. Here $Z^n = Z_1^{n_1} \cdots Z_d^{n_d}$ *if* $Z = (Z_1, \ldots, Z_d) \in \mathcal{L}(\mathcal{C})^d$ *and* $n = (n_1, \ldots, n_d) \in \mathbb{Z}_+^d$.

Similarly one can pose right tangential operator-argument interpolation and bitangential operator-argument Nevanlinna–Pick problems. We note that this and related interpolation problems were studied in [12] by using techniques from reproducing kernel Hilbert spaces, Schur-complements and isometric extensions from the work of [40, 51, 52]. In [19, 42] we showed how the problem can be handled via the Grassmannian approach. We also refer to [22] for a comprehensive survey on the related topics and [14] for discussions of different approaches for bitangential matrix Nevanlinna–Pick interpolation problems.

3. Polynomial vs. non-polynomial multipliers

It is easy to see that all polynomials are multipliers of the Drury–Arveson space: $\mathbb{C}[z_1, \ldots, z_d] \subseteq \mathcal{M}_d$. Naturally we would like to see how differently the non-polynomials multipliers behave compared to polynomial multipliers.

Recall that a commuting tuple of bounded operators $(A_1, \ldots, A_d)$ on a Hilbert space H is said to be a *row contraction* if it satisfies the inequality

$$A_1 A_1^* + \cdots + A_d A_d^* \leq 1.$$

The d-shift $(M_{z_1}, \ldots, M_{z_d})$ on H_d^2 is a natural example of row contraction. In fact, the d-shift is the "master" row contraction in the sense that for each polynomial $p \in \mathbb{C}[z_1, \ldots, z_d]$, the von Neumann inequality

$$\|p(A_1, \ldots, A_d)\| \leq \|p(M_{z_1}, \ldots, M_{z_d})\|$$

holds whenever the commuting tuple $(A_1, \ldots, A_d)$ is a row contraction [9, 39].

In the single variable case, it is well known that the space of multipliers of the classic Hardy space H^2 is the space of bounded holomorphic functions on the unit disk, i.e., $\mathcal{M}_1 = H^\infty(\mathbb{D})$. The multiplier norm of a multiplier f, $\|M_f\|$ is equal to $\|f\|_\infty = \sup_{|z|<1} |f(z)|$. However, for $d \geq 2$, Arveson showed in [9] that $\mathcal{M}_d$ is strictly smaller than $H^\infty(\mathbb{B}_d)$, the space of bounded holomorphic functions on $\mathbb{B}_d$. Moreover, even for polynomials q, $\|q\|_\infty$ in general does not dominate the operator norm of M_q on H_d^2, see [9, 67].

Theorem 3.1. *For $d > 1$ the norms $\|\cdot\|_\infty$ and $\|\cdot\|_{\mathcal{M}_d}$ are not comparable on $\mathcal{M}_d$. There is a strict containment $\mathcal{M}_d \subset H^\infty(\mathbb{B}_d)$, and the d-shift M_z is not subnormal, that is, M_z does not have a joint normal extension.*

Note that it can be shown that if q is a *polynomial*, then

$$\|M_q\|_{\mathcal{Q}} = \|q\|_\infty, \tag{3.1}$$

where $\|M_q\|_{\mathcal{Q}}$ is the essential norm of q. Recall that the *essential norm* of a bounded operator A on a Hilbert space $\mathcal{H}$ is

$$\|A\|_{\mathcal{Q}} = \inf\{\|A + K\| : K \in \mathcal{K}(\mathcal{H})\},$$

where $\mathcal{K}(\mathcal{H})$ is the collection of compact operators on $\mathcal{H}$. Alternately, $\|A\|_{\mathcal{Q}} = \|\pi(A)\|$, where π denotes the quotient homomorphism from $\mathcal{B}(\mathcal{H})$ to the Calkin algebra $\mathcal{Q} = \mathcal{B}(\mathcal{H})/\mathcal{K}(\mathcal{H})$. Indeed by Proposition 5.3 in [9], for each polynomial q, the operator M_q is *essentially normal*, i.e., $[M_q^*, M_q]$ is compact. On the other hand, by Proposition 2.12 in [9], if q is a polynomial, then the spectral radius of M_q equals $\|q\|_\infty$. Since the norm and the spectral radius of any normal element in any C^*-algebra coincide, it follows that $\|M_q\|_{\mathcal{Q}} \leq \|q\|_\infty$ whenever q is a polynomial. The reverse inequality, $\|M_q\|_{\mathcal{Q}} \geq \|q\|_\infty$, can be achieved simply by applying M_q^* to the normalized reproducing kernel of H_d^2.

It turns out that (3.1) in general fails if we consider multipliers which are not polynomials ([44]):

Theorem 3.2. *There exists a sequence $\{\psi_k\} \subset \mathcal{M}_d$ such that*

$$\inf_{k\geq 1} \|M_{\psi_k}\|_{\mathcal{Q}} > 0 \qquad \textit{and} \qquad \lim_{k\to\infty} \|\psi_k\|_\infty = 0.$$

This has implications for other essential properties of multipliers. Recall that an operator T is said to be *hyponormal* if $T^*T - TT^* \geq 0$ and an operator T is said to be *essentially hyponormal* if there is a compact self-adjoint operator K such that

$$T^*T - TT^* + K \geq 0.$$

Obviously, T is essentially hyponormal if and only if $\pi(T)$ is a hyponormal element in the Calkin algebra $\mathcal{Q}$, i.e., $\pi(T^*)\pi(T) - \pi(T)\pi(T^*) \geq 0$. It is well known that the norm of a hyponormal operator coincides with its spectral radius. As we mentioned earlier, by Proposition 2.12 in [9], if q is a polynomial, then the spectral radius of M_q equals $\|q\|_\infty$. Therefore if q is a polynomial such that $\|M_q\| > \|q\|_\infty$, then M_q is not hyponormal. Thus there are plenty of multipliers $f \in \mathcal{M}_d$ for which M_f fails to be hyponormal on H_d^2. This is one phenomenon that sets the Drury–Arveson space H_d^2 apart from the Hardy space and the Bergman space. Note that the phenomenon persists under compact perturbation too.

Theorem 3.3. *There exists a $\psi \in \mathcal{M}_d$ such that the multiplication operator M_ψ on H_d^2 is not essentially hyponormal.*

Let $\mathcal{A}_d$ be the norm closure of the polynomials in $\mathcal{M}_d$. We can see that $\mathcal{A}_d \subset A(\mathbb{B}^d)$, the ball algebra. Thus all multipliers in $\mathcal{A}_d$ are continuous on $\overline{\mathbb{B}^d}$. Note that there are continuous multipliers which are not in $\mathcal{A}_d$. Since the multiplier norm and the supremum norm are not comparable, the image of $\mathcal{A}_d$ inside of $A(\mathbb{B}^d)$ is not closed. It can be shown that the maximal ideal space of $\mathcal{A}_d$ is homeomorphic to $\overline{\mathbb{B}_d}$.

In [31], Clouâtre and Davidson identified $\mathcal{A}_d$ as a direct sum of the preduals of $\mathcal{M}_d$ and of a commutative von Neumann algebra $\mathfrak{W}$,

$$\mathcal{A}_d^* \simeq \mathcal{M}_{d*} \oplus_1 \mathfrak{W}_*.$$

They established analogues of several classical results concerning the dual space of the ball algebra. These developments are deeply intertwined with the problem of peak interpolation for multipliers. It is also worth mentioning that they shed light on the nature of the extreme points of the unit ball of $\mathcal{A}_d^*$. The following results (Theorem 7.5/Theorem 7.7 in [31]) ensure the existence of many extreme points in the closed unit ball of $\mathcal{M}_{d*}$, thus showing a sharp contrast with the more classical situation of the closed unit ball of H^∞ [6].

Theorem 3.4. *Let $f \in \mathcal{A}_d$ with $\|f\|_\infty < \|f\|_{\mathcal{A}_d} = 1$. The set*

$$\mathcal{F} = \{\Psi \in \mathcal{M}_{d*} : \|\Psi\|_{\mathcal{A}_d^*} = 1 = \Psi(f)\}$$

has extreme points, which are also extreme points of the closed unit ball of $\mathcal{M}_{d*}$.

Theorem 3.5. *The following statements hold.*

1. *The set of weak-$*$ exposed points of $\overline{b_1(\mathcal{A}_d^*)}$ that lie in $\mathfrak{W}_*$ is $\{\lambda\tau_\zeta : \lambda \in \mathbb{T}, \zeta \in \mathbb{S}^d\}$, where $\mathbb{T}$ is the unit circle and $\mathbb{S}^d$ is the unit sphere. This set is weak-$*$ compact and it coincides with the extreme points of $\overline{b_1(\mathfrak{W}_*)}$.*
2. *Let $\Phi \in \overline{b_1(\mathcal{M}_{d*})}$ be a weak-$*$ exposed point of $\overline{b_1(\mathcal{A}_d^*)}$, and let $f \in \overline{b_1(\mathcal{A}_d)}$ such that*

$$\mathrm{Re}\Psi(f) < 1 = \mathrm{Re}\Phi(f) \quad \textit{for all} \quad \Psi \in \overline{b_1(\mathcal{A}_d^*)}, \Psi \neq \Phi.$$

   Then, $1 = \|f\|_{\mathcal{A}_d} > \|f\|_\infty$.
3. *If $1 = \|f\|_{\mathcal{A}_d} > \|f\|_\infty$ and $N = \{\xi \in H_d^2 : \|f\xi\|_{H_d^2} = \|\xi\|_{H_d^2}\}$ is one-dimensional, then the functional $[\xi(f\xi)^*]$ is a weak-$*$ exposed point of $\overline{b_1(\mathcal{A}_d^*)}$.*
4. *The extreme points of $\overline{b_1(\mathcal{M}_{d*})}$ are contained in the weak-$*$ closure of the set*

$$\{[\xi(f\xi)^*] : 1 = \|\xi\|_{H_d^2} = \|f\xi\|_{H_d^2} = \|f\|_{\mathcal{A}_d} > \|f\|_\infty\}.$$

4. Corona theorem and spectral theory

Carleson's corona theorem for H^∞ in [29] states that the open unit disk is dense in the maximal ideal space of H^∞. Costea, Sawyer and Wick extended this to multiplier algebras of certain Besov–Sobolev spaces on the unit ball including the multiplier algebra of the Drury–Arveson space. Here is the version of the Corona theorem for $\mathcal{M}_d$ ([33]):

Theorem 4.1. *The corona theorem holds for the multiplier algebra $\mathcal{M}_d$ of the Drury–Arveson space. That is, for $g_1, \dots, g_k \in \mathcal{M}_d$, if there is a $c > 0$ such that*

$$|g_1(z)| + \cdots + |g_k(z)| \geq c$$

for every $z \in \mathbb{B}^d$, then there exist $f_1, \dots, f_k \in \mathcal{M}_d$ such that

$$f_1 g_1 + \cdots + f_k g_k = 1.$$

An immediate consequence of this theorem is the so-called *one function corona theorem.*

Theorem 4.2. *Let $f \in \mathcal{M}_d$. If there is a $c > 0$ such that $|f(z)| \geq c$ for every $z \in \mathbb{B}^d$, then $1/f \in \mathcal{M}_d$.*

There have been several different proofs of this one function corona theorem without invoking the general corona theorem [28, 45, 64]. Also we note that it is true that $1/f \in H_d^2$ for any $f \in H_d^2$ with a lower bound when $d \leq 3$. This is because for any $f \in H_d^2$, $\|f\|_{H_d^2}$ is equivalent to the norm of Rf in the Bergman space if $d = 2$ and in the Hardy space if $d = 3$, where R is the radial derivative. The problem is completely open for $d \geq 4$ since for $d \geq 4$, the norm in the Drury–Arveson space involves higher radial derivatives. There have been related discussions ([4], [64]) but the problem for the general case still requires new ideas.

Recently, in the context of more general Hardy–Sobolev spaces, Cao, He and Zhu developed some spectral theory for multipliers of these spaces ([28]). For $\mathcal{M}_d$ the following results hold. The proofs use one function corona theorem and estimates of higher-order radial derivatives.

Theorem 4.3. *Suppose $f \in \mathcal{M}_d$.*

1. *The spectrum of M_f is the closure of $f(\mathbb{B}^d)$ in the complex plane.*
2. *The essential spectrum of M_f is given by*

$$\sigma_e(M_f) = \bigcap_{r \in (0,1)} \overline{f(\mathbb{B}^d - r\mathbb{B}^d)},$$

   where $r\mathbb{B}^d = \{z \in \mathbb{C}^d : |z| < r\}$
3. *M_f is Fredholm if and only if there exist $r \in (0, 1)$ and $\delta > 0$ such that $|f(z)| \geq \delta$ for all $z \in \mathbb{B}^d - r\mathbb{B}^d$. Moreover, when M_f is Fredholm, its Fredholm index is always 0 for $d > 1$ and is equal to minus the winding number of the mapping $e^{it} \mapsto f(re^{it})$, where $r \in (0, 1)$ is sufficiently close to 1.*

5. Commutators and localization

If we take a list of Hardy-space results and try to determine which ones have analogues on H_d^2 and which ones do not, commutators are certainly very high on any such list. One prominent part of the theory of the Hardy space is the Toeplitz operators on it. Since there is no L^2 associated with H_d^2, the only analogue of

Toeplitz operators on H_d^2 are the multipliers. We can consider the commutators of the form $[M_f^*, M_{z_i}]$, where f is a multiplier for the Drury–Arveson space.

Recall that for each $1 \le p < \infty$, the Schatten class $\mathcal{C}_p$ consists of operators A satisfying the condition $\|A\|_p < \infty$, where the p-norm is given by the formula

$$\|A\|_p = \{\mathrm{tr}((A^*A)^{p/2})\}^{1/p}.$$

Arveson showed in his seminal paper [9] that commutators of the form $[M_{z_j}^*, M_{z_i}]$ on H_d^2 all belong to $\mathcal{C}_p$, $p > d$. As the logical next step, one certainly expects a Schatten class result for commutators on H_d^2 involving multipliers other than the simplest coordinate functions. The following result was proved in [43].

Theorem 5.1. *Let f be a multiplier for the Drury–Arveson space H_d^2. For each $1 \le i \le d$, the commutator $[M_f^*, M_{\zeta_i}]$ belongs to the Schatten class $\mathcal{C}_p$, $p > 2d$. Moreover, for each $2d < p < \infty$, there is a constant C which depends only on p and n such that*

$$\|[M_f^*, M_{z_i}]\|_p \le C\|M_f\|$$

for every multiplier f of H_d^2 and every $1 \le i \le d$.

This Schatten-class result has C^*-algebraic implications.

Let $\mathcal{T}_d$ be the C^*-algebra generated by $M_{z_1}, \dots, M_{z_d}$ on H_d^2. Recall that $\mathcal{T}_d$ was introduced by Arveson in [9]. In more ways than one, $\mathcal{T}_d$ is the analogue of the C^*-algebra generated by Toeplitz operators with *continuous* symbols. Indeed Arveson showed that there is an exact sequence

$$\{0\} \to \mathcal{K} \to \mathcal{T}_d \xrightarrow{\tau} C(\mathbb{S}^d) \to \{0\}, \tag{5.1}$$

where $\mathcal{K}$ is the collection of compact operators on H_d^2. But there is another natural C^*-algebra on H_d^2 which is also related to "Toeplitz operators", where the symbols are not necessarily continuous. We define

$$\mathcal{TM}_d = \text{the } C^*\text{-algebra generated by } \{M_f : fH_d^2 \subset H_d^2\}.$$

Theorem 1.1 tells us that $\mathcal{T}_d$ is contained in the essential center of $\mathcal{TM}_d$, in analogy with the classic situation on the Hardy space of the unit sphere S. This opens the door for us to use the classic localization technique [38] to analyze multipliers.

Let S_w be a class of Schur multipliers defined as follows: For each $w \in \mathbb{B}^d$, let

$$S_w(z) = \frac{1 - |w|}{1 - \langle z, w\rangle}. \tag{5.2}$$

Note that the norm of the operator M_{S_w} on H_d^2 is 1. Here is a localization result shown in [43].

Theorem 5.2. *Let $A \in \mathcal{TM}_d$. Then for each $\xi \in \mathbb{S}^d$, the limit*

$$\lim_{r \uparrow 1} \|AM_{S_{r\xi}}\| \tag{5.3}$$

exists. Moreover, we have

$$\|A\|_{\mathcal{Q}} = \sup_{\xi \in \mathbb{S}^d} \lim_{r \uparrow 1} \|AM_{\mathcal{S}_{r\xi}}\|.$$

Alternatively, we can state this result in a version which may be better suited for applications:

Theorem 5.3. *For each $A \in \mathcal{T}\mathcal{M}_d$, we have*

$$\|A\|_{\mathcal{Q}} = \lim_{r \uparrow 1} \sup_{r \le |w| < 1} \|AM_{S_w}\|.$$

In addition to the analogue of Toeplitz operators and the Toeplitz algebra, it is also interesting to consider possible analogues of Hankel operators in the setting of H_d^2. This is more difficult than the Toeplitz case, since there isn't an L^2 associated with H_d^2. But there are analogous problems. For example, we may ask the following questions: Suppose $f \in \mathcal{M}_d$. Under what condition on f is the commutator $[M_f^*, M_f]$ compact? Under what condition on f does the commutator $[M_f^*, M_f]$ belong to the Schatten class $\mathcal{C}_p$, $p > d$?

The C^*-algebra $\mathcal{T}\mathcal{M}_d$ itself is quite interesting. To see why, let us consider the analogous situation on the Hardy space. Let $H^2(\mathbb{S}^d)$ be the Hardy space on the unit sphere. On $H^2(\mathbb{S}^d)$, we naturally have

$$\mathcal{T}(H^\infty(\mathbb{S}^d)) = C^*\text{-algebra generated by } \{M_f : f \in H^\infty(\mathbb{S}^d)\}.$$

Note that $H^\infty(\mathbb{S}^d)$ is precisely the collection of the multipliers for $H^2(\mathbb{S}^d)$. In this sense, $\mathcal{T}\mathcal{M}_d$ is as close to an analogue of $\mathcal{T}(H^\infty(\mathbb{S}^d))$ as we can get on H_d^2. The significance of this becomes clear when we consider the essential commutants. It is well known [34, 37, 49] that the essential commutant of $\mathcal{T}(H^\infty(\mathbb{S}^d))$ is $\mathcal{T}(\mathrm{QC})$, the C^*-algebra generated by the Toeplitz operators

$$\{T_f : f \in \mathrm{QC} = L^\infty \cap \mathrm{VMO}\}$$

on $H^2(\mathbb{S}^d)$. In this light, it will be interesting to see what the essential commutant of $\mathcal{T}\mathcal{M}_d$ is.

6. Characterizations of multipliers

Due to the importance of multipliers, it is natural to ask whether we have a nice characterization for these multipliers. As we have seen that there have been some work on the characterization of special multipliers such as Schur multipliers and quasi-extreme multipliers. But the determination of which $f \in H_d^2$ is a multiplier in general is still very challenging.

Let m be an integer such that $2m \geq d$. Then given any $f \in H_d^2$, one can define the measure $d\mu_f$ on $\mathbf{B}$ by the formula

$$d\mu_f(z) = |(R^m f)(z)|^2(1-|z|^2)^{2m-d}dv(z), \tag{6.1}$$

where R is the radial derivative and dv is the normalized volume measure on $\mathbb{B}^d$.

In [63] Ortega and Fàbrega proved the following characterization:

Theorem 6.1. *$f \in \mathcal{M}_d$ if and only if $d\mu_f$ is a Carleson measure for H_d^2. That is, $f \in \mathcal{M}_d$ if and only if there is a C such that*

$$\int |h(z)|^2 d\mu_f(z) \leq C\|h\|^2$$

for every $h \in H_d^2$.

In [7] Arcozzi, Rochberg and Sawyer gave a characterization for all the H_d^2-Carleson measures on $\mathbb{B}^d$. See Theorem 34 in that paper. For a given Borel measure on $\mathbb{B}^d$, the conditions in [7] are not easy to verify. More to the point, Theorem 34 in [7] deals with all Borel measures on $\mathbb{B}^d$, not just the class of measures $d\mu_f$ of the form (6.1). A natural question is to ask the following: Let k_z be the normalized reproducing kernel for H_d^2, i.e., $k_z(w) = \frac{(1-|z|^2)^{1/2}}{1-\langle z,w\rangle}$, $|z|<1, |w|<1$. For $f \in H_d^2$, does the condition

$$\sup_{|z|<1} \|fk_z\| < \infty$$

imply that f is a multiplier for H_d^2?

What makes this question particularly tempting is that an affirmative answer would give a very simple characterization of the membership $f \in \mathcal{M}$. But that would be too simple a characterization, as it turns out. Actually the answer is negative as shown in [46].

Theorem 6.2. *There exists an $f \in H_d^2$ satisfying the conditions $f \notin \mathcal{M}_d$ and $\sup_{|z|<1} \|fk_z\| < \infty$.*

In [4], a more general result of Aleman, Hartz, McCarthy and Richter for complete Pick space implies the following sufficient condition for $f \in \mathcal{M}_d$:

Theorem 6.3. *If $f \in H_d^2$ and satisfies*

$$\sup_{|z|<1} \operatorname{Re} \langle f, K_z f\rangle_{H_d^2} < \infty,$$

then $f \in \mathcal{M}_d$.

In [4] it is proved that the condition is not a necessary condition for some complete Pick space. The characterization problem still remains to be challenging.

Remark. Note that there have been numerous studies on relevant noncommutative generalizations. To list few: [11, 16, 23, 36, 55–59]. Unfortunately we have to omit the discussions here. Interested readers may find information from these papers and the references therein.

References

[1] J. Agler and J.E. McCarthy, *Pick Interpolation and Hilbert Function Spaces*, Graduate Studies in Mathematics Volume **44**, Amer. Mat. Soc., Providence, 2002.

[2] J. Agler and J.E. McCarthy, *Complete Nevanlinna–Pick kernels*, J. Funct. Anal., **175** (2000), 111–124.

[3] D. Alpay and D. Kaliuzhnyĭ-Verbovetzkiĭ, *Matrix-J-unitary non-commutative rational formal power series*, in: The State Space Method Generalizations and Applications (Eds. D. Alpay and I. Gohberg), pp. 49–113, **OT161** Birkhäuser Verlag, Basel, 2006.

[4] A. Aleman, M. Hartz, J. McCarthy and S. Richter, *Factorizations induced by complete Nevanlinna-Pick factors*, Adv. Math. **335** (2018), 372–404.

[5] A. Aleman,M. Hartz, J. McCarthy and S. Richter, *The Smirnov class for spaces with the complete Pick property*, J. London Math. Soc., **96(1)**, 2017, 228–242.

[6] T. Ando, *On the predual of H^∞*, Comment. Math. Special Issue **1** (1978), 33–40.

[7] N. Arcozzi, R. Rochberg and E. Sawyer, *Carleson measures for the Drury–Arveson Hardy space and other Besov–Sobolev spaces on complex balls*, Advances in Math, **218(4)**, 2008, 1107–1180.

[8] A. Arias and G. Popescu, *Noncommutative interpolation and Poisson transforms*, Israel J. Math. **115** (2000), 205–234.

[9] W. Arveson, *Subalgebras of C^*-algebras. III. Multivariable operator theory*, Acta Math. **181** (1998), no. 2, 159–228.

[10] W. Arveson, *The curvature invariant of a Hilbert module over $\mathbb{C}[z_1, \dots, z_d]$*, J. Reine Angew. Math. **522** (2000), 173–236.

[11] J.A. Ball and V. Bolotnikov, *Interpolation in the noncommutative Schur–Agler class*, J. Operator Theory **58** (2007), no. 1, 83–126.

[12] J.A. Ball and V. Bolotnikov, *Interpolation problems for Schur multipliers on the Drury–Arveson space: from Nevanlinna–Pick to Abstract Interpolation Problem*, Integral Equations and Operator Theory **62** (2008), 301–349.

[13] J.A. Ball and V. Bolotnikov, *Canonical transfer-function realization for Schur multipliers on the Drury–Arveson space and models for commuting row contractions*, Indiana Univ. Math. J. **61** (2012), 665–716.

[14] J.A. Ball and V. Bolotnikov, *The bitangential matrix Nevanlinna–Pick interpolation problem revisited*, https://arxiv.org/abs/1611.07097.

[15] J.A. Ball, V. Bolotnikov and Q. Fang, *Multivariable backward-shift invariant subspaces and observability operators*, Multidimens. Syst. Signal Process **18** (2007), 191–248.

[16] J.A. Ball, V. Bolotnikov and Q. Fang, *Schur-class multipliers on the Fock space: de Branges–Rovnyak reproducing kernel spaces and transfer-function realizations*, in: Operator Theory, Structured Matrices, and Dilations, Tiberiu Constantinescu Memorial Volume, Theta Foundation in Advance Mathematics, 2007, 85–114.

[17] J.A. Ball, V. Bolotnikov and Q. Fang, *Transfer-function realization for multipliers of the Arveson space*, J. Math. Anal. Appl. **333** (2007), no. 1, 68–92.

[18] J.A. Ball, V. Bolotnikov and Q. Fang, *Schur-class multipliers on the Arveson space: de Branges–Rovnyak reproducing kernel spaces and commutative transfer-function realizations*, J. Math. Anal. Appl. **341** (2008), no. 1, 519–539.

[19] J.A. Ball and Q. Fang, *Nevanlinna–Pick interpolation via graph spaces and Kreĭn-space geometry: a survey*, in Mathematical Methods in Systems, Optimization, and Control, pp. 43–71, Oper. Theory Adv. Appl. **222**, Birkhäuser, 2012.

[20] J.A. Ball, I. Gohberg and L. Rodman, *Interpolation of Rational Matrix Functions*, **OT45**, Birkhäuser Verlag, Basel-Boston, 1990.

[21] J.A. Ball, G. Groenewald and T. Malakorn, *Structured noncommutative multidimensional linear systems*, SIAM J. Control and Optimization **44**, no. 4 (2005), 1474–1528.

[22] J.A. Ball and S. ter Horst, *Multivariable operator-valued Nevanlinna–Pick interpolation: a survey*, in: Operator Algebras, Operator Theory and Applications (Eds. J.J. Grobler, L.E. Labuschagne, and M. Möller), pages 1–72, **OT195**, Birkhäuser, Basel-Berlin, 2009.

[23] J.A. Ball, G. Marx, and V. Vinnikov. Noncommutative reproducing kernel Hilbert spaces. *J. Funct. Anal.*, **271** (2016), 1844–1920.

[24] J.A. Ball, T.T. Trent and V. Vinnikov, *Interpolation and commutant lifting for multipliers on reproducing kernel Hilbert spaces*, in: Operator Theory and Analysis (Ed. H. Bart, I. Gohberg and A.C.M. Ran), pp. 89–138, **OT122**, Birkhäuser, Basel, 2001.

[25] T. Constantinescu and J.L. Johnson, *A note on noncommutative interpolation*, Can. Math. Bull. **46** (2003) no. 1, 59–70.

[26] L. de Branges and J. Rovnyak, Canonical models in quantum scattering theory, in: *Perturbation Theory and its Applications in Quantum Mechanics* (C. Wilcox, ed.) pp. 295–392, Holt, Rinehart and Winston, New York, 1966.

[27] L. de Branges and J. Rovnyak, *Square summable power series*, Holt, Rinehart and Winston, New York, 1966.

[28] G. Cao, L. He and K. Zhu, *Spectral Theory of Multiplication Operators on Hardy–Sobolev Spaces*, J. Funct. Anal. **275** (2018), no. 5,1259–1279.

[29] L.Carleson, *Interpolations by bounded analytic functions and the corona problem*, Ann. of Math. (2) **76** (1962), 547–559.

[30] R. Clouâtre and K.R. Davidson, *Absolute continuity for commuting row contractions*, J. Funct. Anal. **271**(2016), no. 3, 620–641.

[31] R. Clouâtre and K.R. Davidson, *Duality, convexity and peak interpolation in the Drury–Arveson space*, Adv. Math. **295** (2016), 90–149.

[32] R. Clouâtre and K.R. Davidson, *Ideals in a multiplier algebra on the ball*, Trans. Amer. Math. Soc. **370** (2018), 1509–1527.

[33] S. Costea, E. Sawyer and B. Wick, *The Corona Theorem for the Drury–Arveson Hardy space and other holomorphic Besov–Sobolev spaces on the unit ball in* $\mathbf{C}^d$, Anal. & PDE **4** (2011), 499–550.

[34] K.R. Davidson, *On operators commuting with Toeplitz operators modulo the compact operators*, J. Funct. Anal. **24** (1977), 291–302.

[35] K.R. Davidson and T. Le, *Commutant lifting for commuting row contractions*. Bull. Lond. Math. Soc. **42** (2010) no. 3, 506–516.

[36] K.R. Davidson and D.R. Pitts, *Nevanlinna–Pick interpolation for non-commutative analytic Toeplitz algebras*, Integral Equations & Operator Theory **31** (1998) no. 2, 401–430.

[37] X. Ding and S. Sun, *Essential commutant of analytic Toeplitz operators*, Chinese Sci.Bull. 42 (1997), 548–552.

[38] R.G. Douglas, *Banach algebra techniques in operator theory*, Second edition. Graduate Texts in Mathematics, 179. Springer-Verlag, New York, 1998.

[39] S.W. Drury, *A generalization of von Neumann's inequality to the complex ball*, Proc. Amer. Math. Soc. **68** (1978), 300–304.

[40] H. Dym, *J Contractive Matrix Functions, Reproducing Kernel Hilbert Spaces and Interpolation*, CBMS Regional Conference series **71**, American Mathematical Society, Providence, 1989.

[41] J. Eschmeier and M. Putinar, *Spherical contractions and interpolation problems on the unit ball*, J. Reine Angew. Math. **542** (2002), 219–236.

[42] Q. Fang, *Multivariable Interpolation Problems*, PhD dissertation, Virginia Tech, 2008.

[43] Q. Fang and J. Xia, *Commutators and localization on the Drury–Arveson space*, J. Funct. Anal., **260** (2011) 639–673.

[44] Q. Fang and J. Xia, Multipliers and essential norm on the Drury–Arveson space, *Proc. Amer. Math. Soc.* **139** (2011), 2497–2504.

[45] Q. Fang and J. Xia, *Corrigendum to "Multipliers and essential norm on the Drury–Arveson space"*, Proc. Amer. Math. Soc. 141 (2013), 363–368.

[46] Q. Fang and J. Xia, *On the problem of characterizing multipliers for the Drury–Arveson space*, Indiana Univ. Math. J. **64** No. 3 (2015), 663–696.

[47] C. Foiaş and A.E. Frazho, *The Commutant Lifting Approach to Interpolation Problems*, **OT44** Birkhäuser Verlag, Basel-Boston, 1990.

[48] A.M. Gleason, *Finitely generated ideals in Banach algebras*, J. Math. Mech. **13** (1964), 125–132.

[49] K. Guo and S. Sun, *The essential commutant of the analytic Toeplitz algebra and some problems related to it* (Chinese), Acta Math. Sinica (Chin. Ser.) **39** (1996), 300–313.

[50] D.C. Greene, S. Richter, and C. Sundberg, *The structure of inner multipliers on spaces with complete Nevanlinna–Pick kernels*, J. Funct. Anal. **194** (2002), 311–331.

[51] V. Katsnelson, A. Kheifets and P. Yudiskiĭ, An abstract interpolation problem and extension theory of isometric operators, in: *Operators in Spaces of Functions and Problems in Function Theory* (V.A. Marchenko, ed.), pp. 83–96, **146**, Naukova Dumka, Kiev, 1987; English Transl. in: *Topics in Interpolation Theory* (H. Dym, B. Fritzsche, V. Katsnelson and B. Kirstein, eds.), pp. 283–298, **OT 95**, Birkhäuser, Basel-Berlin-Boston, 1997.

[52] I.V. Kovalishina and V.P. Potapov, *Seven Papers Translated from the Russian*, Amer. Math. Soc. Transl. (2) **138**, Providence, RI, 1988.

[53] S. McCullough and T.T. Trent, *Invariant subspaces and Nevanlinna–Pick kernels*, J. Funct. Anal. **178** (2000), no. 1, 226–249.

[54] P.S. Muhly and B. Solel, *Hardy algebras, W^*-correspondences and interpolation theory*, Math. Ann., **330** (2004), 353–415.

[55] G. Popescu, *Characteristic functions for infinite sequences of noncommuting operators*, J. Operator Theory **22** (1989), 51–71.

[56] G. Popescu, *Isometric dilations for infinite sequences of noncommuting operators*, Trans. Amer. Math. Soc., **316** (1989) no. 2, 523–536.

[57] G. Popescu, *Multi-analytic operators on Fock spaces*, Math. Ann., **303** (1995), 31–46.

[58] G. Popescu, *Multivariable Nehari problem and interpolation*, J. Funct. Anal., **200** (2003) no. 2, 536–581.

[59] G. Popescu, *Operator theory on noncommutative varieties*, Indiana Univ. Math. J., **55** (2006), no. 2, 389–442.

[60] M. Jury, *Clark theory in the Drury–Arveson space,* J. Funct. Anal. **266** (2014), no. 6, 3855–3893.

[61] M. Jury and R.T.W. Martin, *Aleksandrov–Clark theory for the Drury–Arveson space*, Integral Equations & Operator Theory **62** (2018), 301–349.

[62] M. Jury and R.T.W. Martin, *Extremal multipliers of the Drury–Arveson space*, Proc. Amer. Math. Soc. **146** (2018), 4293–4306.

[63] J. Ortega and J. Fàbrega, Pointwise multipliers and decomposition theorems in analytic Besov spaces, Math. Z. **235** (2000), 53–81.

[64] S. Richter and J. Sunkes, *Hankel operators, invariant subspaces, and cyclic vectors in the Drury–Arveson space*, Proc. Amer. Math. Soc. **144** (2016), no. 6, 2575–2586.

[65] W. Rudin, *Function theory in the unit ball of* $\mathbf{C}^n$, Springer-Verlag, New York-Berlin, 1980.

[66] D. Sarason, *Generalized interpolation in H^∞*, Trans. Amer. Math. Soc. **127** (1967), 179–203.

[67] O. Shalit, *Operator theory and function theory in Drury–Arveson space and its quotients*, Daniel Alpay (ed.), Handbook of Operator Theory, pp. 1125–1180. Springer, Basel (2015).

Quanlei Fang
Department of Mathematics & Computer Science
CUNY-BCC
Bronx, NY 10453, USA
e-mail: `quanlei.fang@bcc.cuny.edu`

Operator Theory:
Advances and Applications, Vol. 272, 117–131

Contractively Embedded Invariant Subspaces

Sushil Gorai and Jaydeb Sarkar

To Joseph A. Ball with gratitude and admiration

Abstract. This paper focuses on representations of contractively embedded invariant subspaces in several variables. We present a version of the de Branges theorem for n-tuples of multiplication operators by the coordinate functions on analytic reproducing kernel Hilbert spaces over the unit ball $\mathbb{B}^n$ and the Hardy space over the unit polydics $\mathbb{D}^n$ in $\mathbb{C}^n$.

Mathematics Subject Classification (2010). Primary 46C07, 46E22, 47A13; Secondary 47A15, 47B32.

Keywords. Invariant subspaces, de Branges–Rovnyak spaces, Hardy space, reproducing kernel Hilbert spaces, multipliers, bounded analytic functions.

1. Introduction

The theory of contractively embedded invariant and co-invariant (not necessarily closed) subspaces for the shift operator on the Hardy space was initiated by L. de Branges. This theory was laid out more systematically in the mid 60's by de Branges and Rovnyak (see the monograph by de Branges and Rovnyak [18]). The de Branges and Rovnyak's approach to the theory of contractively embedded invariant and co-invariant subspaces for shift operators on reproducing kernel Hilbert spaces has proved very fruitful in analysing operator and function theoretic problems. As is well known, it was this theory that led de Branges to the affirmative solution of the Bieberbach conjecture [17].

The purpose of this note is to analyze the structure of contractively embedded (not necessarily closed) invariant subspaces for tuples of multiplication operators

The first named author's research work is partially supported by an INSPIRE faculty fellowship (IFA-MA-02) funded by DST, and a MATRICS grant (File no. MTR/2017/000974). The second named author is supported in part by (1) National Board of Higher Mathematics (NBHM), India, grant NBHM/R.P.64/2014, and (2) Mathematical Research Impact Centric Support (MATRICS) grant, File No: MTR/2017/000522, by the Science and Engineering Research Board (SERB), Department of Science & Technology (DST), Government of India.

by the coordinate functions on reproducing kernel Hilbert spaces in several variables. Recall that a Hilbert space $\mathcal{H}$ is said to be contractively embedded in a Hilbert space $\mathcal{K}$ if $\mathcal{H}$ is a vector subspace of $\mathcal{K}$ and the inclusion map $i_{\mathcal{H}} : \mathcal{H} \hookrightarrow \mathcal{K}$ is a contraction. Obviously, the latter condition is equivalent to

$$\|f\|_{\mathcal{K}} \leq \|f\|_{\mathcal{H}},$$

for all $f \in \mathcal{H}$, where $\|\cdot\|_{\mathcal{H}}$ and $\|\cdot\|_{\mathcal{K}}$ denotes the norms on $\mathcal{H}$ and $\mathcal{K}$, respectively. It follows, in particular, that a closed subspace of a Hilbert space is contractively (or isometrically, as an embedding) embedded in the larger Hilbert space.

Now let $\mathcal{K}$ be a Hilbert space, and let $\mathcal{H}$ be a Hilbert space that is contractively embedded in $\mathcal{K}$. Let $(T_1, \ldots, T_n)$ be an n-tuple of commuting bounded linear operators on $\mathcal{K}$, that is,

$$T_i T_j = T_j T_i,$$

for all $i, j = 1, \ldots, n$. Let $\mathcal{H}$ be an *invariant subspace* for $(T_1, \ldots, T_n)$, that is,

$$T_i \mathcal{H} \subseteq \mathcal{H},$$

for all $i = 1, \ldots, n$. Suppose that $T_i|_{\mathcal{H}}$ is bounded on $\mathcal{H}$, that is, there exists $M > 0$ such that

$$\|T_i f\|_{\mathcal{H}} \leq M \|f\|_{\mathcal{H}},$$

for all $f \in \mathcal{H}$ and $i = 1, \ldots, n$. Then clearly $(T_1|_{\mathcal{H}}, \ldots, T_n|_{\mathcal{H}})$ is an n-tuple of commuting bounded linear operators on $\mathcal{H}$. The question of interest here is to represent $\mathcal{H}$ in terms of the (algebraic or analytic properties of the) tuple $(T_1, \ldots, T_n)$.

We pause now to examine one concrete example of the above invariant subspace problem. Following standard notation, let $H^2(\mathbb{D})$ denote the Hardy space over the unit disc $\mathbb{D}$. Let M_z on $H^2(\mathbb{D})$ be the multiplication operator by the independent variable z, that is,

$$(M_z f)(w) = w f(w),$$

for all $f \in H^2(\mathbb{D})$ and $w \in \mathbb{D}$. It follows that M_z is a shift of multiplicity one (see Section 3). Let $\mathcal{H}$ be a Hilbert space contractively embedded in $H^2(\mathbb{D})$ such that $M_z \mathcal{H} \subseteq \mathcal{H}$. If $M_z|_{\mathcal{H}}$ is an isometry on $\mathcal{H}$, then the celebrated theorem of de Branges says that there is a function $\varphi \in H^\infty(\mathbb{D})$ such that $\|\varphi\|_\infty \leq 1$ and

$$\mathcal{H} = \varphi H^2(\mathbb{D}).$$

Recall that $H^\infty(\mathbb{D})$ is the Banach algebra of all bounded analytic functions on the unit disc $\mathbb{D}$ equipped with the supremum norm [28]. Here the norm on $\mathcal{H}$ is the range norm induced by the injective multiplier M_φ on $H^2(\mathbb{D})$, that is,

$$\|\varphi f\|_{\mathcal{H}} = \|f\|_{H^2(\mathbb{D})},$$

for all $f \in H^2(\mathbb{D})$ (cf. Section 3 in [34] and Theorems 3.5 and 3.7 in [38]). In this context, we refer the reader to the beautiful survey by Ball and Bolotnikov [5] on de Branges–Rovnyak spaces in both one and several variables, the monographs by Fricain and Mashreghi [19], Sarason [33, 34], Nikolʹskiĭ and Vasyunin [29], Sand [31] and Timotin [38]. Also see Singh and Thukral [37] and Sahni and Singh [30].

Another important and relevant piece of work is due to Ball and Kriete [12] and Crofoot [16]. The reader can also see the papers by Chevrot, Guillot and Ransford [14], Costara and Ransford [15] and Sarason [32] in connection with the de Branges–Rovnyak models and (generalized) Dirichlet spaces.

A natural question is now to ask for similar representations of contractively embedded invariant subspaces for (tuples of) multiplication operators by the coordinate function(s) within the framework of analytic reproducing kernel Hilbert spaces [3] in one and several variables.

In Theorems 2.2, 2.3 and 3.3, we present a solution for this problem in the setting of commuting row contractions on Hilbert spaces and analytic Hilbert spaces (see the definition in Section 2) and tuples of shift operators on vector-valued Hardy spaces over the unit polydisc $\mathbb{D}^n$ in $\mathbb{C}^n$, respectively.

The proofs of Theorems 2.2 and 2.3 involve a careful adaptation of techniques used in [35] and [36]. Whereas the setting and the proof of our invariant subspace theorem for the shift on the Hardy space over the polydisc, Theorem 3.3, is closely related to the recently initiated work [23] on the classification of (closed) invariant subspace problem for the Hardy space in several variables.

A somewhat more intriguing and complex problem is the classification of contractively embedded invariant subspaces which admit a co-invariant complemented subspace. Note that an important aspect of the de Branges–Rovnyak theory is the complementations of invariant subspaces of the Hardy space: A contractively embedded invariant subspace for M_z on $H^2(\mathbb{D})$ is complemented in $H^2(\mathbb{D})$ by an M_z^*-invariant (not necessarily closed) subspace (cf. Subsection 3.4 in [38]). We postpone the general discussion on complemented invariant subspaces for a future paper and refer the reader to the papers by Ball, Bolotnikov and Fang [7–9], Ball, Bolotnikov and ter Horst [10, 11] and Benhida and Timotin [13] for related results in the setting of Drury–Arveson space [4].

For the remainder, we adapt the following notations: $\boldsymbol{z}$ denotes the element $(z_1, \dots, z_n)$ in $\mathbb{C}^n$, $z_i \in \mathbb{C}$, $\mathbb{D}^n = \{\boldsymbol{z} \in \mathbb{C}^n : |z_i| < 1, i = 1, \dots, n\}$, $\mathbb{B}^n = \{\boldsymbol{z} \in \mathbb{C}^n : \|\boldsymbol{z}\|_{\mathbb{C}^n} < 1\}$ and

$$\mathbb{Z}_+^n = \{\boldsymbol{k} = (k_1, \dots, k_n) : k_i \in \mathbb{Z}_+, i = 1, \dots, n\}.$$

Also for each multi-index $\boldsymbol{k} \in \mathbb{Z}_+^n$, commuting tuple $T = (T_1, \dots, T_n)$ on a Hilbert space $\mathcal{H}$, and $\boldsymbol{z} \in \mathbb{C}^n$ we denote

$$T^{\boldsymbol{k}} = T_1^{k_1} \cdots T_n^{k_n} \quad \text{and} \quad \boldsymbol{z}^{\boldsymbol{k}} = z_1^{k_1} \cdots z_n^{k_n}.$$

2. Row contractions and reproducing kernel Hilbert spaces

Let n be a natural number, and let $\mathcal{H}$ be a Hilbert space. A commuting tuple of bounded linear operators $(T_1, \dots, T_n)$ acting on $\mathcal{H}$ is called a *row contraction* if the row operator $(T_1, \dots, T_n) : \mathcal{H}^n \to \mathcal{H}$ defined by

$$(T_1, \dots, T_n) \begin{bmatrix} h_1 \\ \vdots \\ h_n \end{bmatrix} = T_1 h_1 + \cdots + T_n h_n,$$

for all $h_1, \dots, h_n \in \mathcal{H}$, is a contraction. Evidently, the tuple $(T_1, \dots, T_n)$ is a row contraction if and only if

$$\|T_1 h_1 + \cdots + T_n h_n\|^2 \le \|h_1\|^2 + \cdots + \|h_n\|^2,$$

for all $h_1, \dots, h_n \in \mathcal{H}$, or equivalently if

$$\sum_{i=1}^{n} T_i T_i^* \le I_{\mathcal{H}}.$$

For a row contraction $T = (T_1, \dots, T_n)$ on a Hilbert space $\mathcal{H}$, we define the *defect operator* and the *defect space* of T as

$$D_T = \left(I_{\mathcal{H}} - \sum_{i=1}^{n} T_i T_i^* \right)^{1/2},$$

and

$$\mathcal{D}_T = \overline{\operatorname{ran}}\, D_T$$

respectively. Consider the map $P_T : \mathcal{B}(\mathcal{H}) \to \mathcal{B}(\mathcal{H})$ defined by

$$P_T(X) = \sum_{i=1}^{n} T_i X T_i^*,$$

for all $X \in \mathcal{B}(\mathcal{H})$. Clearly, P_T is a completely positive map. Moreover, since

$$I_{\mathcal{H}} \ge P_T(I_{\mathcal{H}}) \ge P_T^2(I_{\mathcal{H}}) \ge \cdots \ge 0,$$

it follows that

$$P_\infty(T) = \text{SOT} - \lim_{m \to \infty} P_T^m(I_{\mathcal{H}}),$$

exists and $0 \le P_\infty(T) \le I_{\mathcal{H}}$. We say that T is a *pure row contraction* if

$$P_\infty(T) = 0.$$

Standard examples of pure row contractions are the multiplication operator tuples by the coordinate functions on the Drury–Arveson space, the Hardy space, the Bergman space and the weighted Bergman spaces over $\mathbb{B}^n$. In fact, for each $\lambda \ge 1$, the multiplication operator tuple $(M_{z_1}, \dots, M_{z_n})$ on the reproducing kernel Hilbert space $\mathcal{H}_{K_\lambda}$ is a pure row contraction, where

$$K_\lambda(\boldsymbol{z}, \boldsymbol{w}) = \left(1 - \sum_{i=1}^{n} z_i \bar{w}_i \right)^{-\lambda}, \tag{2.1}$$

for all $\boldsymbol{z}, \boldsymbol{w} \in \mathbb{B}^n$ (cf. Proposition 4.1 in [35]). Note that the Drury–Arveson space H_n^2, the Hardy space $H^2(\mathbb{B}^n)$, the Bergman space $L_a^2(\mathbb{B}^n)$, and the weighted

Bergman space $L^2_{a,\alpha}(\mathbb{B}^n)$, with $\alpha > 0$, are reproducing kernel Hilbert spaces with kernel K_λ for $\lambda = 1, n$, $n+1$ and $n+1+\alpha$, respectively.

Let $\mathcal{E}$ be a Hilbert space. We identify the Hilbert tensor product $H^2_n \otimes \mathcal{E}$ with the $\mathcal{E}$-valued Drury–Arveson space $H^2_n(\mathcal{E})$, or the $\mathcal{E}$-valued reproducing kernel Hilbert space with kernel function

$$\mathbb{B}^n \times \mathbb{B}^n \ni (\boldsymbol{z}, \boldsymbol{w}) \mapsto \left(1 - \sum_{i=1}^n z_i \bar{w}_i\right)^{-1} I_{\mathcal{E}}.$$

Then

$$H^2_n(\mathcal{E}) = \left\{ f \in \mathcal{O}(\mathbb{B}^n, \mathcal{E}) : f(z) = \sum_{\boldsymbol{k} \in \mathbb{Z}^n_+} a_{\boldsymbol{k}} z^{\boldsymbol{k}}, a_{\boldsymbol{k}} \in \mathcal{E}, \|f\|^2 := \sum_{\boldsymbol{k} \in \mathbb{Z}^n_+} \frac{\|a_{\boldsymbol{k}}\|^2}{\gamma_{\boldsymbol{k}}} < \infty \right\},$$

where $\gamma_{\boldsymbol{k}} = \frac{(k_1 + \cdots + k_n)!}{k_1! \cdots k_n!}$ are the multinomial coefficients, $\boldsymbol{k} \in \mathbb{Z}^n_+$ (cf. [4] and [22]).

Now let $\mathcal{K}$ be a Hilbert space, and let $\mathcal{H}$ be a Hilbert space that is contractively embedded in $\mathcal{K}$. Let $T = (T_1, \ldots, T_n)$ be a pure row contraction on $\mathcal{K}$. Let

$$T_j \mathcal{H} \subseteq \mathcal{H},$$

and let

$$R_j = T_j|_{\mathcal{H}},$$

for all $j = 1, \ldots, n$. Suppose that $R = (R_1, \ldots, R_n)$ is a row contraction on $\mathcal{H}$, that is,

$$\left\| \sum_{i=1}^n R_i h_i \right\|^2_{\mathcal{H}} = \left\| \sum_{i=1}^n T_i h_i \right\|^2_{\mathcal{H}} \leq \sum_{i=1}^n \|h_i\|^2_{\mathcal{H}},$$

for all $h_1, \ldots, h_n \in \mathcal{H}$. First we claim that $(R_1, \ldots, R_n)$ is a pure row contraction. Indeed, observe that

$$i_{\mathcal{H}} R_j = T_j i_{\mathcal{H}},$$

for all $j = 1, \ldots, n$. Then

$$i_{\mathcal{H}} R^{\boldsymbol{k}} = T^{\boldsymbol{k}} i_{\mathcal{H}},$$

and so

$$R^{*\boldsymbol{k}} i^*_{\mathcal{H}} = i^*_{\mathcal{H}} T^{*\boldsymbol{k}},$$

for all $\boldsymbol{k} \in \mathbb{Z}^n_+$. This yields

$$i_{\mathcal{H}} R^{\boldsymbol{k}} R^{*\boldsymbol{k}} i^*_{\mathcal{H}} = T^{\boldsymbol{k}} i_{\mathcal{H}} i^*_{\mathcal{H}} T^{*\boldsymbol{k}},$$

for all $\boldsymbol{k} \in \mathbb{Z}^n_+$, and hence

$$i_{\mathcal{H}} P^m_R(I_{\mathcal{H}}) i^*_{\mathcal{H}} = P^m_T(i_{\mathcal{H}} i^*_{\mathcal{H}}),$$

for each $m \geq 0$. Since $P^m_T : \mathcal{B}(\mathcal{K}) \to \mathcal{B}(\mathcal{K})$ is a (completely) positive map and

$$i_{\mathcal{H}} i^*_{\mathcal{H}} \leq I_{\mathcal{K}},$$

(recall that $i_{\mathcal{H}} : \mathcal{H} \hookrightarrow \mathcal{K}$ is a contraction) we obtain that

$$P^m_T(i_{\mathcal{H}} i^*_{\mathcal{H}}) \leq P^m_T(I_{\mathcal{K}}),$$

and hence

$$i_{\mathcal{H}} P_R^m(I_{\mathcal{H}}) i_{\mathcal{H}}^* \leq P_T^m(I_{\mathcal{K}}),$$

for all $m \geq 0$. Now for $f \in \mathcal{K}$ and $m \geq 0$, we compute

$$\begin{aligned} \|P_R^m(I_{\mathcal{H}})^{\frac{1}{2}} i_{\mathcal{H}}^* f\|_{\mathcal{H}}^2 &= \langle P_R^m(I_{\mathcal{H}}) i_{\mathcal{H}}^* f, i_{\mathcal{H}}^* f\rangle_{\mathcal{H}} \\ &= \langle i_{\mathcal{H}} P_R^m(I_{\mathcal{H}}) i_{\mathcal{H}}^* f, f\rangle_{\mathcal{K}} \\ &\leq \langle P_T^m(I_{\mathcal{K}}) f, f\rangle_{\mathcal{K}}. \end{aligned}$$

Since $(T_1, \ldots, T_n)$ is a pure row contraction we see that

$$\lim_{m\to\infty} \|P_R^m(I_{\mathcal{H}})^{\frac{1}{2}} i_{\mathcal{H}}^* f\|_{\mathcal{H}} = 0,$$

for all $f \in \mathcal{K}$. On the other hand, since $i_{\mathcal{H}}$ is one-to-one we see that $i_{\mathcal{H}}^* : \mathcal{K} \to \mathcal{H}$ has dense range, and hence by continuity

$$SOT - \lim_{m\to\infty} P_R^m(I_{\mathcal{H}})^{1/2} = 0.$$

Since the sequence of positive operators $\{P_R^m(I_{\mathcal{H}})\}_{m\geq 0}$ is uniformly bounded (by $\|I_{\mathcal{H}}\| = 1$) we obtain that

$$SOT - \lim_{m\to\infty} P_R^m(I_{\mathcal{H}}) = 0,$$

that is, $(R_1, \ldots, R_n)$ on $\mathcal{H}$ is a pure row contraction.

At this point we pause to recall the dilation result due to Jewell and Lubin [22] and Muller and Vasilescu [27] (also see Arveson [4]) which says that a pure row contraction is jointly unitarily equivalent to the compression of the tuple of multiplication operators by the coordinate functions $\{z_1, \ldots, z_n\}$ on a vector-valued Drury–Arveson space to a joint co-invariant subspace. In other words, the multiplication operator tuple $(M_{z_1}, \ldots, M_{z_n})$ on the Drury–Arveson space plays the role of the model pure row contraction. We state this more formally as follows (see Theorem 3.1 [35] for a proof):

Theorem 2.1. *Let $\mathcal{L}$ be a Hilbert space, and let $X = (X_1, \ldots, X_n)$ be a pure row contraction on $\mathcal{L}$. Then there exists a co-isometry $\Pi_X : H_n^2(\mathcal{D}_X) \to \mathcal{L}$ such that*

$$\Pi_X M_{z_j} = X_j \Pi_X,$$

for all $j = 1, \ldots, n$.

Therefore, by the above dilation theorem applied to the pure row contraction $(R_1, \ldots, R_n)$, we get a co-isometry $\Pi_R : H_n^2(\mathcal{D}_R) \to \mathcal{H}$ such that

$$\Pi_R M_{z_j} = R_j \Pi_R,$$

for all $j = 1, \ldots, n$. Let

$$\Pi = i_{\mathcal{H}} \circ \Pi_R.$$

It follows that $\Pi : H_n^2(\mathcal{D}_R) \to \mathcal{K}$ is a contraction and

$$\operatorname{ran} \Pi = \mathcal{H}.$$

Moreover, since $i_{\mathcal{H}} R_j = T_j i_{\mathcal{H}}$, we have that

$$\Pi M_{z_j} = T_j \Pi,$$

for all $j = 1, \dots, n$. We summarize these results as follows:

Theorem 2.2. *Let $\mathcal{K}$ be a Hilbert space, and let $(T_1, \dots, T_n)$ be a pure row contraction on $\mathcal{K}$. Let $\mathcal{H}$ be a Hilbert space that is contractively embedded in $\mathcal{K}$. Let $T_j \mathcal{H} \subseteq \mathcal{H}$ and*

$$R_j = T_j|_{\mathcal{H}},$$

for all $j = 1, \dots, n$. Let $(R_1, \dots, R_n)$ be a row contraction on $\mathcal{H}$. Then $(R_1, \dots, R_n)$ is a pure row contraction and there exist a Hilbert space $\mathcal{E}_*$ and a contraction $\Pi : H_n^2(\mathcal{E}_*) \to \mathcal{K}$ such that

$$\Pi M_{z_j} = T_j \Pi,$$

for all $j = 1, \dots, n$, and

$$\operatorname{ran} \Pi = \mathcal{H}.$$

Of particular interest is the case where $(T_1, \dots, T_n)$ is the n-tuple of multiplication operators on a Hilbert space of analytic functions in the unit ball. To this end, we first need to introduce analytic Hilbert spaces over $\mathbb{B}^n$ (see [35] and [36] for more details).

Let $K : \mathbb{B}^n \times \mathbb{B}^n \to \mathbb{C}$ be a positive definite kernel such that $K(\boldsymbol{z}, \boldsymbol{w})$ is holomorphic in the $\{z_1, \dots, z_n\}$ variables and anti-holomorphic in $\{w_1, \dots, w_n\}$ variables. Then the corresponding reproducing kernel Hilbert space $\mathcal{H}_K$ is a Hilbert space of holomorphic functions in $\mathbb{B}^n$. We say that $\mathcal{H}_K$ is an *analytic Hilbert space* if $(M_{z_1}, \dots, M_{z_n})$, the n-tuple of multiplication operators by the coordinate functions $\{z_1, \dots, z_n\}$, defines a pure row contraction on $\mathcal{H}_K$. In other words, M_{z_j} on $\mathcal{H}_K$ defined by

$$(M_{z_j} f)(\boldsymbol{w}) = w_j f(\boldsymbol{w}) \qquad (f \in \mathcal{H}_K, \boldsymbol{w} \in \mathbb{B}^n),$$

is bounded for all $j = 1, \dots, n$, the commuting tuple $M_z = (M_{z_1}, \dots, M_{z_n})$ on $\mathcal{H}_K$ satisfies the positivity condition

$$\sum_{i=1}^{n} M_{z_i} M_{z_i}^* \leq I_{\mathcal{H}_K},$$

and

$$P_\infty(M_z) = 0.$$

Let $\mathcal{E}$ be a Hilbert space. Consider the $\mathcal{E}$-valued reproducing kernel Hilbert space $\mathcal{H}_{K_\lambda} \otimes \mathcal{E}$, $\lambda \geq 1$, where K_λ is defined as in (2.1). Then the reproducing kernel Hilbert space $\mathcal{H}_{K_\lambda} \otimes \mathcal{E}$ is analytic, as is well known and also follows, for example, from Proposition 4.1 in [35]. In particular, the vector-valued Drury–Arveson space $H_n^2 \otimes \mathcal{E}$, the Hardy space $H^2(\mathbb{B}^n) \otimes \mathcal{E}$, the Bergman space $L_a^2(\mathbb{B}^n) \otimes \mathcal{E}$, and the vector-valued weighted Bergman spaces $L_{a,\alpha}^2(\mathbb{B}^n) \otimes \mathcal{E}$, with $\alpha > 0$, are analytic Hilbert spaces.

We finally recall a characterization of intertwining maps between vector-valued Drury–Arveson space and analytic Hilbert spaces (cf. Proposition 4.2 in

[35]). Let $\mathcal{E}_1$ and $\mathcal{E}_2$ be Hilbert spaces, $\mathcal{H}_K$ be an analytic Hilbert space and let $X \in \mathcal{B}(H^2_n \otimes \mathcal{E}_1, \mathcal{H}_K \otimes \mathcal{E}_2)$. Then

$$X(M_{z_i} \otimes I_{\mathcal{E}_1}) = (M_{z_i} \otimes I_{\mathcal{E}_2})X,$$

for all $i = 1, \ldots, n$, if and only if there exists a multipler $\Theta \in \mathcal{M}(H^2_n \otimes \mathcal{E}_1, \mathcal{H}_K \otimes \mathcal{E}_2)$ such that

$$X = M_\Theta. \tag{2.2}$$

Recall that the multiplier space $\mathcal{M}(H^2_n \otimes \mathcal{E}_1, \mathcal{H}_K \otimes \mathcal{E}_2)$ is the Banach space of all operator-valued analytic functions $\Theta : \mathbb{B}^n \to \mathcal{B}(\mathcal{E}_1, \mathcal{E}_2)$ such that

$$\Theta f \in \mathcal{H}_K \otimes \mathcal{E}_2,$$

for all $f \in H^2_n \otimes \mathcal{E}_1$. Note that if $\Theta \in \mathcal{M}(H^2_n \otimes \mathcal{E}_1, \mathcal{H}_K \otimes \mathcal{E}_2)$, then the multiplication operator M_Θ defined by

$$(M_\Theta f)(\boldsymbol{w}) = \Theta(\boldsymbol{w}) f(\boldsymbol{w}),$$

for all $f \in H^2_n \otimes \mathcal{E}_1$ and $\boldsymbol{w} \in \mathbb{B}^n$, is a bounded linear operator (by the closed graph theorem) from $H^2_n \otimes \mathcal{E}_1$ to $\mathcal{H}_K \otimes \mathcal{E}_2$ (cf. [20], [26] and [35]). The next corollary now follows directly from Theorem 2.2.

Theorem 2.3. *Let $\mathcal{E}_*$ be a Hilbert space, and let $\mathcal{H}_K$ be an analytic Hilbert space. Let $\mathcal{S}$ be a Hilbert space that is contractively embedded in $\mathcal{H}_K \otimes \mathcal{E}_*$. Let $M_{z_j}\mathcal{S} \subseteq \mathcal{S}$ and*

$$R_j = M_{z_j}|_{\mathcal{S}},$$

for all $j = 1, \ldots, n$, and suppose that $(R_1, \ldots, R_n)$ is a row contraction on $\mathcal{S}$. Then $(R_1, \ldots, R_n)$ is a pure row contraction and there exist a Hilbert space $\mathcal{E}$ and a contractive multiplier $\Theta \in \mathcal{M}(H^2_n \otimes \mathcal{E}, \mathcal{H}_K \otimes \mathcal{E}_*)$ such that

$$\mathcal{S} = \Theta H^2_n(\mathcal{E}).$$

In the case when $\mathcal{H}_K$ is the Drury–Arveson space H^2_n, see the early results in Benhida and Timotin (Theorem 4.2 [13]). In this context we also refer to McCullough and Trent [26] and Greene, Richter and Sundberg [20].

We would like to point out that the theory of contractively embedded backward shift invariant subspaces in reproducing kernel Hilbert spaces and the de Branges–Rovnyak models, in the setting of row contractions, are closely related to the Gleason's problem [1]. In this context, the reader should consult the papers by Alpay and Dubi [2], Ball and Bolotnikov [6], Ball, Bolotnikov and Fang [7, 9], Ball, Bolotnikov and ter Horst [10, 11], Benhida and Timotin [13] and Martin and Ramanantoanina [25].

3. Hardy space over the polydisc

Let n be a natural number. Given a Hilbert space $\mathcal{E}$, we denote by $H^2_{\mathcal{E}}(\mathbb{D}^{n+1})$ the $\mathcal{E}$-valued Hardy space over the polydisc $\mathbb{D}^{n+1}$. In this section we aim to analyze the structure of contractively embedded invariant subspaces for the multiplication tuple on $H^2_{\mathcal{E}}(\mathbb{D}^{n+1})$. The principle of our method is based on the idea [23] that one can represent the tuple of shifts on the Hardy space over $\mathbb{D}^{n+1}$ by a natural $(n+1)$-tuple of multiplication operators on a vector-valued Hardy space over the unit disc. This is the main content of the following theorem (see [23, Theorem 3.1]).

Theorem 3.1. *Let n be a natural number, and let $\mathcal{E}$ be a Hilbert space. Let*

$$\mathcal{E}_n = H^2_{\mathcal{E}}(\mathbb{D}^n).$$

For each $i = 1, \ldots, n$, let $\kappa_i \in H^\infty_{\mathcal{B}(\mathcal{E}_n)}(\mathbb{D})$ denote the $\mathcal{B}(\mathcal{E}_n)$-valued constant function on $\mathbb{D}$ defined by

$$\kappa_i(w) = M_{z_i} \in \mathcal{B}(\mathcal{E}_n),$$

for all $w \in \mathbb{D}$, and let M_{κ_i} denote the multiplication operator on $H^2_{\mathcal{E}_n}(\mathbb{D})$ defined by

$$M_{\kappa_i} f = \kappa_i f,$$

for all $f \in H^2_{\mathcal{E}_n}(\mathbb{D})$. Then $(M_{z_1}, M_{z_2} \ldots, M_{z_{n+1}})$ and $(M_z, M_{\kappa_1}, \ldots, M_{\kappa_n})$ are unitarily equivalent.

Proof. We briefly sketch only the main ideas behind the proof and refer the reader to the proof of Theorem 3.1 in [23] for details. Since the linear spans of

$$\left\{ z_1^{k_1} z_2^{k_2} \cdots z_{n+1}^{k_{n+1}} \eta : k_1, \ldots, k_{n+1} \geq 0, \eta \in \mathcal{E} \right\} \subseteq H^2_{\mathcal{E}}(\mathbb{D}^{n+1}),$$

and

$$\left\{ z^k (z_1^{k_1} \cdots z_n^{k_n} \eta) : k, k_1, \ldots, k_n \geq 0, \eta \in \mathcal{E} \right\} \subseteq H^2_{\mathcal{E}_n}(\mathbb{D}),$$

are dense in $H^2_{\mathcal{E}}(\mathbb{D}^{n+1})$ and $H^2_{\mathcal{E}_n}(\mathbb{D})$, respectively, it follows that the map $U : H^2_{\mathcal{E}}(\mathbb{D}^{n+1}) \to H^2_{\mathcal{E}_n}(\mathbb{D})$ defined by

$$U \left(z_1^{k_1} z_2^{k_2} \cdots z_{n+1}^{k_{n+1}} \eta \right) = z^{k_1} \left(z_1^{k_2} \cdots z_n^{k_{n+1}} \eta \right),$$

for all $k_1, \ldots, k_{n+1} \geq 0$ and $\eta \in \mathcal{E}$, is a unitary operator. Clearly

$$U M_{z_1} = M_z U,$$

and an easy computation yields

$$U M_{z_i} = M_{\kappa_{i-1}} U,$$

for all $i = 2, \ldots, n$. This completes the proof. □

In view of the above theorem, we can now consider the problem of contractively embedded invariant subspaces for the tuple $(M_z, M_{\kappa_1}, \ldots, M_{\kappa_n})$ on $H^2_{\mathcal{E}_n}(\mathbb{D})$ instead of the tuple of multiplication operators $(M_{z_1}, M_{z_2}, \ldots, M_{z_{n+1}})$ on the vector-valued Hardy space $H^2_{\mathcal{E}}(\mathbb{D}^{n+1})$.

Before we proceed to the main result of this section, we need one more result concerning representations of commutators of shift [21] operators. Here our approach follows that of [23] and [24]. Recall that an isometry V on a Hilbert space $\mathcal{H}$ is said to be a *shift* if

$$SOT - \lim_{m \to \infty} V^{*m} = 0,$$

that is, $\|V^{*m}f\| \to 0$ as $m \to \infty$ for all $f \in \mathcal{H}$, or equivalently, if there is no non-trivial reducing subspace of $\mathcal{H}$ on which V is unitary. Now, if V is a shift on $\mathcal{H}$, then

$$\mathcal{H} = \bigoplus_{m=0}^{\infty} V^m \mathcal{W},$$

where $\mathcal{W} = \ker V^* = \mathcal{H} \ominus V\mathcal{H}$ is the wandering subspace for V. By the above decomposition of $\mathcal{H}$, we see that the map $\Pi_V : \mathcal{H} \to H^2_{\mathcal{W}}(\mathbb{D})$ defined by

$$\Pi_V(V^m \eta) = z^m \eta,$$

for all $m \geq 0$ and $\eta \in \mathcal{W}$, is a unitary operator and

$$\Pi_V V = M_z \Pi_V.$$

Following Wold and von Neumann, we call Π_V the *Wold–von Neumann decomposition* of the shift V (see [23] and [24]).

This point of view is very useful in representing the commutators of shifts (see Theorem 2.1 in [24] and Theorem 2.1 in [23]):

Theorem 3.2 (Theorem 2.1 in [24]). *Let $\mathcal{H}$ be a Hilbert space. Let V be a shift on $\mathcal{H}$, and let C be a bounded operator on $\mathcal{H}$. Let Π_V be the Wold–von Neumann decomposition of V, $M = \Pi_V C \Pi_V^*$, and let*

$$\Theta(w) = P_{\mathcal{W}}(I_{\mathcal{H}} - wV^*)^{-1} C|_{\mathcal{W}},$$

for all $w \in \mathbb{D}$. Then $CV = VC$ if and only if $\Theta \in H^\infty_{\mathcal{B}(\mathcal{W})}(\mathbb{D})$ and

$$M = M_\Theta.$$

Since $\|wV^*\| = |w|\|V\| < 1$ for all $w \in \mathbb{D}$, it follows that, given a bounded operator C on $\mathcal{W}$, the function Θ as defined above is a $\mathcal{B}(\mathcal{W})$-valued analytic function on $\mathbb{D}$. It is however not clear that Θ is a bounded function on $\mathbb{D}$, that is, $\Theta \in H^\infty_{\mathcal{B}(\mathcal{W})}(\mathbb{D})$. The above theorem says that this is so if and only if C is in the commutator of V.

Proof of Theorem 3.2. Again we will only sketch the proof and refer the reader to Theorem 2.1 in [24] for a more rigorous proof. Certainly, the sufficient part follows from the representation of C (as $\Pi_V^* M_\Theta \Pi_V = C$) and the fact that $M_z M_\Theta = M_\Theta M_z$. The proof for the necessary part relies on the fact that (cf. [24])

$$I_{\mathcal{H}} = \sum_{m=0}^{\infty} V^m P_{\mathcal{W}} V^{*m},$$

in the strong operator topology. Indeed, if $CV = VC$, then $MM_z = M_z M$, and so

$$M = M_\Theta,$$

for some bounded analytic function $\Theta \in H^\infty_{\mathcal{B}(\mathcal{W})}(\mathbb{D})$ (see, for instance, the equality in (2.2)). Let $w \in \mathbb{D}$ and $\eta \in \mathcal{W}$. Then

$$\Theta(w)\eta = (M_\Theta \eta)(w) = (\Pi_V C \Pi_V^* \eta)(w).$$

Since $\Pi_V^* \eta = \eta$ and

$$C\eta = \sum_{m=0}^{\infty} V^m P_{\mathcal{W}} V^{*m} C\eta,$$

it follows that

$$\begin{aligned}\Theta(w)\eta &= (\Pi_V C\eta)(w)\\ &= \left(\Pi_V\left(\sum_{m=0}^{\infty} V^m P_{\mathcal{W}} V^{*m} C\eta\right)\right)(w)\\ &= \left(\sum_{m=0}^{\infty} M_z^m (P_{\mathcal{W}} V^{*m} C\eta)\right)(w).\end{aligned}$$

Finally, note that $P_{\mathcal{W}} V^{*m} C\eta \in \mathcal{W}$ for all $m \geq 0$, and hence

$$\Theta(w)\eta = \sum_{m=0}^{\infty} w^m (P_{\mathcal{W}} V^{*m} C\eta),$$

from which the result follows. □

We are now ready for the main result concerning contractively embedded invariant subspaces of vector-valued Hardy spaces.

Let n be a natural number, and let $\mathcal{E}$ be a Hilbert space. Let $\mathcal{S}$ be a Hilbert space that is contractively embedded in $H^2_{\mathcal{E}_n}(\mathbb{D})$. Let

$$z\mathcal{S} \subseteq \mathcal{S}, \qquad \text{and} \qquad \kappa_i \mathcal{S} \subseteq \mathcal{S},$$

for all $i = 1, \ldots, n$. Assume that $(R, R_1, \ldots, R_n)$ is an $(n+1)$-tuple of isometries on $\mathcal{S}$, where

$$R = M_z|_{\mathcal{S}}, \qquad \text{and} \qquad R_i = M_{\kappa_i}|_{\mathcal{S}},$$

for all $i = 1, \ldots, n$. We have

$$\bigcap_{m=0}^{\infty} R^m \mathcal{S} = \bigcap_{m=0}^{\infty} z^m \mathcal{S} \subseteq \bigcap_{m=0}^{\infty} z^m H^2_{\mathcal{E}_n}(\mathbb{D}).$$

But M_z on $H^2_{\mathcal{E}_n}(\mathbb{D})$ is a pure isometry (shift), that is,

$$\bigcap_{m=0}^{\infty} z^m H^2_{\mathcal{E}_n}(\mathbb{D}) = \{0\},$$

and so it follows that

$$\bigcap_{m=0}^{\infty} R^m \mathcal{S} = \{0\}.$$

Further, since M_{κ_j} is a shift, it follows that

$$\bigcap_{m=0}^{\infty} \kappa_j^m H^2_{\mathcal{E}_n}(\mathbb{D}) = \{0\},$$

and so

$$\bigcap_{m=0}^{\infty} R_j^m \mathcal{S} = \{0\},$$

for all $j = 1, \ldots, n$, follows in a similar way. In other words, $(R, R_1, \ldots, R_n)$ is an $(n+1)$-tuple of commuting shifts on $\mathcal{S}$. Now we argue essentially as in the proof of Theorem 3.2 in [23]. Let

$$\mathcal{W} = \mathcal{S} \ominus z\mathcal{S},$$

and let $\Pi_R : \mathcal{S} \to H^2_{\mathcal{W}}(\mathbb{D})$ be the Wold–von Neumann decomposition of R on $\mathcal{S}$. In particular, we have

$$R\Pi_R^* = \Pi_R^* M_z. \tag{3.3}$$

Moreover, since $RR_j = R_j R$, applying Theorem 3.2, we have

$$\Pi_R R_j = M_{\Phi_j} \Pi_R, \tag{3.4}$$

where the $\mathcal{B}(\mathcal{W})$-valued analytic function defined by

$$\Phi_j(w) = P_{\mathcal{W}}(I_{\mathcal{S}} - P_{\mathcal{S}} M_z^*)^{-1} M_{\kappa_j}|_{\mathcal{W}},$$

for all $w \in \mathbb{D}$, is in $H^\infty_{\mathcal{B}(\mathcal{W})}(\mathbb{D})$ and $j = 1, \ldots, n$. Now consider the (contractive) inclusion map $i_{\mathcal{S}} : \mathcal{S} \hookrightarrow H^2_{\mathcal{E}_n}(\mathbb{D})$. Set

$$\Pi = i_{\mathcal{S}} \circ \Pi_R^*.$$

Then $\Pi : H^2_{\mathcal{W}}(\mathbb{D}) \to H^2_{\mathcal{E}_n}(\mathbb{D})$ is a contraction. Moreover, since

$$i_{\mathcal{S}} R = M_z i_{\mathcal{S}}, \qquad \text{and} \qquad i_{\mathcal{S}} R_j = M_{\kappa_j} i_{\mathcal{S}},$$

it follows from (3.3) and (3.4) that

$$\Pi M_z = M_z \Pi, \tag{3.5}$$

and

$$\Pi M_{\Phi_j} = M_{\kappa_j} \Pi, \tag{3.6}$$

for all $j = 1, \ldots, n$. Then using (2.2), one sees that

$$\Pi = M_\Theta,$$

for some contractive multiplier $\Theta \in H^\infty_{\mathcal{B}(\mathcal{W}, \mathcal{E}_n)}(\mathbb{D})$, from (3.5), and hence

$$\Theta \Phi_j = \kappa_j \Theta,$$

from (3.6), for all $j = 1, \ldots, n$. Since $\operatorname{ran} i_{\mathcal{S}} = \mathcal{S}$, it follows from the definition of Π that

$$\mathcal{S} = \Theta H^2_{\mathcal{W}}(\mathbb{D}).$$

We can therefore state the following analogue of the de Branges theorem in the setting of Hardy space over the unit polydisc:

Theorem 3.3. *Let n be a natural number, and let $\mathcal{E}$ be a Hilbert space. Let $\mathcal{S}$ be a Hilbert space that is contractively embedded in $H^2_{\mathcal{E}_n}(\mathbb{D})$. Let $z\mathcal{S} \subseteq \mathcal{S}$ and*

$$R = M_z|_{\mathcal{S}}.$$

For each $j = 1, \ldots, n$, let $\kappa_j \mathcal{S} \subseteq \mathcal{S}$ and

$$R_j = M_{\kappa_j}|_{\mathcal{S}}.$$

Set $\mathcal{W} = \mathcal{S} \ominus z\mathcal{S}$ *and*

$$\Phi_j(w) = P_{\mathcal{W}}(I_{\mathcal{S}} - wP_{\mathcal{S}}M_z^*)^{-1}M_{\kappa_j}|_{\mathcal{W}},$$

for all $w \in \mathbb{D}$ *and* $j = 1, \ldots, n$. *If* $(R, R_1, \ldots, R_n)$ *is an* $(n+1)$*-tuple of commuting isometries on* $\mathcal{S}$, *then* $(M_{\Phi_1}, \ldots, M_{\Phi_n})$ *is an* n*-tuple of commuting shifts on* $H^2_{\mathcal{W}}(\mathbb{D})$ *and there exists a contractive multiplier* $\Theta \in H^\infty_{\mathcal{B}(\mathcal{W},\mathcal{E}_n)}(\mathbb{D})$ *such that*

$$\mathcal{S} = \Theta H^2_{\mathcal{W}}(\mathbb{D}), \qquad \text{and} \qquad \kappa_j\Theta = \Theta\Phi_j,$$

for all $j = 1, \ldots, n$.

The preceding result, in view of the de Branges and Rovnyak theory, suggests a very interesting question: How can one characterize those contractively embedded invariant subspaces for $(M_z, M_{\kappa_1}, \ldots, M_{\kappa_n})$ on $H^2_{\mathcal{E}_n}(\mathbb{D})$ which are complemented by invariant subspaces for $(M_z^*, M_{\kappa_1}^*, \ldots, M_{\kappa_n}^*)$ on $H^2_{\mathcal{E}_n}(\mathbb{D})$? The answer to this question is not known.

References

[1] D. Alpay and C. Dubi, *On commuting operators solving Gleason's problem*, Proc. Amer. Math. Soc. 133 (2005), 3285–3293.

[2] D. Alpay and C. Dubi, *Backward shift operator and finite dimensional de Branges Rovnyak spaces in the ball*, Linear Algebra Appl. 371 (2003), 277–285.

[3] N. Aronszajn, *Theory of reproducing kernels*, Trans. Am. Math. Soc. 68 (1950), 337–404.

[4] W. Arveson, *Subalgebras of C^*-algebras III: Multivariable operator theory*, Acta Math. 181 (1998), 159–228.

[5] J. Ball and V. Bolotnikov, *de Branges–Rovnyak spaces: Basics and theory*, Operator Theory, 2015.

[6] J. Ball and V. Bolotnikov, *Canonical de Branges–Rovnyak model transfer-function realization for multivariable Schur-class functions. Hilbert spaces of analytic functions*, 1–39, CRM Proc. Lecture Notes, 51, Amer. Math. Soc., Providence, RI, 2010.

[7] J. Ball, V. Bolotnikov and Q. Fang, *Schur-class multipliers on the Arveson space: de Branges–Rovnyak reproducing kernel spaces and commutative transfer-function realizations*, J. Math. Anal. Appl. 341 (2008), 519–539.

[8] J. Ball, V. Bolotnikov and Q. Fang, *Multivariable backward-shift-invariant subspaces and observability operators*, Multidimens. Syst. Signal Process. 18, (2007) 191–248.

[9] J. Ball, V. Bolotnikov and Q. Fang, *Schur-class multipliers on the Fock space: de Branges–Rovnyak reproducing kernel spaces and transfer-function realizations*, Operator theory, structured matrices, and dilations, 85–114, Theta Ser. Adv. Math., 7, Theta, Bucharest, 2007.

[10] J. Ball, V. Bolotnikov and S. ter Horst, *Abstract interpolation in vector-valued de Branges–Rovnyak spaces*, Integral Equations Operator Theory 70 (2011), 227–263.

[11] J. Ball, V. Bolotnikov and S. ter Horst, *Interpolation in de Branges–Rovnyak spaces*, Proc. Amer. Math. Soc. 139 (2011), 609–618.

[12] J. Ball and T.L. Kriete, *Operator-valued Nevanlinna–Pick kernels and the functional models for contraction operators*, Integral Equations Operator Theory 10 (1987), 17–61.

[13] C. Benhida and D. Timotin, *Contractively included subspaces of Pick spaces*, Complex Anal. Oper. Theory 9 (2015), 245–264.

[14] N. Chevrot and D. Guillot, T. Ransford, De Branges–Rovnyak spaces and Dirichlet spaces, J. Funct. Anal. 259 (2010) 2366–2383.

[15] C. Costara and T. Ransford, *Which de Branges–Rovnyak spaces are Dirichlet spaces (and vice versa)?*, J. Funct. Anal. 265 (2013), 3204–3218,

[16] R. Crofoot, *Multipliers between invariant subspaces of the backward shift*, Pacific J. Math. 166 (1994), 225–246.

[17] L. de Branges, *A proof of the Bieberbach conjecture*, Acta Math. 154 (1985), 137–152.

[18] L. de Branges and J. Rovnyak, *Square summable power series*, Holt, Rinehart and Winston, New York (1966).

[19] E. Fricain, and J. Mashreghi, *Theory of $\mathcal{H}(b)$ spaces*, vols. I and II, New Monographs in Mathematics. Cambridge University Press, Cambridge, 2016.

[20] D. Greene, S. Richter and C. Sundberg, *The structure of inner multipliers on spaces with complete Nevanlinna–Pick kernels*, J. Funct. Anal. 194, 311–331 (2002).

[21] P. Halmos, *Shifts on Hilbert spaces*, J. Reine Angew. Math. 208 (1961) 102–112.

[22] N. Jewell and A. Lubin, *Commuting weighted shifts and analytic function theory in several variables*, J. Operator Theory 1 (1979), 207–223.

[23] A. Maji, A. Mundayadan, J. Sarkar and Sankar T.R., *Characterization of Invariant subspaces in the polydisc*, arXiv:1710.09853. To appear in Journal of Operator Theory.

[24] A. Maji, J. Sarkar and Sankar T.R., *Pairs of Commuting Isometries – I*, arXiv: 1708.02609. To appear in Studia Mathematica.

[25] R.T.W. Martin and A. Ramanantoanina, *A Gleason solution model for row contractions*, Oper. Theory Adv. Appl. 272 (2019), 249–305 (this volume).

[26] S. McCullough and T. Trent, *Invariant subspaces and Nevanlinna–Pick kernels*, J. Funct. Anal. 178, 226–249 (2000).

[27] V. Muller and F.-H. Vasilescu, *Standard models for some commuting multioperators*, Proc. Amer. Math. Soc. 117 (1993), no. 4, 979–989.

[28] B. Sz.-Nagy and C. Foiaş, *Harmonic Analysis of Operators on Hilbert Space. North-Holland*, Amsterdam-London, 1970.

[29] N. Nikoĺskiĭ and V. Vasyunin, *Notes on two function models*, The Bieberbach conjecture (West Lafayette, Ind., 1985), 113–141, Math. Surveys Monogr., 21, Amer. Math. Soc., Providence, RI, 1986.

[30] N. Sahni and D. Singh, *Invariant subspaces of certain sub Hilbert spaces of H^2*, Proc. Japan Acad. Ser. A Math. Sci. 87 (2011), 56–59.

[31] M. Sand, *Spaces contractively invariant for the backward shift*, J. Operator Theory 34 (1995), 125–144.

[32] D. Sarason, *Local Dirichlet spaces as de Branges–Rovnyak spaces*, Proc. Amer. Math. Soc. 125 (1997) 2133–2139.

[33] D. Sarason, *Sub-Hardy Hilbert spaces in the unit disk*, University of Arkansas Lecture Notes in the Mathematical Sciences. John Wiley & Sons Inc., New York (1994).

[34] D. Sarason, *Shift-invariant spaces from the Brangesian point of view*, The Bieberbach conjecture (West Lafayette, Ind., 1985), 153–166, Math. Surveys Monogr., 21, Amer. Math. Soc., Providence, RI, 1986.

[35] J. Sarkar, *An invariant subspace theorem and invariant subspaces of analytic reproducing kernel Hilbert spaces – II*, Complex Anal. Oper. Theory 10 (2016), 769–782.

[36] J. Sarkar, *An invariant subspace theorem and invariant subspaces of analytic reproducing kernel Hilbert spaces. I*, J. Operator Theory 73 (2015), 433–441.

[37] D. Singh and V. Thukral, *Multiplication by finite Blaschke factors on de Branges spaces*, J. Operator Theory 37 (1997), 223–245.

[38] D. Timotin, *A short introduction to de Branges–Rovnyak spaces*, Invariant subspaces of the shift operator, 21–38, Contemp. Math., 638, Centre Rech. Math. Proc., Amer. Math. Soc., Providence, RI, 2015.

Sushil Gorai
Department of Mathematics and Statistics
Indian Institute of Science Education and Research Kolkata
Mohanpur 741 246, West Bengal, India
e-mail: sushil.gorai@iiserkol.ac.in

Jaydeb Sarkar
Indian Statistical Institute
Statistics and Mathematics Unit
8th Mile, Mysore Road
Bangalore, 560059, India
e-mail: jay@isibang.ac.in
jaydeb@gmail.com

Operator Theory:
Advances and Applications, Vol. 272, 133–154

A Toeplitz-like Operator with Rational Symbol Having Poles on the Unit Circle II: The Spectrum

G.J. Groenewald, S. ter Horst, J. Jaftha and A.C.M. Ran

Dedicated to Joe Ball on the occasion of his seventieth birthday

Abstract. This paper is a continuation of our study of a class of Toeplitz-like operators with a rational symbol which has a pole on the unit circle. A description of the spectrum and its various parts, i.e., point, residual and continuous spectrum, is given, as well as a description of the essential spectrum. In this case, the essential spectrum need not be connected in $\mathbb{C}$. Various examples illustrate the results.

Mathematics Subject Classification (2010). Primary 47B35, 47A53; Secondary 47A68.

Keywords. Unbounded Toeplitz operator, spectrum, essential spectrum.

1. Introduction

This paper is a continuation of our earlier paper [9] where Toeplitz-like operators with rational symbols which may have poles on the unit circle where introduced. While the aim of [9] was to determine the Fredholm properties of such Toeplitz-like operators, in the current paper we will focus on properties of the spectrum. For this purpose we further analyse this class of Toeplitz-like operators, specifically in the case where the operators are not Fredholm.

We start by recalling the definition of our Toeplitz-like operators. Let Rat denote the space of rational complex functions. Write $\mathrm{Rat}(\mathbb{T})$ and $\mathrm{Rat}_0(\mathbb{T})$ for the

This work is based on the research supported in part by the National Research Foundation of South Africa (Grant Number 90670 and 93406).
Part of the research was done during a sabbatical of the third author, in which time several research visits to VU Amsterdam and North-West University were made. Support from University of Cape Town and the Department of Mathematics, VU Amsterdam is gratefully acknowledged.

subspaces of Rat consisting of the rational functions in Rat with all poles on $\mathbb{T}$ and the strictly proper rational functions in Rat with all poles on the unit circle $\mathbb{T}$, respectively. For $\omega \in \text{Rat}$, possibly having poles on $\mathbb{T}$, we define a Toeplitz-like operator $T_\omega(H^p \to H^p)$, for $1 < p < \infty$, as follows:

$$\text{Dom}(T_\omega)=\{g \in H^p \,|\, \omega g = f + \rho \text{ with } f \in L^p, \rho \in \text{Rat}_0(\mathbb{T})\}, \ T_\omega g = \mathbb{P}f. \quad (1.1)$$

Here $\mathbb{P}$ is the Riesz projection of L^p onto H^p.

In [9] it was established that this operator is a densely defined, closed operator which is Fredholm if and only if ω has no zeroes on $\mathbb{T}$. In case the symbol ω of T_ω is in $\text{Rat}(\mathbb{T})$ with no zeroes on $\mathbb{T}$, i.e., T_ω Fredholm, explicit formulas for the domain, kernel, range and a complement of the range were also obtained in [9]. Here we extend these results to the case that ω is allowed to have zeroes on $\mathbb{T}$, cf., Theorem 2.2 below. By a reduction to the case of symbols in $\text{Rat}(\mathbb{T})$, we then obtain for general symbols in Rat, in Proposition 2.4 below, necessary and sufficient conditions for T_ω to be injective or have dense range, respectively.

Main results. Using the fact that $\lambda I_{H^p} - T_\omega = T_{\lambda-\omega}$, our extended analysis of the operator T_ω enables us to describe the spectrum of T_ω, and its various parts. Our first main result is a description of the essential spectrum of T_ω, i.e., the set of all $\lambda \in \mathbb{C}$ for which $\lambda I_{H^p} - T_\omega$ is not Fredholm.

Theorem 1.1. *Let $\omega \in \text{Rat}$. Then the essential spectrum $\sigma_{\text{ess}}(T_\omega)$ of T_ω is an algebraic curve in $\mathbb{C}$ which is given by*

$$\sigma_{\text{ess}}(T_\omega) = \omega(\mathbb{T}) := \{\omega(e^{i\theta}) \mid 0 \leq \theta \leq 2\pi, \ e^{i\theta} \text{ not a pole of } \omega\}.$$

Furthermore, the map $\lambda \mapsto \text{Index}(T_{\lambda-\omega})$ is constant on connected components of $\mathbb{C}\backslash\omega(\mathbb{T})$ and the intersection of the point spectrum, residual spectrum and resolvent set of T_ω with $\mathbb{C}\backslash\omega(\mathbb{T})$ coincides with sets of $\lambda \in \mathbb{C}\backslash\omega(\mathbb{T})$ with $\text{Index}(T_{\lambda-\omega})$ being strictly positive, strictly negative and zero, respectively.

Various examples, specifically in Section 5, show that the algebraic curve $\omega(\mathbb{T})$, and thus the essential spectrum of T_ω, need not be connected in $\mathbb{C}$.

Our second main result provides a description of the spectrum of T_ω and its various parts. Here and throughout the paper $\mathcal{P}$ stands for the subspace of H^p consisting of all polynomials and $\mathcal{P}_k$ for the subspace of $\mathcal{P}$ consisting of all polynomials of degree at most k.

Theorem 1.2. *Let $\omega \in \text{Rat}$, say $\omega = s/q$ with $s, q \in \mathcal{P}$ co-prime. Define*

$$\begin{aligned} k_q &= \sharp\{\text{roots of } q \text{ inside } \overline{\mathbb{D}}\} &&= \sharp\{\text{poles of } \lambda - \omega \text{ inside } \overline{\mathbb{D}}\}, \\ k_\lambda^- &= \sharp\{\text{roots of } \lambda q - s \text{ inside } \mathbb{D}\} &&= \sharp\{\text{zeroes of } \lambda - \omega \text{ inside } \mathbb{D}\}, \\ k_\lambda^0 &= \sharp\{\text{roots of } \lambda q - s \text{ on } \mathbb{T}\} &&= \sharp\{\text{zeroes of } \lambda - \omega \text{ on } \mathbb{T}\}, \end{aligned} \quad (1.2)$$

where in all these sets multiplicities of the roots, poles and zeroes are to be taken into account. Then the resolvent set $\rho(T_\omega)$, *point spectrum* $\sigma_{\rm p}(T_\omega)$, *residual spectrum* $\sigma_{\rm r}(T_\omega)$ *and continuous spectrum* $\sigma_{\rm c}(T_\omega)$ *of* T_ω *are given by*

$$\begin{aligned} \rho(T_\omega) &= \{\lambda \in \mathbb{C} \mid k_\lambda^0 = 0 \text{ and } k_q = k_\lambda^-\}, \\ \sigma_{\rm p}(T_\omega) &= \{\lambda \in \mathbb{C} \mid k_q > k_\lambda^- + k_\lambda^0\}, \quad \sigma_{\rm r}(T_\omega) = \{\lambda \in \mathbb{C} \mid k_q < k_\lambda^-\}, \\ \sigma_{\rm c}(T_\omega) &= \{\lambda \in \mathbb{C} \mid k_\lambda^0 > 0 \text{ and } k_\lambda^- \leq k_q \leq k_\lambda^- + k_\lambda^0\}. \end{aligned} \tag{1.3}$$

Furthermore, $\sigma_{\rm ess}(T_\omega) = \omega(\mathbb{T}) = \{\lambda \in \mathbb{C} \mid k_\lambda^0 > 0\}$.

Again, in subsequent sections various examples are given that illustrate these results. In particular, examples are given where T_ω has a bounded resolvent set, even with an empty resolvent set. This is in sharp contrast to the case where ω has no poles on the unit circle $\mathbb{T}$. For in this case the operator is bounded, the resolvent set is a nonempty unbounded set and the spectrum a compact set, and the essential spectrum is connected.

Both Theorems 1.1 and 1.2 are proven in Section 3.

Discussion of the literature. In the case of a bounded selfadjoint Toeplitz operator on ℓ^2, Hartman and Wintner in [11] showed that the point spectrum is empty when the symbol is real and rational and posed the problem of specifying the spectral properties of such a Toeplitz operator. Gohberg in [7], and more explicitly in [8], showed that a bounded Toeplitz operator with continuous symbol is Fredholm exactly when the symbol has no zeroes on $\mathbb{T}$, and in this case the index of the operator coincides with the negative of the winding number of the symbol with respect to zero. This implies immediately that the essential spectrum of a Toeplitz operator with continuous symbol is the image of the unit circle.

Hartman and Wintner in [12] followed up on their earlier question by showing that in the case where the symbol, φ, is a bounded real-valued function on $\mathbb{T}$, the spectrum of the Toeplitz operator on H^2 is contained in the interval bounded by the essential lower and upper bounds of φ on $\mathbb{T}$ as well as that the point spectrum is empty whenever φ is not a constant. Halmos, after posing in [10] the question whether the spectrum of a Toeplitz operator is connected, with Brown in [1] showed that the spectrum cannot consist of only two points. Widom, in [16], established that bounded Toeplitz operators on H^2 have connected spectrum, and later extended the result for general H^p, with $1 \leq p \leq \infty$. That the essential (Fredholm) spectrum of a bounded Toeplitz operator in H^2 is connected was shown by Douglas in [5]. For the case of bounded Toeplitz operators in H^p it is posed as an open question in Böttcher and Silbermann in [2, Page 70] whether the essential (Fredholm) spectrum of a Toeplitz operator in H^p is necessarily connected. Clark, in [3], established conditions on the argument of the symbol φ in the case $\varphi \in L^q, q \geq 2$ that would give the kernel index of the Toeplitz operator with symbol φ on L^p, where $\frac{1}{p} + \frac{1}{q} = 1$, to be $m \in \mathbb{N}$.

Janas, in [13], discussed unbounded Toeplitz operators on the Bargmann–Segal space and showed that $\sigma_{\rm ess}(T_\varphi) \subset \cap_{R>0}$ closure $\{\varphi(z) : |z| \geq R\}$.

Overview. The paper is organized as follows. Besides the current introduction, the paper consists of five sections. In Section 2 we extend a few results concerning the operator T_ω from [9] to the case where T_ω need not be Fredholm. These results are used in Section 3 to compute the spectrum of T_ω and various of its subparts, and by doing so we prove the main results, Theorems 1.1 and 1.2. The remaining three sections contain examples that illustrate our main results and show in addition that the resolvent set can be bounded, even empty, and that the essential spectrum can be disconnected in $\mathbb{C}$.

Figures. We conclude this introduction with a remark on the figures in this paper illustrating the spectrum and essential spectrum for several examples. The color coding in these figures is as follows: the white region is the resolvent set, the black curve is the essential spectrum, and the colors in the other regions codify the Fredholm index, where red indicates index 2, blue indicates index 1, cyan indicates index -1, magenta indicates index -2.

2. Review and new results concerning T_ω

In this section we recall some results concerning the operator T_ω defined in (1.1) that were obtained in [9] and will be used in the present paper to determine spectral properties of T_ω. A few new features are added as well, specifically relating to the case where T_ω is not Fredholm.

The first result provides necessary and sufficient conditions for T_ω to be Fredholm, and gives a formula for the index of T_ω in case T_ω is Fredholm.

Theorem 2.1 (Theorems 1.1 and 5.4 in [9]). *Let $\omega \in \mathrm{Rat}$. Then T_ω is Fredholm if and only if ω has no zeroes on $\mathbb{T}$. In case T_ω is Fredholm, the Fredholm index of T_ω is given by*

$$\mathrm{Index}(T_\omega) = \sharp\left\{\begin{matrix}\textit{poles of } \omega \textit{ in } \overline{\mathbb{D}} \textit{ multi.}\\ \textit{taken into account}\end{matrix}\right\} - \sharp\left\{\begin{matrix}\textit{zeroes of } \omega \textit{ in } \mathbb{D} \textit{ multi.}\\ \textit{taken into account}\end{matrix}\right\},$$

and T_ω is either injective or surjective. In particular, T_ω is injective, invertible or surjective if and only if $\mathrm{Index}(T_\omega) \leq 0$, $\mathrm{Index}(T_\omega) = 0$ *or* $\mathrm{Index}(T_\omega) \geq 0$, *respectively.*

Special attention is given in [9] to the case where ω is in $\mathrm{Rat}(\mathbb{T})$, since in that case the kernel, domain and range can be computed explicitly; for the domain and range this was done under the assumption that T_ω is Fredholm. In the following result we collect various statements from Proposition 4.5 and Theorems 1.2 and 4.7 in [9] and extend to or improve some of the claims regarding the case that T_ω is not Fredholm.

Theorem 2.2. *Let $\omega \in \mathrm{Rat}(\mathbb{T})$, say $\omega = s/q$ with $s, q \in \mathcal{P}$ co-prime. Factor $s = s_- s_0 s_+$ with s_-, s_0 and s_+ having roots only inside, on, or outside $\mathbb{T}$. Then*

$$\begin{aligned} \mathrm{Ker}(T_\omega) &= \{r_0/s_+ \mid \deg(r_0) < \deg(q) - \deg(s_- s_0)\}; \\ \mathrm{Dom}(T_\omega) &= qH^p + \mathcal{P}_{\deg(q)-1}; \quad \mathrm{Ran}(T_\omega) = sH^p + \widetilde{\mathcal{P}}, \end{aligned} \tag{2.1}$$

where $\widetilde{\mathcal{P}}$ *is the subspace of* $\mathcal{P}$ *given by*

$$\widetilde{\mathcal{P}} = \{r \in \mathcal{P} \mid rq = r_1 s + r_2 \text{ for } r_1, r_2 \in \mathcal{P}_{\deg(q)-1}\} \subset \mathcal{P}_{\deg(s)-1}. \tag{2.2}$$

Furthermore, $H^p = \overline{\mathrm{Ran}(T_\omega)} + \widetilde{\mathcal{Q}}$ *forms a direct sum decomposition of* H^p*, where*

$$\widetilde{\mathcal{Q}} = \mathcal{P}_{k-1} \quad \text{with} \quad k = \max\{\deg(s_-) - \deg(q), 0\}, \tag{2.3}$$

following the convention $\mathcal{P}_{-1} := \{0\}$.

The following result will be useful in the proof of Theorem 2.2.

Lemma 2.3. *Factor* $s \in \mathcal{P}$ *as* $s = s_- s_0 s_+$ *with* s_-*,* s_0 *and* s_+ *having roots only inside, on, or outside* $\mathbb{T}$*. Then* $sH^p = s_- s_0 H^p$ *and* $\overline{sH^p} = s_- H^p$*.*

Proof. Since s_+ has no roots inside $\overline{\mathbb{D}}$, we have $s_+ H^p = H^p$. Furthermore, s_0 is an H^∞ outer function (see, e.g., [14], Example 4.2.5) so that $\overline{s_0 H^p} = H^p$. Since s_- has all its roots inside $\mathbb{D}$, $T_{s_-} : H^p \to H^p$ is an injective operator with closed range. Consequently, we have

$$\overline{sH^p} = \overline{s_- s_0 s_+ H^p} = \overline{s_- s_0 H^p} = s_- \overline{s_0 H^p} = s_- H^p,$$

as claimed. □

Proof of Theorem 2.2. In case T_ω is Fredholm, i.e., s_0 constant, all statements follow from Theorem 1.2 in [9]. Without the Fredholm condition, the formula for $\mathrm{Ker}(T_\omega)$ follows from [9, Lemma 4.1] and for $\mathrm{Dom}(T_\omega)$ and $\mathrm{Ran}(T_\omega)$ Proposition 4.5 of [9] provides

$$\begin{aligned} qH^p + \mathcal{P}_{\deg(q)-1} &\subset \mathrm{Dom}(T_\omega); \\ T_\omega(qH^p + \mathcal{P}_{\deg(q)-1}) &= sH^p + \widetilde{\mathcal{P}} \subset \mathrm{Ran}(T_\omega). \end{aligned} \tag{2.4}$$

Thus in order to prove (2.1), it remains to show that $\mathrm{Dom}(T_\omega) \subset qH^p + \mathcal{P}_{\deg(q)-1}$.

Assume $g \in \mathrm{Dom}(T_\omega)$. Thus there exist $h \in H^p$ and $r \in \mathcal{P}_{\deg(q)-1}$ so that $sg = qh + r$. Since s and q are co-prime, there exist $a, b \in \mathcal{P}$ such that $sa + qb \equiv 1$. Next write $ar = qr_1 + r_2$ for $r_1, r_2 \in \mathcal{P}$ with $\deg(r_2) < \deg(q)$. Thus $sg = qh + r = qh + qbr + sar = q(h + br + sr_1) + sr_2$. Hence $g = q(h + br + sr_1)/s + r_2$. We are done if we can show that $\widetilde{h} := (h + br + sr_1)/s$ is in H^p.

The case where g is rational is significantly easier, but still gives an idea of the complications that arise, so we include a proof. Hence assume $g \in \mathrm{Rat} \cap H^p$. Then $h = (sg - r)/q$ is also in $\mathrm{Rat} \cap H^p$, and $\widetilde{h}$ is also rational. It follows that $q(h + br + sr_1)/s = q\widetilde{h} = g - r_2 \in \mathrm{Rat} \cap H^p$ and thus cannot have poles in $\overline{\mathbb{D}}$. Since q and s are co-prime and h cannot have poles inside $\overline{\mathbb{D}}$, it follows that $\widetilde{h} = (h + br + sr_1)/s$ cannot have poles in $\overline{\mathbb{D}}$. Thus $\widetilde{h}$ is a rational function with no poles in $\overline{\mathbb{D}}$, which implies $\widetilde{h} \in H^p$.

Now we prove the claim for the general case. Assume $q\widetilde{h} + r_2 = g \in H^p$, but $\widetilde{h} = (h + br + sr_1)/s \notin H^p$, i.e., $\widetilde{h}$ is not analytic on $\mathbb{D}$ or $\int_{\mathbb{T}} |\widetilde{h}(z)|^p \mathrm{d}z = \infty$. Set $\widehat{h} = h + br + sr_1 \in H^p$, so that $\widetilde{h} = \widehat{h}/s$. We first show $\widetilde{h}$ must be analytic on $\mathbb{D}$. Since $\widetilde{h} = \widehat{h}/s$ and $\widehat{h} \in H^p$, $\widetilde{h}$ is analytic on $\mathbb{D}$ except possibly at the roots of s.

However, if $\widetilde{h}$ would not be analytic at a root $z_0 \in \mathbb{D}$ of s, then also $g = q\widetilde{h} + r_2$ should not be analytic at z_0, since q is bounded away from 0 on a neighborhood of z_0, using that s and q are co-prime. Thus $\widetilde{h}$ is analytic on $\mathbb{D}$. It follows that $\int_{\mathbb{T}} |\widetilde{h}(z)|^p \mathrm{d}z = \infty$.

Since s and q are co-prime, we can divide $\mathbb{T}$ as $\mathbb{T}_1 \cup \mathbb{T}_2$ with $\mathbb{T}_1 \cap \mathbb{T}_2 = \emptyset$ and each of $\mathbb{T}_1$ and $\mathbb{T}_2$ being nonempty unions of circular arcs, with $\mathbb{T}_1$ containing all roots of s on $\mathbb{T}$ as interior points and $\mathbb{T}_2$ containing all roots of q on $\mathbb{T}$ as interior points. Then there exist $N_1, N_2 > 0$ such that $|q(z)| > N_1$ on $\mathbb{T}_1$ and $|s(z)| > N_2$ on $\mathbb{T}_2$. Note that

$$\int_{\mathbb{T}_2} |\widetilde{h}(z)|^p \mathrm{d}z = \int_{\mathbb{T}_2} |\widehat{h}(z)/s(z)|^p \mathrm{d}z \leq N_2^{-p} \int_{\mathbb{T}_2} |\widehat{h}(z)|^p \mathrm{d}z \leq 2\pi N_2^{-p} \|\widehat{h}\|_{H^p}^p < \infty.$$

Since $\int_{\mathbb{T}} |\widetilde{h}(z)|^p \mathrm{d}z = \infty$ and $\int_{\mathbb{T}_2} |\widetilde{h}(z)|^p \mathrm{d}z < \infty$, it follows that $\int_{\mathbb{T}_1} |\widetilde{h}(z)|^p \mathrm{d}z = \infty$. However, since $|q(z)| > N_1$ on $\mathbb{T}_1$, this implies that

$$\begin{aligned} \|g - r_2\|_{H^p}^p &= \frac{1}{2\pi} \int_{\mathbb{T}} |g(z) - r_2(z)|^p \mathrm{d}z = \frac{1}{2\pi} \int_{\mathbb{T}} |q(z)\widetilde{h}(z)|^p \mathrm{d}z \\ &\geq \frac{1}{2\pi} \int_{\mathbb{T}_1} |q(z)\widetilde{h}(z)|^p \mathrm{d}z \geq \frac{N_1^p}{2\pi} \int_{\mathbb{T}_1} |\widetilde{h}(z)|^p \mathrm{d}z = \infty, \end{aligned}$$

in contradiction with the assumption that $g \in H^p$. Thus we can conclude that $\widetilde{h} \in H^p$ so that $g = q\widetilde{h} + r_2$ is in $qH^p + \mathcal{P}_{\deg(q)-1}$.

It remains to show that $H^p = \overline{\mathrm{Ran}(T_\omega)} + \widetilde{\mathcal{Q}}$ is a direct sum decomposition of H^p. Again, for the case that T_ω is Fredholm this follows from [9, Theorem 1.2]. By the preceding part of the proof we know, even in the non-Fredholm case, that $\mathrm{Ran}(T_\omega) = sH^p + \widetilde{\mathcal{P}}$. Since $\widetilde{\mathcal{P}}$ is finite-dimensional, and thus closed, we have

$$\overline{\mathrm{Ran}(T_\omega)} = \overline{sH^p} + \widetilde{\mathcal{P}} = s_- H^p + \widetilde{\mathcal{P}},$$

using Lemma 2.3 in the last identity. We claim that

$$\overline{\mathrm{Ran}(T_\omega)} = s_- H^p + \widetilde{\mathcal{P}} = s_- H^p + \widetilde{\mathcal{P}}_-,$$

where $\widetilde{\mathcal{P}}_-$ is defined by

$$\widetilde{\mathcal{P}}_- := \{r \in \mathcal{P} \mid qr = r_1 s_- + r_2 \text{ for } r_1, r_2 \in \mathcal{P}_{\deg(q)-1}\} \subset \mathcal{P}_{\deg(s_-)-1}.$$

Once the above identity for $\overline{\mathrm{Ran}(T_\omega)}$ is established, the fact that $\widetilde{\mathcal{Q}}$ is a complement of $\overline{\mathrm{Ran}(T_\omega)}$ follows directly by applying Lemma 4.8 of [9] to $s = s_-$.

We first show that $\overline{\mathrm{Ran}(T_\omega)} = s_- H^p + \widetilde{\mathcal{P}}$ is contained in $s_- H^p + \widetilde{\mathcal{P}}_-$. Let $g = s_- h + r$ with $h \in H^p$ and $r \in \widetilde{\mathcal{P}}$, say $qr = r_1 s + r_2$ with $r_1, r_2 \in \mathcal{P}_{\deg(q)-1}$. Write $r_1 s_0 s_+ = \widetilde{r}_1 q + \widetilde{r}_2$ with $\deg(\widetilde{r}_2) < \deg(q)$. Then

$$qr = r_1 s_- s_0 s_+ + r_2 = q\widetilde{r}_1 s_- + \widetilde{r}_2 s_- + r_2, \text{ so that } q(r - \widetilde{r}_1 s_-) = \widetilde{r}_2 s_- + r_2,$$

with $r_2, \widetilde{r}_2 \in \mathcal{P}_{\deg(q)-1}$. Thus $r - \widetilde{r}_1 s_- \in \widetilde{\mathcal{P}}_-$. Therefore, we have

$$g = s_-(h + \widetilde{r}_1) + (r - \widetilde{r}_1 s_-) \in s_- H^p + \widetilde{\mathcal{P}}_-,$$

proving that $\overline{\mathrm{Ran}(T_\omega)} \subset s_- H^p + \widetilde{\mathcal{P}}_-$.

For the reverse inclusion, assume $g = s_-h+r \in s_-H^p+\widetilde{\mathcal{P}}_-$. Say $qr = r_1s_-+r_2$ with $r_1, r_2 \in \mathcal{P}_{\deg(q)-1}$. Since s_0s_+ and q are co-prime and $\deg(r_1) < \deg(q)$ there exist polynomials $\widetilde{r}_1$ and $\widetilde{r}_2$ with $\deg(\widetilde{r}_1) < \deg(q)$ and $\deg(\widetilde{r}_2) < \deg(s_0s_+)$ that satisfy the Bézout equation $\widetilde{r}_1s_0s_+ + \widetilde{r}_2q = r_1$. Then

$$\widetilde{r}_1s + r_2 = \widetilde{r}_1s_0s_+s_- + r_2 = (r_1 - \widetilde{r}_2q)s_- + r_2 = r_1s_- + r_2 - q\widetilde{r}_2s_- = q(r - \widetilde{r}_2s_-).$$

Hence $r - \widetilde{r}_2s_-$ is in $\widetilde{\mathcal{P}}$, so that

$$g = s_-h + r = s_-(h + \widetilde{r}_2) + (r - \widetilde{r}_2s_-) \in s_-H^p + \widetilde{\mathcal{P}}.$$

This proves the reverse inclusion, and completes the proof of Theorem 2.2. □

The following result makes precise when T_ω is injective and when T_ω has dense range, even in the case where T_ω is not Fredholm.

Proposition 2.4. *Let $\omega \in \mathrm{Rat}$. Then T_ω is injective if and only if*

$$\sharp\left\{\begin{matrix}\textit{poles of } \omega \textit{ in } \overline{\mathbb{D}} \textit{ multi.}\\ \textit{taken into account}\end{matrix}\right\} \leq \sharp\left\{\begin{matrix}\textit{zeroes of } \omega \textit{ in } \overline{\mathbb{D}} \textit{ multi.}\\ \textit{taken into account}\end{matrix}\right\}.$$

Moreover, T_ω has dense range if and only if

$$\sharp\left\{\begin{matrix}\textit{poles of } \omega \textit{ in } \overline{\mathbb{D}} \textit{ multi.}\\ \textit{taken into account}\end{matrix}\right\} \geq \sharp\left\{\begin{matrix}\textit{zeroes of } \omega \textit{ in } \mathbb{D} \textit{ multi.}\\ \textit{taken into account}\end{matrix}\right\}.$$

In particular, T_ω is injective or has dense range.

Proof. First assume $\omega \in \mathrm{Rat}(\mathbb{T})$. By Corollary 4.2 in [9], T_ω is injective if and only if the number of zeroes of ω inside $\overline{\mathbb{D}}$ is greater than or equal to the number of poles of ω, in both cases with multiplicity taken into account. By Theorem 2.2, T_ω has dense range precisely when $\widetilde{\mathcal{Q}}$ in (2.3) is trivial. The latter happens if and only if the number of poles of ω is greater than or equal to the number of zeroes of ω inside $\mathbb{D}$, again taking multiplicities into account. Since in this case all poles of ω are in $\mathbb{T}$, our claim follows for $\omega \in \mathrm{Rat}(\mathbb{T})$.

Now we turn to the general case, i.e., we assume $\omega \in \mathrm{Rat}$. In the remainder of the proof, whenever we speak of numbers of zeroes or poles, this always means that the respective multiplicities are to be taken into account. Recall from [9, Lemma 5.1] that we can factor $\omega(z) = \omega_-(z)z^\kappa\omega_0(z)\omega_+(z)$ with $\omega_-, \omega_0, \omega_+ \in \mathrm{Rat}$, ω_- having no poles or zeroes outside $\mathbb{D}$, ω_+ having no poles or zeroes inside $\overline{\mathbb{D}}$ and ω_0 having poles and zeroes only on $\mathbb{T}$, and κ the difference between the number of zeroes of ω in $\mathbb{D}$ and the number of poles of ω in $\mathbb{D}$. Moreover, we have $T_\omega = T_{\omega_-}T_{z^\kappa\omega_0}T_{\omega_+}$ and T_{ω_-} and T_{ω_+} are boundedly invertible on H^p. Thus T_ω is injective or has closed range if and only it $T_{z^\kappa\omega_0}$ is injective or has closed range, respectively.

Assume $\kappa \geq 0$. Then $z^\kappa\omega_0 \in \mathrm{Rat}(\mathbb{T})$ and the results for the case that the symbol is in $\mathrm{Rat}(\mathbb{T})$ apply. Since the zeroes and poles of ω_0 coincide with the zeroes and poles of ω on $\mathbb{T}$, it follows that the number of poles of $z^\kappa\omega_0$ is equal to the number of poles of ω on $\mathbb{T}$ while the number of zeroes of $z^\kappa\omega_0$ is equal to κ plus the number of zeroes of ω on $\mathbb{T}$ which is equal to the number of zeroes of ω in $\overline{\mathbb{D}}$

minus the number of poles of ω in $\mathbb{D}$. It thus follows that $T_{z^\kappa\omega_0}$ is injective, and equivalently T_ω is injective, if and only if the number of zeroes of ω in $\overline{\mathbb{D}}$ is greater than or equal to the number of poles of ω in $\overline{\mathbb{D}}$, as claimed.

Next, we consider the case where $\kappa < 0$. In that case $T_{z^\kappa\omega_0} = T_{z^\kappa}T_{\omega_0}$, by Lemma 5.3 of [9]. We prove the statements regarding injectivity and T_ω having closed range separately.

First we prove the injectivity claim for the case where $\kappa < 0$. Write $\omega_0 = s_0/q_0$ with $s_0, q_0 \in \mathcal{P}$ co-prime. Note that all the roots of s_0 and q_0 are on $\mathbb{T}$. We need to show that $T_{z^\kappa\omega_0}$ is injective if and only if $\deg(s_0) \geq \deg(q_0) - \kappa$ (recall, κ is negative).

Assume $\deg(s_0) + \kappa \geq \deg(q_0)$. Then $\deg(s_0) > \deg(q_0)$, since $\kappa < 0$, and thus T_{ω_0} is injective. We have $\operatorname{Ker}(T_{z^\kappa}) = \mathcal{P}_{|\kappa|-1}$. So it remains to show $\mathcal{P}_{|\kappa|-1} \cap \operatorname{Ran}(T_{\omega_0}) = \{0\}$. Assume $r \in \mathcal{P}_{|\kappa|-1}$ is also in $\operatorname{Ran}(T_{\omega_0})$. So, by Lemma 2.3 in [9], there exist $g \in H^p$ and $r' \in \mathcal{P}_{\deg(q_0)-1}$ so that $s_0 g = q_0 r + r'$, i.e., $g = (q_0 r + r')/s_0$. This shows that g is in $\operatorname{Rat}(\mathbb{T}) \cap H^p$, which can only happen in case g is a polynomial. Thus, in the fraction $(q_0 r + r')/s_0$, all roots of s_0 must cancel against roots of $q_0 r + r'$. However, since $\deg(s_0) + \kappa \geq \deg(q_0)$, with $\kappa < 0$, $\deg(r) < \deg|\kappa| - 1$ and $\deg(r') < \deg(q_0)$, we have $\deg(q_0 r + r') < \deg(s_0)$ and it is impossible that all roots of s_0 cancel against roots of $q_0 r + r'$, leading to a contradiction. This shows $\mathcal{P}_{|\kappa|-1} \cap \operatorname{Ran}(T_{\omega_0}) = \{0\}$, which implies $T_{z^\kappa\omega_0}$ is injective. Hence also T_ω is injective.

Conversely, assume $\deg(s_0) + \kappa < \deg(q_0)$, i.e., $\deg(s_0) < \deg(q_0) + |\kappa| =: b$, since $\kappa < 0$. Then

$$s_0 \in \mathcal{P}_{b-1} = q_0\mathcal{P}_{|\kappa|-1} + \mathcal{P}_{\deg(q_0)-1}.$$

This shows there exist $r \in \mathcal{P}_{|\kappa|-1}$ and $r' \in \mathcal{P}_{\deg(q_0)-1}$ so that $s_0 = q_0 r + r'$. In other words, the constant function $g \equiv 1 \in H^p$ is in $\operatorname{Dom}(T_{\omega_0})$ and $T_{\omega_0} g = r \in \mathcal{P}_{|\kappa|-1} = \operatorname{Ker}(T_{z^\kappa})$, so that $g \in \operatorname{Ker}(T_{z^\kappa\omega_0})$. This implies T_ω is not injective.

Finally, we turn to the proof of the dense range claim for the case $\kappa < 0$. Since $\kappa < 0$ by assumption, ω has more poles in $\overline{\mathbb{D}}$ (and even in $\mathbb{D}$) than zeroes in $\mathbb{D}$. Thus to prove the dense range claim in this case, it suffices to show that $\kappa < 0$ implies that $T_{z^\kappa\omega_0}$ has dense range. We have $T_{z^\kappa\omega_0} = T_{z^\kappa}T_{\omega_0}$ and T_{z^κ} is surjective. Also, $\omega_0 \in \operatorname{Rat}(\mathbb{T})$ has no zeroes inside $\mathbb{D}$. So the proposition applies to ω_0, as shown in the first paragraph of the proof, and it follows that T_{ω_0} has dense range. But then also $T_{z^\kappa\omega_0} = T_{z^\kappa}T_{\omega_0}$ has dense range, and our claim follows. □

3. The spectrum of T_ω

In this section we determine the spectrum and various subparts of the spectrum of T_ω for the general case, $\omega \in \operatorname{Rat}$, as well as some refinements for the case where $\omega \in \operatorname{Rat}(\mathbb{T})$ is proper. In particular, we prove our main results, Theorems 1.1 and 1.2.

Note that for $\omega \in \operatorname{Rat}$ and $\lambda \in \mathbb{C}$ we have $\lambda I - T_\omega = T_{\lambda-\omega}$. Thus we can relate questions on the spectrum of T_ω to question on injectivity, surjectivity, closed rangeness, etc. for Toeplitz-like operators with an additional complex parameter. By this observation, the spectrum of T_ω, and its various subparts, can be determined using the results of Section 2.

Proof of Theorem 1.1. Since $\lambda I - T_\omega = T_{\lambda-\omega}$ and $T_{\lambda-\omega}$ is Fredholm if and only if $\lambda - \omega$ has no zeroes on $\mathbb{T}$, by Theorem 2.1, it follows that λ is in the essential spectrum if and only if $\lambda = \omega(e^{i\theta})$ for some $0 \leq \theta \leq 2\pi$. This shows that $\sigma_{\text{ess}}(T_\omega)$ is equal to $\omega(\mathbb{T})$.

To see that $\omega(\mathbb{T})$ is an algebraic curve, let $\omega = s/q$ with $s, q \in \mathcal{P}$ co-prime. Then $\lambda = u+iv = \omega(z)$ for $z = x+iy$ with $x^2+y^2 = 1$ if and only if $\lambda q(z)-s(z) = 0$. Denote $q(z) = q_1(x,y) + iq_2(x,y)$ and $s(z) = s_1(x,y) + is_2(x,y)$, where $z = x + iy$ and the functions q_1, q_2, s_1, s_2 are real polynomials in two variables. Then $\lambda = u+iv$ is on the curve $\omega(\mathbb{T})$ if and only if

$$\begin{aligned} q_1(x,y)u - q_2(x,y)v &= s_1(x,y), \\ q_2(x,y)u + q_1(x,y)v &= s_2(x,y), \\ x^2 + y^2 &= 1. \end{aligned}$$

Solving for u and v, this is equivalent to

$$\begin{aligned} (q_1(x,y)^2 + q_2(x,y)^2)u - (q_1(x,y)s_1(x,y) + q_2(x,y)s_2(x,y)) &= 0, \\ (q_1(x,y)^2 + q_2(x,y)^2)v - (q_1(x,y)s_2(x,y) - q_2(x,y)s_1(x,y)) &= 0, \\ x^2 + y^2 &= 1. \end{aligned}$$

This describes an algebraic curve in the plane.

For λ in the complement of the curve $\omega(\mathbb{T})$ the operator $\lambda I - T_\omega = T_{\lambda-\omega}$ is Fredholm, and according to Theorem 2.1 the index is given by

$$\text{Index}(\lambda - T_\omega) = \sharp\{\text{ poles of } \omega \text{ in } \overline{\mathbb{D}}\} - \sharp\{\text{zeroes of } \omega - \lambda \text{ inside } \mathbb{D}\},$$

taking the multiplicities of the poles and zeroes into account. Indeed, $\lambda - \omega = \frac{\lambda q - s}{q}$ and since q and s are co-prime, $\lambda q - s$ and q are also co-prime. Thus Theorem 2.1 indeed applies to $T_{\lambda-\omega}$. Furthermore, $\lambda - \omega$ has the same poles as ω, i.e., the roots of q. Likewise, the zeroes of $\lambda - \omega$ coincide with the roots of the polynomial $\lambda q - s$. Since the roots of this polynomial depend continuously on the parameter λ the number of them is constant on connected components of the complement of the curve $\omega(\mathbb{T})$.

That the index is constant on connected components of the complement of the essential spectrum in fact holds for any unbounded densely defined operator (see [15, Theorem VII.5.2]; see also [4, Proposition XI.4.9] for the bounded case; for a much more refined analysis of this point see [6]).

Finally, the relation between the index of $T_{\lambda-\omega}$ and λ being in the resolvent set, point spectrum or residual spectrum follows directly by applying the last part of Theorem 2.1 to $T_{\lambda-\omega}$. □

Next we prove Theorem 1.2 using some of the new results on T_ω derived in Section 2.

Proof of Theorem 1.2. That the two formulas for the numbers k_q, k_λ^- and k_λ^0 coincides follows from the analysis in the proof of Theorem 1.1, using the co-primeness of $\lambda q - s$ and q. By Theorem 2.1, $T_{\lambda-\omega}$ is Fredholm if and only if $k_\lambda^0 = 0$, proving the formula for $\sigma_{\text{ess}}(T_\omega)$. The formula for the resolvent set follows directly from the fact that the resolvent set is contained in the complement of $\sigma_{\text{ess}}(T_\omega)$, i.e., $k_\lambda^0 = 0$, and that it there coincides with the set of λ's for which the index of $T_{\lambda-\omega}$ is zero, together with the formula for $\text{Index}(T_{\lambda-\omega})$ obtained in Theorem 2.1.

The formulas for the point spectrum and residual spectrum follow by applying the criteria for injectivity and closed rangeness of Proposition 2.4 to $T_{\lambda-\omega}$ together with the fact that $T_{\lambda-\omega}$ must be either injective or have dense range.

For the formula for the continuous spectrum, note that $\sigma_c(T_\omega)$ must be contained in the essential spectrum, i.e., $k_\lambda^0 > 0$. The condition $k_\lambda^- \leq k_q \leq k_\lambda^- + k_\lambda^0$ excludes precisely that λ is in the point or residual spectrum. □

For the case where $\omega \in \text{Rat}(\mathbb{T})$ is proper we can be a bit more precise.

Theorem 3.1. *Let $\omega \in \text{Rat}(\mathbb{T})$ be proper, say $\omega = s/q$ with $s, q \in \mathcal{P}$ co-prime. Thus $\deg(s) \leq \deg(q)$ and all roots of q are on $\mathbb{T}$. Let a be the leading coefficient of q and b the coefficient of s corresponding to the monomial $z^{\deg(q)}$, hence $b = 0$ if and only if ω is strictly proper. Then $\sigma_r(T_\omega) = \emptyset$, and the point spectrum is given by*

$$\sigma_p(T_\omega) = \omega(\mathbb{C}\backslash\overline{\mathbb{D}}) \cup \{b/a\}.$$

Here $\omega(\mathbb{C}\backslash\overline{\mathbb{D}}) = \{\omega(z) \mid z \in \mathbb{C}\backslash\overline{\mathbb{D}}\}$. In particular, if ω is strictly proper, then $0 = b/a$ is in $\sigma_p(T_\omega)$. Finally,

$$\sigma_c(T_\omega) = \{\lambda \in \mathbb{C} \mid k_\lambda^0 > 0 \text{ and all roots of } \lambda q - s \text{ are in } \overline{\mathbb{D}}\}.$$

Proof. Let $\omega = s/q \in \text{Rat}(\mathbb{T})$ be proper with $s, q \in \mathcal{P}$ co-prime. Then $k_q = \deg(q)$. Since $\deg(s) \leq \deg(q)$, for any $\lambda \in \mathbb{C}$ we have

$$k_\lambda^- + k_\lambda^0 \leq \deg(\lambda q - s) \leq \deg(q) = k_q.$$

It now follows directly from (1.3) that $\sigma_r(T_\omega) = \emptyset$ and $\sigma_c(T_\omega) = \{\lambda \in \mathbb{C} \mid k_\lambda^0 > 0, k_\lambda^- + k_\lambda^0 = \deg(q)\}$. To determine the point spectrum, again using (1.3), one has to determine when strict inequality occurs. We have $\deg(\lambda q - s) < \deg(q)$ precisely when the leading coefficient of λq is cancelled in $\lambda q - s$ or if $\lambda = 0$ and $\deg(s) < \deg(q)$. Both cases correspond to $\lambda = b/a$. For the other possibility of having strict inequality, $k_\lambda^- + k_\lambda^0 < \deg(\lambda q - s)$, note that this happens precisely when $\lambda q - s$ has a root outside $\overline{\mathbb{D}}$, or equivalently $\lambda = \omega(z)$ for a $z \notin \overline{\mathbb{D}}$. □

4. The spectrum may be unbounded, the resolvent set empty

In this section we present some first examples, showing that the spectrum can be unbounded and the resolvent set may be empty.

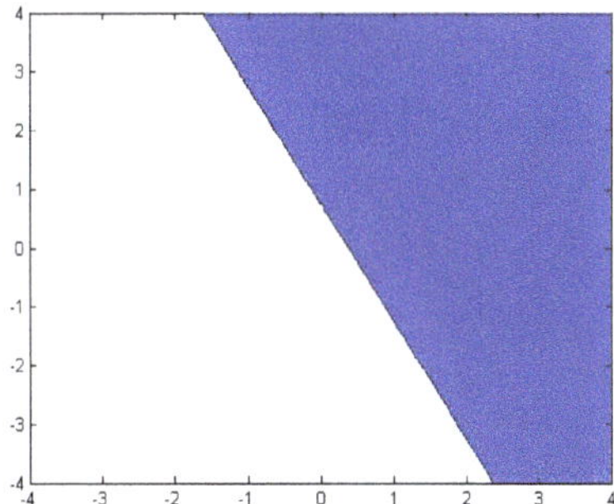

FIGURE 1. Spectrum of T_ω where $\omega(z) = \frac{z-\alpha}{z-1}$, with $\alpha = -\frac{i}{2}$.

Example 4.1. Let $\omega(z) = \frac{z-\alpha}{z-1}$ for some $1 \neq \alpha \in \mathbb{C}$, say $\alpha = a + ib$, with a and b real. Let $L \subset \mathbb{C}$ be the line given by

$$L = \{z = x + iy \in \mathbb{C} \mid 2by = (a^2 + b^2 - 1) + (2 - 2a)x\}. \tag{4.1}$$

Then we have

$$\rho(T_\omega) = \omega(\mathbb{D}), \quad \sigma_{\text{ess}}(T_\omega) = \omega(\mathbb{T}) = L = \sigma_{\text{c}}(T_\omega),$$
$$\sigma_{\text{p}}(T_\omega) = \omega(\mathbb{C}\backslash\overline{\mathbb{D}}), \quad \sigma_{\text{r}}(T_\omega) = \emptyset.$$

Moreover, the point spectrum of T_ω is the open half-plane determined by L that contains 1 and the resolvent set of T_ω is the other open half-plane determined by L, see Figure 1.

To see that these claims are true note that for $\lambda \neq 1$

$$\lambda - \omega(z) = \frac{z(\lambda - 1) + \alpha - \lambda}{z - 1} = \frac{1}{\lambda - 1} \frac{z + \frac{\alpha - \lambda}{\lambda - 1}}{z - 1},$$

while for $\lambda = 1$ we have $\lambda - \omega(z) = \frac{\alpha - \lambda}{z-1}$. Thus $\lambda = 1 \in \sigma_{\text{p}}(T_\omega)$ for every $1 \neq \alpha \in \mathbb{C}$ as in that case $k_q = 1 > 0 = k_\lambda^- + k_\lambda^0$. For $\lambda \neq 1$, $\lambda - \omega$ has a zero at $\frac{\alpha - \alpha}{\lambda - 1}$ of multiplicity one. For $\lambda = x + iy$ we have $|\alpha - \lambda| = |\lambda - 1|$ if and only if $(a-x)^2 + (b-y)^2 = (x-1)^2 + y^2$, which in turn is equivalent to $2by = (a^2 + b^2 - 1) + (2 - 2a)x$. Hence the zero of $\lambda - \omega$ is on $\mathbb{T}$ precisely when λ is on the line L. This shows $\sigma_{\text{ess}}(T_\omega) = L$. One easily verifies that the point spectrum and resolvent set correspond to the two half-planes indicated above and that these coincide with the images of ω under $\mathbb{C}\backslash\overline{\mathbb{D}}$ and $\mathbb{D}$, respectively. Since $\lambda - \omega$ can have at most one zero, it is clear from Theorem 1.2 that $\sigma_{\text{r}}(T_\omega) = \emptyset$, so that $\sigma_{\text{c}}(T_\omega) = L = \sigma_{\text{ess}}(T_\omega)$, as claimed. □

Example 4.2. Let $\omega(z) = \frac{1}{(z-1)^k}$ for some positive integer $k > 1$. Then

$$\sigma_{\text{p}}(T_\omega) = \sigma(T_\omega) = \mathbb{C}, \quad \sigma_r(T_\omega) = \sigma_{\text{c}}(T_\omega) = \rho(T_\omega) = \emptyset,$$

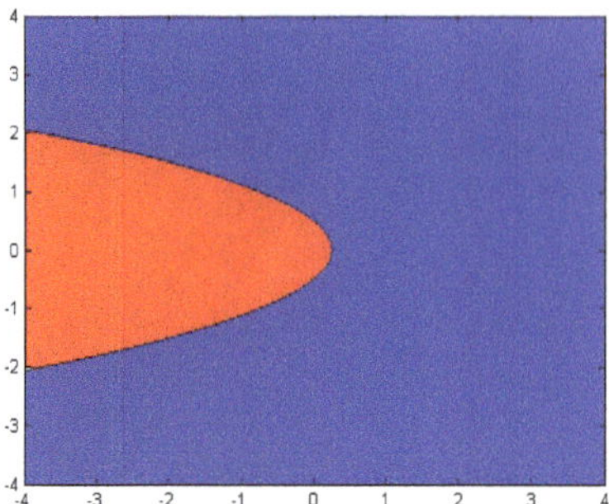

FIGURE 2. Spectrum of T_ω where $\omega(z) = \frac{1}{(z-1)^2}$.

and the essential spectrum is given by

$$\sigma_{\rm ess}(T_\omega) = \omega(\mathbb{T}) = \{(it - \tfrac{1}{2})^k \mid t \in \mathbb{R}\}.$$

For $k = 2$ the situation is as in Figure 2; one can check that the curve $\omega(\mathbb{T})$ is the parabola $\mathrm{Re}(z) = \frac{1}{4} - \mathrm{Im}(z)^2$. (Recall that different colors indicate different Fredholm index, as explained at the end of the introduction.)

To prove the statements, we start with the observation that for $|z| = 1$, $\frac{1}{z-1}$ is of the form $it - \frac{1}{2}, t \in \mathbb{R}$. Thus for $z \in \mathbb{T}$ with $\frac{1}{z-1} = it - \frac{1}{2}$ we have

$$\omega(z) = \frac{1}{(z-1)^k} = (z-1)^{-k} = (it - \tfrac{1}{2})^k.$$

This proves the formula for $\sigma_{\rm ess}(T_\omega)$. For $\lambda = re^{i\theta} \neq 0$ we have

$$\lambda - \omega(z) = \frac{\lambda(z-1)^k - 1}{(z-1)^k}.$$

Thus $\lambda - \omega(z) = 0$ if and only if $(z-1)^k = \lambda^{-1}$, i.e., $z = 1 + r^{-1/k}e^{i(\theta+2\pi l)/k}$ for $l = 0, \ldots, k-1$. Thus the zeroes of $\lambda - \omega$ are k equally spaced points on the circle with center 1 and radius $r^{-1/k}$. Clearly, since $k > 1$, not all zeroes can be inside $\overline{\mathbb{D}}$, so $k_q > k_\lambda^0 + k_\lambda^-$, and thus $\lambda \in \sigma_{\rm p}(T_\omega)$. It follows directly from Theorem 1.2 that $0 \in \sigma_{\rm p}(T_\omega)$. Thus $\sigma_{\rm p}(T_\omega) = \mathbb{C}$, as claimed. The curve $\omega(\mathbb{T})$ divides the plane into several regions on which the index is a positive constant integer, but the index may change between different regions. □

5. The essential spectrum need not be connected

For a continuous function ω on the unit circle it is obviously the case that the curve $\omega(\mathbb{T})$ is a connected and bounded curve in the complex plane, and hence the essential spectrum of T_ω is connected in this case. It was proved by Widom [16] that also for ω piecewise continuous the essential spectrum of T_ω is connected, and it is the image of a curve related to $\omega(\mathbb{T})$ (roughly speaking, filling the jumps with line segments). Douglas [5] proved that even for $\omega \in L^\infty$ the essential spectrum of T_ω as an operator on H^2 is connected. In [2] the question is raised whether or not the essential spectrum of T_ω as an operator on H^p is always connected when $\omega \in L^\infty$.

Returning to our case, where ω is a rational function possibly with poles on the unit circle, clearly when ω does have poles on the unit circle it is not a-priori necessary that $\sigma_{\text{ess}}(T_\omega) = \omega(\mathbb{T})$ is connected. We shall present examples that show that indeed the essential spectrum need not be connected, in contrast with the case where $\omega \in L^\infty$.

Consider $\omega = s/q \in \text{Rat}(\mathbb{T})$ with $s, q \in \mathcal{P}$ with real coefficients. In that case $\overline{\omega(z)} = \omega(\overline{z})$, so that the essential spectrum is symmetric with respect to the real axis. In particular, if $\omega(\mathbb{T}) \cap \mathbb{R} = \emptyset$, then the essential spectrum is disconnected. The converse direction need not be true, since the essential spectrum can consist of several disconnected parts on the real axis, as the following example shows.

Example 5.1. Consider $\omega(z) = \frac{z}{z^2+1}$. Then

$$\sigma_{\text{ess}}(T_\omega) = \omega(\mathbb{T}) = (-\infty, -1] \cup [1, \infty) = \sigma_{\text{c}}(T_\omega), \quad \sigma_{\text{p}}(T_\omega) = \mathbb{C} \backslash \omega(\mathbb{T}),$$

and thus $\sigma_{\text{r}}(T_\omega) = \rho(T_\omega) = \emptyset$. Further, for $\lambda \notin \omega(\mathbb{T})$ the Fredholm index is 1.

Indeed, note that for $z = e^{i\theta} \in \mathbb{T}$ we have

$$\omega(z) = \frac{1}{z + z^{-1}} = \frac{1}{2\,\text{Re}(z)} = \frac{1}{2\cos(\theta)} \in \mathbb{R}.$$

Letting θ run from 0 to 2π, one finds that $\omega(\mathbb{T})$ is equal to the union of $(-\infty, -1]$ and $[1, \infty)$, as claimed. Since ω is strictly proper, $\sigma_{\text{r}}(T_\omega) = \emptyset$ by Theorem 3.1. Applying Theorem 2.1 to T_ω we obtain that T_ω is Fredholm with index 1. Hence T_ω is not injective, so that $0 \in \sigma_{\text{p}}(T_\omega)$. However, since $\mathbb{C}\backslash\omega(\mathbb{T})$ is connected, it follows from Theorem 1.1 that the index of $T_{\lambda-\omega}$ is equal to 1 on $\mathbb{C}\backslash\omega(\mathbb{T})$, so that $\mathbb{C}\backslash\omega(\mathbb{T}) \subset \sigma_{\text{p}}(T_\omega)$. However, for λ on $\omega(\mathbb{T})$ the function $\lambda - \omega$ has two zeroes on $\mathbb{T}$ as well as two poles on $\mathbb{T}$. It follows that $\omega(\mathbb{T}) = \sigma_{\text{c}}(T_\omega)$, which shows all the above formulas for the spectral parts hold.

As a second example we specify q to be $z^2 - 1$ and determine a condition on s that guarantees $\sigma_{\text{ess}}(T_\omega) = \omega(\mathbb{T})$ in not connected.

Example 5.2. Consider $\omega(z) = \frac{s(z)}{z^2-1}$ with $s \in \mathcal{P}$ a polynomial with real coefficients. Then for $z \in \mathbb{T}$ we have

$$\omega(z) = \frac{\overline{z}s(z)}{z - \overline{z}} = \frac{\overline{z}s(z)}{-2i\,\text{Im}(z)} = \frac{i\overline{z}s(z)}{2\,\text{Im}(z)}, \quad \text{so that} \quad \text{Im}(\omega(z)) = \frac{\text{Re}(\overline{z}s(z))}{2\,\text{Im}(z)}.$$

Hence $\text{Im}(\omega(z)) = 0$ if and only if $\text{Re}(\overline{z}s(z)) = 0$. Say $s(z) = \sum_{j=0}^k a_j z^j$. Then for $z \in \mathbb{T}$ we have

$$\text{Re}(\overline{z}s(z)) = \sum_{j=0}^{k} a_j \text{Re}(z^{j-1}).$$

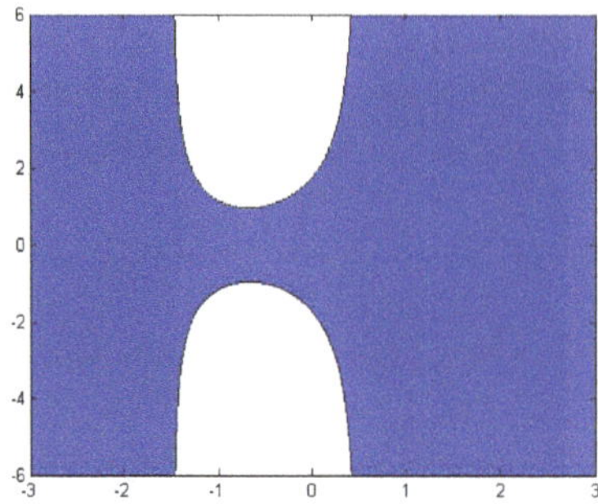

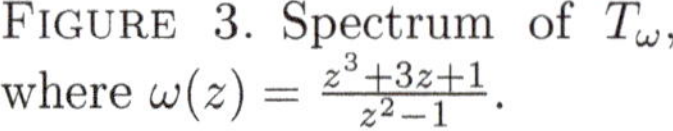

FIGURE 3. Spectrum of T_ω, where $\omega(z) = \frac{z^3+3z+1}{z^2-1}$.

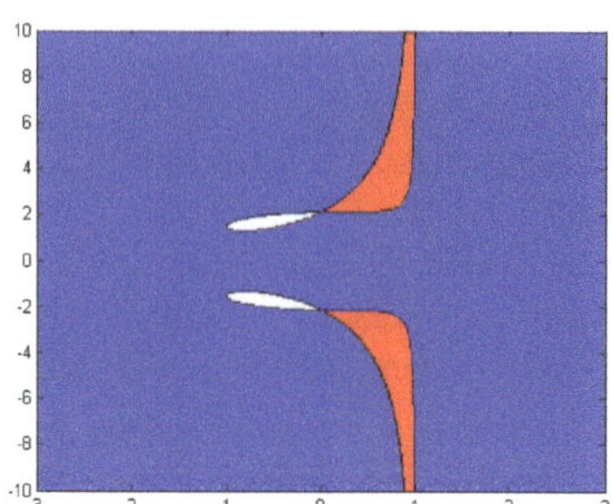

FIGURE 4. Spectrum of T_ω, where $\omega(z) = \frac{z^4+3z+1}{z^2-1}$.

Since $|\mathrm{Re}(z^j)| \leq 1$, we obtain that $|\mathrm{Re}(\overline{z}s(z))| > 0$ for all $z \in \mathbb{T}$ in case $2|a_1| > \sum_{j=0}^{k} |a_j|$. Hence in that case $\omega(\mathbb{T}) \cap \mathbb{R} = \emptyset$ and we find that the essential spectrum is disconnected in $\mathbb{C}$.

We consider two concrete examples, where this criteria is satisfied.

Firstly, take $\omega(z) = \frac{z^3+3z+1}{z^2-1}$. Then

$$\omega(e^{i\theta}) = \frac{1}{2}(2\cos\theta - 1) - \frac{i}{2}\frac{2(\cos\theta + 1/4)^2 + 7/4}{\sin\theta},$$

which is the curve given in Figure 3, that also shows the spectrum and resolvent as well as the essential spectrum.

Secondly, take $\omega(z) = \frac{z^4+3z+1}{z^2-1}$. Figure 4 shows the spectrum and resolvent and the essential spectrum. Observe that this is also a case where the resolvent is a bounded set.

6. A parametric example

In this section we take $\omega_k(z) = \frac{z^k+\alpha}{(z-1)^2}$ for $\alpha \in \mathbb{C}, \alpha \neq -1$ and for various integers $k \geq 1$. Note that the case $k = 0$ was dealt with in Example 4.2 (after scaling with the factor $1 + \alpha$). The zeroes of $\lambda - \omega_k$ are equal to the roots of

$$p_{\lambda,\alpha,k}(z) = \lambda q(z) - s(z) = \lambda(z-1)^2 - (z^k + \alpha).$$

Thus, λ is in the resolvent set $\rho(T_{\omega_k})$ whenever $p_{\lambda,\alpha,k}$ has at least two roots in $\mathbb{D}$ and no roots on $\mathbb{T}$. Note that Theorem 3.1 applies in case $k = 1, 2$. We discuss the first of these two cases in detail, and then conclude with some figures that contain possible configurations of other cases.

Example 6.1. Let $\omega(z) = \omega_1(z) = \frac{z+\alpha}{(z-1)^2}$ for $\alpha \neq -1$. Then

$$\sigma_{\mathrm{ess}}(T_\omega) = \omega(\mathbb{T}) = \{(it - \tfrac{1}{2}) + (1+\alpha)(it - \tfrac{1}{2})^2 \mid t \in \mathbb{R}\}. \tag{6.1}$$

Define the circle

$$\mathbb{T}(-\tfrac{1}{2},\tfrac{1}{2}) = \{z \in \mathbb{C} \mid |z+\tfrac{1}{2}| = \tfrac{1}{2}\},$$

and write $\mathbb{D}(-\frac{1}{2},\frac{1}{2})$ for the open disc formed by the interior of $\mathbb{T}(-\frac{1}{2},\frac{1}{2})$ and $\mathbb{D}^c(-\frac{1}{2},\frac{1}{2})$ for the open exterior of $\mathbb{T}(-\frac{1}{2},\frac{1}{2})$.

For $\alpha \notin \mathbb{T}(-\frac{1}{2},\frac{1}{2})$ the curve $\omega(\mathbb{T})$ is equal to the parabola in $\mathbb{C}$ given by

$$\omega(\mathbb{T}) = \{-(\alpha+1)(x(y)+iy) \mid y \in \mathbb{R}\}, \qquad \text{where}$$
$$x(y) = \frac{|\alpha+1|^4}{(|\alpha|^2+\mathrm{Re}(\alpha))^2}y^2 + \frac{(\mathrm{Re}(\alpha)+1)|\alpha+1|^2\mathrm{Im}(\alpha)}{(|\alpha|^2+\mathrm{Re}(\alpha))^2}y + \frac{|\alpha|^2(1-|\alpha|^2)}{(|\alpha|^2+\mathrm{Re}(\alpha))^2},$$

while for $\alpha \in \mathbb{T}(-\frac{1}{2},\frac{1}{2})$ the curve $\omega(\mathbb{T})$ becomes the half-line given by

$$\omega(\mathbb{T}) = \left\{-(\alpha+1)r - \frac{(\alpha+1)(1+2\overline{\alpha})}{4(1-|\alpha|^2)} \mid r \geq 0\right\}.$$

As ω is strictly proper, we have $\sigma_{\mathrm{r}}(T_\omega) = \emptyset$. For the remaining parts of the spectrum we consider three cases.

(i) For $\alpha \in \mathbb{D}(-\frac{1}{2},\frac{1}{2})$ the points $-\frac{1}{2}$ and 0 are separated by the parabola $\omega(\mathbb{T})$ and the connected component of $\mathbb{C}\backslash\omega(\mathbb{T})$ that contains $-\frac{1}{2}$ is equal to $\rho(T_\omega)$, while the connected component that contains 0 is equal to $\sigma_{\mathrm{p}}(T_\omega)$. Finally, $\sigma_{\mathrm{ess}}(T_\omega) = \omega(\mathbb{T}) = \sigma_{\mathrm{c}}(T_\omega)$.

(ii) For $\alpha \in \mathbb{T}(-\frac{1}{2},\frac{1}{2})$ we have

$$\rho(T_\omega) = \emptyset, \quad \sigma_{\mathrm{c}}(T_\omega) = \omega(\mathbb{T}) = \sigma_{\mathrm{ess}}(T_\omega), \quad \sigma_{\mathrm{p}}(T_\omega) = \mathbb{C}\backslash\omega(\mathbb{T}),$$

and for each $\lambda \in \omega(\mathbb{T})$, $\lambda - \omega$ has two zeroes on $\mathbb{T}$.

(iii) For $\alpha \in \mathbb{D}^c(-\frac{1}{2},\frac{1}{2})$ we have $\sigma_{\mathrm{p}}(T_\omega) = \mathbb{C}$, and hence $\rho(T_\omega) = \sigma_{\mathrm{c}}(T_\omega) = \emptyset$.

The proof of these statements will be separated into three steps.

Step 1. We first determine the formula of $\omega(\mathbb{T})$ and show this is a parabola. Note that

$$\omega(z) = \frac{z+\alpha}{(z-1)^2} = \frac{z-1}{(z-1)^2} + \frac{1+\alpha}{(z-1)^2} = \frac{1}{z-1} + (\alpha+1)\frac{1}{(z-1)^2}.$$

Let $|z| = 1$. Then $\frac{1}{z-1}$ is of the form $it - \frac{1}{2}$ with $t \in \mathbb{R}$. So $\omega(\mathbb{T})$ is the curve

$$\omega(\mathbb{T}) = \{(it-\tfrac{1}{2}) + (\alpha+1)(it-\tfrac{1}{2})^2 \mid t \in \mathbb{R}\}.$$

Thus (6.1) holds. Now observe that

$$\begin{aligned}
&(it-\tfrac{1}{2}) + (\alpha+1)(it-\tfrac{1}{2})^2\\
&= -t^2(\alpha+1) + t(i-(\alpha+1)i) + (-\tfrac{1}{2}+\tfrac{1}{4}(\alpha+1))\\
&= -t^2(\alpha+1) + (-\alpha i)t + (-\tfrac{1}{4}+\tfrac{1}{4}\alpha)\\
&= -(\alpha+1)\left(t^2 + t\frac{\alpha i}{\alpha+1} - \frac{1}{4}\left(\frac{\alpha-1}{\alpha+1}\right)\right).
\end{aligned}$$

The prefactor $-(1+\alpha)$ acts as a rotation combined with a real scalar multiplication, so $\omega(\mathbb{T})$ is also given by

$$\omega(\mathbb{T}) = -(\alpha+1)\left\{t^2 + t\left(\frac{\alpha i}{\alpha+1}\right) - \frac{1}{4}\left(\frac{\alpha-1}{\alpha+1}\right) \mid t \in \mathbb{R}\right\}. \tag{6.2}$$

Thus if the above curve is a parabola, so is $\omega(\mathbb{T})$. Write

$$x(t) = \mathrm{Re}\left(t^2 + t\frac{\alpha i}{1+\alpha} - \frac{1}{4}\left(\frac{\alpha-1}{\alpha+1}\right)\right),$$

$$y(t) = \mathrm{Im}\left(t^2 + t\frac{\alpha i}{1+\alpha} - \frac{1}{4}\left(\frac{\alpha-1}{\alpha+1}\right)\right).$$

Since

$$\frac{\alpha i}{\alpha+1} = \frac{-\mathrm{Im}(\alpha) + i(|\alpha|^2 + \mathrm{Re}(\alpha))}{|\alpha+1|^2} \quad \text{and} \quad \frac{\alpha-1}{\alpha+1} = \frac{(|\alpha|^2-1) + 2i\mathrm{Im}(\alpha)}{|\alpha+1|^2}$$

we obtain that

$$x(t) = t^2 - \frac{\mathrm{Im}(\alpha)}{|\alpha+1|^2}t - \frac{|\alpha|^2-1}{4|\alpha+1|^2}, \quad y(t) = \frac{|\alpha|^2 + \mathrm{Re}(\alpha)}{|\alpha+1|^2}t - \frac{\mathrm{Im}(\alpha)}{2|\alpha+1|^2}.$$

Note that $|\alpha + \frac{1}{2}|^2 = |\alpha|^2 + \mathrm{Re}(\alpha) + \frac{1}{4}$. Therefore, we have $|\alpha|^2 + \mathrm{Re}(\alpha) = 0$ if and only if $|\alpha + \frac{1}{2}| = \frac{1}{2}$. Thus $|\alpha|^2 + \mathrm{Re}(\alpha) = 0$ holds if and only if α is on the circle $\mathbb{T}(-\frac{1}{2}, \frac{1}{2})$.

In case $\alpha \notin \mathbb{T}(-\frac{1}{2}, \frac{1}{2})$, i.e., $|\alpha|^2 + \mathrm{Re}(\alpha) \neq 0$, we can express t in terms of y, and feed this into the formula for x. One can then compute that

$$x = \frac{|\alpha+1|^4}{(|\alpha|^2 + \mathrm{Re}(\alpha))^2}y^2 + \frac{(\mathrm{Re}(\alpha)+1)|\alpha+1|^2\mathrm{Im}(\alpha)}{(|\alpha|^2 + \mathrm{Re}(\alpha))^2}y + \frac{|\alpha|^2(1-|\alpha|^2)}{(|\alpha|^2 + \mathrm{Re}(\alpha))^2}.$$

Inserting this formula into (6.2), we obtain the formula for $\omega(\mathbb{T})$ for the case where $\alpha \notin \mathbb{T}(-\frac{1}{2}, \frac{1}{2})$.

In case $\alpha \in \mathbb{T}(-\frac{1}{2}, \frac{1}{2})$, i.e., $|\alpha|^2 + \mathrm{Re}(\alpha) = 0$, we have

$$|\alpha+1|^2 = 1 - |\alpha|^2 = 1 + \mathrm{Re}(\alpha), \quad \mathrm{Im}(\alpha)^2 = |\alpha|^2(1-|\alpha|^2)$$

and using these identities one can compute that

$$y(t) = \frac{-2\mathrm{Im}(\alpha)}{4(1-|\alpha|^2)} \quad \text{and} \quad x(t) = \left(t - \frac{\mathrm{Im}(\alpha)}{2(1-|\alpha|^2)}\right)^2 + \frac{1+2\mathrm{Re}(\alpha)}{4(1-|\alpha|^2)}.$$

Thus $\{x(t) + iy(t) \mid t \in \mathbb{R}\}$ determines a half-line in $\mathbb{C}$, parallel to the real axis and starting in $\frac{1+2\overline{\alpha}}{4(1-|\alpha|^2)}$ and moving in positive direction. It follows that $\omega(\mathbb{T})$ is the half-line

$$\omega(\mathbb{T}) = \left\{-(\alpha+1)r - \frac{(\alpha+1)(1+2\overline{\alpha})}{4(1-|\alpha|^2)} \mid r \geq 0\right\},$$

as claimed.

Step 2. Next we determine the various parts of the spectrum in $\mathbb{C}\backslash\omega(\mathbb{T})$. Since ω is strictly proper, Theorem 3.1 applies, and we know $\sigma_{\mathrm{r}}(T_\omega) = \emptyset$ and $\sigma_{\mathrm{p}} = \omega(\mathbb{C}\backslash\overline{\mathbb{D}}) \cup \{0\}$.

For $k = 1$, the polynomial $p_{\lambda,\alpha}(z) = p_{\lambda,\alpha,1}(z) = \lambda z^2 - (1 + 2\lambda)z + \lambda - \alpha$ has roots

$$\frac{-(1 + 2\lambda) \pm \sqrt{1 + 4\lambda(1 + \alpha)}}{2\lambda}.$$

We consider three cases, depending on whether α is inside, on or outside the circle $\mathbb{T}(-\frac{1}{2}, \frac{1}{2})$.

Assume $\alpha \in \mathbb{D}(-\frac{1}{2}, \frac{1}{2})$. Then $\omega(\mathbb{T})$ is a parabola in $\mathbb{C}$. For $\lambda = -\frac{1}{2}$ we find that $\lambda - \omega$ has zeroes $\pm i\sqrt{1 + 2\alpha}$, which are both inside $\mathbb{D}$, because of our assumption. Thus $-\frac{1}{2} \in \rho(T_\omega)$, so that $\rho(T_\omega) \neq \emptyset$. Therefore the connected component of $\mathbb{C}\backslash\omega(\mathbb{T})$ that contains $-\frac{1}{2}$ is contained in $\rho(T_\omega)$, which must also contain $\omega(\mathbb{D})$. Note that $0 \in \omega(\mathbb{T})$ if and only if $|\alpha| = 1$. However, there is no intersection of the disc $\alpha \in \mathbb{D}(-\frac{1}{2}, \frac{1}{2})$ and the unit circle $\mathbb{T}$. Thus 0 is in $\sigma_{\mathrm{p}}(T_\omega)$, but not on $\omega(\mathbb{T})$. Hence 0 is contained in the connected component of $\mathbb{C}\backslash\omega(\mathbb{T})$ that does not contain $-\frac{1}{2}$. This implies that the connected component containing 0 is included in $\sigma_{\mathrm{p}}(T_\omega)$. This proves our claims for the case $\alpha \in \mathbb{D}(-\frac{1}{2}, \frac{1}{2})$.

Now assume $\alpha \in \mathbb{T}(-\frac{1}{2}, \frac{1}{2})$. Then $\omega(\mathbb{T})$ is a half-line, and thus $\mathbb{C}\backslash\omega(\mathbb{T})$ consists of one connected component. Note that the intersection of the disc determined by $|\alpha + \frac{1}{2}| < \frac{1}{2}$ and the unit circle consists of -1 only. But $\alpha \neq -1$, so it again follows that $0 \notin \omega(\mathbb{T})$. Therefore the $\mathbb{C}\backslash\omega(\mathbb{T}) = \sigma_{\mathrm{p}}(T_\omega)$. Moreover, the reasoning in the previous case shows that $\lambda = -\frac{1}{2}$ is in $\sigma_{\mathrm{c}}(T_\omega)$ since both zeroes of $-\frac{1}{2} - \omega$ are on $\mathbb{T}$.

Finally, consider that case where α is in the exterior of $\mathbb{T}(-\frac{1}{2}, \frac{1}{2})$, i.e., $|\alpha + \frac{1}{2}| > \frac{1}{2}$. In this case, $|\alpha| = 1$ is possible, so that $0 \in \sigma_{\mathrm{p}}(T_\omega)$ could be on $\omega(\mathbb{T})$. We show that $\alpha = \omega(0) \in \omega(\mathbb{D})$ is in $\sigma_{\mathrm{p}}(T_\omega)$. If $\alpha = 0$, this is clearly the case. So assume $\alpha \neq 0$. The zeroes of $\alpha - \omega$ are then equal to 0 and $\frac{1+2\alpha}{\alpha}$. Note that $|\frac{1+2\alpha}{\alpha}| > 1$ if and only if $|1 + 2\alpha|^2 - |\alpha|^2 > 0$. Moreover, we have

$$|1 + 2\alpha|^2 - |\alpha|^2 = 3|\alpha|^2 + 4\mathrm{Re}(\alpha) + 1 = 3|\alpha + \tfrac{2}{3}|^2 - \tfrac{1}{3}.$$

Thus, the second zero of $\alpha - \omega$ is outside $\overline{\mathbb{D}}$ if and only if $|\alpha + \frac{2}{3}|^2 > \frac{1}{9}$. Since the disc indicated by $|\alpha + \frac{2}{3}| \leq \frac{1}{3}$ is contained in the interior of $\mathbb{T}(-\frac{1}{2}, \frac{1}{2})$, it follows that for α satisfying $|\alpha + \frac{1}{2}| > \frac{1}{2}$ one zero of $\alpha - \omega$ is outside $\overline{\mathbb{D}}$, and thus $\omega(0) = \alpha \in \sigma_{\mathrm{p}}(T_\omega)$. Note that

$$\mathbb{C} = \omega(\mathbb{C}) = \omega(\mathbb{D}) \cup \omega(\mathbb{T}) \cup \omega(\mathbb{C}\backslash\overline{\mathbb{D}}),$$

and that $\omega(\mathbb{D})$ and $\omega(\mathbb{C}\backslash\overline{\mathbb{D}})$ are connected components, both contained in $\sigma_{\mathrm{p}}(T_\omega)$. This shows that $\mathbb{C}\backslash\omega(\mathbb{T})$ is contained in $\sigma_{\mathrm{p}}(T_\omega)$.

Step 3. In the final part we prove the claim regarding the essential spectrum $\sigma_{\mathrm{ess}}(T_\omega) = \omega(\mathbb{T})$. Let $\lambda \in \omega(\mathbb{T})$ and write z_1 and z_2 for the zeroes of $\lambda - \omega$. One of the zeroes must be on $\mathbb{T}$, say $|z_1| = 1$. Then $\lambda \in \sigma_{\mathrm{p}}(\mathbb{T})$ if and only if $|z_1 z_2| = |z_2| > 1$. From the form of $p_{\lambda,\alpha}$ determined above we obtain that

$$\lambda z^2 - (1 + 2\lambda)z + \lambda - \alpha = \lambda(z - z_1)(z - z_2).$$

Determining the constant term on the right-hand sides shows that $\lambda z_1 z_2 = \lambda - \alpha$. Thus

$$|z_2| = |z_1 z_2| = \frac{|\lambda - \alpha|}{|\lambda|}.$$

This shows that $\lambda \in \sigma_{\rm p}(T_\omega)$ if and only if $|\lambda - \alpha| > |\lambda|$, i.e., λ is in the half-plane containing zero determined by the line through $\frac{1}{2}\alpha$ perpendicular to the line segment from zero to α.

Consider the line given by $|\lambda - \alpha| = |\lambda|$ and the parabola $\omega(\mathbb{T})$, which is a half-line in case $\alpha \in \mathbb{T}(-\frac{1}{2}, \frac{1}{2})$. We show that $\omega(\mathbb{T})$ and the line intersect only for $\alpha \in \mathbb{T}(-\frac{1}{2}, \frac{1}{2})$, and that in the latter case $\omega(\mathbb{T})$ is contained in the line. Hence for each value of $\alpha \neq -1$, the essential spectrum consists of either point spectrum or of continuous spectrum, and for $\alpha \in \mathbb{T}(-\frac{1}{2}, \frac{1}{2})$ both zeroes of $\lambda - \omega$ are on $\mathbb{T}$, so that $\omega(\mathbb{T})$ is contained in $\sigma_{\rm c}(T_\omega)$.

As observed in (6.1), the parabola $\omega(\mathbb{T})$ is given by the parametrization $(it - \frac{1}{2})^2(\alpha + 1) + (it - \frac{1}{2})$ with $t \in \mathbb{R}$, while the line is given by the parametrization $\frac{1}{2}\alpha + si\alpha$ with $s \in \mathbb{R}$. Fix a $t \in \mathbb{R}$ and assume the point on $\omega(\mathbb{T})$ parameterized by t intersects with the line, i.e., assume there exists a $s \in \mathbb{R}$ such that:

$$\left(it - \tfrac{1}{2}\right)^2 (\alpha + 1) + \left(- \tfrac{1}{2}\right) = \tfrac{1}{2}\alpha + si\alpha,$$

Thus

$$\left(-t^2 - it + \tfrac{1}{4}\right)(\alpha + 1) + \left(it - \tfrac{1}{2}\right) = \tfrac{1}{2}\alpha + si\alpha,$$

and rewrite this as

$$i(-t(\alpha + 1) + t - \alpha s) + \left(\left(-t^2 + \tfrac{1}{4}\right)(\alpha + 1) - \tfrac{1}{2} - \tfrac{1}{2}\alpha\right) = 0,$$

which yields

$$-\alpha i(t + s) + (\alpha + 1)\left(-t^2 - \tfrac{1}{4}\right) = 0.$$

Since $t^2 + \frac{1}{4} > 0$, this certainly cannot happen in case $\alpha = 0$. So assume $\alpha \neq 0$. Multiply both sides by $-\overline{\alpha}$ to arrive at

$$|\alpha|^2 i(t + s) + (|\alpha|^2 + \overline{\alpha})\left(t^2 + \tfrac{1}{4}\right) = 0.$$

Separate the real and imaginary part to arrive at

$$(|\alpha|^2 + \mathrm{Re}(\alpha))\left(t^2 + \tfrac{1}{4}\right) + i(|\alpha|^2(t + s) - \left(t^2 + \tfrac{1}{4}\right)\mathrm{Im}(\alpha)) = 0.$$

Thus

$$(|\alpha|^2 + \mathrm{Re}(\alpha))\left(t^2 + \tfrac{1}{4}\right) = 0 \quad \text{and} \quad |\alpha|^2(t + s) = \left(t^2 + \tfrac{1}{4}\right)\mathrm{Im}(\alpha).$$

Since $t^2 + \frac{1}{4} > 0$, the first identity yields $|\alpha|^2 + \mathrm{Re}(\alpha) = 0$, which happens precisely when $\alpha \in \mathbb{T}(-\frac{1}{2}, \frac{1}{2})$. Thus there cannot be an intersection when $\alpha \notin \mathbb{T}(-\frac{1}{2}, \frac{1}{2})$. On the other hand, for $\alpha \in \mathbb{T}(-\frac{1}{2}, \frac{1}{2})$ the first identity always holds, while there always exists an $s \in \mathbb{R}$ that satisfies the second equation. Thus, in that case, for any $t \in \mathbb{R}$, the point on $\omega(\mathbb{T})$ parameterized by t intersects the line, and thus $\omega(\mathbb{T})$ must be contained in the line.

We conclude by showing that $\omega(\mathbb{T}) \subset \sigma_{\rm p}(T_\omega)$ when $|\alpha + \frac{1}{2}| > \frac{1}{2}$ and that $\omega(\mathbb{T}) \subset \sigma_{\rm c}(T_\omega)$ when $|\alpha + \frac{1}{2}| < \frac{1}{2}$. Recall that the two cases correspond to $|\alpha|^2 +$

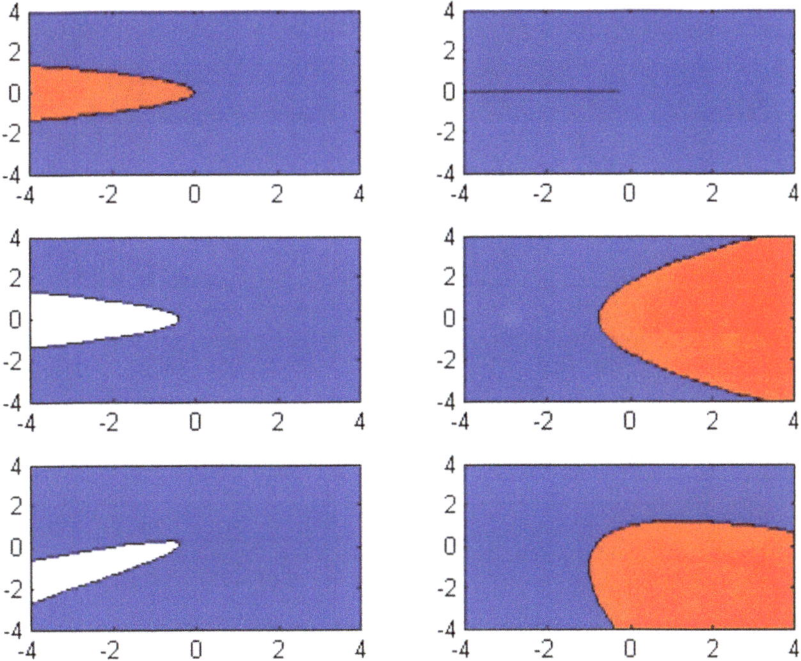

FIGURE 5. Spectrum of T_ω, where $\omega(z) = \frac{z+\alpha}{(z-1)^2}$ for some values of α, with $\alpha = 1$, and $\alpha = 0$ (top row left and right), $\alpha = 1/2$ and $\alpha = -2$ (middle row left and right), $\alpha = -\frac{1}{2} + \frac{1}{4}i$ and $\alpha = -2 + i$ (bottom row).

$\mathrm{Re}(\alpha) > 0$ and $|\alpha|^2 + \mathrm{Re}(\alpha) < 0$, respectively. To show that this is the case, we take the point on the parabola parameterized by $t = 0$, i.e., take $\lambda = \frac{1}{4}(\alpha+1) - \frac{1}{2} = \frac{1}{4}(\alpha - 1)$. Then $\lambda - \alpha = -\frac{1}{4}(3\alpha + 1)$. So

$$|\lambda - \alpha|^2 = \tfrac{1}{16}(9|\alpha|^2 + 6\mathrm{Re}(\alpha) + 1) \quad \text{and} \quad |\lambda|^2 = \tfrac{1}{16}(|\alpha^2| - 2\mathrm{Re}(\alpha) + 1).$$

It follows that $|\lambda - \alpha| > |\lambda|$ if and only if

$$\tfrac{1}{16}(9|\alpha|^2 + 6\mathrm{Re}(\alpha) + 1) > |\lambda|^2 = \tfrac{1}{16}(|\alpha^2| - 2\mathrm{Re}(\alpha) + 1),$$

or equivalently,

$$8(|\alpha|^2 + \mathrm{Re}(\alpha)) > 0.$$

This proves our claim for the case $|\lambda + \frac{1}{2}| > \frac{1}{2}$. The other claim follows by reversing the directions in the above inequalities.

Figure 5 presents some illustrations of the possible situations. □

The case $k = 2$ can be dealt with using the same techniques, and very similar results are obtained in that case.

The next examples deal with other cases of ω_k, now with $k > 2$.

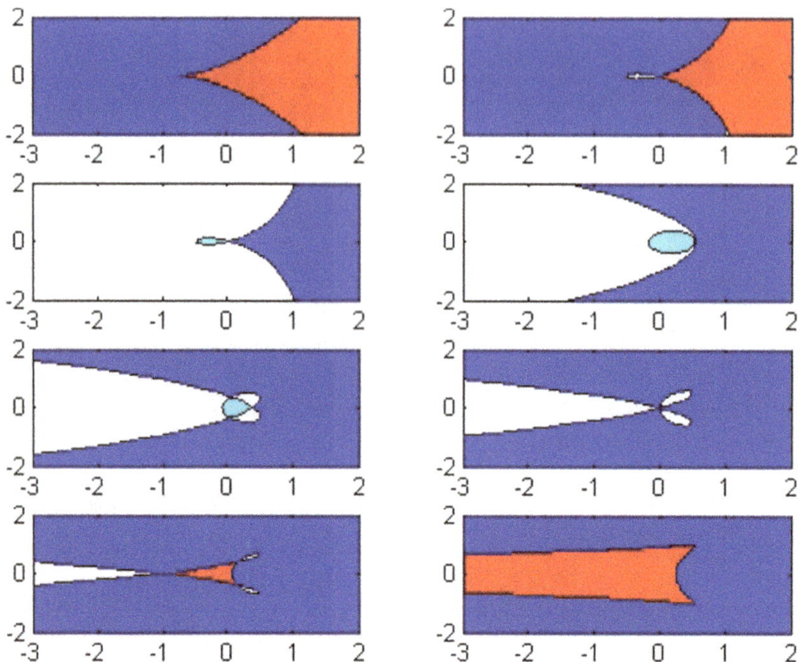

FIGURE 6. Spectrum of T_ω where $\omega(z) = \frac{z^3+\alpha}{(z-1)^2}$ for several values of α, with α being (left to right and top to bottom) respectively, -2, -1.05, -0.95, 0.3, 0.7, 1, 1.3, 2.

Example 6.2. Let $\omega = \frac{z^3+\alpha}{(z-1)^2}$. Then

$$\omega(z) = \frac{z^3+\alpha}{(z-1)^2} = (z-1) + 3 + \frac{3}{z-1} + \frac{1+\alpha}{(z-1)^2}.$$

For $z \in \mathbb{T}$, $\frac{1}{z-1}$ has the form $-\frac{1}{2} + ti, t \in \mathbb{R}$ and so $\omega(\mathbb{T})$ has the form

$$\omega(\mathbb{T}) = \left\{ \frac{1}{-\frac{1}{2}+ti} + 3 + 3\left(-\frac{1}{2}+ti\right) + (1+\alpha)\left(-\frac{1}{2}+ti\right)^2, \mid t \in \mathbb{R} \right\}.$$

Also $\lambda - \omega(z) = \frac{\lambda(z-1)^2 - z^3 - \alpha}{(z-1)^2}$ and so for invertibility we need the polynomial $p_{\lambda,\alpha}(z) = \lambda(z-1)^2 - z^3 - \alpha$ to have exactly two roots in $\mathbb{D}$. Since this is a polynomial of degree 3 the number of roots inside $\mathbb{D}$ can be zero, one, two or three, and the index of $\lambda - T_\omega$ correspondingly can be two, one, zero or minus one. Examples are given in Figure 6.

Example 6.3. To get some idea of possible other configurations we present some examples with other values of k.

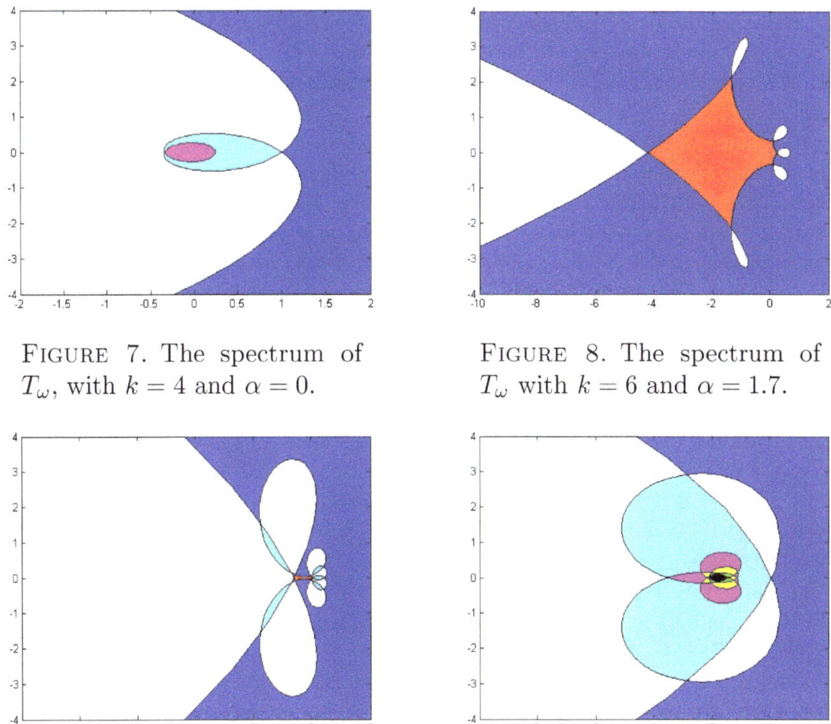

FIGURE 7. The spectrum of T_ω, with $k = 4$ and $\alpha = 0$.

FIGURE 8. The spectrum of T_ω with $k = 6$ and $\alpha = 1.7$.

FIGURE 9. The spectrum of T_ω for $k = 7$ and $\alpha = 1.1$ (left) and $k = 7$, $\alpha = 0.8$ (right)

For $\omega(z) = \frac{z^4}{(z-1)^2}$ (so $k = 4$ and $\alpha = 0$) the essential spectrum of T_ω is the curve in Figure 7, the white region is the resolvent set, and color coding for the Fredholm index is as earlier in the paper. For $\omega(z) = \frac{z^6+1.7}{(z-1)^2}$ (so $k = 6$ and $\alpha = 1.7$) see Figure 8, and as a final example Figure 9 presents the essential spectrum and spectrum for $\omega(z) = \frac{z^7+1.1}{(z-1)^2}$ and $\omega(z) = \frac{z^7+0.8}{(z-1)^2}$. In the latter figure color coding is as follows: the Fredholm index is -3 in the yellow region, -4 in the green region and -5 in the black region.

Acknowledgement. The present work is based on research supported in part by the National Research Foundation of South Africa. Any opinion, finding and conclusion or recommendation expressed in this material is that of the authors and the NRF does not accept any liability in this regard.

References

[1] A. Brown and P.R. Halmos, Algebraic properties of Toeplitz operators, *J. Reine Angw. Math.* **213** (1964), 89–102.

[2] A. Böttcher and B. Silbermann, *Analysis of Toeplitz operators. Second edition*, Springer Monographs in Mathematics, Springer–Verlag, Berlin, 2006.
[3] D.N. Clark, On the point spectrum on a Toeplitz operator, *Trans. Amer. Math. Soc.* **126** (1967), 251–266.
[4] J.B. Conway, *A course in Functional analysis. Second edition*, Springer, 1990.
[5] R.G. Douglas, *Banach algebra techniques in Operator Theory. Second Edition*, Graduate Texts in Mathematics, 179, Springer, New York, 1988.
[6] K.H. Förster and M.A. Kaashoek. The asymptotic behaviour of the reduced minumum modulus of a Fredholm operator. *Proc. Amer. Math. Soc.* **49** (1975), 123–131.
[7] I. Gohberg, On an application of the theory of normed rings to singular integral equations, *Uspehi. Matem. Nauk* **7** (1952), 149–156 [Russian].
[8] I. Gohberg, Toeplitz matrices composed of the Fourier coefficients of piecewise continuous functions, *Funktcional. Anal. Priložen.* **1** (1967), 91–92 [Russian].
[9] G.J. Groenewald, S. ter Horst, J. Jaftha and A.C.M. Ran, A Toeplitz-like operator with rational symbol having poles on the unit circle I: Fredholm properties, *Oper. Theory Adv. Appl.* **271** (2018), 239–268.
[10] P.R Halmos, A glimpse into Hilbert space. 1963 *Lectures on Modern Mathematics*, Vol. I pp. 1–22, Wiley, New York
[11] P. Hartman and A. Wintner, On the spectra of Toeplitz's Matrices, *Amer. J. Math.* **72** (1950), 359–366.
[12] P. Hartman and A. Wintner, The spectra of Toeplitz's matrices, *Amer. J. Math.* **76** (1954), 887–864.
[13] J. Janas, Unbounded Toeplitz operators in the Bargmann–Segal space, *Studia Math.* **99** (1991), 87–99.
[14] N.K. Nikolskii, *Operators, functions and systems: An easy reading. Vol. I: Hardy, Hankel and Toeplitz*, American Mathematical Society, Providence, RI, 2002.
[15] M. Schechter, *Principles of functional analysis*, Academic Press inc., New York, 1971.
[16] H. Widom, On the spectrum of a Toeplitz operator, *Pacific J. Math.* **14** (1964), 365–375.

G.J. Groenewald and S. ter Horst
Department of Mathematics
Unit for BMI
North-West University
Potchefstroom, 2531 South Africa
e-mail: Gilbert.Groenewald@nwu.ac.za
Sanne.TerHorst@nwu.ac.za

J. Jaftha
Numeracy Centre
University of Cape Town
Rondebosch 7701
Cape Town; South Africa
e-mail:Jacob.Jaftha@uct.ac.za

A.C.M. Ran
Department of Mathematics
Faculty of Science, VU Amsterdam
De Boelelaan 1081a
1081 HV Amsterdam, The Netherlands
and
Unit for BMI
North-West University
Potchefstroom, 2531 South Africa
e-mail: a.c.m.ran@vu.nl

Operator Theory:
Advances and Applications, Vol. 272, 155–212

The Twofold Ellis–Gohberg Inverse Problem in an Abstract Setting and Applications

S. ter Horst, M.A. Kaashoek and F. van Schagen

Dedicated to Joe Ball on the occasion of his seventieth birthday

Abstract. In this paper we consider a twofold Ellis–Gohberg type inverse problem in an abstract $*$-algebraic setting. Under natural assumptions, necessary and sufficient conditions for the existence of a solution are obtained, and it is shown that in case a solution exists, it is unique. The main result relies strongly on an inversion formula for a 2×2 block operator matrix whose off diagonal entries are Hankel operators while the diagonal entries are identity operators. Various special cases are presented, including the cases of matrix-valued L^1-functions on the real line and matrix-valued Wiener functions on the unit circle of the complex plane. For the latter case, it is shown how the results obtained in an earlier publication by the authors can be recovered.

Mathematics Subject Classification (2010). Primary: 47A56; Secondary: 15A29, 47B35, 47G10.

Keywords. Inverse problem, operator inversion, Wiener functions, abstract Toeplitz and Hankel operators, integral operators.

1. Introduction

In the present paper we consider a twofold inverse problem related to orthogonal matrix function equations considered by R.J. Ellis and I. Gohberg for the scalar-valued case and mainly in discrete time; see [4] and the book [5]. The problem is referred to as the twofold EG inverse problem for short. Solutions of the onefold version of the problem, both in discrete and continuous time setting, have been obtained in [14, 15]. For the discrete time setting a solution of the twofold problem is given in [10]. One of our aims is to solve the twofold problem for the case of L^1-matrix functions on the real line which has not be done yet. More generally,

This work is based on the research supported in part by the National Research Foundation of South Africa (Grant Numbers 90670 and 93406).

we will solve an abstract $*$-algebraic version of the twofold EG inverse problem that contains various special cases, including the case of L^1-matrix functions on the real line. Our abstract setting will include an abstract inversion theorem which plays an important role in various concrete cases as well.

The abstract version of the twofold EG inverse problem we shall be dealing with is presented in Section 2. Here, for convenience of the reader, we consider the twofold EG inverse problem for L^1-matrix functions on the real line, and present the two main theorems for this case, Theorem 1.1 and Theorem 1.2 below. This requires some notation and terminology.

Throughout $\mathbb{C}^{r\times s}$ denotes the linear space of all $r \times s$ matrices with complex entries and $L^1(\mathbb{R})^{r\times s}$ denotes the space of all $r \times s$ matrices of which the entries are Lebesgue integrable functions on the real line $\mathbb{R}$. Furthermore

$$L^1(\mathbb{R}_+)^{r\times s} = \{f \in L^1(\mathbb{R})^{r\times s} \mid \mathrm{supp}(f) \subset \mathbb{R}_+ = [0,\infty)\},$$
$$L^1(\mathbb{R}_-)^{r\times s} = \{f \in L^1(\mathbb{R})^{r\times s} \mid \mathrm{supp}(f) \subset \mathbb{R}_- = (-\infty,0]\}.$$

Here $\mathrm{supp}(f)$ indicates the support of the function f. Now assume we are given

$$a \in L^1(\mathbb{R}_+)^{p\times p}, \quad c \in L^1(\mathbb{R}_-)^{q\times p}, \tag{1.1}$$
$$b \in L^1(\mathbb{R}_+)^{p\times q}, \quad d \in L^1(\mathbb{R}_-)^{q\times q}. \tag{1.2}$$

Given these data the *twofold EG inverse problem* referred to in the title is the problem to find $g \in L^1(\mathbb{R}_+)^{p\times q}$ satisfying

$$a + g \star c \in L^1(\mathbb{R}_-)^{p\times p}, \quad g^* + g^* \star a + c \in L^1(\mathbb{R}_+)^{q\times p}, \tag{1.3}$$
$$d + g^* \star b \in L^1(\mathbb{R}_+)^{q\times q}, \qquad g + g \star d + b \in L^1(\mathbb{R}_-)^{p\times q}. \tag{1.4}$$

Here $g^*(t) = g(-t)^*$ for each $t \in \mathbb{R}$, and as usual $f \star h$ denotes the convolution product of $L^1(\mathbb{R})$ matrix functions f and h.

The onefold version of the problem, when only a and c in (1.1) have been given and the problem is to find g such that (1.3) is satisfied, has been dealt with in [15].

To see the EG inverse problem from an operator point of view, let $g \in L^1(\mathbb{R}_+)^{p\times q}$, and let G and G_* be the Hankel operators defined by

$$G : L^1(\mathbb{R}_-)^q \to L^1(\mathbb{R}_+)^p, \quad (Gf)(t) = \int_{-\infty}^0 g(t-s)f(s)\,\mathrm{d}s\,, \quad t \geq 0; \tag{1.5}$$

$$G_* : L^1(\mathbb{R}_+)^p \to L^1(\mathbb{R}_-)^q, \quad (G_*h)(t) = \int_0^\infty g^*(t-s)h(s)\,\mathrm{d}s\,, \quad t \leq 0. \tag{1.6}$$

Here $L^1(\mathbb{R}_\pm)^r = L^1(\mathbb{R}_\pm)^{r\times 1}$. Using these Hankel operators, the conditions in (1.3) and (1.4) are equivalent to

$$a + Gc = 0, \qquad G_*a + c = -g^*, \tag{1.7}$$
$$d + G_*b = 0, \qquad Gd + b = -g. \tag{1.8}$$

To understand the above identities let us mention that we follow the convention that an operator acting on columns can be extended in a canonical way to an operator acting on matrices. We do this without changing the notation. For instance, in the first identity in (1.7) the operator G acts on each of the p columns of the $q \times p$ matrix function c, and Gc is the resulting $p \times p$ matrix function. Thus the first condition in (1.7) is equivalent to the first condition in (1.3). Similarly, the second condition in (1.3) is equivalent to the second condition in (1.7) and so on.

Hence the four conditions in (1.7) and (1.8) can be summarized by

$$\begin{bmatrix} I & G \\ G_* & I \end{bmatrix} \begin{bmatrix} a \\ c \end{bmatrix} = \begin{bmatrix} 0 \\ -g^* \end{bmatrix} \quad \text{and} \quad \begin{bmatrix} I & G \\ G_* & I \end{bmatrix} \begin{bmatrix} b \\ d \end{bmatrix} = \begin{bmatrix} -g \\ 0 \end{bmatrix}. \tag{1.9}$$

In other words, in this context the inverse problem is to reconstruct, if possible, a (block) Hankel operator and its associate, from the given data $\{a, b, c, d\}$.

To describe the main theorems in the present context, we need some further preliminaries about Laurent, Hankel and Wiener–Hopf operators. Let ρ be a function on $\mathbb{R}$ given by

$$\rho(t) = r_0 + r(t), t \in \mathbb{R}, \quad \text{where } r \in L^1(\mathbb{R})^{k\times m} \text{ and } r_0 \in \mathbb{C}^{k\times m}. \tag{1.10}$$

With ρ in (1.10) we associate the Laurent operator $L_\rho : L^1(\mathbb{R})^m \to L^1(\mathbb{R})^k$ which is defined by

$$(L_\rho f)(t) = r_0 f(t) + \int_{-\infty}^{\infty} r(t-s) f(s)\, ds, \quad (t \in \mathbb{R}). \tag{1.11}$$

Furthermore, we write L_ρ as a 2×2 operator matrix relative to the direct sum decompositions $L^1(\mathbb{R})^\ell = L^1(\mathbb{R}_-)^\ell \dot{+} L^1(\mathbb{R}_+)^\ell$, $\ell = m, k$, as follows:

$$L_\rho = \begin{bmatrix} T_{-,\rho} & H_{-,\rho} \\ H_{+,\rho} & T_{+,\rho} \end{bmatrix} : \begin{bmatrix} L^1(\mathbb{R}_-)^m \\ L^1(\mathbb{R}_+)^m \end{bmatrix} \to \begin{bmatrix} L^1(\mathbb{R}_-)^k \\ L^1(\mathbb{R}_+)^k \end{bmatrix}.$$

Thus $T_{-,\rho}$ and $T_{+,\rho}$ are the Wiener–Hopf operators given by

$$(T_{-,\rho} f)(t) = r_0 f(t) + \int_{-\infty}^{0} r(t-s) f(s)\, ds, \quad t \le 0, \quad f \in L^1(\mathbb{R}_-)^m, \tag{1.12}$$

$$(T_{+,\rho} f)(t) = r_0 f(t) + \int_{0}^{\infty} r(t-s) f(s)\, ds, \quad t \ge 0, \quad f \in L^1(\mathbb{R}_+)^m, \tag{1.13}$$

and $H_{-,\rho}$ and $H_{+,\rho}$ are the Hankel operators given by

$$(H_{-,\rho} f)(t) = \int_{0}^{\infty} r(t-s) f(s)\, ds, \quad t \le 0, \quad f \in L^1(\mathbb{R}_+)^m, \tag{1.14}$$

$$(H_{+,\rho} f)(t) = \int_{-\infty}^{0} r(t-s) f(s)\, ds, \quad t \ge 0 \quad f \in L^1(\mathbb{R}_-)^m. \tag{1.15}$$

In particular, the Hankel operators G and G_* appearing in (1.9) are equal to $G = H_{+,g}$ and $G_* = H_{-,g^*}$, respectively.

In what follows, instead of the data set $\{a, b, c, d\}$ we will often use the equivalent data set $\{\alpha, \beta, \gamma, \delta\}$, where

$$\alpha = e_p + a, \quad \beta = b, \quad \gamma = c, \quad \delta = e_q + d. \tag{1.16}$$

Here e_p and e_q are the functions on $\mathbb{R}$ identically equal to the unit matrix I_p and I_q, respectively. Using the data in (1.16) and the definitions of Toeplitz and Hankel operators in (1.12)–(1.15), we define the following operators:

$$M_{11} = T_{+,\alpha}T_{+,\alpha^*} - T_{+,\beta}T_{+,\beta^*} : L^1(\mathbb{R}_+)^p \to L^1(\mathbb{R}_+)^p, \tag{1.17}$$
$$M_{21} = H_{-,\gamma}T_{+,\alpha^*} - H_{-,\delta}T_{+,\beta^*} : L^1(\mathbb{R}_+)^p \to L^1(\mathbb{R}_-)^q, \tag{1.18}$$
$$M_{12} = H_{+,\beta}T_{-,\delta^*} - H_{+,\alpha}T_{-,\gamma^*} : L^1(\mathbb{R}_-)^q \to L^1(\mathbb{R}_+)^p, \tag{1.19}$$
$$M_{22} = T_{-,\delta}T_{-,\delta^*} - T_{-,\gamma}T_{-,\gamma^*} : L^1(\mathbb{R}_-)^q \to L^1(\mathbb{R}_-)^q. \tag{1.20}$$

Notice that these four operators are uniquely determined by the data.

We are now ready to state, in the present context, our two main theorems. In the abstract setting these theorems appear in Sections 6 and 7, respectively. The first is an inversion theorem and the second presents the solution of the EG inverse problem.

Theorem 1.1. *Let $g \in L^1(\mathbb{R}_+)^{p\times q}$, and let W be the operator given by*

$$W := \begin{bmatrix} I & H_{+,g} \\ H_{-,g^*} & I \end{bmatrix} : \begin{bmatrix} L^1(\mathbb{R}_+)^p \\ L^1(\mathbb{R}_-)^q \end{bmatrix} \to \begin{bmatrix} L^1(\mathbb{R}_+)^p \\ L^1(\mathbb{R}_-)^q \end{bmatrix}. \tag{1.21}$$

Then W is invertible if and only if g is a solution to a twofold EG inverse problem for some data set $\{a, b, c, d\}$ as in (1.1) and (1.2), that is, if and only if the following two equations are solvable:

$$W \begin{bmatrix} a \\ c \end{bmatrix} = \begin{bmatrix} 0 \\ -g^* \end{bmatrix} \quad \text{and} \quad W \begin{bmatrix} b \\ d \end{bmatrix} = \begin{bmatrix} -g \\ 0 \end{bmatrix}. \tag{1.22}$$

In that case the inverse of W is given by

$$W^{-1} = \begin{bmatrix} M_{11} & M_{12} \\ M_{21} & M_{22} \end{bmatrix}, \tag{1.23}$$

where M_{ij}, $1 \le i, j \le 2$, are the operators defined by (1.17)–(1.20) with α, β, γ, δ being given by (1.16) where a, b, c, d are given by (1.22). Furthermore, the operators M_{11} and M_{22} are invertible and

$$H_{+,g} = -M_{11}^{-1}M_{12} = -M_{12}M_{22}^{-1}, \tag{1.24}$$
$$H_{-,g^*} = -M_{21}M_{11}^{-1} = -M_{22}^{-1}M_{21}, \tag{1.25}$$
$$g = -M_{11}^{-1}b, \quad g^* = -M_{22}^{-1}c. \tag{1.26}$$

For the second theorem we need a generalization of the convolution product $\star$, which we shall denote by the symbol $\diamond$. In fact, given the data set $\{a, b, c, d\}$

and the equivalent data set $\{\alpha, \beta, \gamma, \delta\}$ given by (1.16), we define the following $\diamond$-products:

$$\begin{aligned} \alpha^* \diamond \alpha &:= e_p + a^* + a + a^* \star a, & \gamma^* \diamond \gamma &:= c^* \star c, \\ \delta^* \diamond \delta &:= e_q + d^* + d + d^* \star d, & \beta^* \diamond \beta &:= b^* \star b, \\ \alpha^* \diamond \beta &:= b + a^* \star b, & \gamma^* \diamond \delta &:= c^* + c^* \star d. \end{aligned}$$

Theorem 1.2. *Let $\{a, b, c, d\}$ be the functions given by (1.1) and (1.2), let α, β, γ, δ be the functions given by (1.16), and let e_p and e_q be the functions on $\mathbb{R}$ identically equal to the unit matrix I_p and I_q, respectively. Then the twofold EG inverse problem associated with the data set $\{a, b, c, d\}$ has a solution if and only if the following two conditions are satisfied:*

(L1) $\alpha^* \diamond \alpha - \gamma^* \diamond \gamma = e_p$, $\delta^* \diamond \delta - \beta^* \diamond \beta = e_q$, $\alpha^* \diamond \beta = \gamma^* \diamond \delta$;

(L2) *the operators M_{11} and M_{22} defined by (1.17) and (1.20) are one-to-one.*

In that case M_{11} and M_{22} are invertible, and the (unique) solution g and its adjoint g^* are given by

$$g = -M_{11}^{-1} b \quad \text{and} \quad g^* = -M_{22}^{-1} c. \tag{1.27}$$

Here b and c are the matrix functions appearing in (1.2) and (1.1), respectively.

Assuming that condition (L1) above is satisfied, the invertibility of the operator M_{11} is equivalent to the injectivity of the operator M_{11}, and the invertibility M_{22} is equivalent to the injectivity of the operator M_{22}. To prove these equivalences we use the fact (cf. formulas (4.18) and (4.19) in [12, Section 4.3]) that M_{11} and M_{22} are also given by

$$M_{11} = I + H_{+,\beta} H_{-,\beta^*} - H_{+,\alpha} H_{-,\alpha^*}, \tag{1.28}$$

$$M_{22} = I + H_{-,\gamma} H_{+,\gamma^*} - H_{-,\delta} H_{+,\delta^*}. \tag{1.29}$$

Since the Hankel operators appearing in these formulas are all compact operators, M_{11} and M_{22} are Fredholm operators with index zero, and thus invertible if and only if they are one-to-one.

We shall see in Lemma 8.5, again assuming that condition (L1) above is satisfied, that the operators M_{21} and M_{12} are also given by

$$M_{21} = T_{-,\delta} H_{-,\beta^*} - T_{-,\gamma} H_{-,\alpha^*}, \tag{1.30}$$

$$M_{12} = T_{+,\alpha} H_{+,\gamma^*} - T_{+,\beta} H_{+,\delta^*}. \tag{1.31}$$

Since the functions a, b, c, d are $L^1(\mathbb{R})$ matrix functions, the operators M_{ij}, $1 \le i, j \le 2$, are also well-defined as bounded linear operators on the corresponding L^2 spaces. It follows that Theorems 1.1 and 1.2 remain true if the L^1 spaces in (1.21) are replaced by corresponding L^2 spaces. In this L^2-setting, Theorems 1.1 and 1.2 are the continuous analogs of Theorems 3.1 and 4.1 in [10]. Furthermore, in this L^2-setting the adjoints of the operators M_{ij}, $1 \le i, j \le 2$, as operators on L^2-spaces, are well-defined as well. In fact, assuming condition (L1) is satisfied and using (1.17)–(1.20) and the identities (1.30), (1.31), we see that in the L^2 setting we have

$$\begin{aligned} M_{11}^* &= M_{11}, \quad M_{21}^* = M_{12}, \\ M_{12}^* &= M_{21}, \quad M_{22}^* = M_{22}. \end{aligned} \tag{1.32}$$

Theorem 1.1 belongs to the wide class of inversion theorems for structured operators. In particular, the theorem can be viewed as an analogue of the Gohberg–Heinig inversion theorem for convolution operators on a finite interval [7]. In its present form, Theorem 1.1 can be seen as an addition to Theorem 12.2.4 in [5], where, using a somewhat different notation, the invertibility of W is proved. The formula for the inverse of W could be obtained from [9, Theorem 0.1], where the formula for M_{11} appears in a somewhat different notation. Note that [9, Theorem 0.1] also solves the asymmetric version of the inversion problem. Formulas (1.24)–(1.26) seem to be new.

As mentioned before, in the present paper we put the twofold EG problem in an abstract $*$-algebraic setting. This allows us to consider and solve non-stationary twofold EG problems (see Subsection 3.2 for an example). Furthermore, Theorems 1.1 and 1.2 are obtained as corollaries of the two abstract theorems, Theorem 6.1 and Theorem 7.1, derived in this paper. Also, as we shall prove in Section 10, Theorems 3.1 and 4.1 in [10] appear as corollaries of our main theorems.

The paper consists of ten sections (including the present introduction) and an appendix. In Section 2 we introduce the abstract $*$-algebraic setting and state the main problem. Section 3 presents a numerical example and a number of illustrative special cases, including various Wiener algebra examples. Sections 4 and 5 have a preliminary character. Here we introduce Toeplitz-like and Hankel-like operators, which play an important role in the abstract setting, and we derive a number of identities and lemmas that are used in the proofs of the main results. In Section 6 the abstract inversion theorem (Theorem 6.1) is proved, and in Section 7 the solution to the abstract twofold EG inverse problem (Theorem 7.1) is presented and proved. Theorems 1.1 and 1.2 are proved in Section 8 using the results of Section 6 and Section 7. In Section 9 we further specify Theorem 7.1 for the case when there are additional invertibility conditions on the underlying data. The proof in this section is direct and does not use Theorem 7.1. As mentioned in the previous paragraph, Theorems 3.1 and 4.1 in [10] are derived in Section 10 as corollaries of our main theorems in Sections 6 and 7.

Finally, in the Appendix we review a number of results that play an important role in Section 8, where we have to relate Hankel-type and Toeplitz-type operators used in Section 6 and Section 7 to classical Hankel and Wiener–Hopf integral operators. The Appendix consists of three subsections. In Subsection A.1 we recall the definition of a Hankel operator on $L^2(\mathbb{R}_+)$ and review some basic facts. In Subsection A.2 we present a theorem (partially new) characterizing classical Hankel integral operators mapping $L^1(\mathbb{R}_+)^p$ into $L^1(\mathbb{R}_+)^q$. Two auxiliary results are presented in the final subsection.

2. General setting and main problem

We first describe the general $*$-algebraic setting that we will be working with. To do this we use the notation introduced on pages 109 and 110 of [13]; see also the first two pages of [8, Section II.1]. Throughout $\mathcal{A}$, $\mathcal{B}$, $\mathcal{C}$ and $\mathcal{D}$ are complex linear vector spaces such that the following set of 2×2 block matrices form an algebra:

$$\mathcal{M} = \mathcal{M}_{\mathcal{A},\mathcal{B},\mathcal{C},\mathcal{D}} = \left\{ f = \begin{bmatrix} a & b \\ c & d \end{bmatrix} \mid a \in \mathcal{A},\ b \in \mathcal{B},\ c \in \mathcal{C},\ d \in \mathcal{D} \right\}. \tag{2.1}$$

Furthermore, we assume $\mathcal{A}$ and $\mathcal{D}$ are $*$-algebras (see [18, Chapter IV] for the definition of this notion) with units $e_{\mathcal{A}}$ and $e_{\mathcal{D}}$, respectively, and endowed with involutions *. The diagonal

$$e_{\mathcal{M}} = \begin{bmatrix} e_{\mathcal{A}} & 0 \\ 0 & e_{\mathcal{D}} \end{bmatrix}$$

is the unit element of $\mathcal{M}$. Moreover, $\mathcal{C}$ is a linear space isomorphic to $\mathcal{B}$ via a conjugate linear transformation * whose inverse is also denoted by *. We require $\mathcal{M}$ to be a $*$-algebra with respect to the usual matrix multiplication and with the involution given by

$$\begin{bmatrix} a & b \\ c & d \end{bmatrix}^* = \begin{bmatrix} a^* & c^* \\ b^* & d^* \end{bmatrix}.$$

The algebras $\mathcal{A}$ and $\mathcal{D}$ are assumed to admit direct sum decompositions:

$$\mathcal{A} = \mathcal{A}_-^0 \dot{+} \mathcal{A}_d \dot{+} \mathcal{A}_+^0, \quad \mathcal{D} = \mathcal{D}_-^0 \dot{+} \mathcal{D}_d \dot{+} \mathcal{D}_+^0. \tag{2.2}$$

In these two direct sum decompositions the summands are assumed to be subalgebras of $\mathcal{A}$ and $\mathcal{D}$, respectively. Furthermore, we require

$$\begin{aligned} &e_{\mathcal{A}} \in \mathcal{A}_d, \quad (\mathcal{A}_-^0)^* = \mathcal{A}_+^0, \quad (\mathcal{A}_d)^* = \mathcal{A}_d, \\ &e_{\mathcal{D}} \in \mathcal{D}_d, \quad (\mathcal{D}_-^0)^* = \mathcal{D}_+^0, \quad (\mathcal{D}_d)^* = \mathcal{D}_d. \end{aligned} \tag{2.3}$$

Set

$$\mathcal{A}_- = \mathcal{A}_-^0 \dot{+} \mathcal{A}_d, \quad \mathcal{A}_+ = \mathcal{A}_d \dot{+} \mathcal{A}_+^0, \quad \mathcal{D}_- = \mathcal{D}_-^0 \dot{+} \mathcal{D}_d, \quad \mathcal{D}_+ = \mathcal{D}_d \dot{+} \mathcal{D}_+^0.$$

We also assume that $\mathcal{B}$ and $\mathcal{C}$ admit direct sum decompositions:

$$\mathcal{B} = \mathcal{B}_- \dot{+} \mathcal{B}_+, \quad \mathcal{C} = \mathcal{C}_- \dot{+} \mathcal{C}_+, \quad \text{such that} \quad \mathcal{C}_- = \mathcal{B}_+^*, \quad \mathcal{C}_+ = \mathcal{B}_-^*. \tag{2.4}$$

These direct sum decompositions yield a direct sum decomposition of $\mathcal{M}$, namely $\mathcal{M} = \mathcal{M}_-^0 \dot{+} \mathcal{M}_d \dot{+} \mathcal{M}_+^0$, where

$$\mathcal{M}_-^0 = \begin{bmatrix} \mathcal{A}_-^0 & \mathcal{B}_- \\ \mathcal{C}_- & \mathcal{D}_-^0 \end{bmatrix}, \quad \mathcal{M}_d = \begin{bmatrix} \mathcal{A}_d & 0 \\ 0 & \mathcal{D}_d \end{bmatrix}, \quad \mathcal{M}_+^0 = \begin{bmatrix} \mathcal{A}_+^0 & \mathcal{B}_+ \\ \mathcal{C}_+ & \mathcal{D}_+^0 \end{bmatrix}. \tag{2.5}$$

Note that

$$(\mathcal{M}_-^0)^* = \mathcal{M}_+^0, \quad (\mathcal{M}_+^0)^* = \mathcal{M}_-^0, \quad \mathcal{M}_d^* = \mathcal{M}_d.$$

Finally, we assume that the products of elements from the summands in $\mathcal{M} = \mathcal{M}_-^0 \dot{+} \mathcal{M}_d \dot{+} \mathcal{M}_+^0$ satisfy the rules of the following table.

Multiplication table

$\times$	$\mathcal{M}_-^0$	$\mathcal{M}_d$	$\mathcal{M}_+^0$
$\mathcal{M}_-^0$	$\mathcal{M}_-^0$	$\mathcal{M}_-^0$	$\mathcal{M}$
$\mathcal{M}_d$	$\mathcal{M}_-^0$	$\mathcal{M}_d$	$\mathcal{M}_+^0$
$\mathcal{M}_+^0$	$\mathcal{M}$	$\mathcal{M}_+^0$	$\mathcal{M}_+^0$

We say that the algebra $\mathcal{M} = \mathcal{M}_{\mathcal{A},\mathcal{B},\mathcal{C},\mathcal{D}}$ defined by (2.1) is *admissible* if all the conditions listed in the above paragraph are satisfied.

Main problem. We are now ready to state the main problem that we shall be dealing with. Let $\alpha \in \mathcal{A}_+$, $\beta \in \mathcal{B}_+$, $\gamma \in \mathcal{C}_-$ and $\delta \in \mathcal{D}_-$ be given. We call $g \in \mathcal{B}_+$ a *solution to the twofold EG inverse problem associated with* α, β, γ, *and* δ whenever

$$\alpha + g\gamma - e_{\mathcal{A}} \in \mathcal{A}_-^0 \quad \text{and} \quad g^*\alpha + \gamma \in \mathcal{C}_+, \tag{2.6}$$

$$g\delta + \beta \in \mathcal{B}_- \quad \text{and} \quad \delta + g^*\beta - e_{\mathcal{D}} \in \mathcal{D}_+^0. \tag{2.7}$$

Our main aim is to determine necessary and sufficient conditions for this inverse problem to be solvable and to derive explicit formulas for its solution. We shall show that the solution, if it exists, is unique. The following result is a special case of [13, Theorem 1.2].

Proposition 2.1. *If the twofold EG inverse problem associated with* α, β, γ *and* δ *has a solution, then*

(C1) $\quad \alpha^*\alpha - \gamma^*\gamma = P_{\mathcal{A}_d}\alpha,$
(C2) $\quad \delta^*\delta - \beta^*\beta = P_{\mathcal{D}_d}\delta,$
(C3) $\quad \alpha^*\beta = \gamma^*\delta.$

Here $P_{\mathcal{A}_d}$ *and* $P_{\mathcal{D}_d}$ *denote the projections of* $\mathcal{A}$ *and* $\mathcal{D}$ *onto* $\mathcal{A}_d$ *and* $\mathcal{D}_d$, *respectively, along* $\mathcal{A}^0 = \mathcal{A}_-^0 \dot{+} \mathcal{A}_+^0$ *and* $\mathcal{D}^0 = \mathcal{D}_-^0 \dot{+} \mathcal{D}_+^0$, *respectively.*

Notice that (C1) and (C2) imply that

$$a_0 := P_{\mathcal{A}_d}\alpha = (P_{\mathcal{A}_d}\alpha)^* = a_0^* \quad \text{and} \quad d_0 := P_{\mathcal{D}_d}\delta = (P_{\mathcal{D}_d}\delta)^* = d_0^*. \tag{2.8}$$

Furthermore, together the three conditions (C1)–(C3) are equivalent to

$$\begin{bmatrix} \alpha^* & \gamma^* \\ \beta^* & \delta^* \end{bmatrix} \begin{bmatrix} e_{\mathcal{A}} & 0 \\ 0 & -e_{\mathcal{D}} \end{bmatrix} \begin{bmatrix} \alpha & \beta \\ \gamma & \delta \end{bmatrix} = \begin{bmatrix} a_0 & 0 \\ 0 & -d_0 \end{bmatrix}. \tag{2.9}$$

Remark 2.2. Since a_0 and d_0 belong to $\mathcal{A}_d$ and $\mathcal{D}_d$, respectively, invertibility of a_0 in $\mathcal{A}$ and of d_0 in $\mathcal{D}$ imply that $a_0^{-1} \in \mathcal{A}_d$ and $d_0^{-1} \in \mathcal{D}_d$. In other words, a_0 and d_0 are invertible in $\mathcal{A}_d$ and $\mathcal{D}_d$, respectively.

Remark 2.3. In the sequel it will often be assumed that a_0 and d_0 are invertible. In that case the following three conditions are well-defined.

(C4) $\quad \alpha a_0^{-1}\alpha^* - \beta d_0^{-1}\beta^* = e_{\mathcal{A}},$
(C5) $\quad \delta d_0^{-1}\delta^* - \gamma a_0^{-1}\gamma^* = e_{\mathcal{D}},$
(C6) $\quad \alpha a_0^{-1}\gamma^* = \beta d_0^{-1}\delta^*.$

In solving the twofold EG inverse problem referred to above we shall always assume that a_0 and d_0 are invertible and that the six conditions (C1)–(C6) are fulfilled.

The next lemma shows that in many cases (C4)–(C6) are satisfied whenever conditions (C1)–(C3) are satisfied.

Lemma 2.4. *Let* $\alpha \in \mathcal{A}_+$, $\beta \in \mathcal{B}_+$, $\gamma \in \mathcal{C}_-$, $\delta \in \mathcal{D}_-$, *and let*

$$Q = \begin{bmatrix} \alpha & \beta \\ \gamma & \delta \end{bmatrix}.$$

Assume that a_0 *and* α *are invertible in* $\mathcal{A}$, *and that* d_0 *and* δ *are invertible in* $\mathcal{D}$. *If, in addition,* α, β, γ *and* δ *satisfy conditions* (C1)–(C3), *then* Q *is invertible, and conditions* (C4)–(C6) *are satisfied.*

Proof. Since δ is invertible, a classical Schur complement argument (see, e.g., formula (2.3) in [1, Chapter 2]) shows that

$$Q = \begin{bmatrix} \alpha & \beta \\ \gamma & \delta \end{bmatrix} = \begin{bmatrix} e_{\mathcal{A}} & \beta\delta^{-1} \\ 0 & e_{\mathcal{D}} \end{bmatrix} \begin{bmatrix} \Delta & 0 \\ 0 & \delta \end{bmatrix} \begin{bmatrix} e_{\mathcal{A}} & 0 \\ \gamma\delta^{-1} & e_{\mathcal{D}} \end{bmatrix}, \quad \text{with } \Delta = \alpha - \beta\delta^{-1}\gamma.$$

Using the invertibility of α and δ, we can rewrite (C3) as $\beta\delta^{-1} = \alpha^{-*}\gamma^*$. The latter identity together with (C1) yields:

$$\Delta = \alpha - \beta\delta^{-1}\gamma = \alpha - \alpha^{-*}\gamma^*\gamma = \alpha^{-*}\left(\alpha^*\alpha - \gamma^*\gamma\right) = \alpha^{-*}a_0.$$

It follows that the Schur complement Δ is invertible. But then Q is invertible too, and the identity (2.9) shows that the inverse Q^{-1} of Q is given by

$$Q^{-1} = \begin{bmatrix} a_0^{-1} & 0 \\ 0 & -d_0^{-1} \end{bmatrix} \begin{bmatrix} \alpha^* & \gamma^* \\ \beta^* & \delta^* \end{bmatrix} \begin{bmatrix} e_{\mathcal{A}} & 0 \\ 0 & -e_{\mathcal{D}} \end{bmatrix}.$$

Since QQ^{-1} is a 2×2 block identity matrix, we conclude that

$$\begin{bmatrix} \alpha & \beta \\ \gamma & \delta \end{bmatrix} \begin{bmatrix} a_0^{-1} & 0 \\ 0 & -d_0^{-1} \end{bmatrix} \begin{bmatrix} \alpha^* & \gamma^* \\ \beta^* & \delta^* \end{bmatrix} \begin{bmatrix} e_{\mathcal{A}} & 0 \\ 0 & -e_{\mathcal{D}} \end{bmatrix} = \begin{bmatrix} e_{\mathcal{A}} & 0 \\ 0 & e_{\mathcal{D}} \end{bmatrix}$$

This yields

$$\begin{bmatrix} \alpha & \beta \\ \gamma & \delta \end{bmatrix} \begin{bmatrix} a_0^{-1} & 0 \\ 0 & -d_0^{-1} \end{bmatrix} \begin{bmatrix} \alpha^* & \gamma^* \\ \beta^* & \delta^* \end{bmatrix} = \begin{bmatrix} e_{\mathcal{A}} & 0 \\ 0 & -e_{\mathcal{D}} \end{bmatrix}, \tag{2.10}$$

and hence (C4)–(C6) are satisfied. □

3. A numerical example and some illustrative special cases

In this section we present a few inverse problems which are special cases of the abstract problem presented in the previous section.

3.1. A numerical example

As a first illustration we consider a simple example of a problem for 3×3 matrices. Given

$$\alpha = \tfrac{1}{8}\begin{bmatrix} -2 & 2 & 0 \\ 0 & -3 & -4 \\ 0 & 0 & 6 \end{bmatrix}, \qquad \beta = -\tfrac{1}{8}\begin{bmatrix} -2 & -6 & 0 \\ 0 & 1 & -4 \\ 0 & 0 & -2 \end{bmatrix}, \tag{3.1}$$

$$\gamma = -\tfrac{1}{8}\begin{bmatrix} -2 & 0 & 0 \\ -4 & 1 & 0 \\ 0 & -6 & -2 \end{bmatrix}, \qquad \delta = \tfrac{1}{8}\begin{bmatrix} 6 & 0 & 0 \\ -4 & -3 & 0 \\ 0 & 2 & -2 \end{bmatrix}, \tag{3.2}$$

we seek a 3×3 upper triangular matrix g such that

$$\alpha + g\gamma = \begin{bmatrix} 1 & 0 & 0 \\ \star & 1 & 0 \\ \star & \star & 1 \end{bmatrix}, \quad g^*\alpha + \gamma = \begin{bmatrix} 0 & \star & \star \\ 0 & 0 & \star \\ 0 & 0 & 0 \end{bmatrix}, \tag{3.3}$$

$$\beta + g\delta = \begin{bmatrix} 0 & 0 & 0 \\ \star & 0 & 0 \\ \star & \star & 0 \end{bmatrix}, \quad g^*\beta + \delta = \begin{bmatrix} 1 & \star & \star \\ 0 & 1 & \star \\ 0 & 0 & 1 \end{bmatrix}. \tag{3.4}$$

Here the symbols $\star$ denote unspecified entries. By direct checking it is easy to see that the matrix $g_\circ$ given by

$$g_\circ = \begin{bmatrix} 1 & 2 & 0 \\ 0 & 1 & 2 \\ 0 & 0 & 1 \end{bmatrix}$$

is upper triangular and satisfies (3.3) and (3.4). From the general results about existence of solutions and methods to determine solutions, which will be presented in this paper, it follows that $g_\circ$ is the only solution. For this example it also straightforward to check that conditions (C1)–(C3) presented in the previous section are satisfied.

3.2. A class of finite-dimensional matrix examples

We will put the problem considered in the preceding example into the general setting considered in the previous section. Let $p \geq 1$ be an integer (in the example above we took $p = 3$), and let

$$\mathcal{A} = \mathcal{B} = \mathcal{C} = \mathcal{D} = \mathbb{C}^{p \times p}. \tag{3.5}$$

The involution * is given by the usual transposed conjugate of a matrix. Let $\mathcal{A}_+^0 = \mathcal{B}_+^0 = \mathcal{C}_+^0 = \mathcal{D}_+^0$ be the subspace of $\mathbb{C}^{p \times p}$ of the strictly upper triangular matrices and $\mathcal{A}_-^0 = \mathcal{B}_-^0 = \mathcal{C}_-^0 = \mathcal{D}_-^0$ the subspace of the strictly lower triangular matrices. Furthermore, let $\mathcal{A}_d = \mathcal{B}_d = \mathcal{C}_d = \mathcal{D}_d$ be the subspace consisting of the $p \times p$ diagonal matrices, that is, matrices with all entries off the main diagonal being equal to zero. We set

$$\mathcal{A}_- = \mathcal{A}_-^0 \dot{+} \mathcal{A}_d, \quad \mathcal{A}_+ = \mathcal{A}_d \dot{+} \mathcal{A}_+^0, \quad \mathcal{D}_- = \mathcal{D}_-^0 \dot{+} \mathcal{D}_d, \quad \mathcal{D}_+ = \mathcal{D}_d \dot{+} \mathcal{D}_+^0, \tag{3.6}$$

$$\mathcal{B}_- = \mathcal{B}_-^0, \qquad \mathcal{B}_+ = \mathcal{B}_d \dot{+} \mathcal{B}_+^0, \quad \mathcal{C}_- = \mathcal{C}_-^0 \dot{+} \mathcal{C}_d, \quad \mathcal{C}_+ = \mathcal{C}_+^0. \tag{3.7}$$

The problem we consider in this setting is the following. Let α, β, γ, δ be given $p \times p$ matrices, and assume that $\alpha \in \mathcal{A}_+$, $\beta \in \mathcal{B}_+$, $\gamma \in \mathcal{C}_-$ and $\delta \in \mathcal{D}_-$. Then a $p \times p$ matrix $g \in \mathcal{B}_+$ is said to be a solution to the EG inverse problem for the given data α, β, γ, δ whenever the four inclusions in (2.6) and (2.7) are satisfied. In the numerical example considered above this amounts to the conditions (3.3) and (3.4) being fulfilled.

If a solution exists, then the conditions (C1)–(C3) are satisfied, what in this setting means that

$$\begin{bmatrix} \alpha^* & \gamma^* \\ \beta^* & \gamma^* \end{bmatrix} \begin{bmatrix} I_p & 0 \\ 0 & -I_p \end{bmatrix} \begin{bmatrix} \alpha & \beta \\ \gamma & \delta \end{bmatrix} = \begin{bmatrix} \alpha_d & 0 \\ 0 & -\delta_d \end{bmatrix}. \tag{3.8}$$

Here I_p is the $p \times p$ identity matrix, α_d is the diagonal matrix whose main diagonal coincides with the one of α, and δ_d is the diagonal matrix whose main diagonal coincides with the one of δ. If α_d and δ_d are invertible, then α and δ are invertible as $p \times p$ (lower or upper) triangular matrices. In this case, as we shall see in Theorem 9.1, the twofold EG inverse problem is solvable, the solution is unique, and the solution is given by $g = -P_{\mathcal{B}_+}\left((\alpha^*)^{-1}\gamma^*\right)$.

The above special case is an example of a non-stationary EG inverse problem. We intend to deal with other non-stationary problems in a later publication, using elements of [8]; see also [11, Section 5].

3.3. Wiener algebra examples

Let $\mathcal{N}$ be a unital $*$-algebra with unit $e_{\mathcal{N}}$ and involution *. We assume that $\mathcal{N}$ admits a direct sum decomposition:

$$\mathcal{N} = \mathcal{N}_{-,0} \dot{+} \mathcal{N}_d \dot{+} \mathcal{N}_{+,0}.$$

In this direct sum decomposition the summands are subalgebras of $\mathcal{N}$, and we require

$$e_{\mathcal{N}} \in \mathcal{N}_d, \quad (\mathcal{N}_d)^* = \mathcal{N}_d, \quad (\mathcal{N}_{-,0})^* = \mathcal{N}_{+,0},$$
$$\mathcal{N}_d \mathcal{N}_{\pm,0} \subset \mathcal{N}_{\pm,0}, \quad \mathcal{N}_{\pm,0} \mathcal{N}_d \subset \mathcal{N}_{\pm,0}.$$

Given $\mathcal{N}$ we construct two admissible algebras $\mathcal{M}_{\mathcal{A},\mathcal{B},\mathcal{C},\mathcal{D}}$ using the following two translation tables:

Table 1

$\mathcal{A}$	$\mathcal{A}_+^0$	$\mathcal{A}_d$	$\mathcal{A}_-^0$	$\mathcal{B}$	$\mathcal{B}_+$	$\mathcal{B}_-$
$\mathcal{N}^{p\times p}$	$\mathcal{N}_{+,0}^{p\times p}$	$\mathcal{N}_d^{p\times p}$	$\mathcal{N}_{-,0}^{p\times p}$	$\mathcal{N}^{p\times q}$	$\mathcal{N}_+^{p\times q}$	$\mathcal{N}_{-,0}^{p\times q}$
$\mathcal{C}$	$\mathcal{C}_-$	$\mathcal{C}_+$	$\mathcal{D}$	$\mathcal{D}_+^0$	$\mathcal{D}_d$	$\mathcal{D}_-^0$
$\mathcal{N}^{q\times p}$	$\mathcal{N}_-^{q\times p}$	$\mathcal{N}_{+,0}^{q\times p}$	$\mathcal{N}^{q\times q}$	$\mathcal{N}_{+,0}^{q\times q}$	$\mathcal{N}_d^{q\times q}$	$\mathcal{N}_{-,0}^{q\times q}$

Table 2

$\mathcal{A}$	$\mathcal{A}_+^0$	$\mathcal{A}_d$	$\mathcal{A}_-^0$	$\mathcal{B}$	$\mathcal{B}_+$	$\mathcal{B}_-$
$\mathcal{N}^{p\times p}$	$\mathcal{N}_{+,0}^{p\times p}$	$\mathcal{N}_d^{p\times p}$	$\mathcal{N}_{-,0}^{p\times p}$	$\mathcal{N}_0^{p\times q}$	$\mathcal{N}_{+,0}^{p\times q}$	$\mathcal{N}_{-,0}^{p\times q}$
$\mathcal{C}$	$\mathcal{C}_-$	$\mathcal{C}_+$	$\mathcal{D}$	$\mathcal{D}_+^0$	$\mathcal{D}_d$	$\mathcal{D}_-^0$
$\mathcal{N}_0^{q\times p}$	$\mathcal{N}_{-,0}^{q\times p}$	$\mathcal{N}_{+,0}^{q\times p}$	$\mathcal{N}^{q\times q}$	$\mathcal{N}_{+,0}^{q\times q}$	$\mathcal{N}_d^{q\times q}$	$\mathcal{N}_{-,0}^{q\times q}$

Here $\mathcal{N}_0$ is the space defined by $\mathcal{N}_0 = \mathcal{N}_{-,0} + \mathcal{N}_{+,0}$.
In subsequent special cases we make these examples more concrete.

3.3.1. The Wiener algebra on the real line. Recall that the Wiener algebra on the real line $\mathcal{W}(\mathbb{R})$ consists of the functions φ of the form

$$\varphi(\lambda) = f_0 + \int_{-\infty}^{\infty} e^{i\lambda t} f(t)\,\mathrm{d}t\,, \quad \lambda \in \mathbb{R}, \tag{3.9}$$

with $f_0 \in \mathbb{C}$ and $f \in L^1(\mathbb{R})$. The subspaces $\mathcal{W}(\mathbb{R})_{\pm,0}$ consist of the functions φ in $\mathcal{W}(\mathbb{R})$ for which in the representation (3.9) the constant $f_0 = 0$ and $f \in L^1(\mathbb{R}_\pm)$. A function φ belongs to the subspace $\mathcal{W}(\mathbb{R})_0$ if and only if $f = 0$ in the representation in (3.9). With $\mathcal{N} = \mathcal{W}(\mathbb{R})$, it is straightforward to check that the spaces $\mathcal{A}, \mathcal{B}, \mathcal{C}$, and $\mathcal{D}$ and their subspaces defined by Table 2 have all the properties listed in the first two paragraphs of Section 2, that is, $\mathcal{M} = \mathcal{M}_{\mathcal{A},\mathcal{B},\mathcal{C},\mathcal{D}}$ is admissible. Now let

$$\alpha \in e_p + \mathcal{W}(\mathbb{R})_{+,0}^{p\times p}, \quad \beta \in \mathcal{W}(\mathbb{R})_{+,0}^{p\times q}, \quad \gamma \in \mathcal{W}(\mathbb{R})_{-,0}^{q\times p}, \quad \delta \in e_q + \mathcal{W}(\mathbb{R})_{-,0}^{q\times q},$$

where e_p and e_q are the functions identically equal to the unit matrix I_p and I_q, respectively. Then the twofold EG inverse problem is to find $g \in \mathcal{W}(\mathbb{R})_{+,0}^{p\times q}$ such that the following four inclusions are satisfied:

$$\alpha + g\gamma - e_p \in \mathcal{W}(\mathbb{R})_{-,0}^{p\times p} \quad \text{and} \quad g^*\alpha + \gamma \in \mathcal{W}(\mathbb{R})_{+,0}^{q\times p};$$
$$g\delta + \beta \in \mathcal{W}(\mathbb{R})_{-,0}^{p\times q} \quad \text{and} \quad \delta + g^*\beta - e_q \in \mathcal{W}(\mathbb{R})_{+,0}^{q\times q}.$$

Notice that these inclusions are just the same as the inclusions in (2.6) and (2.7). In this way this twofold EG inverse problem is put in the abstract setting of the twofold EG inverse problem defined in Section 2.

Remark 3.1. The version of the twofold EG inverse problem considered in this subsubsection is isomorphic to the twofold EG inverse problem considered in the introduction. This follows from the definition of the Wiener algebra $\mathcal{W}(\mathbb{R})$ in (3.9). The solution of the twofold EG inverse problem as described in this subsubsection follows from Theorems 1.1 and 1.2. The latter two theorems will be proved in Section 8.

Note that in this special case the algebra $\mathcal{M} = \mathcal{M}_{\mathcal{A},\mathcal{B},\mathcal{C},\mathcal{D}}$ appearing in (2.1) can be considered as a subalgebra of $\mathcal{W}(\mathbb{R})^{(p+q)\times(p+q)}$. Indeed,

$$\mathcal{M} = \mathcal{W}(\mathbb{R})_{-,0}^{(p+q)\times(p+q)} \dot{+} \mathcal{M}_d \dot{+} \mathcal{W}(\mathbb{R})_{+,0}^{(p+q)\times(p+q)}$$

with

$$\mathcal{M}_d = \left\{ \begin{bmatrix} a_0 & 0 \\ 0 & d_0 \end{bmatrix} \mid a_0 \in \mathbb{C}^{p\times p}, d_0 \in \mathbb{C}^{q\times q} \right\}.$$

The case $\mathcal{N} = \mathcal{RW}(\mathbb{R})$. Let $\mathcal{RW}(\mathbb{R})$ be the subalgebra of $\mathcal{W}(\mathbb{R})$ consisting of all rational functions in $\mathcal{W}(\mathbb{R})$. With $\mathcal{N} = \mathcal{RW}(\mathbb{R})$ it is straightforward to check that the resulting $\mathcal{A}, \mathcal{B}, \mathcal{C}, \mathcal{D}$ defined in Table 2 have all the properties listed in the first two paragraphs of Section 2, that is, $\mathcal{M} = \mathcal{M}_{\mathcal{A},\mathcal{B},\mathcal{C},\mathcal{D}}$ is admissible. Let

$$\begin{aligned} &\alpha \in e_p + \mathcal{RW}(\mathbb{R})^{p\times p}_{+,0}, \quad && \beta \in \mathcal{RW}(\mathbb{R})^{p\times q}_{+,0}, \\ &\gamma \in \mathcal{RW}(\mathbb{R})^{q\times p}_{-,0}, && \delta \in e_q + \mathcal{RW}(\mathbb{R})^{q\times q}_{-,0}. \end{aligned}$$

The twofold EG inverse problem is to find $g \in \mathcal{RW}(\mathbb{R})^{p\times q}_{+,0}$ such that the following four inclusions are satisfied:

$$\begin{aligned} \alpha + g\gamma - e_p \in \mathcal{RW}(\mathbb{R})^{p\times p}_{-,0}, \quad &\text{and} \quad g^*\alpha + \gamma \in \mathcal{RW}(\mathbb{R})^{q\times p}_{+,0}; \\ g\delta + \beta \in \mathcal{RW}(\mathbb{R})^{p\times q}_{-,0} \quad &\text{and} \quad \delta + g^*\beta - e_q \in \mathcal{RW}(\mathbb{R})^{q\times q}_{+,0}. \end{aligned}$$

In a forthcoming paper we plan to deal with the twofold EG inverse problem for rational functions in $\mathcal{W}(\mathbb{R})$, using minimal realizations of the rational functions involved and related state space techniques. The latter will lead to new explicit formulas for the solution.

The case $\mathcal{N} = \mathcal{FW}(\mathbb{R})$. Let $\mathcal{FW}(\mathbb{R})$ denote the subalgebra of $\mathcal{W}(\mathbb{R})$ of functions in $\mathcal{W}(\mathbb{R})$ whose inverse Fourier transforms are elements in $L^1(\mathbb{R})$ with finite support. Hence if $\rho \in \mathcal{W}(\mathbb{R})$ is given by

$$\rho(\lambda) = r_0 + \int_{-\infty}^{\infty} e^{i\lambda t} r(t)\, dt \quad (\lambda \in \mathbb{R}),$$

with $r \in L^1(\mathbb{R})$ and $r_0 \in \mathbb{C}$, then $\rho \in \mathcal{FW}(\mathbb{R})$ in case there are real numbers $\tau_1 < \tau_2$ so that $r(t) = 0$ for all $t \notin [\tau_1, \tau_2]$. In this case, one easily verifies that $\mathcal{M} = \mathcal{M}_{\mathcal{A},\mathcal{B},\mathcal{C},\mathcal{D}}$ with $\mathcal{A}, \mathcal{B}, \mathcal{C}, \mathcal{D}$ as in Table 2 is admissible. The twofold EG inverse problem specified for these choices can be stated as follows. Let

$$\begin{aligned} &\alpha \in e_p + \mathcal{FW}(\mathbb{R})^{p\times p}_{+,0}, \quad && \beta \in \mathcal{FW}(\mathbb{R})^{p\times q}_{+,0}, \\ &\gamma \in \mathcal{FW}(\mathbb{R})^{q\times p}_{-,0}, && \delta \in e_q + \mathcal{FW}(\mathbb{R})^{q\times q}_{-,0}. \end{aligned}$$

The twofold EG inverse problem now is to find $g \in \mathcal{RW}(\mathbb{R})^{p\times q}_{+,0}$ such that the following four inclusions are satisfied:

$$\begin{aligned} \alpha + g\gamma - e_p \in \mathcal{FW}(\mathbb{R})^{p\times p}_{-,0} \quad &\text{and} \quad g^*\alpha + \gamma \in \mathcal{FW}(\mathbb{R})^{q\times p}_{+,0}; \\ g\delta + \beta \in \mathcal{FW}(\mathbb{R})^{p\times q}_{-,0} \quad &\text{and} \quad \delta + g^*\beta - e_q \in \mathcal{FW}(\mathbb{R})^{q\times q}_{+,0}. \end{aligned}$$

We plan to return to this case in a forthcoming paper.

3.3.2. The Wiener algebra on the unit circle. Let $\mathcal{N} = \mathcal{W}(\mathbb{T})$, where $\mathcal{W}(\mathbb{T})$ is the Wiener algebra of functions on the unit circle $\mathbb{T}$, that is, the algebra of all functions on $\mathbb{T}$ with absolutely converging Fourier series. Define $\mathcal{A}, \mathcal{B}, \mathcal{C}, \mathcal{D}$ as in Table 1. In this case $\mathcal{M} = \mathcal{M}_{\mathcal{A},\mathcal{B},\mathcal{C},\mathcal{D}}$ is admissible too. Note that the Fourier transform defines an isomorphism between $\mathcal{W}(\mathbb{T})$ and the algebra ℓ^1 of absolutely converging complex sequences. The version of the twofold EG inverse problem for ℓ^1 has been solved in [10]. In Section 10 we give a new proof of the main theorems in [10] by putting the inversion theorem, [10, Theorem 3.1], and the solution of the twofold EG inverse problem, [10, Theorem 4.1], into the general setting of Section 2 and using the results of Sections 4–7.

The case $\mathcal{N} = \mathcal{RW}(\mathbb{T})$. Let $\mathcal{RW}(\mathbb{T})$ be the subalgebra of $\mathcal{W}(\mathbb{T})$ consisting of all rational functions in $\mathcal{W}(\mathbb{T})$. With $\mathcal{N} = \mathcal{RW}(\mathbb{T})$ and $\mathcal{A}, \mathcal{B}, \mathcal{C}, \mathcal{D}$ as in Table 1, the resulting algebra $\mathcal{M} = \mathcal{M}_{\mathcal{A},\mathcal{B},\mathcal{C},\mathcal{D}}$ is admissible. Let

$$\alpha \in \mathcal{RW}(\mathbb{T})_+^{p\times p}, \quad \beta \in \mathcal{RW}(\mathbb{T})_+^{p\times q},$$
$$\gamma \in \mathcal{RW}(\mathbb{T})_-^{q\times p}, \quad \delta \in \mathcal{RW}(\mathbb{T})_-^{q\times q}.$$

The twofold EG inverse problem is to find $g \in \mathcal{RW}(\mathbb{T})_+^{p\times q}$ such that the following four inclusions are satisfied:

$$\alpha + g\gamma - e_p \in \mathcal{RW}(\mathbb{T})_{-,0}^{p\times p} \quad \text{and} \quad g^*\alpha + \gamma \in \mathcal{RW}(\mathbb{T})_{+,0}^{q\times p};$$
$$g\delta + \beta \in \mathcal{RW}(\mathbb{T})_{-,0}^{p\times q} \quad \text{and} \quad \delta + g^*\beta - e_q \in \mathcal{RW}(\mathbb{T})_{+,0}^{q\times q}.$$

The onefold EG inverse problem for rational matrix functions on $\mathbb{T}$ is treated in [15, Section 6]. Minimal realizations of the functions involved play an important role in the approach in [15]. We intend to work on the twofold EG inverse problem for rational matrix functions on the unit circle in a later publication, again using minimal state space realizations of the functions involved.

The case $\mathcal{N} = \mathcal{TP}$. Let $\mathcal{TP}$ be the set consisting of the trigonometric polynomials in z viewed as a subalgebra of $\mathcal{W}(\mathbb{T})$, and write $\mathcal{TP}_+$ and $\mathcal{TP}_-$ for the subalgebras of polynomials in z and in z^{-1}, respectively. With $\mathcal{TP}_{+,0}$ and $\mathcal{TP}_{-,0}$ we denote the corresponding spaces with the constant functions left out. Again, $\mathcal{A}, \mathcal{B}, \mathcal{C}, \mathcal{D}$ are defined as in Table 1, and the algebra $\mathcal{M} = \mathcal{M}_{\mathcal{A},\mathcal{B},\mathcal{C},\mathcal{D}}$ is admissible. Let

$$\alpha \in \mathcal{TP}_+^{p\times p}, \quad \beta \in \mathcal{TP}_+^{p\times q}, \quad \gamma \in \mathcal{TP}_-^{q\times p}, \quad \delta \in \mathcal{TP}_-^{q\times q}.$$

In this context the twofold EG inverse problem is to find $g \in \mathcal{TP}_+^{p\times q}$ such that the following four inclusions are satisfied:

$$\alpha + g\gamma - e_p \in \mathcal{TP}_{-,0}^{p\times p} \quad \text{and} \quad g^*\alpha + \gamma \in \mathcal{TP}_{+,0}^{q\times p};$$
$$g\delta + \beta \in \mathcal{TP}_{-,0}^{p\times q} \quad \text{and} \quad \delta + g^*\beta - e_q \in \mathcal{TP}_{+,0}^{q\times q}.$$

For this case, a solution to the twofold EG inverse problem has been obtained in [10, Section 9].

4. Preliminaries about Toeplitz-like and Hankel-like operators

In this section we define Toeplitz-like and Hankel-like operators and derive some of their properties. First some notation. In what follows the direct sum of two linear spaces $\mathcal{N}$ and $\mathcal{L}$ will be denoted by $\mathcal{N}\dot{+}\mathcal{L}$. Thus (see [3, pages 37, 38]) the space $\mathcal{N}\dot{+}\mathcal{L}$ consists of all (n,ℓ) with $n\in\mathcal{N}$ and $\ell\in\mathcal{L}$ and its the linear structure is given by

$$(n_1,\ell_1)+(n_2,\ell_2)=(n_1+n_2,\ell_1+\ell_2),\quad \lambda(n,\ell)=(\lambda n,\lambda\ell)\quad (\lambda\in\mathbb{C}).$$

In a canonical way $\mathcal{N}$ and $\mathcal{L}$ can be identified with the linear spaces

$$\{(n,\ell)\mid n\in\mathcal{N},\ell=0\in\mathcal{L}\}\quad\text{and}\quad\{(n,\ell)\mid n=0\in\mathcal{N},\ell\in\mathcal{L}\},$$

respectively. We will use these identifications without further explanation.

Throughout this section $\mathcal{A}$, $\mathcal{B}$, $\mathcal{C}$ and $\mathcal{D}$ are as in Section 2, and we assume that $\mathcal{M}_{\mathcal{A},\mathcal{B},\mathcal{C},\mathcal{D}}$ is admissible. Put $\mathcal{X}=\mathcal{A}\dot{+}\mathcal{B}$ and $\mathcal{Y}=\mathcal{C}\dot{+}\mathcal{D}$. Thus $\mathcal{X}$ is the direct sum of $\mathcal{A}$ and $\mathcal{B}$, and $\mathcal{Y}$ is the direct sum of $\mathcal{C}$ and $\mathcal{D}$. Furthermore, let

$$\mathcal{X}_+=\mathcal{A}_+\dot{+}\mathcal{B}_+,\quad \mathcal{X}_-=\mathcal{A}_-^0\dot{+}\mathcal{B}_-,\tag{4.1}$$

$$\mathcal{Y}_+=\mathcal{C}_+\dot{+}\mathcal{D}_+^0,\quad \mathcal{Y}_-=\mathcal{C}_-\dot{+}\mathcal{D}_-.\tag{4.2}$$

With these direct sums we associate four projections, denoted by

$$\mathbf{P}_{\mathcal{X}_+},\quad \mathbf{P}_{\mathcal{X}_-},\quad \mathbf{P}_{\mathcal{Y}_+},\quad \mathbf{P}_{\mathcal{Y}_-}.$$

By definition, $\mathbf{P}_{\mathcal{X}_+}$ is the projection of $\mathcal{X}$ onto $\mathcal{X}_+$ along $\mathcal{X}_-$, and $\mathbf{P}_{\mathcal{X}_-}$ is the projection of $\mathcal{X}$ onto $\mathcal{X}_-$ along $\mathcal{X}_+$. The two other projections $\mathbf{P}_{\mathcal{Y}_+}$ and $\mathbf{P}_{\mathcal{Y}_-}$ are defined in a similar way, replacing $\mathcal{X}$ by $\mathcal{Y}$.

We proceed with defining multiplication (or Laurent-like) operators and related Toeplitz-like and Hankel-like operators distinguishing four cases. In each case the Toeplitz- and Hankel-like operators are compressions of the multiplication operators. Our terminology differs from the one used in [19] and [20]. Intertwining relations with shift-like operators appear later in the end of Section 7, in Section 8, and in the Appendix.

1. The case when $\rho\in\mathcal{A}$. Assume $\rho\in\mathcal{A}$. Then $\rho\mathcal{A}\subset\mathcal{A}$ and $\rho\mathcal{B}\subset\mathcal{B}$ and therefore we have for $x=(\alpha,\beta)\in\mathcal{X}$ that $\rho x=(\rho\alpha,\rho\beta)\in\mathcal{X}$, i.e., $\rho\mathcal{X}\subset\mathcal{X}$. We define the multiplication operator $\mathbf{L}_\rho:\mathcal{X}\to\mathcal{X}$ by putting $\mathbf{L}_\rho x=\rho x$ for $x\in\mathcal{X}$. With respect to the decomposition $\mathcal{X}=\mathcal{X}_-\dot{+}\mathcal{X}_+$ we write $\mathbf{L}_\rho$ as a 2×2 operator matrix as follows

$$\mathbf{L}_\rho=\begin{bmatrix}\mathbf{T}_{-,\rho} & \mathbf{H}_{-,\rho}\\ \mathbf{H}_{+,\rho} & \mathbf{T}_{+,\rho}\end{bmatrix}:\begin{bmatrix}\mathcal{X}_-\\ \mathcal{X}_+\end{bmatrix}\to\begin{bmatrix}\mathcal{X}_-\\ \mathcal{X}_+\end{bmatrix}.\tag{4.3}$$

Thus for $x_-\in\mathcal{X}_-$ and $x_+\in\mathcal{X}_+$ we have

$$\mathbf{T}_{-,\rho}x_-=\mathbf{P}_{\mathcal{X}_-}(\rho x_-),\quad \mathbf{T}_{+,\rho}x_+=\mathbf{P}_{\mathcal{X}_+}(\rho x_+),$$
$$\mathbf{H}_{+,\rho}x_-=\mathbf{P}_{\mathcal{X}_+}(\rho x_-),\quad \mathbf{H}_{-,\rho}x_+=\mathbf{P}_{\mathcal{X}_-}(\rho x_+).$$

We have $\mathbf{L}_\rho[\mathcal{A}] \subset \mathcal{A}$ and $\mathbf{L}_\rho[\mathcal{B}] \subset \mathcal{B}$. Similarly, one has the inclusions

$$\mathbf{T}_{\pm,\rho}[\mathcal{A}_\pm] \subset \mathcal{A}_\pm, \quad \mathbf{T}_{\pm,\rho}[\mathcal{B}_\pm] \subset \mathcal{B}_\pm, \tag{4.4}$$

$$\mathbf{H}_{\pm,\rho}[\mathcal{A}_\mp] \subset \mathcal{A}_\pm \quad \mathbf{H}_{\pm,\rho}[\mathcal{B}_\mp] \subset \mathcal{B}_\pm. \tag{4.5}$$

Furthermore, as expected from the classical theory of Hankel operators, we have

$$\rho \in \mathcal{A}_+ \Rightarrow \mathbf{H}_{-,\rho} = 0 \quad \text{and} \quad \phi \in \mathcal{A}_- \Rightarrow \mathbf{H}_{+,\phi} = 0. \tag{4.6}$$

2. The case when $\boldsymbol{\rho \in \mathcal{B}}$. For $\rho \in \mathcal{B}$ we have $\rho\mathcal{C} \subset \mathcal{A}$ and $\rho\mathcal{D} \subset \mathcal{B}$, and therefore $\rho\mathcal{Y} \subset \mathcal{X}$. We define the multiplication operator $\mathbf{L}_\rho : \mathcal{Y} \to \mathcal{X}$ by putting $\mathbf{L}_\rho y = \rho y$ for $y \in \mathcal{Y}$. With respect to the decompositions $\mathcal{Y} = \mathcal{Y}_- \dot{+} \mathcal{Y}_+$ and $\mathcal{X} = \mathcal{X}_- \dot{+} \mathcal{X}_+$ we write $\mathbf{L}_\rho$ as a 2×2 operator matrix as follows

$$\mathbf{L}_\rho = \begin{bmatrix} \mathbf{T}_{-,\rho} & \mathbf{H}_{-,\rho} \\ \mathbf{H}_{+,\rho} & \mathbf{T}_{+,\rho} \end{bmatrix} : \begin{bmatrix} \mathcal{Y}_- \\ \mathcal{Y}_+ \end{bmatrix} \to \begin{bmatrix} \mathcal{X}_- \\ \mathcal{X}_+ \end{bmatrix}. \tag{4.7}$$

Thus for $y_- \in \mathcal{Y}_-$ and $y_+ \in \mathcal{Y}_+$ we have

$$\mathbf{T}_{-,\rho} y_- = \mathbf{P}_{\mathcal{X}_-}(\rho y_-), \quad \mathbf{T}_{+,\rho} y_+ = \mathbf{P}_{\mathcal{X}_+}(\rho y_+),$$
$$\mathbf{H}_{+,\rho} y_- = \mathbf{P}_{\mathcal{X}_+}(\rho y_-), \quad \mathbf{H}_{-,\rho} y_+ = \mathbf{P}_{\mathcal{X}_-}(\rho y_+).$$

We have $\mathbf{L}_\rho[\mathcal{C}] \subset \mathcal{A}$ and $\mathbf{L}_\rho[\mathcal{D}] \subset \mathcal{B}$. Similarly, one has

$$\mathbf{T}_{\pm,\rho}[\mathcal{C}_\pm] \subset \mathcal{A}_\pm, \quad \mathbf{T}_{\pm,\rho}[\mathcal{D}_\pm] \subset \mathcal{B}_\pm, \tag{4.8}$$

$$\mathbf{H}_{\pm,\rho}[\mathcal{C}_\mp] \subset \mathcal{A}_\pm, \quad \mathbf{H}_{\pm,\rho}[\mathcal{D}_\mp] \subset \mathcal{B}_\pm. \tag{4.9}$$

Furthermore, we have

$$\rho \in \mathcal{B}_+ \Rightarrow \mathbf{H}_{-,\rho} = 0 \quad \text{and} \quad \phi \in \mathcal{B}_- \Rightarrow \mathbf{H}_{+,\phi} = 0. \tag{4.10}$$

3. The case when $\boldsymbol{\rho \in \mathcal{C}}$. Let $\rho \in \mathcal{C}$. Then $\rho\mathcal{A} \subset \mathcal{C}$ and $\rho\mathcal{B} \subset \mathcal{D}$, and therefore $\rho\mathcal{X} \subset \mathcal{Y}$. We define the multiplication operator $\mathbf{L}_\rho : \mathcal{X} \to \mathcal{Y}$ by putting $\mathbf{L}_\rho x = \rho x$ for $x \in \mathcal{X}$. With respect to the decomposition $\mathcal{X} = \mathcal{X}_- \dot{+} \mathcal{X}_+$ and $\mathcal{Y} = \mathcal{Y}_- \dot{+} \mathcal{Y}_+$ we write $\mathbf{L}_\rho$ as a 2×2 operator matrix as follows

$$\mathbf{L}_\rho = \begin{bmatrix} \mathbf{T}_{-,\rho} & \mathbf{H}_{-,\rho} \\ \mathbf{H}_{+,\rho} & \mathbf{T}_{+,\rho} \end{bmatrix} : \begin{bmatrix} \mathcal{X}_- \\ \mathcal{X}_+ \end{bmatrix} \to \begin{bmatrix} \mathcal{Y}_- \\ \mathcal{Y}_+ \end{bmatrix}. \tag{4.11}$$

Thus for $x_- \in \mathcal{X}_-$ and $x_+ \in \mathcal{X}_+$ we have

$$\mathbf{T}_{-,\rho} x_- = \mathbf{P}_{\mathcal{Y}_-}(\rho x_-), \quad \mathbf{T}_{+,\rho} x_+ = \mathbf{P}_{\mathcal{Y}_+}(\rho x_+),$$
$$\mathbf{H}_{+,\rho} x_- = \mathbf{P}_{\mathcal{Y}_+}(\rho x_-), \quad \mathbf{H}_{-,\rho} x_+ = \mathbf{P}_{\mathcal{Y}_-}(\rho x_+).$$

We have $\mathbf{L}_\rho[\mathcal{A}] \subset \mathcal{C}$ and $\mathbf{L}_\rho[\mathcal{B}] \subset \mathcal{D}$. Similarly, one has the inclusions

$$\mathbf{T}_{\pm,\rho}[\mathcal{A}_\pm] \subset \mathcal{C}_\pm, \quad \mathbf{T}_{\pm,\rho}[\mathcal{B}_\pm] \subset \mathcal{D}_\pm, \tag{4.12}$$

$$\mathbf{H}_{\pm,\rho}[\mathcal{A}_\mp] \subset \mathcal{C}_\pm, \quad \mathbf{H}_{\pm,\rho}[\mathcal{B}_\mp] \subset \mathcal{D}_\pm. \tag{4.13}$$

Furthermore, we have

$$\rho \in \mathcal{C}_+ \Rightarrow \mathbf{H}_{-,\rho} = 0 \quad \text{and} \quad \phi \in \mathcal{C}_- \Rightarrow \mathbf{H}_{+,\phi} = 0. \tag{4.14}$$

4. The case when $\rho \in \mathcal{D}$. For $\rho \in \mathcal{D}$ we have $\rho\mathcal{C} \subset \mathcal{C}$ and $\rho\mathcal{D} \subset \mathcal{D}$, and therefore $\rho\mathcal{Y} \subset \mathcal{Y}$. We define the multiplication operator $\mathbf{L}_\rho : \mathcal{Y} \to \mathcal{Y}$ by putting $\mathbf{L}_\rho y = \rho y$ for $y \in \mathcal{Y}$. With respect to the decomposition $\mathcal{Y} = \mathcal{Y}_- \dot{+} \mathcal{Y}_+$ we write $\mathbf{L}_\rho$ as a 2×2 operator matrix as follows

$$\mathbf{L}_\rho = \begin{bmatrix} \mathbf{T}_{-,\rho} & \mathbf{H}_{-,\rho} \\ \mathbf{H}_{+,\rho} & \mathbf{T}_{+,\rho} \end{bmatrix} : \begin{bmatrix} \mathcal{Y}_- \\ \mathcal{Y}_+ \end{bmatrix} \to \begin{bmatrix} \mathcal{Y}_- \\ \mathcal{Y}_+ \end{bmatrix}. \tag{4.15}$$

Thus for $y_- \in \mathcal{Y}_-$ and $y_+ \in \mathcal{Y}_+$ we have

$$\mathbf{T}_{-,\rho} y_- = \mathbf{P}_{\mathcal{Y}_-}(\rho y_-), \quad \mathbf{T}_{+,\rho} y_+ = \mathbf{P}_{\mathcal{Y}_+}(\rho y_+),$$
$$\mathbf{H}_{+,\rho} y_- = \mathbf{P}_{\mathcal{Y}_+}(\rho y_-), \quad \mathbf{H}_{-,\rho} y_+ = \mathbf{P}_{\mathcal{Y}_-}(\rho y_+).$$

We have $\mathbf{L}_\rho[\mathcal{C}] \subset \mathcal{C}$ and $\mathbf{L}_\rho[\mathcal{D}] \subset \mathcal{D}$. Similarly, we have the inclusions

$$\mathbf{T}_{\pm,\rho}[\mathcal{C}_\pm] \subset \mathcal{C}_\pm, \quad \mathbf{T}_{\pm,\rho}[\mathcal{D}_\pm] \subset \mathcal{D}_\pm, \tag{4.16}$$
$$\mathbf{H}_{\pm,\rho}[\mathcal{C}_\mp] \subset \mathcal{C}_\pm, \quad \mathbf{H}_{\pm,\rho}[\mathcal{D}_\mp] \subset \mathcal{D}_\pm. \tag{4.17}$$

Furthermore, we have

$$\rho \in \mathcal{D}_+ \Rightarrow \mathbf{H}_{-,\rho} = 0, \qquad \phi \in \mathcal{D}_- \Rightarrow \mathbf{H}_{+,\phi} = 0. \tag{4.18}$$

5. Multiplicative identities. Let $\mathcal{U}$, $\mathcal{V}$ and $\mathcal{Z}$ each be one of the spaces $\mathcal{X}$ or $\mathcal{Y}$ defined above. The corresponding decomposition of the spaces we denote as $\mathcal{U} = \mathcal{U}_- \dot{+} \mathcal{U}_+$ and similarly for $\mathcal{V}$ and $\mathcal{Z}$. Let ϕ be such that for $u \in \mathcal{U}$ we have $\phi u \in \mathcal{V}$ and ρ be such that for $v \in \mathcal{V}$ we have $\rho v \in \mathcal{Z}$. Then we have that $\mathbf{L}_{\rho\phi} = \mathbf{L}_\rho \mathbf{L}_\phi$, which gives

$$\begin{bmatrix} \mathbf{T}_{-,\rho\phi} & \mathbf{H}_{-,\rho\phi} \\ \mathbf{H}_{+,\rho\phi} & \mathbf{T}_{+,\rho\phi} \end{bmatrix} = \begin{bmatrix} \mathbf{T}_{-,\rho} & \mathbf{H}_{-,\rho} \\ \mathbf{H}_{+,\rho} & \mathbf{T}_{+,\rho} \end{bmatrix} \begin{bmatrix} \mathbf{T}_{-,\phi} & \mathbf{H}_{-,\phi} \\ \mathbf{H}_{+,\phi} & \mathbf{T}_{+,\phi} \end{bmatrix} : \begin{bmatrix} \mathcal{U}_- \\ \mathcal{U}_+ \end{bmatrix} \to \begin{bmatrix} \mathcal{Z}_- \\ \mathcal{Z}_+ \end{bmatrix}. \tag{4.19}$$

In particular, we have the following identities:

$$\mathbf{T}_{+,\rho\phi} = \mathbf{T}_{+,\rho}\mathbf{T}_{+,\phi} + \mathbf{H}_{+,\rho}\mathbf{H}_{-,\phi} : \mathcal{U}_+ \to \mathcal{Z}_+, \tag{4.20}$$
$$\mathbf{H}_{+,\rho\phi} = \mathbf{T}_{+,\rho}\mathbf{H}_{+,\phi} + \mathbf{H}_{+,\rho}\mathbf{T}_{-,\phi} : \mathcal{U}_- \to \mathcal{Z}_+, \tag{4.21}$$
$$\mathbf{H}_{-,\rho\phi} = \mathbf{H}_{-,\rho}\mathbf{T}_{+,\phi} + \mathbf{T}_{-,\rho}\mathbf{H}_{-,\phi} : \mathcal{U}_+ \to \mathcal{Z}_-, \tag{4.22}$$
$$\mathbf{T}_{-,\rho\phi} = \mathbf{H}_{-,\rho}\mathbf{H}_{+,\phi} + \mathbf{T}_{-,\rho}\mathbf{T}_{-,\phi} : \mathcal{U}_- \to \mathcal{Z}_-. \tag{4.23}$$

5. Further notations and auxiliary results

In this section we bring together a number of identities and lemmas that will be used in the proofs of the main results. Throughout this section $\alpha \in \mathcal{A}_+$, $\beta \in \mathcal{B}_+$, $\gamma \in \mathcal{C}_-$, and $\delta \in \mathcal{D}_-$. Furthermore, g is an arbitrary element in $\mathcal{B}_+$. We split this section into two parts.

PART 1. With g we associate the operator $\mathbf{\Omega}$ given by

$$\mathbf{\Omega} = \begin{bmatrix} \mathbf{I}_{\mathcal{X}_+} & \mathbf{H}_{+,g} \\ \mathbf{H}_{-,g^*} & \mathbf{I}_{\mathcal{Y}_-} \end{bmatrix} : \begin{bmatrix} \mathcal{X}_+ \\ \mathcal{Y}_- \end{bmatrix} \to \begin{bmatrix} \mathcal{X}_+ \\ \mathcal{Y}_- \end{bmatrix}. \tag{5.1}$$

Here $\mathcal{X}_\pm$ and $\mathcal{Y}_\pm$ are as in (4.1) and (4.2), respectively. Using the properties of Hankel-like operators given in the previous section we see that

$$(2.6) \iff \mathbf{\Omega}\begin{bmatrix}\alpha\\ \gamma\end{bmatrix} = \begin{bmatrix}e_{\mathcal{A}}\\ 0\end{bmatrix}, \tag{5.2}$$

$$(2.7) \iff \mathbf{\Omega}\begin{bmatrix}\beta\\ \delta\end{bmatrix} = \begin{bmatrix}0\\ e_{\mathcal{D}}\end{bmatrix}. \tag{5.3}$$

Summarizing this yields the following corollary.

Corollary 5.1. *The element $g \in \mathcal{B}_+$ is a solution to the twofold EG inverse problem associated with α, β, γ, and δ if and only if*

$$\mathbf{\Omega}\begin{bmatrix}\alpha\\ \gamma\end{bmatrix} = \begin{bmatrix}e_{\mathcal{A}}\\ 0\end{bmatrix} \quad \textit{and} \quad \mathbf{\Omega}\begin{bmatrix}\beta\\ \delta\end{bmatrix} = \begin{bmatrix}0\\ e_{\mathcal{D}}\end{bmatrix}, \tag{5.4}$$

We also have the following implications:

$$\alpha + g\gamma \in \mathcal{A}_- \implies \mathbf{H}_{+,g\gamma} = -\mathbf{H}_{+,\alpha}, \tag{5.5}$$
$$g^*\alpha + \gamma \in \mathcal{C}_+ \iff \mathbf{H}_{-,g^*\alpha} = -\mathbf{H}_{-,\gamma}, \tag{5.6}$$
$$\beta + g\delta \in \mathcal{B}_- \iff \mathbf{H}_{+,g\delta} = -\mathbf{H}_{+,\beta}, \tag{5.7}$$
$$g^*\beta + \delta \in \mathcal{D}_+ \implies \mathbf{H}_{-,g^*\beta} = -\mathbf{H}_{-,\delta}. \tag{5.8}$$

After taking adjoints in the left-hand inclusions above we obtain

$$\alpha + g\gamma \in \mathcal{A}_- \implies \mathbf{H}_{-,\gamma^*g^*} = -\mathbf{H}_{-,\alpha^*}, \tag{5.9}$$
$$g^*\alpha + \gamma \in \mathcal{C}_+ \iff \mathbf{H}_{+,\alpha^*g} = -\mathbf{H}_{+,\gamma^*}, \tag{5.10}$$
$$\beta + g\delta \in \mathcal{B}_- \iff \mathbf{H}_{-,\delta^*g^*} = -\mathbf{H}_{-,\beta^*}, \tag{5.11}$$
$$g^*\beta + \delta \in \mathcal{D}_+ \implies \mathbf{H}_{+,\beta^*g} = -\mathbf{H}_{+,\delta^*}. \tag{5.12}$$

Notice that the first inclusion in (2.6) implies that $\alpha + g\gamma \in \mathcal{A}_-$ and the second inclusion in (2.7) implies $g^*\beta + \delta \in \mathcal{D}_+$. The implications from left to right are obvious. To prove the implications from right to left in (5.6), (5.7), (5.10) and (5.11) one reasons as follows. For example, for (5.6) one uses that $e_{\mathcal{A}} \in \mathcal{X}_+$, such that

$$0 = \mathbf{H}_{-,g^*\alpha+\gamma}e_{\mathcal{A}} = \mathbf{P}_{\mathcal{X}_-}(g^*\alpha + \gamma)e_{\mathcal{A}} = \mathbf{P}_{\mathcal{C}_-}(g^*\alpha + \gamma).$$

Hence $g^*\alpha + \gamma \in \mathcal{C}_+$, as claimed. Since $e_{\mathcal{A}} \notin \mathcal{X}_-$ and $e_{\mathcal{D}} \notin \mathcal{Y}_+$, the reverse implications in (5.5), (5.8), (5.9) and (5.12) cannot be derived in this way.

Note that $\alpha \in \mathcal{A}_+$, $\beta \in \mathcal{B}_+$, $\gamma \in \mathcal{C}_-$, $\delta \in \mathcal{D}_-$ implies that

$$\mathbf{H}_{-,\alpha} = 0, \quad \mathbf{H}_{-,\beta} = 0, \quad \mathbf{H}_{+,\gamma} = 0, \quad \mathbf{H}_{+,\delta} = 0, \tag{5.13}$$
$$\mathbf{H}_{+,\alpha^*} = 0, \quad \mathbf{H}_{+,\beta^*} = 0, \quad \mathbf{H}_{-,\gamma^*} = 0, \quad \mathbf{H}_{-,\delta^*} = 0. \tag{5.14}$$

Using the identities (5.13) and (5.14) together with the product formulas at the end of Section 4 we obtain the following eight identities:

$$\mathbf{H}_{+,\alpha^* g} = \mathbf{T}_{+,\alpha^*}\mathbf{H}_{+,g}, \qquad \mathbf{H}_{+,\beta^* g} = \mathbf{T}_{+,\beta^*}\mathbf{H}_{+,g}, \tag{5.15}$$

$$\mathbf{H}_{-,\gamma^* g^*} = \mathbf{T}_{-,\gamma^*}\mathbf{H}_{-,g^*}, \quad \mathbf{H}_{-,\delta^* g^*} = \mathbf{T}_{-,\delta^*}\mathbf{H}_{-,g^*}, \tag{5.16}$$

$$\mathbf{H}_{+,g\gamma} = \mathbf{H}_{+,g}\mathbf{T}_{-,\gamma}, \qquad \mathbf{H}_{+,g\delta} = \mathbf{H}_{+,g}\mathbf{T}_{-,\delta}, \tag{5.17}$$

$$\mathbf{H}_{-,g^*\alpha} = \mathbf{H}_{-,g^*}\mathbf{T}_{+,\alpha}, \qquad \mathbf{H}_{-,g^*\beta} = \mathbf{H}_{-,g^*}\mathbf{T}_{+,\beta}. \tag{5.18}$$

The next lemma is an immediate consequence of the definitions.

Lemma 5.2. *For $g \in \mathcal{B}_+$ and $h \in \mathcal{C}_-$ we have*

$$\mathbf{H}_{+,g} e_{\mathcal{D}} = g \quad \text{and} \quad \mathbf{H}_{-,h} e_{\mathcal{A}} = h.$$

We conclude this part with the following lemma.

Lemma 5.3. *Assume that conditions* (C1)–(C3) *are satisfied. Then*

$$\begin{bmatrix} \mathbf{T}_{+,\alpha^*} \\ \mathbf{T}_{+,\beta^*} \end{bmatrix} \begin{bmatrix} \mathbf{H}_{+,\beta} & \mathbf{H}_{+,\alpha} \end{bmatrix} = \begin{bmatrix} \mathbf{H}_{+,\gamma^*} \\ \mathbf{H}_{+,\delta^*} \end{bmatrix} \begin{bmatrix} \mathbf{T}_{-,\delta} & \mathbf{T}_{-,\gamma} \end{bmatrix}. \tag{5.19}$$

and

$$\begin{bmatrix} \mathbf{T}_{-,\delta^*} \\ \mathbf{T}_{-,\gamma^*} \end{bmatrix} \begin{bmatrix} \mathbf{H}_{-,\gamma} & \mathbf{H}_{-,\delta} \end{bmatrix} = \begin{bmatrix} \mathbf{H}_{-,\beta^*} \\ \mathbf{H}_{-,\alpha^*} \end{bmatrix} \begin{bmatrix} \mathbf{T}_{+,\alpha} & \mathbf{T}_{+,\beta} \end{bmatrix}. \tag{5.20}$$

Proof. In the course of the proof we repeatedly use the product rules (4.20)–(4.23). Using the first identity in (5.14), condition (C3) and the fourth identity in (5.13) we see that

$$\begin{aligned} \mathbf{T}_{+,\alpha^*}\mathbf{H}_{+,\beta} &= \mathbf{H}_{+,\alpha^*\beta} - \mathbf{H}_{+,\alpha^*}\mathbf{T}_{-,\beta} \\ &= \mathbf{H}_{+,\alpha^*\beta} = \mathbf{H}_{+,\gamma^*\delta} \\ &= \mathbf{H}_{+,\gamma^*}\mathbf{T}_{-,\delta} + \mathbf{T}_{+,\gamma^*}\mathbf{H}_{+,\delta} = \mathbf{H}_{+,\gamma^*}\mathbf{T}_{-,\delta}. \end{aligned}$$

It follows that

$$\mathbf{T}_{+,\alpha^*}\mathbf{H}_{+,\beta} = \mathbf{H}_{+,\gamma^*}\mathbf{T}_{-,\delta}. \tag{5.21}$$

Next, using the first identity in (5.14) and the third in (5.13) we obtain

$$\begin{aligned} \mathbf{T}_{+,\alpha^*}\mathbf{H}_{+,\alpha} &= \mathbf{H}_{+,\alpha^*\alpha} - \mathbf{H}_{+,\alpha^*}\mathbf{T}_{-,\alpha} = \mathbf{H}_{+,\alpha^*\alpha}, \\ \mathbf{H}_{+,\gamma^*}\mathbf{T}_{-,\gamma} &= \mathbf{H}_{+,\gamma^*\gamma} - \mathbf{T}_{+,\gamma^*}\mathbf{H}_{+,\gamma} = \mathbf{H}_{+,\gamma^*\gamma}. \end{aligned}$$

On the other hand, using condition (C1) and the second identity in (4.6) with $\phi = a_0 = \mathbf{P}_{\mathcal{A}_d}\alpha$, we see that $\mathbf{H}_{+,\alpha^*\alpha} - \mathbf{H}_{+,\gamma^*\gamma} = \mathbf{H}_{+,a_0} = 0$. We proved

$$\mathbf{T}_{+,\alpha^*}\mathbf{H}_{+,\alpha} = \mathbf{H}_{+,\gamma^*}\mathbf{T}_{-,\gamma}. \tag{5.22}$$

The next two equalities are proved in a similar way as the previous two:

$$\mathbf{T}_{+,\beta^*}\mathbf{H}_{+,\beta} = \mathbf{H}_{+,\delta^*}\mathbf{T}_{-,\delta}, \tag{5.23}$$

$$\mathbf{T}_{+,\beta^*}\mathbf{H}_{+,\alpha} = \mathbf{H}_{+,\delta^*}\mathbf{T}_{-,\gamma}. \tag{5.24}$$

Observe that (5.21), (5.22), (5.23) and (5.24) can be rewritten as (5.19).

The equality (5.20) is proved similarly, using (C2) instead of (C1). □

PART 2. In the second part of this section we assume that $a_0 = \mathbf{P}_{\mathcal{A}_d}\alpha$ and $d_0 = \mathbf{P}_{\mathcal{A}_d}\delta$ are invertible in $\mathcal{A}_d$ and $\mathcal{D}_d$, respectively. Using the notations introduced in the previous section we associate with the elements $\alpha, \beta, \gamma, \delta$ the following operators:

$$\mathbf{R}_{11} = \mathbf{T}_{+,\alpha}a_0^{-1}\mathbf{T}_{+,\alpha^*} - \mathbf{T}_{+,\beta}d_0^{-1}\mathbf{T}_{+,\beta^*} : \mathcal{X}_+ \to \mathcal{X}_+, \tag{5.25}$$

$$\mathbf{R}_{21} = \mathbf{H}_{-,\gamma}a_0^{-1}\mathbf{T}_{+,\alpha^*} - \mathbf{H}_{-,\delta}d_0^{-1}\mathbf{T}_{+,\beta^*} : \mathcal{X}_+ \to \mathcal{Y}_-, \tag{5.26}$$

$$\mathbf{R}_{12} = \mathbf{H}_{+,\beta}d_0^{-1}\mathbf{T}_{-,\delta^*} - \mathbf{H}_{+,\alpha}a_0^{-1}\mathbf{T}_{-,\gamma^*} : \mathcal{Y}_- \to \mathcal{X}_+, \tag{5.27}$$

$$\mathbf{R}_{22} = \mathbf{T}_{-,\delta}d_0^{-1}\mathbf{T}_{-,\delta^*} - \mathbf{T}_{-,\gamma}a_0^{-1}\mathbf{T}_{-,\gamma^*} : \mathcal{Y}_- \to \mathcal{Y}_-. \tag{5.28}$$

Lemma 5.4. *Assume that conditions* (C1) *and* (C2) *are satisfied and that a_0 and d_0 are invertible in $\mathcal{A}_d$ and $\mathcal{D}_d$, respectively. Then the following identities hold true:*

$$\mathbf{R}_{11}e_{\mathcal{A}} = \alpha, \quad \mathbf{R}_{12}e_{\mathcal{D}} = \beta, \quad \mathbf{R}_{21}e_{\mathcal{A}} = \gamma, \quad \mathbf{R}_{22}e_{\mathcal{D}} = \delta. \tag{5.29}$$

Proof. Note that $\beta^* \in \mathcal{C}_-$. Thus

$$\mathbf{T}_{+,\beta^*}e_{\mathcal{A}} = \mathbf{P}_{\mathcal{Y}_+}(\beta^* e_{\mathcal{A}}) = \mathbf{P}_{\mathcal{Y}_+}\beta^* = 0.$$

Since $\alpha^* \in \mathcal{A}_-^*$, we have

$$\mathbf{T}_{+,\alpha^*}e_{\mathcal{A}} = \mathbf{P}_{\mathcal{X}_+}(\alpha^* e_{\mathcal{A}}) = \mathbf{P}_{\mathcal{X}_+}\alpha^* = a_0^*.$$

Using $a_0 = a_0^*$ (by the first part of (2.8)), it follows that

$$\mathbf{R}_{11}e_{\mathcal{A}} = \mathbf{T}_{+,\alpha}a_0^{-1}\mathbf{T}_{+,\alpha^*}e_{\mathcal{A}} = \mathbf{T}_{+,\alpha}a_0^{-1}a_0^* = \mathbf{T}_{+,\alpha}e_{\mathcal{A}} = P_{\mathcal{X}_+}(\alpha e_{\mathcal{A}}) = \alpha.$$

Notice that we used condition (C1). This proves the first identity (5.29).

Next, using $\gamma \in \mathcal{C}_-$, $\mathbf{T}_{+,\alpha^*}e_{\mathcal{A}} = a_0^*$, and $\mathbf{T}_{+,\beta^*}e_{\mathcal{A}} = \mathbf{P}_{\mathcal{Y}_+}\beta^* = 0$ we obtain

$$\mathbf{R}_{21}e_{\mathcal{A}} = \mathbf{H}_{-,\gamma}a_0^{-1}\mathbf{T}_{+,\alpha^*}e_{\mathcal{A}} = \mathbf{H}_{-,\gamma}e_{\mathcal{A}} = \mathbf{P}_{\mathcal{Y}_-}(\gamma e_{\mathcal{A}}) = \mathbf{P}_{\mathcal{Y}_-}\gamma = \gamma,$$

which proves the third identity in (5.29). The two other identities in (5.29), involving $\mathbf{R}_{12}$ and $\mathbf{R}_{22}$, are obtained in a similar way, using (C2), (2.8) and

$$\mathbf{T}_{-,\gamma^*}e_{\mathcal{D}} = 0, \quad \mathbf{T}_{-,\delta^*}e_{\mathcal{D}} = d_0^*, \quad \mathbf{H}_{+,\beta}e_{\mathcal{D}} = \beta, \quad \mathbf{T}_{-,\delta}e_{\mathcal{D}} = \delta.$$

This proves the lemma. □

The next lemma presents alternative formulas for the operators $\mathbf{R}_{ij}$, $1 \leq i, j \leq 2$, given by (5.25)–(5.28), assuming conditions (C4)–(C6) are satisfied.

Lemma 5.5. *Assume that a_0 and d_0 are invertible in $\mathcal{A}_d$ and $\mathcal{D}_d$, respectively, and that conditions* (C4), (C5), *and* (C6) *are satisfied. Then*

$$\mathbf{R}_{11} = \mathbf{I}_{\mathcal{X}_+} - \mathbf{H}_{+,\alpha}a_0^{-1}\mathbf{H}_{-,\alpha^*} + \mathbf{H}_{+,\beta}d_0^{-1}\mathbf{H}_{-,\beta^*} : \mathcal{X}_+ \to \mathcal{X}_+, \tag{5.30}$$

$$\mathbf{R}_{21} = \mathbf{T}_{-,\delta}d_0^{-1}\mathbf{H}_{-,\beta^*} - \mathbf{T}_{-,\gamma}a_0^{-1}\mathbf{H}_{-,\alpha^*} : \mathcal{X}_+ \to \mathcal{Y}_-, \tag{5.31}$$

$$\mathbf{R}_{12} = \mathbf{T}_{+,\alpha}a_0^{-1}\mathbf{H}_{+,\gamma^*} - \mathbf{T}_{+,\beta}d_0^{-1}\mathbf{H}_{+,\delta^*} : \mathcal{Y}_- \to \mathcal{X}_+, \tag{5.32}$$

$$\mathbf{R}_{22} = \mathbf{I}_{\mathcal{Y}_-} - \mathbf{H}_{-,\delta}d_0^{-1}\mathbf{H}_{+,\delta^*} + \mathbf{H}_{-,\gamma}a_0^{-1}\mathbf{H}_{+,\gamma^*} : \mathcal{Y}_- \to \mathcal{Y}_-. \tag{5.33}$$

Proof. First notice that $a_0^{-1} \in \mathcal{A}_d$ and $d_0^{-1} \in \mathcal{D}_d$ yield the following identities

$$\mathbf{T}_{+,\alpha a_0^{-1}} = \mathbf{T}_{+,\alpha} a_0^{-1}, \quad \mathbf{H}_{+,\alpha a_0^{-1}} = \mathbf{H}_{+,\alpha} a_0^{-1}, \tag{5.34}$$

$$\mathbf{T}_{+,\beta d_0^{-1}} = \mathbf{T}_{+,\beta} d_0^{-1}, \quad \mathbf{H}_{+,\beta d_0^{-1}} = \mathbf{H}_{+,\beta} d_0^{-1}, \tag{5.35}$$

$$\mathbf{T}_{-,\delta d_0^{-1}} = \mathbf{T}_{-,\delta} d_0^{-1}, \quad \mathbf{H}_{-,\delta d_0^{-1}} = \mathbf{H}_{-,\delta} d_0^{-1}, \tag{5.36}$$

$$\mathbf{T}_{-,\gamma a_0^{-1}} = \mathbf{T}_{-,\gamma} a_0^{-1}, \quad \mathbf{H}_{-,\gamma a_0^{-1}} = \mathbf{H}_{-,\gamma} a_0^{-1}. \tag{5.37}$$

Next, note that condition (C4) implies that $\mathbf{T}_{+,\alpha a_0^{-1}\alpha^* - \beta d_0^{-1}\beta^* - e_{\mathcal{A}}} = 0$. It follows that

$$\mathbf{T}_{+,\alpha a_0^{-1}\alpha^*} - \mathbf{T}_{+,\beta d_0^{-1}\beta^*} - \mathbf{I}_{\mathcal{X}_+} = 0.$$

Applying the product rule (4.20) and the identities in (5.34) and (5.35) we see that

$$\mathbf{T}_{+,\alpha} a_0^{-1} \mathbf{T}_{+,\alpha^*} - \mathbf{T}_{+,\beta} d_0^{-1} \mathbf{T}_{+,\beta^*} = \mathbf{I}_{\mathcal{X}_+} - \mathbf{H}_{+,\alpha} a_0^{-1} \mathbf{H}_{-,\alpha^*} + \mathbf{H}_{+,\beta} d_0^{-1} \mathbf{H}_{-,\beta^*}.$$

It follows that the operator $\mathbf{R}_{11}$ defined by (5.25) is also given by (5.30). In a similar way one shows that condition (C5) yields the identity (5.33).

Since (C6) states $\alpha a_0^{-1}\gamma^* = \beta d_0^{-1}\delta^*$, we have the equality $\mathbf{H}_{+,\alpha a_0^{-1}\gamma^*} = \mathbf{H}_{+,\beta d_0^{-1}\delta^*}$. Applying the product rule (4.21) and the identities in (5.34) and (5.35) it follows that

$$\mathbf{H}_{+,\alpha} a_0^{-1} \mathbf{T}_{-,\gamma^*} + \mathbf{T}_{+,\alpha} a_0^{-1} \mathbf{H}_{+,\gamma^*} = \mathbf{H}_{+,\beta} d_0^{-1} \mathbf{T}_{-,\delta^*} + \mathbf{T}_{+,\beta} d_0^{-1} \mathbf{H}_{+,\delta^*}.$$

This yields

$$\mathbf{R}_{12} = \mathbf{H}_{+,\beta} d_0^{-1} \mathbf{T}_{-,\delta^*} - \mathbf{H}_{+,\alpha} a_0^{-1} \mathbf{T}_{-,\gamma^*} = \mathbf{T}_{+,\alpha} a_0^{-1} \mathbf{H}_{+,\gamma^*} - \mathbf{T}_{+,\beta} d_0^{-1} \mathbf{H}_{+,\delta^*},$$

which proves (5.32).

Finally, to prove the identity (5.31), note that, by taking adjoints, condition (C6) yields that $\delta d_0^{-1}\beta^* = \gamma a_0^{-1}\alpha^*$. But then using the identities in (5.36) and (5.37), arguments similar to the ones used in the previous paragraph, yield the identity (5.31). □

The following lemma contains some useful formulas that we will prove by direct verification.

Lemma 5.6. *Assume that* a_0 *and* d_0 *are invertible in* $\mathcal{A}_d$ *and* $\mathcal{D}_d$*, respectively, and that the conditions* (C1)–(C6) *are satisfied. Let* $\mathbf{R}_{ij}$*,* $i,j = 1,2$*, be given by* (5.25)–(5.28). *Then*

$$\begin{bmatrix} \mathbf{R}_{11} & \mathbf{R}_{12} \\ \mathbf{R}_{21} & \mathbf{R}_{22} \end{bmatrix} \begin{bmatrix} \mathbf{I}_{\mathcal{X}_+} & 0 \\ 0 & -\mathbf{I}_{\mathcal{Y}_-} \end{bmatrix} \begin{bmatrix} \mathbf{R}_{11} & \mathbf{R}_{12} \\ \mathbf{R}_{21} & \mathbf{R}_{22} \end{bmatrix} = \begin{bmatrix} \mathbf{R}_{11} & 0 \\ 0 & -\mathbf{R}_{22} \end{bmatrix}. \tag{5.38}$$

This implies that

$$\mathbf{R} = \begin{bmatrix} \mathbf{R}_{11} & \mathbf{R}_{12} \\ \mathbf{R}_{21} & \mathbf{R}_{22} \end{bmatrix} : \begin{bmatrix} \mathcal{X}_+ \\ \mathcal{Y}_- \end{bmatrix} \to \begin{bmatrix} \mathcal{X}_+ \\ \mathcal{Y}_- \end{bmatrix} \tag{5.39}$$

is invertible if and only if $\mathbf{R}_{11}$ *and* $\mathbf{R}_{22}$ *are invertible. Furthermore, in that case*

$$\mathbf{R}^{-1} = \begin{bmatrix} \mathbf{I}_{\mathcal{X}_+} & -\mathbf{R}_{12}\mathbf{R}_{22}^{-1} \\ -\mathbf{R}_{21}\mathbf{R}_{11}^{-1} & \mathbf{I}_{\mathcal{Y}_-} \end{bmatrix} = \begin{bmatrix} \mathbf{I}_{\mathcal{X}_+} & -\mathbf{R}_{11}^{-1}\mathbf{R}_{12} \\ -\mathbf{R}_{22}^{-1}\mathbf{R}_{21} & \mathbf{I}_{\mathcal{Y}_-} \end{bmatrix}. \tag{5.40}$$

Proof. To check (5.38) we will prove the four identities

$$\mathbf{R}_{11}\mathbf{R}_{12} = \mathbf{R}_{12}\mathbf{R}_{22}, \qquad \mathbf{R}_{22}\mathbf{R}_{21} = \mathbf{R}_{21}\mathbf{R}_{11}, \tag{5.41}$$

$$\mathbf{R}_{11}\mathbf{R}_{11} - \mathbf{R}_{12}\mathbf{R}_{21} = \mathbf{R}_{11}, \qquad \mathbf{R}_{22}\mathbf{R}_{22} - \mathbf{R}_{21}\mathbf{R}_{12} = \mathbf{R}_{22}. \tag{5.42}$$

From (5.25)–(5.28) and (5.30)–(5.33) it follows that

$$\begin{aligned} \mathbf{R}_{11}\mathbf{R}_{12} = &\begin{bmatrix} \mathbf{T}_{+,\alpha} & \mathbf{T}_{+,\beta} \end{bmatrix} \begin{bmatrix} a_0^{-1} & 0 \\ 0 & -d_0^{-1} \end{bmatrix} \begin{bmatrix} \mathbf{T}_{+,\alpha^*} \\ \mathbf{T}_{+,\beta^*} \end{bmatrix} \\ &\times \begin{bmatrix} \mathbf{H}_{+,\beta} & \mathbf{H}_{+,\alpha} \end{bmatrix} \begin{bmatrix} d_0^{-1} & 0 \\ 0 & -a_0^{-1} \end{bmatrix} \begin{bmatrix} \mathbf{T}_{-,\delta^*} \\ \mathbf{T}_{-,\gamma^*} \end{bmatrix}, \end{aligned}$$

and

$$\begin{aligned} \mathbf{R}_{12}\mathbf{R}_{22} = &\begin{bmatrix} \mathbf{T}_{+,\alpha} & \mathbf{T}_{+,\beta} \end{bmatrix} \begin{bmatrix} a_0^{-1} & 0 \\ 0 & -d_0^{-1} \end{bmatrix} \begin{bmatrix} \mathbf{H}_{+,\gamma^*} \\ \mathbf{H}_{+,\delta*} \end{bmatrix} \\ &\times \begin{bmatrix} \mathbf{T}_{-,\delta} & \mathbf{T}_{-,\gamma} \end{bmatrix} \begin{bmatrix} d_0^{-1} & 0 \\ 0 & -a_0^{-1} \end{bmatrix} \begin{bmatrix} \mathbf{T}_{-,\delta^*} \\ \mathbf{T}_{-,\gamma^*} \end{bmatrix}. \end{aligned}$$

But then (5.19) shows that $\mathbf{R}_{11}\mathbf{R}_{12} = \mathbf{R}_{12}\mathbf{R}_{22}$. In a similar way, using (5.20) one proves that $\mathbf{R}_{22}\mathbf{R}_{21} = \mathbf{R}_{21}\mathbf{R}_{11}$.

Next observe that

$$\begin{aligned} \mathbf{R}_{11}(\mathbf{R}_{11} - \mathbf{I}_{\mathcal{X}_+}) = &- \begin{bmatrix} \mathbf{T}_{+,\alpha} & \mathbf{T}_{+,\beta} \end{bmatrix} \begin{bmatrix} a_0^{-1} & 0 \\ 0 & -d_0^{-1} \end{bmatrix} \begin{bmatrix} \mathbf{T}_{+,\alpha^*} \\ \mathbf{T}_{+,\beta^*} \end{bmatrix} \\ &\times \begin{bmatrix} \mathbf{H}_{+,\alpha} & \mathbf{H}_{+,\beta} \end{bmatrix} \begin{bmatrix} a_0^{-1} & 0 \\ 0 & -d_0^{-1} \end{bmatrix} \begin{bmatrix} \mathbf{H}_{-,\alpha^*} \\ \mathbf{H}_{-,\beta^*} \end{bmatrix}, \end{aligned}$$

and

$$\begin{aligned} \mathbf{R}_{12}\mathbf{R}_{21} = &\begin{bmatrix} \mathbf{T}_{+,\alpha} & \mathbf{T}_{+,\beta} \end{bmatrix} \begin{bmatrix} a_0^{-1} & 0 \\ 0 & -d_0^{-1} \end{bmatrix} \begin{bmatrix} \mathbf{H}_{+,\gamma^*} \\ \mathbf{H}_{+,\delta*} \end{bmatrix} \\ &\times \begin{bmatrix} \mathbf{T}_{-,\delta} & \mathbf{T}_{-,\gamma} \end{bmatrix} \begin{bmatrix} d_0^{-1} & 0 \\ 0 & -a_0^{-1} \end{bmatrix} \begin{bmatrix} \mathbf{H}_{-,\beta^*} \\ \mathbf{H}_{-,\alpha^*} \end{bmatrix} \\ = &- \begin{bmatrix} \mathbf{T}_{+,\alpha} & \mathbf{T}_{+,\beta} \end{bmatrix} \begin{bmatrix} a_0^{-1} & 0 \\ 0 & -d_0^{-1} \end{bmatrix} \begin{bmatrix} \mathbf{H}_{+,\gamma^*} \\ \mathbf{H}_{+,\delta*} \end{bmatrix} \\ &\times \begin{bmatrix} \mathbf{T}_{-,\gamma} & \mathbf{T}_{-,\delta} \end{bmatrix} \begin{bmatrix} a_0^{-1} & 0 \\ 0 & -d_0^{-1} \end{bmatrix} \begin{bmatrix} \mathbf{H}_{-,\alpha^*} \\ \mathbf{H}_{-,\beta^*} \end{bmatrix}. \end{aligned}$$

But then (5.19) implies that $\mathbf{R}_{11}\mathbf{R}_{11} - \mathbf{R}_{12}\mathbf{R}_{21} = \mathbf{R}_{11}$. Similarly, using (5.20) one proves that $\mathbf{R}_{22}\mathbf{R}_{22} - \mathbf{R}_{21}\mathbf{R}_{12} = \mathbf{R}_{22}$.

The final statements (5.39) and (5.40) are immediate from (5.41) and (5.42). □

6. An abstract inversion theorem

Let $\mathcal{M} = \mathcal{M}_{\mathcal{A},\mathcal{B},\mathcal{C},\mathcal{D}}$ be an admissible algebra. Fix $g \in \mathcal{B}_+$, and let $\mathbf{\Omega}$ be the operator given by

$$\mathbf{\Omega} := \begin{bmatrix} I_{\mathcal{X}_+} & \mathbf{H}_{+,g} \\ \mathbf{H}_{-,g^*} & \mathbf{I}_{\mathcal{Y}_-} \end{bmatrix} : \begin{bmatrix} \mathcal{X}_+ \\ \mathcal{Y}_- \end{bmatrix} \to \begin{bmatrix} \mathcal{X}_+ \\ \mathcal{Y}_- \end{bmatrix}. \tag{6.1}$$

We shall prove the following inversion theorem.

Theorem 6.1. *Let $\mathcal{M} = \mathcal{M}_{\mathcal{A},\mathcal{B},\mathcal{C},\mathcal{D}}$ be an admissible algebra and let $g \in \mathcal{B}_+$. Then the operator $\mathbf{\Omega}$ defined by (6.1) is invertible if there exist $\alpha \in \mathcal{A}_+$, $\beta \in \mathcal{B}_+$, $\gamma \in \mathcal{C}_-$, $\delta \in \mathcal{D}_-$ such that*

$$\mathbf{\Omega} \begin{bmatrix} \alpha \\ \gamma \end{bmatrix} = \begin{bmatrix} e_{\mathcal{A}} \\ 0 \end{bmatrix} \quad \textit{and} \quad \mathbf{\Omega} \begin{bmatrix} \beta \\ \delta \end{bmatrix} = \begin{bmatrix} 0 \\ e_{\mathcal{D}} \end{bmatrix}, \tag{6.2}$$

and the following two conditions are satisfied:

(a) *$a_0 := P_{\mathcal{A}_d}\alpha$ and $d_0 := P_{\mathcal{D}_d}\delta$ are invertible in $\mathcal{A}_d$ and $\mathcal{D}_d$, respectively;*
(b) *conditions* (C4)–(C6) *are satisfied*

In that case the inverse of $\mathbf{\Omega}$ is given by

$$\mathbf{\Omega}^{-1} = \begin{bmatrix} \mathbf{R}_{11} & \mathbf{R}_{12} \\ \mathbf{R}_{21} & \mathbf{R}_{22} \end{bmatrix}, \tag{6.3}$$

where $\mathbf{R}_{ij}$, $1 \le i,j \le 2$, are the operators defined by (5.25)–(5.28). *Furthermore, the operators $\mathbf{R}_{11}$ and $\mathbf{R}_{22}$ are invertible and*

$$\mathbf{H}_{+,g} = -\mathbf{R}_{11}^{-1}\mathbf{R}_{12} = -\mathbf{R}_{12}\mathbf{R}_{22}^{-1}, \tag{6.4}$$

$$\mathbf{H}_{-,g^*} = -\mathbf{R}_{21}\mathbf{R}_{11}^{-1} = -\mathbf{R}_{22}^{-1}\mathbf{R}_{21}, \tag{6.5}$$

$$g = -\mathbf{R}_{11}^{-1}\beta, \quad g^* = -\mathbf{R}_{22}^{-1}\gamma. \tag{6.6}$$

Remark 6.2. In contrast to Theorem 1.1, the above theorem is not an "if and only if" statement. In this general setting we only have the following partial converse: if the operator $\mathbf{\Omega}$ given by (6.1) is invertible, then there exist $\alpha \in \mathcal{A}_+$, $\beta \in \mathcal{B}_+$, $\gamma \in \mathcal{C}_-$, $\delta \in \mathcal{D}_-$ such that the equations (6.2) are satisfied. It can happen that the operator $\mathbf{\Omega}$ is invertible and item (a) is not satisfied; see Example 6.5 given at the end of the present section.

Note that the operators $\mathbf{R}_{ij}$, $1 \le i,j \le 2$, appearing in (6.3) do not depend on the particular choice of g, but on α, β, γ, δ only. It follows that Theorem 6.1 yields the following corollary.

Corollary 6.3. *Let $\alpha \in \mathcal{A}_+$, $\beta \in \mathcal{B}_+$, $\gamma \in \mathcal{C}_-$, $\delta \in \mathcal{D}_-$, and assume that*

(a) *$a_0 = P_{\mathcal{A}_d}\alpha$ and $d_0 = P_{\mathcal{D}_d}\delta$ are invertible in $\mathcal{A}_d$ and $\mathcal{D}_d$, respectively;*
(b) *conditions* (C4)–(C6) *are satisfied.*

Under these conditions, if the twofold EG inverse problem associated with $\alpha \in \mathcal{A}_+$, $\beta \in \mathcal{B}_+$, $\gamma \in \mathcal{C}_-$, $\delta \in \mathcal{D}_-$ has a solution, then the solution is unique.

Proof. Assume that the twofold EG inverse problem associated with $\alpha \in \mathcal{A}_+$, $\beta \in \mathcal{B}_+$, $\gamma \in \mathcal{C}_-$, $\delta \in \mathcal{D}_-$ has a solution, g say. Then (see Corollary 5.1) the two identities in (6.2) are satisfied. Furthermore, by assumption, items (a) and (b) in Theorem 6.1 are satisfied too. We conclude that (6.3) holds, and hence $\mathbf{\Omega}$ is uniquely determined by the operators $\mathbf{R}_{ij}$, $1 \le i, j \le 2$. But these $\mathbf{R}_{ij}$, $1 \le i, j \le 2$, do not depend on g, but on α, β, γ, δ only. It follows that the same is true for $\mathbf{H}_{+,g}$. But $\mathbf{H}_{+,g}e_{\mathcal{D}} = P_{\mathcal{X}_+}ge_{\mathcal{D}} = P_{\mathcal{X}_+}g = g$. Thus g is uniquely determined by the data. □

The following lemma will be useful in the proof of Theorem 6.1.

Lemma 6.4. *Let $g \in \mathcal{B}_+$ satisfy the inclusions* (2.6) *and* (2.7). *Then the following identities hold:*

$$\mathbf{T}_{+,\alpha^*}\mathbf{H}_{+,g} = -\mathbf{H}_{+,\gamma^*}, \quad \mathbf{T}_{+,\beta^*}\mathbf{H}_{+,g} = -\mathbf{H}_{-,\delta^*}, \tag{6.7}$$

$$\mathbf{T}_{-,\gamma^*}\mathbf{H}_{-,g^*} = -\mathbf{H}_{-,\alpha^*}, \quad \mathbf{T}_{-,\delta^*}\mathbf{H}_{-,g^*} = -\mathbf{H}_{-,\beta^*}, \tag{6.8}$$

$$\mathbf{H}_{+,g}\mathbf{T}_{-,\delta} = -\mathbf{H}_{+,\beta}, \quad \mathbf{H}_{+,g}\mathbf{T}_{-,\gamma} = -\mathbf{H}_{+,\alpha}, \tag{6.9}$$

$$\mathbf{H}_{-,g^*}\mathbf{T}_{+,\alpha} = -\mathbf{H}_{-,\gamma}, \quad \mathbf{H}_{-,g^*}\mathbf{T}_{+,\beta} = -\mathbf{H}_{-,\delta}. \tag{6.10}$$

Proof. The above identities follow by using the implications in (5.5)–(5.8) and (5.9)–(5.12) together with the identities in (5.15)–(5.18). Let us illustrate this by proving the first identity in (6.7).

From the first identity in (5.15) we know that $\mathbf{T}_{+,\alpha^*}\mathbf{H}_{+,g} = \mathbf{H}_{+,\alpha^*g}$. Since $g \in \mathcal{B}_+$ satisfies the first inclusion in (2.6), the equivalence in (5.10) tells us that $\mathbf{H}_{+,\alpha^*g} = -\mathbf{H}_{+,\gamma^*}$. Hence $\mathbf{T}_{+,\alpha^*}\mathbf{H}_{+,g} = -\mathbf{H}_{+,\gamma^*}$, and the first identity (6.7) is proved. □

Proof of Theorem 6.1. Recall that the two identities in (6.2) together are equivalent to $g \in \mathcal{B}_+$ being a solution to the twofold EG inverse problem associated with $\alpha \in \mathcal{A}_+$, $\beta \in \mathcal{B}_+$, $\gamma \in \mathcal{C}_-$, $\delta \in \mathcal{D}_-$, and hence the two identities in (6.2) imply that the conditions (C1)–(C3) are satisfied. Given item (b) in Theorem 6.1 we conclude that all conditions (C1)–(C6) are satisfied.

The remainder of the proof is divided into three parts.

PART 1. First we will prove that

$$\begin{bmatrix} \mathbf{R}_{11} & \mathbf{R}_{12} \\ \mathbf{R}_{21} & \mathbf{R}_{22} \end{bmatrix} \begin{bmatrix} I_{\mathcal{X}_+} & \mathbf{H}_{+,g} \\ \mathbf{H}_{-,g^*} & I_{\mathcal{Y}_-} \end{bmatrix} = \begin{bmatrix} I_{\mathcal{X}_+} & 0 \\ 0 & I_{\mathcal{Y}_-} \end{bmatrix}. \tag{6.11}$$

We start with the identity $\mathbf{R}_{11} + \mathbf{R}_{12}\mathbf{H}_{-,g^*} = I_{\mathcal{X}_+}$. Using the two identities in (6.8) we have

$$\begin{aligned} \mathbf{R}_{12}\mathbf{H}_{-,g^*} &= \mathbf{H}_{+,\beta}d_0^{-1}\mathbf{T}_{-,\delta^*}\mathbf{H}_{-,g^*} - \mathbf{H}_{+,\alpha}a_0^{-1}\mathbf{T}_{-,\gamma^*}\mathbf{H}_{-,g^*} \\ &= -\mathbf{H}_{+,\beta}d_0^{-1}\mathbf{H}_{-,\beta^*} + \mathbf{H}_{+,\alpha}a_0^{-1}\mathbf{H}_{-,\alpha^*} \\ &= -\mathbf{R}_{11} + I_{\mathcal{X}_+}, \end{aligned}$$

which proves $\mathbf{R}_{11} + \mathbf{R}_{12}\mathbf{H}_{-,g^*} = I_{\mathcal{X}_+}$.

Similarly, using the two identities in (6.7), we obtain

$$\begin{aligned}\mathbf{R}_{11}\mathbf{H}_{+,g} &= \mathbf{T}_{+,\alpha}a_0^{-1}\mathbf{T}_{+,\alpha^*}\mathbf{H}_{+,g} - \mathbf{T}_{+,\beta}d_0^{-1}\mathbf{T}_{+,\beta^*}\mathbf{H}_{+,g}\\ &= -\mathbf{T}_{+,\alpha}a_0^{-1}\mathbf{H}_{+,\gamma^*} + \mathbf{T}_{+,\beta}d_0^{-1}\mathbf{H}_{+,\delta^*}\\ &= -\mathbf{R}_{12}.\end{aligned}$$

Thus $\mathbf{R}_{11}\mathbf{H}_{+,g} + \mathbf{R}_{12} = 0$.

The equalities $\mathbf{R}_{21}\mathbf{H}_{+,g} + \mathbf{R}_{22} = I_{\mathcal{Y}}$ and $\mathbf{R}_{21} + \mathbf{R}_{22}\mathbf{H}_{-,g^*} = 0$ are proved in a similar way.

PART 2. In this part we prove that

$$\begin{bmatrix} I_{\mathcal{X}_+} & \mathbf{H}_{+,g}\\ \mathbf{H}_{-,g^*} & I_{\mathcal{Y}_-}\end{bmatrix}\begin{bmatrix}\mathbf{R}_{11} & \mathbf{R}_{12}\\ \mathbf{R}_{21} & \mathbf{R}_{22}\end{bmatrix} = \begin{bmatrix} I_{\mathcal{X}_+} & 0\\ 0 & I_{\mathcal{Y}_-}\end{bmatrix}. \tag{6.12}$$

To see this we first show that $\mathbf{R}_{11} + \mathbf{H}_{+,g}\mathbf{R}_{21} = I_{\mathcal{X}_+}$. We use (5.31) and the two identities in (6.9). This yields

$$\begin{aligned}\mathbf{H}_{+,g}\mathbf{R}_{21} &= \mathbf{H}_{+,g}\mathbf{T}_{-,\delta}d_0^{-1}\mathbf{H}_{-,\beta^*} - \mathbf{H}_{+,g}\mathbf{T}_{-,\gamma}a_0^{-1}\mathbf{H}_{-,\alpha^*}\\ &= -\mathbf{H}_{+,\beta}d_0^{-1}\mathbf{H}_{-,\beta^*} + \mathbf{H}_{+,\alpha}a_0^{-1}\mathbf{H}_{-,\alpha^*}\\ &= I_{\mathcal{X}_+} - \mathbf{R}_{11}.\end{aligned}$$

where the last equality follows from (5.30). We proved $\mathbf{R}_{11} + \mathbf{H}_{+,g}\mathbf{R}_{21} = I_{\mathcal{X}_+}$.

Next we will prove that $\mathbf{R}_{12} + \mathbf{H}_{+,g}\mathbf{R}_{22} = 0$. Using (5.28) and the identities in (6.9) we obtain

$$\begin{aligned}\mathbf{H}_{+,g}\mathbf{R}_{22} &= \mathbf{H}_{+,g}\mathbf{T}_{-,\delta}d_0^{-1}\mathbf{T}_{-,\delta^*} - \mathbf{H}_{+,g}\mathbf{T}_{-,\gamma}a_0^{-1}\mathbf{T}_{-,\gamma^*}\\ &= -\mathbf{H}_{+,\beta}d_0^{-1}\mathbf{T}_{-,\delta^*} + \mathbf{H}_{+,\alpha}a_0^{-1}\mathbf{T}_{-,\gamma^*}\\ &= -\mathbf{R}_{12}.\end{aligned}$$

We proved that $\mathbf{R}_{12} + \mathbf{H}_{+,g}\mathbf{R}_{22} = 0$.

The equalities $\mathbf{H}_{-,g^*}\mathbf{R}_{11} + \mathbf{R}_{21} = 0$ and $\mathbf{H}_{-,g^*}\mathbf{R}_{12} + \mathbf{R}_{22} = I_{\mathcal{Y}_-}$ are proved in a similar way.

PART 3. To finish the proof we note that (6.11) and (6.12) imply that the operator $\boldsymbol{\Omega}$ is invertible and that its inverse is given by

$$\boldsymbol{\Omega}^{-1} = \begin{bmatrix}\mathbf{R}_{11} & \mathbf{R}_{12}\\ \mathbf{R}_{21} & \mathbf{R}_{22}\end{bmatrix},$$

which completes the proof. □

As mentioned in the introduction, Theorem 6.1 has many predecessors. See also Sections 8 and 10.

Example 6.5. We conclude this section with an example of the type announced in Remark 6.2, i.e., the operator $\boldsymbol{\Omega}$ is invertible and item (a) in Theorem 6.1 is not satisfied. We use a special case of the example in Subsection 3.2. Let $p = 2$ and

$$g = \begin{bmatrix} 1 & 1\\ 0 & 1\end{bmatrix}.$$

Then $\mathbf{\Omega}$ is invertible. To see this we choose bases for the upper triangular and the lower triangular matrices and determine the matrix of $\mathbf{\Omega}$ with respect to these bases. The basis we choose for the upper triangular matrices is

$$\left\{\begin{bmatrix}1 & 0\\ 0 & 0\end{bmatrix}, \begin{bmatrix}0 & 1\\ 0 & 0\end{bmatrix}, \begin{bmatrix}0 & 0\\ 0 & 1\end{bmatrix}\right\}$$

and the basis we choose for the lower triangular matrices is

$$\left\{\begin{bmatrix}1 & 0\\ 0 & 0\end{bmatrix}, \begin{bmatrix}0 & 0\\ 1 & 0\end{bmatrix}, \begin{bmatrix}0 & 0\\ 0 & 1\end{bmatrix}\right\}.$$

Then it follows that the matrix for $\mathbf{\Omega}$ with respect to these bases is

$$\begin{bmatrix}1 & 0 & 0 & 1 & 1 & 0\\ 0 & 1 & 0 & 0 & 0 & 1\\ 0 & 0 & 1 & 0 & 0 & 1\\ 1 & 0 & 0 & 1 & 0 & 0\\ 1 & 0 & 0 & 0 & 1 & 0\\ 0 & 1 & 1 & 0 & 0 & 1\end{bmatrix},$$

which is an invertible matrix. The solution of the two equations (6.2) is

$$\alpha = \begin{bmatrix}-1 & -1\\ 0 & 0\end{bmatrix}, \quad \gamma = \begin{bmatrix}1 & 0\\ 1 & 1\end{bmatrix}, \quad \beta = \begin{bmatrix}1 & 1\\ 0 & 1\end{bmatrix}, \quad \delta = \begin{bmatrix}0 & 0\\ -1 & -1\end{bmatrix}.$$

We see that α and δ are not invertible and then the diagonals α_d and δ_d are also not invertible. It is also easy to check that α, β, γ and δ satisfy the inclusions (2.6) and (2.7).

7. Solution to the abstract twofold EG inverse problem

The next theorem is the main result of this section.

Theorem 7.1. *Let $\alpha \in \mathcal{A}_+$, $\beta \in \mathcal{B}_+$, $\gamma \in \mathcal{C}_-$, $\delta \in \mathcal{D}_-$, and assume that*

(a) *$a_0 = P_{\mathcal{A}_d}\alpha$ and $d_0 = P_{\mathcal{D}_d}\delta$ are invertible in $\mathcal{A}_d$ and $\mathcal{D}_d$, respectively;*
(b) *conditions (C1)–(C6) are satisfied.*

Furthermore, let $\mathbf{R}_{11}$, $\mathbf{R}_{12}$, $\mathbf{R}_{21}$, $\mathbf{R}_{22}$ be the operators defined by (5.25)–(5.28). Then the twofold EG inverse problem associated with the data set $\{\alpha, , \beta, \gamma, \delta\}$ has a solution if and only if

(i) *$\mathbf{R}_{11} : \mathcal{X}_+ \to \mathcal{X}_+$ and $\mathbf{R}_{22} : \mathcal{Y}_- \to \mathcal{Y}_-$ are invertible;*
(ii) *$\left(\mathbf{R}_{11}^{-1}\beta\right)^* = \mathbf{R}_{22}^{-1}\gamma$;*
(iii) *$\mathbf{R}_{11}^{-1}\mathbf{R}_{12} = \mathbf{H}_{+,\rho}$ for some $\rho \in \mathcal{B}$ and $\mathbf{R}_{22}^{-1}\mathbf{R}_{21} = \mathbf{H}_{-,\eta}$ for some $\eta \in \mathcal{C}$.*

In that case the solution g of the twofold EG inverse problem associated with α, β, γ and δ is unique and is given by

$$g = -\mathbf{R}_{11}^{-1}\beta = -(\mathbf{R}_{22}^{-1}\gamma)^*. \tag{7.1}$$

Proof. The proof is divided into two parts. Note that the uniqueness statement is already covered by Corollary 6.3. In the first part of the proof we prove the necessity of the conditions (i), (ii), (iii).

PART 1. Assume $g \in \mathcal{B}_+$ is a solution to the twofold EG inverse problem associated with the data set $\{\alpha, \beta, \gamma, \delta\}$. Note that conditions (a) and (b) in Theorem 7.1 imply conditions (a) and (b) in Theorem 6.1. Furthermore, from Corollary 5.1 we know that the identities in (6.2) are satisfied. Thus Theorem 6.1 tells us that operator $\mathbf{\Omega}$ defined by (6.1) is invertible and its inverse is given by (6.3). In particular, the operator $\mathbf{R}$ defined by

$$\mathbf{R} = \begin{bmatrix} \mathbf{R}_{11} & \mathbf{R}_{12} \\ \mathbf{R}_{21} & \mathbf{R}_{22} \end{bmatrix}$$

is invertible. But then the second part of Lemma 5.6 tells us that the operators $\mathbf{R}_{11}$ and $\mathbf{R}_{22}$ are invertible, i.e., condition (i) is fulfilled. Furthermore, again using the second part of Lemma 5.6, we have

$$\mathbf{\Omega} = \mathbf{R}^{-1} = \begin{bmatrix} \mathbf{I}_{\mathcal{X}_+} & -\mathbf{R}_{12}\mathbf{R}_{22}^{-1} \\ -\mathbf{R}_{21}\mathbf{R}_{11}^{-1} & \mathbf{I}_{\mathcal{Y}_-} \end{bmatrix} = \begin{bmatrix} \mathbf{I}_{\mathcal{X}_+} & -\mathbf{R}_{11}^{-1}\mathbf{R}_{12} \\ -\mathbf{R}_{22}^{-1}\mathbf{R}_{21} & \mathbf{I}_{\mathcal{Y}_-} \end{bmatrix}.$$

In particular, we have

$$\mathbf{H}_{+,g} = -\mathbf{R}_{11}^{-1}\mathbf{R}_{12} = -\mathbf{R}_{12}\mathbf{R}_{22}^{-1}, \quad \mathbf{H}_{-,g^*} = -\mathbf{R}_{21}\mathbf{R}_{11}^{-1} = -\mathbf{R}_{22}^{-1}\mathbf{R}_{21}.$$

The preceding two identities show that item (iii) holds with $\rho = -g$ and $\eta = -g^*$. Finally, since $\rho = -g$ and $\eta = -g^*$, the identities in (7.1) imply that item (ii) is satisfied.

PART 2. In this part we assume that conditions (i), (ii), (iii) are satisfied and we show that the twofold EG inverse problem associated with the data set $\{\alpha, \beta, \gamma, \delta\}$ has a solution.

Put $g = -P_{\mathcal{B}_+}\rho$ and $h = -P_{\mathcal{C}_-}\eta$. We shall show that $h = g^*$ and for this choice of g the inclusions (2.6) and (2.7) are fulfilled. Note that $P_{\mathcal{B}_+}(g + \rho) = 0$, so that $g + \rho \in \mathcal{B}_-$. From the second part of (4.10) we then obtain that $\mathbf{H}_{+,g} = -\mathbf{H}_{+,\rho}$, and, by a similar argument, from the first part of (4.14) it follows that $\mathbf{H}_{-,h} = -\mathbf{H}_{-,\eta}$. Using these identities and those given by Lemma 5.2 together with the second and third identity in (5.29) we see that condition (iii) yields

$$\begin{aligned} g &= \mathbf{H}_{+,g}e_{\mathcal{D}} = -\mathbf{H}_{+,\rho}e_{\mathcal{D}} = -\mathbf{R}_{11}^{-1}\mathbf{R}_{12}e_{\mathcal{D}} = -\mathbf{R}_{11}^{-1}\beta, \\ h &= \mathbf{H}_{-,h}e_{\mathcal{A}} = -\mathbf{H}_{-,\eta}e_{\mathcal{A}} = -\mathbf{R}_{22}^{-1}\mathbf{R}_{21}e_{\mathcal{A}} = -\mathbf{R}_{22}^{-1}\gamma. \end{aligned}$$

But then (ii) implies that $h = g^*$. Furthermore, (iii) tells us that

$$\mathbf{R}_{11}^{-1}\mathbf{R}_{12} = -\mathbf{H}_{+,g} \quad \text{and} \quad \mathbf{R}_{22}^{-1}\mathbf{R}_{21} = -\mathbf{H}_{-,g^*}. \tag{7.2}$$

According to Lemma 5.6 condition (i) implies that the operator $\mathbf{R}$ given by (5.39) is invertible, and its inverse is given by (5.40). This together with the identities in (7.2) implies that

$$\mathbf{R}^{-1} = \begin{bmatrix} \mathbf{R}_{11} & \mathbf{R}_{12} \\ \mathbf{R}_{21} & \mathbf{R}_{22} \end{bmatrix}^{-1} = \begin{bmatrix} I_{\mathcal{X}_+} & -\mathbf{R}_{11}^{-1}\mathbf{R}_{12} \\ -\mathbf{R}_{22}^{-1}\mathbf{R}_{21} & I_{\mathcal{Y}_-} \end{bmatrix} = \begin{bmatrix} I_{\mathcal{X}_+} & \mathbf{H}_{+g} \\ \mathbf{H}_{-,g^*} & I_{\mathcal{Y}_-} \end{bmatrix}.$$

Next note that the identities in (5.29) can be rephrased as

$$\begin{bmatrix} \mathbf{R}_{11} & \mathbf{R}_{12} \\ \mathbf{R}_{21} & \mathbf{R}_{22} \end{bmatrix} \begin{bmatrix} e_{\mathcal{A}} & 0 \\ 0 & e_{\mathcal{D}} \end{bmatrix} = \begin{bmatrix} \alpha & \beta \\ \gamma & \delta \end{bmatrix}.$$

But then

$$\begin{bmatrix} I_{\mathcal{X}_+} & \mathbf{H}_{+g} \\ \mathbf{H}_{-,g^*} & I_{\mathcal{Y}_-} \end{bmatrix} \begin{bmatrix} \alpha & \beta \\ \gamma & \delta \end{bmatrix} = \begin{bmatrix} e_{\mathcal{A}} & 0 \\ 0 & e_{\mathcal{D}} \end{bmatrix},$$

and the equivalences in (5.2) and (5.3) tell us that with our choice of g the inclusions (2.6) and (2.7) are fulfilled. Hence g is a solution to the twofold EG inverse problem associated with the data set $\{\alpha, \beta, \gamma, \delta\}$. Since $g = -\mathbf{R}_{11}^{-1}\beta$, the proof is complete. □

A variation on condition (iii) in Theorem 7.1 does not appear in the solution to the twofold EG inverse problem in $L^1(\mathbb{R})$ as formulate in the introduction, e.g., in Theorem 1.2, and neither in the solution to the discrete twofold EG inverse problem in [10]. This is because in the abstract setting presented in this paper we do not have a characterization of Hankel-type operators via an intertwining condition as in the discrete case as well as in the continuous case (where an extra condition is needed, as shown in the Appendix). Lemma 7.2 below provides, at the abstract level, a result that will be useful in proving that condition (iii) is implied by the assumptions made for the special cases we consider.

Assume we have operators $\mathbf{V}_{\mathcal{Z},\pm} : \mathcal{Z}_\pm \to \mathcal{Z}_\pm$ and $\mathbf{V}_{*,\mathcal{Z},\pm} : \mathcal{Z}_\pm \to \mathcal{Z}_\pm$, with $\mathcal{Z}$ either $\mathcal{X}$ or $\mathcal{Y}$, that are such that $\mathbf{V}_{*,\mathcal{Z},\pm}\mathbf{V}_{\mathcal{Z},\pm} = I_{\mathcal{Z}_\pm}$ and

$$\begin{aligned}
&\text{for any } \phi \in \mathcal{A}\text{:} && \mathbf{V}_{*,\mathcal{X},\pm}\mathbf{H}_{\pm,\phi} = \mathbf{H}_{\pm,\phi}\mathbf{V}_{\mathcal{X},\mp}, && \mathbf{V}_{*,\mathcal{X},\pm}\mathbf{T}_{\pm,\phi}\mathbf{V}_{\mathcal{X},\pm} = \mathbf{T}_{\pm,\phi};\\
&\text{for any } \phi \in \mathcal{B}\text{:} && \mathbf{V}_{*,\mathcal{X},\pm}\mathbf{H}_{\pm,\phi} = \mathbf{H}_{\pm,\phi}\mathbf{V}_{\mathcal{Y},\mp}, && \mathbf{V}_{*,\mathcal{X},\pm}\mathbf{T}_{\pm,\phi}\mathbf{V}_{\mathcal{Y},\pm} = \mathbf{T}_{\pm,\phi};\\
&\text{for any } \psi \in \mathcal{D}\text{:} && \mathbf{V}_{*,\mathcal{Y},\pm}\mathbf{H}_{\pm,\psi} = \mathbf{H}_{\pm,\psi}\mathbf{V}_{\mathcal{Y},\mp}, && \mathbf{V}_{*,\mathcal{Y},\pm}\mathbf{T}_{\pm,\phi}\mathbf{V}_{\mathcal{Y},\pm} = \mathbf{T}_{\pm,\phi};\\
&\text{for any } \psi \in \mathcal{C}\text{:} && \mathbf{V}_{*,\mathcal{Y},\pm}\mathbf{H}_{\pm,\psi} = \mathbf{H}_{\pm,\psi}\mathbf{V}_{\mathcal{X},\mp}, && \mathbf{V}_{*,\mathcal{Y},\pm}\mathbf{T}_{\pm,\phi}\mathbf{V}_{\mathcal{X},\pm} = \mathbf{T}_{\pm,\phi},
\end{aligned}$$

and

$$\begin{aligned}
&\text{for any } \phi \in \mathcal{A}_\pm\text{:} && \mathbf{T}_{\pm,\phi}\mathbf{V}_{\mathcal{X},\pm} = \mathbf{V}_{\mathcal{X},\pm}\mathbf{T}_{\pm,\phi};\\
&\text{for any } \phi \in \mathcal{B}_\pm\text{:} && \mathbf{T}_{\pm,\phi}\mathbf{V}_{\mathcal{Y},\pm} = \mathbf{V}_{\mathcal{X},\pm}\mathbf{T}_{\pm,\phi};\\
&\text{for any } \psi \in \mathcal{C}_\pm\text{:} && \mathbf{T}_{\pm,\psi}\mathbf{V}_{\mathcal{Y},\pm} = \mathbf{V}_{\mathcal{Y},\pm}\mathbf{T}_{\pm,\psi};\\
&\text{for any } \phi \in \mathcal{B}_\pm\text{:} && \mathbf{T}_{\pm,\psi}\mathbf{V}_{\mathcal{X},\pm} = \mathbf{V}_{\mathcal{Y},\pm}\mathbf{T}_{\pm,\psi}.
\end{aligned}$$

Lemma 7.2. *With $\mathbf{R}_{ij}$ defined as above one has the equalities*

$$\mathbf{R}_{11}\mathbf{V}_{*,\mathcal{X},+}\mathbf{R}_{12} = \mathbf{R}_{12}\mathbf{V}_{\mathcal{Y},-}\mathbf{R}_{22} \quad \text{and} \quad \mathbf{R}_{22}\mathbf{V}_{*,\mathcal{Y},-}\mathbf{R}_{21} = \mathbf{R}_{21}\mathbf{V}_{\mathcal{X},+}\mathbf{R}_{11}.$$

Moreover, if $\mathbf{R}_{11}$ and $\mathbf{R}_{22}$ are invertible, then

$$\mathbf{V}_{*,\mathcal{X},+}\mathbf{R}_{12}\mathbf{R}_{22}^{-1} = \mathbf{R}_{11}^{-1}\mathbf{R}_{12}\mathbf{V}_{\mathcal{Y},-}. \tag{7.3}$$

Proof. First we will prove that $\mathbf{R}_{11}\mathbf{V}_{*,\mathcal{X},+}\mathbf{R}_{12} = \mathbf{R}_{12}\mathbf{V}_{\mathcal{Y},-}\mathbf{R}_{22}$. We start with deriving the equality

$$\begin{bmatrix}\mathbf{T}_{+,\alpha^*}\\ \mathbf{T}_{+,\beta^*}\end{bmatrix}\mathbf{V}_{*,\mathcal{X},+}\begin{bmatrix}\mathbf{H}_{+,\beta} & \mathbf{H}_{+,\alpha}\end{bmatrix} = \begin{bmatrix}\mathbf{H}_{+,\gamma^*}\\ \mathbf{H}_{+,\delta*}\end{bmatrix}\mathbf{V}_{\mathcal{Y},-}\begin{bmatrix}\mathbf{T}_{-,\delta} & \mathbf{T}_{-,\gamma}\end{bmatrix}. \tag{7.4}$$

To obtain (7.4), first notice that

$$\mathbf{V}_{*,\mathcal{X},+}\begin{bmatrix}\mathbf{H}_{+,\beta} & \mathbf{H}_{+,\alpha}\end{bmatrix} = \begin{bmatrix}\mathbf{H}_{+,\beta} & \mathbf{H}_{+,\alpha}\end{bmatrix}\begin{bmatrix}\mathbf{V}_{\mathcal{Y},-} & 0\\ 0 & \mathbf{V}_{\mathcal{X},-}\end{bmatrix};$$
$$\mathbf{V}_{\mathcal{Y},-}\begin{bmatrix}\mathbf{T}_{-,\delta} & \mathbf{T}_{-,\gamma}\end{bmatrix} = \begin{bmatrix}\mathbf{T}_{-,\delta} & \mathbf{T}_{-,\gamma}\end{bmatrix}\begin{bmatrix}\mathbf{V}_{\mathcal{Y},-} & 0\\ 0 & \mathbf{V}_{\mathcal{X},-}\end{bmatrix}.$$

Then use (5.19) to get that

$$\begin{aligned}\begin{bmatrix}\mathbf{T}_{+,\alpha^*}\\ \mathbf{T}_{+,\beta^*}\end{bmatrix}\mathbf{V}_{*,\mathcal{X},+}\begin{bmatrix}\mathbf{H}_{+,\beta} & \mathbf{H}_{+,\alpha}\end{bmatrix} &= \begin{bmatrix}\mathbf{T}_{+,\alpha^*}\\ \mathbf{T}_{+,\beta^*}\end{bmatrix}\begin{bmatrix}\mathbf{H}_{+,\beta} & \mathbf{H}_{+,\alpha}\end{bmatrix}\begin{bmatrix}\mathbf{V}_{\mathcal{Y},-} & 0\\ 0 & \mathbf{V}_{\mathcal{X},-}\end{bmatrix}\\ &= \begin{bmatrix}\mathbf{H}_{+,\gamma^*}\\ \mathbf{H}_{+,\delta*}\end{bmatrix}\begin{bmatrix}\mathbf{T}_{-,\delta} & \mathbf{T}_{-,\gamma}\end{bmatrix}\begin{bmatrix}\mathbf{V}_{\mathcal{Y},-} & 0\\ 0 & \mathbf{V}_{\mathcal{X},-}\end{bmatrix}\\ &= \begin{bmatrix}\mathbf{H}_{+,\gamma^*}\\ \mathbf{H}_{+,\delta*}\end{bmatrix}\mathbf{V}_{\mathcal{Y},-}\begin{bmatrix}\mathbf{T}_{-,\delta} & \mathbf{T}_{-,\gamma}\end{bmatrix}.\end{aligned}$$

We proved (7.4). By multiplying (7.4) on the left and the right by

$$\begin{bmatrix}\mathbf{T}_{+,\alpha} & \mathbf{T}_{+,\beta}\end{bmatrix} \quad\text{and}\quad \begin{bmatrix}\mathbf{T}_{-,\delta^*}\\ \mathbf{T}_{-,\gamma^*}\end{bmatrix},$$

respectively, one gets $\mathbf{R}_{11}\mathbf{V}_{*,\mathcal{X},+}\mathbf{R}_{12} = \mathbf{R}_{12}\mathbf{V}_{\mathcal{Y},-}\mathbf{R}_{22}$. Furthermore, the equality $\mathbf{R}_{22}\mathbf{V}_{*,\mathcal{Y},-}\mathbf{R}_{21} = \mathbf{R}_{21}\mathbf{V}_{\mathcal{X},+}\mathbf{R}_{11}$ can be proved in a similar way.

Given the invertibility of $\mathbf{R}_{11}$ and $\mathbf{R}_{22}$ the preceding two identities yield the identity (7.3) trivially. □

8. Proof of Theorems 1.1 and 1.2

In this section we will prove Theorems 1.1 and 1.2. Recall that in this case the data are given by (1.1) and (1.2), and the twofold EG inverse problem is to find $g \in L^1(\mathbb{R}_+)^{p\times q}$ such that (1.3) and (1.4) are satisfied.

As a first step, the above problem will be put into the general setting introduced in Section 2 using a particular choice for $\mathcal{A}$, $\mathcal{B}$, $\mathcal{C}$, $\mathcal{D}$, namely as follows:

$$\mathcal{A} = \{f \mid f = \eta e_p + f_0, \text{ where } \eta \in \mathbb{C}^{p\times p},\ f_0 \in L^1(\mathbb{R})^{p\times p}\}, \tag{8.1}$$
$$\mathcal{B} = L^1(\mathbb{R})^{p\times q}, \qquad \mathcal{C} = L^1(\mathbb{R})^{q\times p}, \tag{8.2}$$
$$\mathcal{D} = \{h \mid h = \zeta e_q + h_0, \text{ where } \zeta \in \mathbb{C}^{q\times q},\ h_0 \in L^1(\mathbb{R})^{q\times q}\}. \tag{8.3}$$

Furthermore, $\mathcal{A}$, $\mathcal{B}$, $\mathcal{C}$, $\mathcal{D}$ admit decompositions as in (2.2) and (2.4) using

$$\begin{aligned} &\mathcal{A}_+^0 = L^1(\mathbb{R}_+)^{p\times p}, && \mathcal{A}_-^0 = L^1(\mathbb{R}_-)^{p\times p}, && \mathcal{A}_d = \{\eta e_p \mid \eta \in \mathbb{C}^{p\times p}\},\\ &\mathcal{B}_+ = L^1(\mathbb{R}_+)^{p\times q}, && \mathcal{B}_- = L^1(\mathbb{R}_-)^{p\times q}, &&\\ &\mathcal{C}_+ = L^1(\mathbb{R}_+)^{q\times p}, && \mathcal{C}_- = L^1(\mathbb{R}_-)^{q\times p}, &&\\ &\mathcal{D}_+^0 = L^1(\mathbb{R}_+)^{q\times q}, && \mathcal{D}_-^0 = L^1(\mathbb{R}_-)^{q\times q}, && \mathcal{D}_d = \{\zeta e_q \mid \zeta \in \mathbb{C}^{q\times q}\}. \end{aligned}$$

Here e_m, for $m = p, q$, is the constant $m \times m$ matrix function on $\mathbb{R}$ whose value is the $m \times m$ identity matrix I_m. Thus given $\eta \in \mathbb{C}^{m\times m}$, the symbol ηe_m denotes the constant matrix function on $\mathbb{R}$ identically equal to η.

We proceed by defining the algebraic structure. The addition is the usual addition of functions and is denoted by $+$. For the product we use the symbol $\diamond$ which in certain cases is just the usual convolution product $\star$. If $f = \eta_f e_p + f_0 \in \mathcal{A}$ and $\tilde{f} = \eta_{\tilde{f}} e_p + \tilde{f}_0 \in \mathcal{A}$, then the $\diamond$ product is defined by

$$f \diamond \tilde{f} := \eta_f \eta_{\tilde{f}} e_p + \big(\eta_f \tilde{f}_0 + f_0 \eta_{\tilde{f}} e_p + f_0 \star \tilde{f}_0\big).$$

Thus for $f \in L^1(\mathbb{R})^{n\times m}$ and $h \in L^1(\mathbb{R})^{m\times k}$ the product $f \diamond h$ is the convolution product $f \star h$. The product of elements $f = \eta_f e_p + f_0 \in \mathcal{A}$ and $h_0 \in \mathcal{B}$ is defined as $f \diamond h_0 = \eta_f h_0 + f_0 \star h_0$. Other products are defined likewise. One only needs the matrix dimension to allow the multiplication. The units $e_{\mathcal{A}}$ and $e_{\mathcal{D}}$ in $\mathcal{A}$ and $\mathcal{D}$ are given by $e_{\mathcal{A}} = e_p$ and $e_{\mathcal{D}} = e_q$, respectively. Finally, the adjoint f^* for $f \in L^1(\mathbb{R})^{r\times s}$ is defined by $f^*(\lambda) = f(-\lambda)^*$, $\lambda \in \mathbb{R}$, so that $f^* \in L^1(\mathbb{R})^{s\times r}$. For $f = \eta e_s + f_0$ with $\eta \in \mathbb{C}^{s\times s}$ and $f_0 \in L^1(\mathbb{R})^{s\times s}$ we define f^* by $f^* = \eta^* e_s + f_0^*$, where η^* is the adjoint of the matrix η. It easily follows that all conditions of the first paragraph of Section 2 are satisfied. We conclude that $\mathcal{M}_{\mathcal{A},\mathcal{B},\mathcal{C},\mathcal{D}}$ is admissible.

Remark 8.1. Observe that for a data set $\{a, b, c, d\}$ as in (1.1) with α, β, γ, δ the functions given by (1.16), the inclusions for a, b, c and d in (1.3) and (1.4) are equivalent to the inclusions (2.6) and (2.7) for α, β, γ and δ. Thus the solutions $g \in L^1(\mathbb{R}_+)^{p\times q}$ for the twofold EG inverse problem formulated in the introduction coincide with the solutions of the abstract twofold EG inverse problem of Section 2 using the specification given in the present section. Furthermore, in this case

$$\text{(C1)} \iff \alpha^* \diamond \alpha - \gamma^* \diamond \gamma = e_p; \tag{8.4}$$

$$\text{(C2)} \iff d^* \diamond \delta - \beta^* \diamond \beta = e_q; \tag{8.5}$$

$$\text{(C3)} \iff \alpha^* \diamond \beta = \gamma^* \diamond \delta. \tag{8.6}$$

Thus (C1)–(C3) are satisfied if and only if the following three identities hold true:

$$\alpha^* \diamond \alpha - \gamma^* \diamond \gamma = e_p, \quad d^* \diamond \delta - \beta^* \diamond \beta = e_q, \quad \alpha^* \diamond \beta = \gamma^* \diamond \delta. \tag{8.7}$$

8.1. Proof of Theorem 1.1

Note that Theorem 1.1 is an "if and only if" theorem. We first proof the "only if" part. Let $g \in L^1(\mathbb{R}_+)^{p\times q}$, and assume that the operator W given by (1.21) is invertible. Note that

$$\begin{bmatrix} 0 \\ -g^* \end{bmatrix} \in \begin{bmatrix} L^1(\mathbb{R}_+)^{p\times p} \\ L^1(\mathbb{R}_-)^{q\times p} \end{bmatrix} \quad \text{and} \quad \begin{bmatrix} -g \\ 0 \end{bmatrix} \in \begin{bmatrix} L^1(\mathbb{R}_+)^{p\times q} \\ L^1(\mathbb{R}_-)^{q\times q} \end{bmatrix}.$$

Since W is invertible, we see that there exist

$$a \in L^1(\mathbb{R}_+)^{p\times p}, \quad c \in L^1(\mathbb{R}_-)^{q\times p}, \quad b \in L^1(\mathbb{R}_+)^{p\times q}, \quad d \in L^1(\mathbb{R}_-)^{q\times q}$$

such that

$$W \begin{bmatrix} a \\ c \end{bmatrix} = \begin{bmatrix} 0 \\ -g^* \end{bmatrix} \quad \text{and} \quad W \begin{bmatrix} b \\ d \end{bmatrix} = \begin{bmatrix} -g \\ 0 \end{bmatrix}.$$

But this implies that g is a solution to the twofold EG inverse problem defined by the data set $\{a, b, c, d\}$. Thus the "only if" part of Theorem 1.1 is proved.

Next we prove the "if" part of Theorem 1.1. We assume that $g \in L^1(\mathbb{R}_+)^{p\times q}$ is a solution to the twofold EG inverse problem defined by the data set $\{a, b, c, d\}$ given by (1.1) and (1.2). Furthermore, α, β, γ, and δ are given by (1.16), and $\mathcal{M} = \mathcal{M}_{\mathcal{A},\mathcal{B},\mathcal{C},\mathcal{D}}$ is the admissible algebra defined in the beginning of this section. Our aim is to obtain the "if" part of Theorem 1.1 as a corollary of Theorem 6.1. For that purpose various results of Section 2 and Sections 4–6 have to be specified further for the case when $\mathcal{A}$, $\mathcal{B}$, $\mathcal{C}$, $\mathcal{D}$ are given by (8.1)–(8.3) in the beginning of this section. This will be done in four steps.

Step 1. Results from Section 2. Since $g \in L^1(\mathbb{R}_+)^{p\times q} = \mathcal{B}_+$ is a solution to the twofold EG inverse problem associated with the data $\{\alpha, \beta, \gamma, \delta\}$, we know from Proposition 2.1 that conditions (C1)–(C3) are satisfied. Furthermore,

$$a_0 = P_{\mathcal{A}_d}\alpha = e_p \quad \text{and} \quad d_0 = P_{\mathcal{D}_d}\delta = e_q. \tag{8.8}$$

But then the fact that α, β, γ, and δ are matrix functions implies that conditions (C4)–(C6) are also satisfied. Indeed, using the identities in (8.8), we see from (2.9) that

$$\begin{bmatrix} \alpha^*(\lambda) & \gamma^*(\lambda) \\ \beta^*(\lambda) & \delta^*(\lambda) \end{bmatrix} \begin{bmatrix} I_p & 0 \\ 0 & -I_q \end{bmatrix} \begin{bmatrix} \alpha(\lambda) & \beta(\lambda) \\ \gamma(\lambda) & \delta(\lambda) \end{bmatrix} = \begin{bmatrix} I_p & 0 \\ 0 & -I_q \end{bmatrix}, \quad \lambda \in \mathbb{R}. \tag{8.9}$$

In particular, the first matrix in the left-hand side of (8.9) is surjective and third matrix in the left-hand side of (8.9) is injective. But all matrices in (8.9) are finite square matrices. It follows that all these matrices are invertible. Hence

$$\begin{bmatrix} \alpha & \beta \\ \gamma & \delta \end{bmatrix}^{-1} \begin{bmatrix} e_p & 0 \\ 0 & -e_q \end{bmatrix} \begin{bmatrix} \alpha^* & \gamma^* \\ \beta^* & \delta^* \end{bmatrix}^{-1} = \begin{bmatrix} e_p & 0 \\ 0 & -e_q \end{bmatrix},$$

which yields

$$\begin{bmatrix} \alpha & \beta \\ \gamma & \delta \end{bmatrix} \begin{bmatrix} e_p & 0 \\ 0 & -e_q \end{bmatrix} \begin{bmatrix} \alpha^* & \gamma^* \\ \beta^* & \delta^* \end{bmatrix} = \begin{bmatrix} e_p & 0 \\ 0 & -e_q \end{bmatrix}.$$

The latter implies that conditions (1.4)–(1.6) are satisfied. In particular, we have proved that

(i) $a_0 = P_{\mathcal{A}_d}\alpha$ and $d_0 = P_{\mathcal{D}_d}\delta$ are invertible in $\mathcal{A}_d$ and $\mathcal{D}_d$, respectively;
(ii) conditions (C1)–(C6) are satisfied.

STEP 2. RESULTS FROM SECTION 4. In the present context the spaces $\mathcal{X}$ and $\mathcal{Y}$, $\mathcal{X}_+$ and $\mathcal{Y}_+$, and $\mathcal{X}_-$ and $\mathcal{Y}_-$ defined in the first paragraph of Section 4 are given by

$$\begin{aligned}
\mathcal{X} &= \mathcal{A}\dot{+}\mathcal{B} = \left(\mathbb{C}^{p\times p}e_p + L^1(\mathbb{R})^{p\times p}\right)\dot{+}L^1(\mathbb{R})^{p\times q},\\
\mathcal{X}_+ &= \mathcal{A}_+\dot{+}\mathcal{B}_+ = \left(\mathbb{C}^{p\times p}e_p + L^1(\mathbb{R}_+)^{p\times p}\right)\dot{+}L^1(\mathbb{R}_+)^{p\times q},\\
\mathcal{X}_- &= \mathcal{A}_-^0\dot{+}\mathcal{B}_- = L^1(\mathbb{R}_-)^{p\times p}\dot{+}L^1(\mathbb{R}_-)^{p\times q},
\end{aligned}$$

and

$$\begin{aligned}
\mathcal{Y} &= \mathcal{C}\dot{+}\mathcal{D} = L^1(\mathbb{R})^{q\times p}\dot{+}\left(L^1(\mathbb{R})^{q\times q} + \mathbb{C}^{q\times q}e_q\right),\\
\mathcal{Y}_+ &= \mathcal{C}_+\dot{+}\mathcal{D}_+^0 = L^1(\mathbb{R}_+)^{q\times p}\dot{+}L^1(\mathbb{R}_+)^{q\times q},\\
\mathcal{Y}_- &= \mathcal{C}_-\dot{+}\mathcal{D}_- = L^1(\mathbb{R}_-)^{q\times p}\dot{+}\left(L^1(\mathbb{R}_-)^{q\times q} + \mathbb{C}^{q\times q}e_q\right).
\end{aligned}$$

In the sequel we write $x \in \mathcal{X}$ as $x = (f, g)$, where $f = \eta_f e_p + f_0 \in \mathcal{A}$ and $g \in \mathcal{B}$. In a similar way vectors $x_+ \in \mathcal{X}_+$ and $x_- \in \mathcal{X}_-$ will be written as

$$\begin{aligned}
x_+ &= (f_+, g_+), \text{ where } f_+ = \eta_{f_+}e_p + f_{+,0} \in \mathcal{A}_+ \text{ and } g_+ \in \mathcal{B}_+,\\
x_- &= (f_-, g_-), \text{ where } f_- \in \mathcal{A}_-^0 \text{ and } g_- \in \mathcal{B}_-.
\end{aligned}$$

Analogous notations will be used for vectors $y \in \mathcal{Y}$, $y_+ \in \mathcal{Y}_+$, and $y_- \in \mathcal{Y}_-$. Indeed, $y \in \mathcal{Y}$ will be written as (h, k), where $h \in \mathcal{C}$, and $k = \zeta e_q + k_0 \in \mathcal{D}$, and

$$\begin{aligned}
y_+ &= (h_+, k_+), \text{ where } h_+ \in \mathcal{C}_+ \text{ and } k_+ \in \mathcal{D}_+^0,\\
y_- &= (h_-, k_-), \text{ where } h_- \in \mathcal{C}_- \text{ and } k_- = \zeta_{k_-}e_q + k_{-,0} \in \mathcal{D}_-.
\end{aligned}$$

Furthermore, in what follows $0_{p\times q}$ and $0_{q\times p}$ denote the linear spaces consisting only of the zero $p \times q$ and zero $q \times p$ matrix, respectively.

Using the above notation we define the following operators:

$$J_{\mathcal{X}_+} : \mathcal{X}_+ \to \begin{bmatrix} \mathbb{C}^{p\times p}\dot{+}0_{p\times q} \\ L^1(\mathbb{R}_+)^{p\times p}\dot{+}L^1(\mathbb{R}_+)^{p\times q} \end{bmatrix}, \quad J_{\mathcal{X}_+}x_+ = \begin{bmatrix} (\eta_{f_+}, 0) \\ (f_{+,0}, g_+) \end{bmatrix},$$

$$J_{\mathcal{X}_-} : \mathcal{X}_- \to L^1(\mathbb{R}_-)^{p\times p}\dot{+}L^1(\mathbb{R}_-)^{p\times q}, \quad J_{\mathcal{X}_-}x_- = (f_-, g_-),$$

and

$$J_{\mathcal{Y}_+} : \mathcal{Y}_+ \to L^1(\mathbb{R}_+)^{q\times p}\dot{+}L^1(\mathbb{R}_+)^{q\times q}, \quad J_{\mathcal{Y}_+}y_+ = (h_+, k_+),$$

$$J_{\mathcal{Y}_-} : \mathcal{Y}_- \to \begin{bmatrix} 0_{q\times p}\dot{+}\mathbb{C}^{q\times q} \\ L^1(\mathbb{R}_-)^{q\times p}\dot{+}L^1(\mathbb{R}_-)^{q\times q} \end{bmatrix}, \quad J_{\mathcal{Y}_-}y_- = \begin{bmatrix} (0, \zeta_{k_-}) \\ (h_-, k_{-,0}) \end{bmatrix}.$$

Note that all four operators defined above are invertible operators.

Next, in our present setting where $\alpha, \beta, \gamma, \delta$ are given by given by (1.16), we relate the Toeplitz-like and Hankel-like operators introduced in Section 4 to ordinary Wiener–Hopf and Hankel integral operators.

Let $\alpha = a_+ + e_p + a_- \in \mathcal{A}_+ + \mathcal{A}_d + \mathcal{A}_- = \mathcal{A}$. Then

$$J_{\mathcal{X}_+}\mathbf{T}_{+,\alpha} = \begin{bmatrix} I_p & 0 \\ a_+ & T_{+,\alpha} \end{bmatrix} J_{\mathcal{X}_+}, \quad J_{\mathcal{X}_-}\mathbf{T}_{-,\alpha} = T_{-,\alpha}J_{\mathcal{X}_-},$$

$$J_{\mathcal{X}_+}\mathbf{H}_{+,\alpha} = \begin{bmatrix} 0 \\ H_{+,\alpha} \end{bmatrix} J_{\mathcal{X}_-}, \quad J_{\mathcal{X}_-}\mathbf{H}_{-,\alpha} = \begin{bmatrix} a_- & H_{-,\alpha} \end{bmatrix} J_{\mathcal{X}_+}.$$

For $\beta = b_+ + b_- \in \mathcal{B}_+ + \mathcal{B}_- = \mathcal{B}$ we get

$$J_{\mathcal{X}_+}\mathbf{T}_{+,\beta} = \begin{bmatrix} 0 \\ T_{+,\beta} \end{bmatrix} J_{\mathcal{Y}_+}, \quad J_{\mathcal{X}_-}\mathbf{T}_{-,\beta} = \begin{bmatrix} b_- & T_{-,\beta} \end{bmatrix} J_{\mathcal{Y}_-},$$

$$J_{\mathcal{X}_+}\mathbf{H}_{+,\beta} = \begin{bmatrix} 0 & 0 \\ b_+ & H_{+,\beta} \end{bmatrix} J_{\mathcal{Y}_-}, \quad J_{\mathcal{X}_-}\mathbf{H}_{-,\beta} = H_{-,\beta}J_{\mathcal{Y}_+}.$$

Let $\gamma = c_+ + c_- \in \mathcal{C}_+ + \mathcal{C}_- = \mathcal{C}$. We have the equalities

$$J_{\mathcal{Y}_-}\mathbf{T}_{-,\gamma} = \begin{bmatrix} 0 \\ T_{-,\gamma} \end{bmatrix} J_{\mathcal{X}_-}, \quad J_{\mathcal{Y}_+}\mathbf{T}_{+,\gamma} = \begin{bmatrix} c_+ & T_{+,\gamma} \end{bmatrix} J_{\mathcal{X}_+},$$

$$J_{\mathcal{Y}_-}\mathbf{H}_{-,\gamma} = \begin{bmatrix} 0 & 0 \\ c_- & H_{-,\gamma} \end{bmatrix} J_{\mathcal{X}_+}, \quad J_{\mathcal{Y}_+}\mathbf{H}_{+,\gamma} = H_{+,\gamma}J_{\mathcal{X}_-}.$$

Let $\delta = d_+ + e_q + d_- \in \mathcal{D}_+ + \mathcal{D}_d + \mathcal{D}_- = \mathcal{D}$. Then

$$J_{\mathcal{Y}_-}\mathbf{T}_{-,\delta} = \begin{bmatrix} I_q & 0 \\ d_- & T_{-,\delta} \end{bmatrix} J_{\mathcal{Y}_-}, \quad J_{\mathcal{Y}_+}\mathbf{T}_{+,\delta} = T_{+,\delta}J_{\mathcal{X}_-},$$

$$J_{\mathcal{Y}_-}\mathbf{H}_{-,\delta} = \begin{bmatrix} 0 \\ H_{-,\delta} \end{bmatrix} J_{\mathcal{Y}_+}, \quad J_{\mathcal{Y}_+}\mathbf{H}_{+,\delta} = \begin{bmatrix} d_+ & H_{+,\delta} \end{bmatrix} J_{\mathcal{Y}_-}.$$

The following lemma is an immediate consequence of the above relations.

Lemma 8.2. *For $g \in \mathcal{B}_+$ one has*

$$\begin{bmatrix} J_{\mathcal{X}_+} & 0 \\ 0 & J_{\mathcal{Y}_-} \end{bmatrix} \begin{bmatrix} \mathbf{I}_{\mathcal{X}_+} & \mathbf{H}_{+,g} \\ \mathbf{H}_{-,g^*} & \mathbf{I}_{\mathcal{Y}_-} \end{bmatrix} = \begin{bmatrix} I_p & 0 & 0 & 0 \\ 0 & I & g & H_{+,g} \\ 0 & 0 & I_q & 0 \\ g^* & H_{-,g^*} & 0 & I \end{bmatrix} \begin{bmatrix} J_{\mathcal{X}_+} & 0 \\ 0 & J_{\mathcal{Y}_-} \end{bmatrix}. \tag{8.10}$$

Furthermore, if $\mathbf{\Omega}$ is the operator defined by (6.1) *using the present data, and if W is the operator defined by* (1.21)*, then* (8.10) *shows that $\mathbf{\Omega}$ is invertible if and only if W is invertible.*

STEP 3. RESULTS FROM SECTION 5. As before we assume that a, b, c and d are given by (1.1) and (1.2) and α, b, γ and δ by (1.16), and that $g \in L^1(\mathbb{R}_+)^{p\times q} = \mathcal{B}_+$ is a solution to the twofold EG inverse problem associated with the data $\{\alpha, \beta, \gamma, \delta\}$. Thus we know from STEP 1 that

(i) $a_0 = P_{\mathcal{A}_d}\alpha$ and $d_0 = P_{\mathcal{D}_d}\delta$ are invertible in $\mathcal{A}_d$ and $\mathcal{D}_d$, respectively;
(ii) conditions (C1)–(C6) are satisfied.

In particular, all conditions underlying the lemmas proved in Section 5 are fulfilled.

The following lemma is an immediate consequence of Lemma 5.3.

Lemma 8.3. *Since conditions* (C1)–(C3) *are satisfied, we have*

$$\begin{bmatrix} T_{+,\alpha^*} \\ T_{+,\beta^*} \end{bmatrix} \begin{bmatrix} H_{+,\beta} & H_{+,\alpha} \end{bmatrix} = \begin{bmatrix} H_{+,\gamma^*} \\ H_{+,\delta^*} \end{bmatrix} \begin{bmatrix} T_{-,\delta} & T_{-,\gamma} \end{bmatrix}. \tag{8.11}$$

and

$$\begin{bmatrix} T_{-,\delta^*} \\ T_{-,\gamma^*} \end{bmatrix} \begin{bmatrix} H_{-,\gamma} & H_{-,\delta} \end{bmatrix} = \begin{bmatrix} H_{-,\beta^*} \\ H_{-,\alpha^*} \end{bmatrix} \begin{bmatrix} T_{+,\alpha} & T_{+,\beta} \end{bmatrix}. \tag{8.12}$$

Proof. The above equalities (8.11) and (8.12) follow from the equalities (5.19) and (5.20) and the representations of the Hankel-like and Toeplitz-like operators given in the paragraph preceding Lemma 8.2. For example to prove (8.11) note that

$$\begin{bmatrix} J_{\mathcal{X}_+} & 0 \\ 0 & J_{\mathcal{Y}_+} \end{bmatrix} \begin{bmatrix} \mathbf{T}_{+,\alpha^*} \\ \mathbf{T}_{+,\beta^*} \end{bmatrix} = \begin{bmatrix} I_p & 0 \\ 0 & T_{+,\alpha^*} \\ 0 & T_{+,\beta^*} \end{bmatrix} J_{\mathcal{X}_+},$$

and

$$J_{\mathcal{X}_+} \begin{bmatrix} \mathbf{H}_{+,\beta} & \mathbf{H}_{+,\alpha} \end{bmatrix} = \begin{bmatrix} 0 & 0 & 0 \\ b & H_{+,\beta} & H_{+,\alpha} \end{bmatrix} \begin{bmatrix} J_{\mathcal{Y}_-} & 0 \\ 0 & J_{\mathcal{X}_-} \end{bmatrix}.$$

On the other hand

$$\begin{bmatrix} J_{\mathcal{X}_+} & 0 \\ 0 & J_{\mathcal{Y}_+} \end{bmatrix} \begin{bmatrix} \mathbf{H}_{+,\gamma^*} \\ \mathbf{H}_{+,\delta^*} \end{bmatrix} = \begin{bmatrix} 0 & 0 \\ c^* & H_{+,\gamma^*} \\ d^* & H_{+,\delta^*} \end{bmatrix} J_{\mathcal{Y}_-},$$

and

$$J_{\mathcal{Y}_-} \begin{bmatrix} \mathbf{T}_{-,\delta} & \mathbf{T}_{-,\gamma} \end{bmatrix} = \begin{bmatrix} I_q & 0 & 0 \\ d & T_{-,\delta} & T_{-,\gamma} \end{bmatrix} \begin{bmatrix} J_{\mathcal{Y}_-} & 0 \\ 0 & J_{\mathcal{X}_-} \end{bmatrix}.$$

The equality (8.11) now follows from (5.19). The equality (8.12) can be verified in the same manner. □

In what follows M is the operator given by

$$M = \begin{bmatrix} M_{11} & M_{12} \\ M_{21} & M_{22} \end{bmatrix} : \begin{bmatrix} L^1(\mathbb{R}_+)^p \\ L^1(\mathbb{R}_-)^q \end{bmatrix} \to \begin{bmatrix} L^1(\mathbb{R}_+)^p \\ L^1(\mathbb{R}_-)^q \end{bmatrix}, \tag{8.13}$$

where M_{11}, M_{12}, M_{21}, and M_{22} are the operators defined by (1.17)–(1.20).

Lemma 8.4. *Let* $\mathbf{R}_{11}$, $\mathbf{R}_{12}$, $\mathbf{R}_{21}$, *and* $\mathbf{R}_{22}$ *be defined by* (5.25)–(5.28), *and let* M_{11}, M_{12}, M_{21}, *and* M_{22} *be defined by* (1.17)–(1.20). *Then*

$$\begin{bmatrix} J_{\mathcal{X}_+} & 0 \\ 0 & J_{\mathcal{Y}_-} \end{bmatrix} \begin{bmatrix} \mathbf{R}_{11} & \mathbf{R}_{12} \\ \mathbf{R}_{21} & \mathbf{R}_{22} \end{bmatrix} = \begin{bmatrix} I_p & 0 & 0 & 0 \\ a & M_{11} & b & M_{12} \\ 0 & 0 & I_q & 0 \\ c & M_{21} & d & M_{22} \end{bmatrix} \begin{bmatrix} J_{\mathcal{X}_+} & 0 \\ 0 & J_{\mathcal{Y}_-} \end{bmatrix}. \tag{8.14}$$

In particular, $\mathbf{R}$ *given by* (5.39) *is invertible if and only if* M *defined by* (8.13) *is invertible. Moreover,* M_{11} *is invertible if and only if* $\mathbf{R}_{11}$ *is invertible and* M_{22} *is invertible if and only if* $\mathbf{R}_{22}$ *is invertible.*

Proof. From the relations between Hankel-like operators and Toeplitz-like operators on the one hand and Hankel integral operators and Wiener–Hopf operators on the other hand we have the identities:

$$J_{\mathcal{X}_+}\mathbf{R}_{11} = \begin{bmatrix} I_p & 0 \\ a & M_{11} \end{bmatrix} J_{\mathcal{X}_+}, \quad J_{\mathcal{X}_+}\mathbf{R}_{12} = \begin{bmatrix} 0 & 0 \\ b & M_{12} \end{bmatrix} J_{\mathcal{Y}_-}, \tag{8.15}$$

$$J_{\mathcal{Y}_-}\mathbf{R}_{21} = \begin{bmatrix} 0 & 0 \\ c & M_{21} \end{bmatrix} J_{\mathcal{X}_+}, \quad J_{\mathcal{Y}_-}\mathbf{R}_{22} = \begin{bmatrix} I_q & 0 \\ d & M_{22} \end{bmatrix} J_{\mathcal{Y}_-}. \tag{8.16}$$

Putting together these equalities gives the equality (8.14). The equality (8.14) implies that $\mathbf{R}$ is invertible it and only if M is invertible. The final statement follows from the first equality in (8.15) and the second equality in (8.16). □

We continue with specifying two other lemmas from Section 5.

Lemma 8.5. *Since conditions* (C4)–(C6) *are satisfied, we have*

$$M_{11} = I_p - H_{+,\alpha}H_{-,\alpha^*} + H_{+,\beta}H_{-,\beta^*} : L^1(\mathbb{R}_+)^p \to L^1(\mathbb{R}_+)^p, \tag{8.17}$$

$$M_{21} = T_{-,\delta}H_{-,\beta^*} - T_{-,\gamma}H_{-,\alpha^*} : L^1(\mathbb{R}_+)^p \to L^1(\mathbb{R}_-)^q, \tag{8.18}$$

$$M_{12} = T_{+,\alpha}H_{+,\gamma^*} - T_{+,\beta}H_{+,\delta^*} : L^1(\mathbb{R}_-)^q \to L^1(\mathbb{R}_+)^p, \tag{8.19}$$

$$M_{22} = I_q - H_{-,\delta}H_{+,\delta^*} + H_{-,\gamma}H_{+,\gamma^*} : L^1(\mathbb{R}_-)^q \to L^1(\mathbb{R}_-)^q. \tag{8.20}$$

Proof. The result is an immediate consequence of Lemma 5.5 and the relations between the $\mathbf{R}_{ij}$ and M_{ij} in (8.15) and (8.16). □

Lemma 8.6. *Let M_{ij}, $i,j = 1,2$, be given by* (1.17)–(1.20), *and let M be given by* (8.13). *Since conditions* (C1)–(C6) *are satisfied, we have*

$$\begin{bmatrix} M_{11} & M_{12} \\ M_{21} & M_{22} \end{bmatrix} \begin{bmatrix} I_p & 0 \\ 0 & -I_q \end{bmatrix} \begin{bmatrix} M_{11} & M_{12} \\ M_{21} & M_{22} \end{bmatrix} = \begin{bmatrix} M_{11} & 0 \\ 0 & -M_{22} \end{bmatrix}. \tag{8.21}$$

In particular, M is invertible if and only if M_{11} and M_{22} are invertible. Furthermore, in that case

$$M^{-1} = \begin{bmatrix} I_p & -M_{12}M_{22}^{-1} \\ -M_{21}M_{11}^{-1} & I_q \end{bmatrix} = \begin{bmatrix} I_p & -M_{11}^{-1}M_{12} \\ -M_{22}^{-1}M_{21} & I_q \end{bmatrix}. \tag{8.22}$$

Proof. The result is an immediate consequence of Lemma 5.6 and the relations between $\mathbf{R}_{ij}$ and M_{ij} given in (8.15) and (8.16). □

STEP 4. RESULTS OF SECTION 6. We use Theorem 6.1 to prove the "if" part of Theorem 1.1 and the identities (1.23), (1.24), (1.25), and (1.26).

First we check that the various conditions appearing in Theorem 6.1 are satisfied given our data. Since $g \in L^1(\mathbb{R}_+)^{p\times q} = \mathcal{B}_+$ is a solution to the twofold EG inverse problem, Proposition 5.1 tells us that

$$\mathbf{\Omega}\begin{bmatrix} \alpha \\ \gamma \end{bmatrix} = \begin{bmatrix} e_{\mathcal{A}} \\ 0 \end{bmatrix} \quad \text{and} \quad \mathbf{\Omega}\begin{bmatrix} \beta \\ \delta \end{bmatrix} = \begin{bmatrix} 0 \\ e_{\mathcal{D}} \end{bmatrix}.$$

Thus the identities in (6.2) are satisfied. Next, note that the final conclusion of STEP 1 tells us that of items (a) and (b) in Theorem 6.1 are also satisfied.

Thus Theorem 6.1 tells us that the operator $\mathbf{\Omega}$ is invertible. But then we can use Lemma 8.2 to conclude that the operator W defined by (1.21) is invertible too. This concludes the proof of the "if" part of Theorem 1.1.

Theorem 6.1 also tells us that the inverse $\mathbf{R}$ of $\mathbf{\Omega}$ is given by (6.3). From (8.10) and (8.14) it then follows that the inverse of W is the operator M defined by (8.13). This proves identity (1.23).

From Lemma 8.6 we know that M_{11} and M_{22} are invertible. The identities in (1.24) and (1.25) are obtained by comparing the off diagonal entries of $W = M^{-1}$ in (1.23) and (8.22).

Finally to see that the identities in (1.26) hold true, note that from $\mathbf{R}\mathbf{\Omega} = I$ it follows that

$$\begin{bmatrix} I_p & 0 & 0 & 0 \\ a & M_{11} & b & M_{12} \\ 0 & 0 & I & 0 \\ c & M_{21} & d & M_{22} \end{bmatrix} \begin{bmatrix} I_p & 0 & 0 & 0 \\ 0 & I & g & H_{+,g} \\ 0 & 0 & I & 0 \\ g^* & H_{-,g^*} & 0 & I \end{bmatrix} = I.$$

In particular, $M_{11}g + b = 0$ and $M_{22}g^* + c = 0$. Using the invertibility of M_{11} and M_{22} we obtain the formulas for g and g^* in (1.26). This completes the proof. □

8.2. Proof of Theorem 1.2

Throughout this subsection, as in Theorem 1.2, $\{a, b, c, d\}$ are the functions given by (1.1) and (1.2), and α, β, γ, δ are the functions given by (1.16). Furthermore, e_p and e_q are the functions on $\mathbb{R}$ identically equal to the unit matrix I_p and I_q, respectively. Finally $\mathcal{M} = \mathcal{M}_{\mathcal{A},\mathcal{B},\mathcal{C},\mathcal{D}}$ is the admissible algebra constructed in the beginning three paragraphs of the present section. In what follows we split the proof of Theorem 1.2 into two parts.

PART 1. In this part we assume that the twofold EG inverse problem associated with the data set $\{a, b, c, d\}$ has a solution, $g \in L^1(\mathbb{R}_+)^{p\times q} = \mathcal{B}_+$ say. Then we know from Proposition 2.1 that conditions (C1) – (C3) are satisfied. But the latter, using the final part of Remark 8.1, implies that condition (L1) is satisfied. Furthermore, the second part of Theorem 1.1 tells us that the operators M_{11} and M_{22} are invertible, and hence condition (L2) is satisfied too. Finally, the two identities in (1.26) yield the two identities in (1.27). This concludes the first part of the proof.

PART 2. In this part we assume that (L1) and (L2) are satisfied. Our aim is to show that the twofold EG inverse problem associated with the data set $\{a, b, c, d\}$ has a solution.

We begin with some preliminaries. Recall that

$$a_0 = P_{\mathcal{A}_d}\alpha = e_p \quad \text{and} \quad d_0 = P_{\mathcal{D}_d}\delta = e_q.$$

Furthermore, from the identities in (8.7) we know that (C1)–(C3) are satisfied. But then we can repeat the arguments in Step 1 of the proof of Theorem 1.1 to show that conditions (C4)–(C6) are also satisfied. Thus all conditions (C1)–(C6) are fulfilled. Finally, note that in the paragraph directly after Theorem 1.2 we showed that (L1) and (L2) imply that M_{11} and M_{22} are invertible, and hence we can apply Lemma 8.6 to see that the inverse of

$$M := \begin{bmatrix} M_{11} & M_{12} \\ M_{21} & M_{22} \end{bmatrix}$$

is given by (8.22), i.e.,

$$M^{-1} = \begin{bmatrix} I_p & -M_{12}M_{22}^{-1} \\ -M_{21}M_{11}^{-1} & I_q \end{bmatrix} = \begin{bmatrix} I_p & -M_{11}^{-1}M_{12} \\ -M_{22}^{-1}M_{21} & I_q \end{bmatrix}. \tag{8.23}$$

It remains to show that there exists a $g \in L^1(\mathbb{R}_+)^{p\times q}$ such that

$$-M_{11}^{-1}M_{12} = H_{+,g} \quad \text{and} \quad -M_{22}^{-1}M_{21} = H_{-,g^*}. \tag{8.24}$$

To do this we need (in the context of the present setting) a more general version of Lemma 7.2. We cannot apply Lemma 7.2 directly because of the role of the constant functions in $\mathbb{C}^{p\times p}e_p$ and $\mathbb{C}^{q\times q}e_q$. The more general version of Lemma 7.2 will be given and proved in the following intermezzo.

Intermezzo. First we introduce the required transition operators. Let $\tau \geq 0$. Define $V_{r,\tau} : L^2(\mathbb{R}_+)^r \to L^2(\mathbb{R}_+)^r$ by

$$(V_{r,\tau}f)(t) = \begin{cases} f(t-\tau), & t \geq \tau, \\ 0, & 0 \leq t \leq \tau. \end{cases}$$

Note that its adjoint $V_{r,\tau}^*$ is given by $\left(V_{r,\tau}^*f\right)(t) = f(t+\tau)$ for $t \geq 0$. We also need the flip over operator J_r from $L^2(\mathbb{R}_+)^r$ to $L^2(\mathbb{R}_-)^r$ given by $(J_rf)(t) = f(-t)$. With some abuse of notation we also consider $V_{r,\tau}$, J_r and their adjoints as operators acting on L^1-spaces. For $\varphi \in \mathbb{C}^{k\times m}\dot{+}L^1(\mathbb{R}_+)^{k\times m}$ we then have

$$\begin{aligned} V_{k,\tau}^*H_{+,\varphi} &= H_{+,\varphi}J_mV_{m,\tau}J_m, \quad \text{and} \quad J_kV_{k,\tau}^*J_kH_{-,\varphi} = H_{-,\varphi}V_{m,\tau}, \\ V_{k,\tau}^*T_{+,\varphi}V_{m,\tau} &= T_{+,\varphi}. \quad \text{and} \quad J_kV_{k,\tau}^*J_kT_{-,\varphi}J_mV_{m,\tau}J_m = T_{-,\varphi}. \end{aligned}$$

The following lemma is the more general version of Lemma 7.2 mentioned above. The result will be used to show that $M_{11}^{-1}M_{12}$ and $M_{22}^{-1}M_{21}$ are classical Hankel integral operators.

Lemma 8.7. *With M_{ij}, $1 \leq i,j \leq 2$, defined by* (1.17)–(1.20) *we have the following equalities:*

$$M_{11}V_{p,\tau}^*M_{12} = M_{12}J_qV_{q,\tau}J_qM_{22} \quad \textit{and} \quad M_{22}j_qV_{q,\tau}^*J_qM_{21} = M_{21}V_{p,\tau}M_{11}.$$

Moreover, if M_{11} and M_{22} are invertible, then

$$V_{p,\tau}^*M_{12}M_{22}^{-1} = M_{11}^{-1}M_{12}J_qV_{q,\tau}J_q, \quad J_qV_{q-,\tau}^*J_qM_{21}M_{11}^{-1} = M_{22}^{-1}M_{21}V_{p,\tau}.$$

Proof. First we will prove that $M_{11}V_{p,\tau}^*M_{12} = M_{12}J_qV_{q,\tau}J_qM_{22}$. We start with deriving the equality

$$\begin{bmatrix} T_{+,\alpha^*} \\ T_{+,\beta^*} \end{bmatrix} V_{p,\tau}^* \begin{bmatrix} H_{+,\beta} & H_{+,\alpha} \end{bmatrix} = \begin{bmatrix} H_{+,\gamma^*} \\ H_{+,\delta*} \end{bmatrix} J_qV_{q,\tau}J_q \begin{bmatrix} T_{-,\delta} & T_{-,\gamma} \end{bmatrix}. \tag{8.25}$$

Let $f_- \in L^1(\mathbb{R}_-)^q$ and $h_- \in L^1(\mathbb{R}_-)^p$. To obtain (8.25), first notice that

$$V_{p,\tau}^* \begin{bmatrix} H_{+,\beta} & H_{+,\alpha} \end{bmatrix} \begin{bmatrix} f_- \\ h_- \end{bmatrix} = \begin{bmatrix} H_{+,\beta} & H_{+,\alpha} \end{bmatrix} \begin{bmatrix} J_q V_{q,\tau} J_q & 0 \\ 0 & J_p V_{p,\tau} J_p \end{bmatrix} \begin{bmatrix} f_- \\ h_- \end{bmatrix};$$

$$J_q V_{q,\tau} J_q \begin{bmatrix} T_{-,\delta} & T_{-,\gamma} \end{bmatrix} \begin{bmatrix} f_- \\ h_- \end{bmatrix} = \begin{bmatrix} T_{-,\delta} & T_{-,\gamma} \end{bmatrix} \begin{bmatrix} J_q V_{q,\tau} J_q & 0 \\ 0 & J_p V_{p,\tau} J_p \end{bmatrix} \begin{bmatrix} f_- \\ h_- \end{bmatrix}.$$

Then use (8.11) to get that

$$\begin{aligned} &\begin{bmatrix} T_{+,\alpha^*} \\ T_{+,\beta^*} \end{bmatrix} V_{p,\tau}^* \begin{bmatrix} H_{+,\beta} & H_{+,\alpha} \end{bmatrix} \begin{bmatrix} f_- \\ h_- \end{bmatrix} \\ &\quad = \begin{bmatrix} T_{+,\alpha^*} \\ T_{+,\beta^*} \end{bmatrix} \begin{bmatrix} H_{+,\beta} & H_{+,\alpha} \end{bmatrix} \begin{bmatrix} J_q V_{q,\tau} J_q & 0 \\ 0 & J_p V_{p,\tau} J_p \end{bmatrix} \begin{bmatrix} f_- \\ h_- \end{bmatrix} \\ &\quad = \begin{bmatrix} H_{+,\gamma^*} \\ H_{+,\delta*} \end{bmatrix} \begin{bmatrix} T_{-,\delta} & T_{-,\gamma} \end{bmatrix} \begin{bmatrix} J_q V_{q,\tau} J_q & 0 \\ 0 & J_p V_{p,\tau} J_p \end{bmatrix} \begin{bmatrix} f_- \\ h_- \end{bmatrix} \\ &\quad = \begin{bmatrix} H_{+,\gamma^*} \\ H_{+,\delta*} \end{bmatrix} J_q V_{q,\tau} J_q \begin{bmatrix} T_{-,\delta} & T_{-,\gamma} \end{bmatrix} \begin{bmatrix} f_- \\ h_- \end{bmatrix}. \end{aligned}$$

We proved (8.25). By multiplying (8.25) on the left and the right by

$$\begin{bmatrix} T_{+,\alpha} & T_{+,\beta} \end{bmatrix} \quad \text{and} \quad \begin{bmatrix} T_{-,\delta^*} \\ T_{-,\gamma^*} \end{bmatrix},$$

respectively, one gets $M_{11} V_{p,\tau}^* M_{12} = M_{12} J_q V_{q,\tau} J_q M_{22}$.

The equality $M_{22} J_q V_{q,\tau}^* J_q M_{21} = M_{21} V_{p,\tau} M_{11}$ can be proved in a similar way. The claim regarding the case that M_{11} and M_{22} are invertible follows trivially. $\square$

We continue with the second part of the proof. It remains to show that there exists a $g \in L^1(\mathbb{R}_+)^{p\times q}$ such that the two identities in (8.24) are satisfied. For this purpose we need Lemma 8.7 and various results presented in the Appendix. In particular, in what follows we need the Sobolev space $\text{SB}(\mathbb{R}_+)^n$ which consist of all functions $\varphi \in L^1(\mathbb{R}_+)^n$ such that φ is absolutely continuous on compact intervals of $\mathbb{R}_+$ and $\varphi' \in L^1(\mathbb{R}_+)^n$ (see Subsection A.2). Notice that $M_{12}J_q$ is a sum of products of Wiener–Hopf operators and classical Hankel integral operators. Therefore, by Lemma A.7, the operator $M_{12}J_q$ maps $\text{SB}(\mathbb{R}_+)^q$ into $\text{SB}(\mathbb{R}_+)^p$ and $M_{12}J_q|_{\text{SB}(\mathbb{R}_+)^q}$ is bounded as an operator from $\text{SB}(\mathbb{R}_+)^q$ to $\text{SB}(\mathbb{R}_+)^p$. The operator M_{11} is of the form (A.18). Thus Lemma A.6 tells us that M_{11}^{-1} satisfies the condition (H1) in Theorem A.4. We conclude that the operator $-M_{11}^{-1}M_{12}J_q$ maps $\text{SB}(\mathbb{R}_+)^q$ into $\text{SB}(\mathbb{R}_+)^p$ and $-M_{11}^{-1}M_{12}J_q|_{\text{SB}(\mathbb{R}_+)^q}$ is bounded as an operator from $\text{SB}(\mathbb{R}_+)^q$ to $\text{SB}(\mathbb{R}_+)^p$. Also we know from Lemma 8.7 that

$$V_{p,\tau}(-M_{11}^{-1}M_{12})J_q = (-M_{11}^{-1}M_{12})J_q V_{q,\tau}, \quad \forall \tau \geq 0.$$

According to Theorem A.2 it follows that there exists a $k \in L^\infty(\mathbb{R}_+)^{p\times q}$ such that $-M_{11}^{-1}M_{12}J_q = H(k)$. But then we can apply Corollary A.5 to show that there exists a $g \in L^1(\mathbb{R}_+)^{p\times q}$ such that $-M_{11}^{-1}M_{12} = H_{+,g}$. In a similar way we prove that there exists a $h \in L^1(\mathbb{R}_-)^{q\times p}$ such that $-M_{22}^{-1}M_{21} = H_{-,h}$.

Notice that the operators M_{ij} can be considered to be operators acting between L^2-spaces. This can be done because the Hankel and Wiener–Hopf operators that constitute the M_{ij} can be seen as operators between L^2 spaces. Recall that $H^*_{+,\rho} = H_{-,\rho^*}$ and $T^*_{\pm,\rho} = T_{\pm,\rho^*}$. Using Lemma 8.5 and the definition of M_{12}, we see that $M^*_{12} = M_{21}$, and using the definitions of M_{11} and M_{22} we may conclude that $M^*_{11} = M_{11}$ and $M^*_{22} = M_{22}$. From the equality (8.22) one sees that $H_{-,h} = H^*_{+,g}$. Hence $h = g^*$.

We need to show that for this g the inclusions (1.3) and (1.4) are satisfied, or equivalently that (1.9) is satisfied. Define for this g the operator W by (1.21). We already know that W is invertible and that its inverse is M. Let $\tilde{a}$, $\tilde{b}$, $\tilde{c}$ and $\tilde{d}$ be the solution of

$$W \begin{bmatrix} \tilde{a} & \tilde{b} \\ \tilde{c} & \tilde{d} \end{bmatrix} = \begin{bmatrix} 0 & -g \\ -g^* & 0 \end{bmatrix}.$$

Then put $\tilde{\alpha} = e_p + \tilde{a}$, $\tilde{\beta} = \tilde{b}$, $\tilde{\gamma} = \tilde{c}$ and $\tilde{\delta} = e_q + \tilde{d}$. With the data $\tilde{\alpha}$, $\tilde{\beta}$, $\tilde{\gamma}$, and $\tilde{\delta}$ produce a new $\tilde{M}$ which, according to Theorem 1.1, is also the inverse of W. But then the old M and new $\tilde{M}$ are the same, and hence g is the solution of the EG inverse problem associated with α, β, γ and δ. All the conditions of Theorem 1.1 are now satisfied and we conclude that $g = -M_{11}^{-1}b$ and $g^* = -M_{22}^{-1}c$. □

9. The EG inverse problem with additional invertibility conditions

As before $\alpha \in \mathcal{A}_+$, $\beta \in \mathcal{B}_+$, $\gamma \in \mathcal{C}_-$, and $\delta \in \mathcal{D}_-$. In this section we consider the case when α is invertible in $\mathcal{A}_+$ and δ is invertible in $\mathcal{D}_-$. Notice that in the example discussed in Subsection 3.2 this condition is satisfied whenever $a_0 := P_{\mathcal{A}_d}\alpha$ and $d_0 := P_{\mathcal{D}_d}\delta$ are invertible.

Theorem 9.1. *Let $\alpha \in \mathcal{A}_+$, $\beta \in \mathcal{B}_+$, $\gamma \in \mathcal{C}_-$ and $\delta \in \mathcal{D}_-$ and assume that α and δ are invertible in $\mathcal{A}_+$ and $\mathcal{D}_-$, respectively. If, in addition, α, β, γ and δ satisfy the conditions* (C1) *and* (C2)*, then $g_1 = -P_{\mathcal{B}_+}(\alpha^{-*}\gamma^*)$ is the unique element of $\mathcal{B}_+$ that satisfies* (2.6) *and $g_2 = -P_{\mathcal{B}_+}(\beta\delta^{-1})$ is the unique element of $\mathcal{B}_+$ that satisfies* (2.7)*. Moreover, in that case, $g_1 = g_2$ if and only if condition* (C3) *is satisfied. In particular, if* (C1)*–*(C3) *hold, then $g = g_1 = g_2$ is the unique solution to the twofold EG inverse problem associated with the data set $\{\alpha, \beta, \gamma, \delta\}$.*

Proof. The inclusion $\alpha^{-1} \in \mathcal{A}_+$ implies that a_0 is invertible with inverse in $\mathcal{A}_d$. Similarly, $\delta^{-1} \in \mathcal{D}_-$ implies that d_0 is invertible with inverse in $\mathcal{D}_d$.

Let $g_1 := -P_{\mathcal{B}_+}(\alpha^{-*}\gamma^*)$. First we prove that g_1 satisfies the second inclusion in (2.6). From the definition of g_1 it follows that $g_1 + \alpha^{-*}\gamma^* = \beta_1$ for some $\beta_1 \in \mathcal{B}_-$. Taking adjoints we see that $g_1^* + \gamma\alpha^{-1} = \beta_1^* \in \mathcal{C}_+$. Multiplying from the right by α and using the multiplication table in Section 2 we see that

$$g_1^*\alpha + \gamma \in \mathcal{C}_+\mathcal{A}_+ \subset \mathcal{C}_+.$$

So g_1 is a solution of the second inclusion in (2.6). Notice that in this paragraph we did not yet use that $\alpha^{-1} \in \mathcal{A}_+$.

The next step is to show that g_1 is the unique element of $\mathcal{B}_+$ that satisfies the second inclusion in (2.6). Assume that $\varphi_1 \in \mathcal{B}_+$ and $P_{\mathcal{C}_-}(\varphi_1^*\alpha + \gamma) = 0$. We will prove that $\varphi_1 = g_1$. Notice that

$$\alpha^*(\varphi_1 - g_1) = (\alpha^*\varphi_1 + \gamma^*) - (\alpha^* g_1 + \gamma^*) \in \mathcal{B}_-.$$

Hence $\varphi_1 - g_1 \in \mathcal{B}_+$ and $\alpha^*(\varphi_1 - g_1) \in \mathcal{B}_-$. Since $\alpha^{-*} \in \mathcal{A}_-$ we have that

$$\varphi_1 - g_1 = \alpha^{-*}\alpha^*(\varphi_1 - g_1) \in \mathcal{A}_-\mathcal{B}_- \subset \mathcal{B}_-$$

and hence $\varphi_1 - g_1 = 0$.

Next remark that $P_{\mathcal{A}_-}(\alpha^{-1} - a_0^{-1}) = 0$. Indeed

$$\alpha^{-1} - a_0^{-1} = \alpha^{-1}(a_0 - \alpha)a_0^{-1} \in \mathcal{A}_+\mathcal{A}_+^0\mathcal{A}_d \subset \mathcal{A}_+^0.$$

To show that g_1 satisfies the first inclusion in (2.6), note that

$$\alpha + g_1\gamma - e_{\mathcal{A}} = \alpha - (\alpha^{-*}\gamma^*)\gamma + \left(P_{\mathcal{B}_-}(\alpha^{-*}\gamma^*)\right)\gamma - e_{\mathcal{A}}.$$

Now use that $\gamma^*\gamma = \alpha^*\alpha - a_0$ to see that

$$\begin{aligned}
\alpha + g_1\gamma - e_{\mathcal{A}} &= \alpha - \alpha^{-*}(\alpha^*\alpha - a_0) + \left(P_{\mathcal{B}_-}(\alpha^{-*}\gamma^*)\right)\gamma - e_{\mathcal{A}} \\
&= \alpha^{-*}a_0 - e_{\mathcal{A}} + \left(P_{\mathcal{B}_-}(\alpha^{-*}\gamma^*)\right)\gamma \\
&= (\alpha^{-*} - a_0^{-1})a_0 + \left(P_{\mathcal{B}_-}(\alpha^{-*}\gamma^*)\right)\gamma.
\end{aligned}$$

Since $\gamma \in \mathcal{C}_-$, we have that $P_{\mathcal{A}_+}\left[\left(P_{\mathcal{B}_-}(\alpha^{-*}\gamma^*)\right)\gamma\right] = 0$, and since $P_{\mathcal{A}_-}(\alpha^{-1} - a_0^{-1}) = 0$, we also have that $P_{\mathcal{A}_+}(\alpha^{-*} - a_0^{-1}) = 0$. We proved the first inclusion in (2.6).

Next, let $g_2 := -P_{\mathcal{B}_+}(\beta\delta^{-1})$. We will show that g_2 is the unique element of $\mathcal{B}_+$ that satisfies the first inclusion in (2.7). To do this, note that $g_2 + \beta\delta^{-1} = \beta_2$ for some $\beta_2 \in \mathcal{B}_-$, which implies that

$$g_2\delta + \beta \in \mathcal{B}_-\mathcal{D}_- \subset \mathcal{B}_-.$$

We proved that $P_{\mathcal{B}_+}(g_2\delta + \beta) = 0$. Assume that $\varphi_2 \in \mathcal{B}_+$ satisfies also the first inclusion in (2.7). Then

$$(\varphi_2 - g_2)\delta = (\varphi_2\delta + \beta) - (g_2\delta + \beta) \in \mathcal{B}_-.$$

Since $\delta \in \mathcal{D}_-$ we have that $(\varphi_2 - g_2) = (\varphi_2 - g_2)\delta\,\delta^{-1} \in \mathcal{B}_-\mathcal{D}_- \subset \mathcal{B}_-$. Hence $\varphi_2 - g_2 = 0$ and g_2 is the unique solution of the first inclusion in (2.7).

We proceed with showing that g_2 also satisfies the second inclusion in (2.7). Indeed

$$\begin{aligned}
\beta^* g_2 + \delta^* - e_{\mathcal{D}} &= \beta^*(-\beta\delta^{-1} + P_{\mathcal{B}_-}(\beta\delta^{-1})) + \delta^* - e_{\mathcal{D}} \\
&= -\beta^*\beta\delta^{-1} + \delta^* - e_{\mathcal{D}} + \beta^* P_{\mathcal{B}_-}(\beta\delta^{-1}) \\
&= -(\delta^*\delta - d_0)\delta^{-1} + \delta^* - e_{\mathcal{D}} + \beta^* P_{\mathcal{B}_-}(\beta\delta^{-1}) \\
&= d_0\delta^{-1} - e_{\mathcal{D}} + \beta^* P_{\mathcal{B}_-}(\beta\delta^{-1}) \in \mathcal{D}_-^0.
\end{aligned}$$

Here we used that $P_{\mathcal{D}_+}(d_0\delta^{-1} - e_{\mathcal{D}}) = P_{\mathcal{D}_+}\left(d_0^{-1}(e_{\mathcal{D}} - \delta)\delta^{-1}\right) = 0$. We proved that g_2 satisfies the second inclusion in (2.7).

If $\alpha^*\beta - \gamma^*\delta = 0$, then $\beta = \alpha^{-*}\gamma^*\delta$. It follows that

$$\begin{aligned} 0 &= P_{\mathcal{B}_+}(\beta - \alpha^{-*}\gamma^*\delta) = P_{\mathcal{B}_+}\left(\beta - P_{\mathcal{B}_+}\left((\alpha^{-*}\gamma^*)\delta\right)\right) \\ &= P_{\mathcal{B}_+}\left(\beta - P_{\mathcal{B}_+}(\alpha^{-*}\gamma^*)\delta\right) = P_{\mathcal{B}_+}(\beta + g_1\delta). \end{aligned}$$

So g_1 also solves the first inclusion in (2.7) and the uniqueness of the solution gives $g_1 = g_2$.

Conversely, if $g_1 = g_2$ then we have a solution of the inclusions (2.6) and (2.7). It follows from [13, Theorem 1.2] that the conditions (C1)–(C3) are satisfied and in particular we get $\alpha^*\beta - \gamma^*\delta = 0$. □

In the next proposition we combine the results of Theorem 9.1 with those of Theorem 6.1.

Proposition 9.2. *Let $\mathcal{M}_{\mathcal{A},\mathcal{B},\mathcal{C},\mathcal{D}}$ be an admissible algebra, and let $\alpha \in \mathcal{A}_+$, $\beta \in \mathcal{B}_+$, $\gamma \in \mathcal{C}_-$, $\delta \in \mathcal{D}_-$ be such that:*

(a) *α is invertible in $\mathcal{A}_+$ and δ is invertible in $\mathcal{D}_-$;*
(b) *conditions* (C1)–(C3) *are satisfied.*

Let

$$\varphi = -P_{\mathcal{B}_+}(a^{-*}\gamma^*) \quad \textit{and} \quad \mathbf{\Omega} = \begin{bmatrix} I_{\mathcal{X}_+} & \mathbf{H}_{+,\varphi} \\ \mathbf{H}_{-,\varphi^*} & I_{\mathcal{Y}_-} \end{bmatrix}. \tag{9.26}$$

Then φ is the solution of the twofold EG inverse problem associated with $\alpha, \beta, \gamma, \delta$, the operator $\mathbf{\Omega}$ is invertible, and

$$\mathbf{\Omega}^{-1} = \begin{bmatrix} \mathbf{R}_{11} & \mathbf{R}_{12} \\ \mathbf{R}_{21} & \mathbf{R}_{22} \end{bmatrix}, \tag{9.27}$$

where the operators $\mathbf{R}_{ij}$, $1 \le i,j \le 2$, are defined by (5.25)–(5.28). *In particular, the operators $\mathbf{R}_{11}$ and $\mathbf{R}_{22}$ are invertible and*

$$\mathbf{H}_{+,\varphi} = -\mathbf{R}_{11}^{-1}\mathbf{R}_{12} = -\mathbf{R}_{12}\mathbf{R}_{22}^{-1}, \quad \mathbf{H}_{-,\varphi^*} = -\mathbf{R}_{21}\mathbf{R}_{11}^{-1} = -\mathbf{R}_{22}^{-1}\mathbf{R}_{21}, \tag{9.28}$$

and

$$\mathbf{R}_{11}\varphi = -\beta \quad \textit{and} \quad \mathbf{R}_{22}\varphi^* = -\gamma. \tag{9.29}$$

Proof. Since the conditions of Theorem 9.1 are satisfied, φ is the unique solution of the twofold EG inverse problem associated with $\alpha, \beta, \gamma, \delta$. Also the fact that α is invertible in $\mathcal{A}_+$ and δ is invertible in $\mathcal{D}_-$ gives that a_0 and d_0 are invertible in $\mathcal{A}_d$ and $\mathcal{D}_d$, respectively, and that according to Lemma 2.4 also the conditions (C4)–(C6) are satisfied. But then all the conditions of Theorem 6.1 are satisfied. The equalities (9.27), (9.28) and (9.29) are now immediate from Theorem 6.1. □

Specifying Theorem 9.1 for the example discussed in Subsection 3.2 yields the following corollary.

Corollary 9.3. *Let $\mathcal{A}$, $\mathcal{B}$, $\mathcal{C}$, $\mathcal{D}$, $\mathcal{A}_\pm$, $\mathcal{B}_\pm$, $\mathcal{C}_\pm$, $\mathcal{D}_\pm$, and $\mathcal{A}_d$, $\mathcal{B}_d$, $\mathcal{C}_d$, $\mathcal{D}_d$ be as in Subsection* 3.2, *and let $\alpha \in \mathcal{A}_+$, $\beta \in \mathcal{B}_+$, $\gamma \in \mathcal{C}_-$, $\delta \in \mathcal{D}_-$ be given. Assume that*

(a) *$\alpha_0 = P_{\mathcal{A}_d}\alpha$ and $\delta_0 = P_{\mathcal{A}_d}\delta$ are invertible.*
(b) *α, β, γ and δ satisfy conditions* (C1)–(C3)

Then $g = -P_{\mathcal{B}_+}(\alpha^{-*}\gamma^*)$ is the unique element of $\mathcal{B}_+$ that satisfies (2.6) *and* (2.7).

Proof. We only need to recall that the invertibility of the diagonal matrices a_0 and d_0 implies invertibility of α and δ in $\mathcal{A}_+$ and $\mathcal{D}_-$, respectively. □

From the above corollary it also follows that in the numerical Example 3.1 the solution g,

$$g = -P_{\mathcal{B}_+}(\alpha^{-*}\gamma^*) = \begin{bmatrix} 1 & 2 & 0 \\ 0 & 1 & 2 \\ 0 & 0 & 1 \end{bmatrix},$$

is the unique solution of equations (3.3) and (3.4).

10. Wiener algebra on the circle

In this section (as announced in Subsubsection 3.3.2) we show how the solution of the discrete twofold EG inverse problem, Theorem 4.1 in [10], can be obtained as a corollary of our abstract Theorem 7.1.

Let us first recall the discrete twofold EG inverse problem as it was presented in [10]. This requires some preliminaries. Throughout $\mathcal{W}^{n\times m}$ denotes the space of $n\times m$ matrix functions with entries in the Wiener algebra on the unit circle which is denoted by $\mathcal{W}$ and not by $\mathcal{W}(\mathbb{T})$ as in Subsubsection 3.3.2. Thus a matrix function φ belongs to $\mathcal{W}^{n\times m}$ if and only if φ is continuous on the unit circle and its Fourier coefficients $\dots\varphi_{-1},\varphi_0,\varphi_1,\dots$ are absolutely summable. We set

$$\begin{aligned} \mathcal{W}_+^{n\times m} &= \{\varphi\in\mathcal{W}^{n\times m} \mid \varphi_j = 0, \quad \text{for } j=-1,-2,\dots\},\\ \mathcal{W}_-^{n\times m} &= \{\varphi\in\mathcal{W}^{n\times m} \mid \varphi_j = 0, \quad \text{for } j=1,2,\dots\},\\ \mathcal{W}_d^{n\times m} &= \{\varphi\in\mathcal{W}^{n\times m} \mid \varphi_j = 0, \quad \text{for } j\neq 0\},\\ \mathcal{W}_{+,0}^{n\times m} &= \{\varphi\in\mathcal{W}^{n\times m} \mid \varphi_j = 0, \quad \text{for } j=0,-1,-2,\dots\},\\ \mathcal{W}_{-,0}^{n\times m} &= \{\varphi\in\mathcal{W}^{n\times m} \mid \varphi_j = 0, \quad \text{for } j=0,1,2,\dots\}. \end{aligned}$$

Given $\varphi\in\mathcal{W}^{n\times m}$ the function φ^* is defined by $\varphi^*(\zeta)=\varphi(\zeta)^*$ for each $\zeta\in\mathbb{T}$. Thus the jth Fourier coefficient of φ^* is given by $(\varphi^*)_j = (\varphi_{-j})^*$. The map $\varphi\mapsto\varphi^*$ defines an involution which transforms $\mathcal{W}^{n\times m}$ into $\mathcal{W}^{m\times n}$, $\mathcal{W}_+^{n\times m}$ into $\mathcal{W}_-^{m\times n}$, $\mathcal{W}_{-,0}^{n\times m}$ into $\mathcal{W}_{+,0}^{m\times n}$, etc.

The data of the discrete EG inverse problem consist of four functions, namely

$$\alpha\in\mathcal{W}_+^{p\times p},\quad \beta\in\mathcal{W}_+^{p\times q},\quad \gamma\in\mathcal{W}_-^{q\times p},\quad \delta\in\mathcal{W}_-^{q\times q}, \tag{10.1}$$

and the problem is to find $g\in\mathcal{W}_+^{p\times q}$ such that

$$\alpha + g\gamma - e_p \in \mathcal{W}_{-,0}^{p\times p} \quad\text{and}\quad g^*\alpha+\gamma\in\mathcal{W}_{+,0}^{q\times p}; \tag{10.2}$$

$$g\delta+\beta\in\mathcal{W}_{-,0}^{p\times q} \quad\text{and}\quad \delta+g^*\beta-e_q\in\mathcal{W}_{+,0}^{q\times q}. \tag{10.3}$$

Here e_p and e_q denote the functions identically equal to the identity matrices I_p and I_q, respectively. If g has these properties, we refer to g as a *solution to the discrete twofold EG inverse problem associated with the data set* $\{\alpha,\beta,\gamma,\delta\}$. If a solution exists, then we know from Theorem 1.2 in [13] that necessarily the following identities hold:

$$\alpha^*\alpha - \gamma^*\gamma = a_0, \quad \delta^*\delta - \beta^*\beta = d_0, \quad \alpha^*\beta = \gamma^*\delta. \tag{10.4}$$

Here a_0 and d_0 are the zeroth Fourier coefficient of α and δ, respectively, and we identify the matrices with a_0 and d_0 with the matrix functions on $\mathbb{T}$ that are identically equal to a_0 and d_0, respectively. In this section we shall assume that a_0 and d_0 are invertible. Then (10.4) is equivalent to

$$\alpha a_0^{-1}\alpha^* - \gamma a_0^{-1}\gamma^* = e_p, \quad \delta d_0^{-1}\delta^* - \beta d_0^{-1}\beta^* = e_q, \quad \alpha a_0^{-1}\gamma^* = \beta d_0^{-1}\delta^*. \tag{10.5}$$

Finally, we associate with the data $\alpha, \beta, \gamma, \delta$ the following operators:

$$R_{11} = T_{+,\alpha}a_0^{-1}T_{+,\alpha^*} - T_{+,\beta}d_0^{-1}T_{+,\beta^*} : \mathcal{W}_+^p \to \mathcal{W}_+^p, \tag{10.6}$$
$$R_{21} = H_{-,\gamma}a_0^{-1}T_{+,\alpha^*} - H_{-,\delta}d_0^{-1}T_{+,\beta^*} : \mathcal{W}_+^p \to \mathcal{W}_-^q, \tag{10.7}$$
$$R_{12} = H_{+,\beta}d_0^{-1}T_{-,\delta^*} - H_{+,\alpha}a_0^{-1}T_{-,\gamma^*} : \mathcal{W}_-^q \to \mathcal{W}_+^p, \tag{10.8}$$
$$R_{22} = T_{-,\delta}d_0^{-1}T_{-,\delta^*} - T_{-,\gamma}a_0^{-1}T_{-,\gamma^*} : \mathcal{W}_-^q \to \mathcal{W}_-^q. \tag{10.9}$$

Here $T_{+,\alpha}$, T_{+,α^*}, $T_{+,\beta}$, $T_{+,\beta}$, $T_{-,\gamma}$, T_{-,γ^*}, $T_{-,\delta}$, T_{-,δ^*} are Toeplitz operators and $H_{+,\alpha}$, $H_{+,\beta}$, $H_{-,\gamma}$, $H_{-,\delta}$ are Hankel operators. The definitions of these operators can be found in the final paragraph of this section.

The next theorem gives the solution of the discrete twofold EG inverse problem. By applying Fourier transforms it is straightforward to check that the theorem is just equivalent to [10, Theorem 4.1].

Theorem 10.1. *Let α, β, γ, δ be the functions given by* (10.1), *with both matrices a_0 and d_0 invertible. Then the discrete twofold EG inverse problem associated with the data set $\{\alpha, \beta, \gamma, \delta\}$ has a solution if and only the following conditions are satisfied:*

(D1) *the identities in* (10.4) *hold true;*
(D2) *the operators R_{11} and R_{22} defined by* (10.6) *and* (10.9) *are one-to-one.*

Furthermore, in that case R_{11} and R_{22} are invertible, the solution is unique and the unique solution g and its adjoint are given by

$$g = -R_{11}^{-1}\beta \quad \text{and} \quad g^* = -R_{22}^{-1}\gamma. \tag{10.10}$$

The next step is to show how the above theorem can be derived as a corollary of our abstract Theorem 7.1. This requires to put the inverse problem in the context of the general scheme of Sections 4–7. To do this (cf., the first paragraph of Subsubsection 3.3.2) we use the following choice of $\mathcal{A}$, $\mathcal{B}$, $\mathcal{C}$, and $\mathcal{D}$:

$$\mathcal{A} = \mathcal{W}^{p\times p}, \quad \mathcal{B} = \mathcal{W}^{p\times q} \quad \mathcal{C} = \mathcal{W}^{q\times p}, \quad \mathcal{D} = \mathcal{W}^{q\times q}. \tag{10.11}$$

The spaces $\mathcal{A}$, $\mathcal{B}$, $\mathcal{C}$, $\mathcal{D}$ admit decompositions as in (2.2) and (2.4) using

$$\begin{aligned}
&\mathcal{A}_+^0 = \mathcal{W}_{+,0}^{p\times p}, \quad \mathcal{A}_-^0 = \mathcal{W}_{-,0}^{p\times p}, \quad \mathcal{A}_d = \{\eta e_p \mid \eta \in \mathbb{C}^{p\times p}\},\\
&\mathcal{B}_+ = \mathcal{W}_{+}^{p\times q}, \quad \mathcal{B}_- = \mathcal{W}_{-,0}^{p\times q},\\
&\mathcal{C}_- = \mathcal{W}_{-}^{q\times p}, \quad \mathcal{C}_+ = \mathcal{W}_{+,0}^{q\times p},\\
&\mathcal{D}_+^0 = \mathcal{W}_{+,0}^{q\times q}, \quad \mathcal{D}_-^0 = \mathcal{W}_{-,0}^{q\times q}, \quad \mathcal{D}_d = \{\zeta e_q \mid \zeta \in \mathbb{C}^{q\times q}\}.
\end{aligned}$$

The algebraic structure is given by the algebraic structure of the Wiener algebra and by the matrices with entries from the Wiener algebra. Note that

$$\alpha \in \mathcal{A}_+, \quad \beta \in \mathcal{B}_+, \quad \gamma \in \mathcal{C}_-, \quad \delta \in \mathcal{D}_-, \tag{10.12}$$

and we are interested (cf., (2.6) and (2.7)) in finding $g \in \mathcal{B}_+$ such that

$$\alpha + g\gamma - e_p \in \mathcal{A}_-^0 \quad \text{and} \quad g^*\alpha + \gamma \in \mathcal{C}_+, \tag{10.13}$$

$$g\delta + \beta \in \mathcal{B}_-^0 \quad \text{and} \quad \delta + g^*\beta - e_q \in \mathcal{D}_+^0. \tag{10.14}$$

Furthermore, $a_0 = P_{\mathcal{A}_d}\alpha$ and $d_0 = P_{\mathcal{A}_d}\delta$ are invertible in $\mathcal{A}_d$ and $\mathcal{D}_d$, respectively.

In the present context the spaces $\mathcal{X}$ and $\mathcal{Y}$, $\mathcal{X}_+$ and $\mathcal{Y}_+$, and $\mathcal{X}_-$ and $\mathcal{Y}_-$ defined in the first paragraph of Section 4 are given by

$$\begin{aligned}
\mathcal{X} &= \mathcal{A}\dot{+}\mathcal{B} = \mathcal{W}^{p\times p}\dot{+}\mathcal{W}^{p\times q}, & \mathcal{Y} &= \mathcal{C}\dot{+}\mathcal{D} = \mathcal{W}^{q\times p}\dot{+}\mathcal{W}^{q\times q},\\
\mathcal{X}_+ &= \mathcal{A}_+\dot{+}\mathcal{B}_+ = \mathcal{W}_+^{p\times p}\dot{+}\mathcal{W}_+^{p\times q}, & \mathcal{Y}_+ &= \mathcal{C}_+\dot{+}\mathcal{D}_+^0 = \mathcal{W}_{+,0}^{q\times p}\dot{+}\mathcal{W}_{+,0}^{q\times q},\\
\mathcal{X}_- &= \mathcal{A}_-^0\dot{+}\mathcal{B}_- = \mathcal{W}_{-,0}^{p\times p}\dot{+}\mathcal{W}_{-,0}^{p\times q}, & \mathcal{Y}_- &= \mathcal{C}_-\dot{+}\mathcal{D}_- = \mathcal{W}_-^{q\times p}\dot{+}\mathcal{W}_-^{q\times q}.
\end{aligned}$$

Remark 10.2. Note that the space $\mathcal{X} = \mathcal{A}\dot{+}\mathcal{B} = \mathcal{W}^{p\times p}\dot{+}\mathcal{W}^{p\times q}$ can be identified in a canonical way with the space $\mathcal{W}^{p\times(p+q)}$, and analogously the subspaces $\mathcal{X}_\pm$ can be identified in a canonical way with subspaces of $\mathcal{W}^{p\times(p+q)}$. For instance, $\mathcal{X}_+$ with $\mathcal{W}_+^{p\times(p+q)}$. Similarly, $\mathcal{Y} = \mathcal{C}\dot{+}\mathcal{D}$ can be identified with $\mathcal{W}^{q\times(p+q)}$, and the spaces $\mathcal{Y}_\pm$ with subspaces of $\mathcal{W}^{q\times(p+q)}$. We will use these identifications in the proof of Theorem 10.1

Remark 10.3. Let $g \in \mathcal{W}_+^{p\times q}$, and let R_{ij}, $1 \le i,j \le 2$, be the operators defined by (10.6)–(10.9). Note that the operators $H_{+,g}$, H_{-,g^*} and R_{ij} act on vector spaces $\mathcal{W}_\pm^m$ with $m = p$ or $m = q$; see the final paragraphs of the present section. As usual we extend the action of these operators to spaces of matrices of the type $\mathcal{W}_\pm^{m\times k}$. In this way (using the preceding remark) we see that the operators $H_{+,g}$, H_{-,g^*} and R_{ij} can be identified with the operators $\mathbf{H}_{+,g}$, $\mathbf{H}_{-,g^*}$ and $\mathbf{R}_{ij}$ as defined in Section 5, respectively.

Proof of Theorem 10.1. We will apply Theorem 7.1 and Lemma 7.2 using $\mathcal{A}$, $\mathcal{B}$, $\mathcal{C}$, $\mathcal{D}$ in (10.11). First we check that the conditions in Theorem 7.1 are satisfied. Condition (a) is satisfied by assumption.

Now assume that there exists a solution g to the twofold EG inverse problem. Then conditions (C1)–(C6) are satisfied too. Next put

$$\Omega = \begin{bmatrix} I_{\mathcal{W}_+^p} & H_{+,g} \\ H_{-,g^*} & I_{\mathcal{W}_-^q} \end{bmatrix} : \begin{bmatrix} \mathcal{W}_+^p \\ \mathcal{W}_-^q \end{bmatrix} \to \begin{bmatrix} \mathcal{W}_+^p \\ \mathcal{W}_-^q \end{bmatrix},$$

$$\boldsymbol{\Omega} = \begin{bmatrix} I_{\mathcal{X}_+} & \mathbf{H}_{+,g} \\ \mathbf{H}_{-,g^*} & \mathbf{I}_{\mathcal{Y}_-} \end{bmatrix} : \begin{bmatrix} \mathcal{X}_+ \\ \mathcal{Y}_- \end{bmatrix} \to \begin{bmatrix} \mathcal{X}_+ \\ \mathcal{Y}_- \end{bmatrix}.$$

Here $\mathcal{X}_\pm$ and $\mathcal{Y}_\pm$ are the spaces defined in the paragraph preceding Remark 10.2. Since the conditions of Theorem 6.1 are satisfied, we know from (6.3) that the operator $\boldsymbol{\Omega}$ is invertible and its inverse is given by $\boldsymbol{\Omega}^{-1} = \mathbf{R}$, where

$$\mathbf{R} = \begin{bmatrix} \mathbf{R}_{11} & \mathbf{R}_{12} \\ \mathbf{R}_{21} & \mathbf{R}_{22} \end{bmatrix}.$$

Thus $\boldsymbol{\Omega}\mathbf{R}$ and $\mathbf{R}\boldsymbol{\Omega}$ are identity operators. Using the similarity mentioned in Remark 10.3 above, we see that ΩR and $R\Omega$ are also identity operators, and hence Ω is invertible. Moreover, the fact that $\mathbf{R}_{11}$ and $\mathbf{R}_{22}$ are invertible implies that R_{11} and R_{22} are invertible. Finally, from (6.4), (6.5) and (6.6) (again using the above Remark 10.3) we obtain the identities

$$H_{+,g} = -R_{11}^{-1}R_{12} = -R_{12}R_{22}^{-1}, \quad H_{-,g^*} = -R_{21}R_{11}^{-1} = -R_{22}^{-1}R_{21}; \tag{10.15}$$

$$g = -R_{11}^{-1}\beta, \quad g^* = -R_{22}^{-1}\gamma. \tag{10.16}$$

It follows that conditions (D1) and (D2) are fulfilled.

Conversely, assume that conditions (D1) and (D2) are satisfied. Then the statements (b) and (i) in Theorem 7.1 follow. To apply Lemma 7.2 we set $\mathbf{V}_{\mathcal{X},\pm} = S_{p,\pm}$ and $\mathbf{V}_{\mathcal{Y},\pm} = S_{q,\pm}$, where $S_{q,\pm}$ and $S_{q,\pm}$ are the shift operators defined in the final paragraph of this section. The intertwining of $\mathbf{V}_{\mathcal{X},\pm}$ and $\mathbf{V}_{\mathcal{Y},\pm}$ with the Hankel-like and Toeplitz-like operators are required for the application of Lemma 7.2. But these intertwining relations correspond with the intertwining of $S_{p,\pm}$ and $S_{q,\pm}$ with the Hankel and Toeplitz operators $H_{\pm,p}$, $H_{\pm,q}$ and $T_{\pm,p}$, $T_{\pm,q}$ appearing in the present section. From the final part of Lemma 7.2 we conclude that

$$S_{*,p,+}R_{12}R_{22}^{-1} = R_{11}^{-1}R_{12}S_{q,-}.$$

Furthermore, Lemma 5.6 tells us that $R_{12}R_{22}^{-1} = R_{11}^{-1}R_{12}$. Hence

$$S_{*,p,+}R_{11}^{-1}R_{12} = R_{11}^{-1}R_{12}S_{q,-}.$$

But then, using Lemma 10.4 below, it follows that there exists a $g \in \mathcal{W}_+^{p\times q}$ such that $H_{+,g} = -R_{11}^{-1}R_{12}$. Similarly we obtain that there exists a $h \in \mathcal{W}_-^{q\times p}$ such that $H_{-,h} = -R_{22}^{-1}R_{21}$. From Lemma 5.4 it then follows that $g = -R_{11}^{-1}\beta$ and $h = -R_{22}^{-1}\gamma$.

It remains to show that $h = g^*$. To do this, we extend some of the operators from (subspaces of) Wiener spaces to subspaces of L^2-spaces over the unit circle $\mathbb{T}$. More specifically, for $m = p, q$ write $L^2(\mathbb{T})^m$ for the space of vectors of size m whose entries are L^2-functions over $\mathbb{T}$, and write $L^2_+(\mathbb{T})^m$ and $L^2_-(\mathbb{T})^m$ for the subspaces of $L^2(\mathbb{T})^m$ consisting of functions in $L^2(\mathbb{T})^m$ such that the Fourier coefficients with strictly negative $(-1,-2,\dots)$ coefficients and positive $(0,1,2,\dots)$ coefficients, respectively, are zero. Then, with some abuse of notation, we extend the operators $H_{+,g}$, $H_{-,h}$ and R_{ij}, $i,j=1,2$, in the following way:

$$\begin{aligned} H_{+,g} &: L^2_-(\mathbb{T})^q \to L^2_+(\mathbb{T})^p, & H_{-,h} &: L^2_+(\mathbb{T})^p \to L^2_-(\mathbb{T})^q,\\ R_{11} &: L^2_+(\mathbb{T})^p \to L^2_+(\mathbb{T})^p, & R_{12} &: L^2_-(\mathbb{T})^q \to L^2_+(\mathbb{T})^p,\\ R_{21} &: L^2_+(\mathbb{T})^p \to L^2_-(\mathbb{T})^q, & R_{22} &: L^2_-(\mathbb{T})^q \to L^2_-(\mathbb{T})^q. \end{aligned}$$

It then follows from the representations (10.6)–(10.9) and (5.30)–(5.33) that $R_{11} = R_{11}^*$, $R_{22} = R_{22}^*$, and $R_{12}^* = R_{21}$. We find that $-H_{+,g} = R_{11}^{-1}R_{12} = \big(R_{22}^{-1}R_{21}\big)^* = -H_{-,h}$, and therefore $h = g^*$. We conclude that the solution of the twofold EG inverse problem is indeed given by (10.10). □

Toeplitz and Hankel operators. Throughout, for a function $\rho \in \mathcal{W}^{n\times m}$, we write $M(\rho)$ for the *multiplication operator* of ρ from $\mathcal{W}^m$ into $\mathcal{W}^n$, that is,

$$M(\rho) : \mathcal{W}^m \to \mathcal{W}^n, \quad (M(\rho)f)(e^{it}) = \rho(e^{it})f(e^{it}) \quad (f \in \mathcal{W}_m,\ t \in [0,2\pi]).$$

We define Toeplitz operators $T_{\pm,\rho}$ and Hankel operators $H_{\pm,\rho}$ as compressions of multiplication operators, as follows. *Fix the dimensions $p \geq 1$ and $q \geq 1$ for the remaining part of this section.*

If $\rho \in \mathcal{W}^{p\times p}$, then

$$\begin{aligned} T_{+,\rho} &= P_{+,p}M(\rho) : \mathcal{W}^p_+ \to \mathcal{W}^p_+, & T_{-,\rho} &= (I-P_{+,p})M(\rho) : \mathcal{W}^p_{-,0} \to \mathcal{W}^p_{-,0},\\ H_{+,\rho} &= P_{+,p}M(\rho) : \mathcal{W}^p_{-,0} \to \mathcal{W}^p_+, & H_{-,\rho} &= (I-P_{+,p})M(\rho) : \mathcal{W}^p_+ \to \mathcal{W}^p_{-,0}. \end{aligned}$$

If $\rho \in \mathcal{W}^{p\times q}$, then

$$\begin{aligned} T_{+,\rho} &= P_{+,p}M(\rho) : \mathcal{W}^q_{+,0} \to \mathcal{W}^p_+, & T_{-,\rho} &= (I-P_{+,p})M(\rho) : \mathcal{W}^q_- \to \mathcal{W}^p_{-,0},\\ H_{+,\rho} &= P_{+,p}M(\rho) : \mathcal{W}^q_- \to \mathcal{W}^p_+, & H_{-,\rho} &= (I-P_{+,p})M(\rho) : \mathcal{W}^q_{+,0} \to \mathcal{W}^p_{-,0}. \end{aligned}$$

If $\rho \in \mathcal{W}^{q\times p}$, then

$$\begin{aligned} T_{+,\rho} &= (I-P_{-,q})M(\rho) : \mathcal{W}^p_+ \to \mathcal{W}^q_{+,0}, & T_{-,\rho} &= P_{-,q}M(\rho) : \mathcal{W}^p_{-,0} \to \mathcal{W}^q_-,\\ H_{+,\rho} &= (I-P_{-,q})M(\rho) : \mathcal{W}^p_{-,0} \to \mathcal{W}^q_{+,0}, & H_{-,\rho} &= P_{-,q}M(\rho) : \mathcal{W}^p_+ \to \mathcal{W}^q_-, \end{aligned}$$

and for $\rho \in \mathcal{W}^{q\times p}$ then

$$\begin{aligned} T_{+,\rho} &= (I-P_{-,q})M(\rho) : \mathcal{W}^q_{+,0} \to \mathcal{W}^q_{+,0}, & T_{-,\rho} &= P_{-,q}M(\rho) : \mathcal{W}^q_- \to \mathcal{W}^q_-,\\ H_{+,\rho} &= (I-P_{-,q})M(\rho) : \mathcal{W}^q_- \to \mathcal{W}^q_{+,0}, & H_{-,\rho} &= P_{-,q}M(\rho) : \mathcal{W}^q_{+,0} \to \mathcal{W}^q_-. \end{aligned}$$

Shift operators. We also define the shift operators used in the present section. Let $\varphi \in \mathcal{W}^{p\times p}$ and $\psi \in \mathcal{W}^{q\times q}$ be defined by $\varphi(z) = ze_p$ and $\psi(z) = ze_q$, with e_p and e_q the constant functions equal to the unity matrix. The shift operators that we need are now defined by

$$\begin{aligned} S_{p,+} &= M(\varphi) : \mathcal{W}^p_+ \to \mathcal{W}^p_+, & S_{q,+} &= M(\psi) : \mathcal{W}^q_{+,0} \to \mathcal{W}^q_{+,0}, \\ S_{q,-} &= M(\psi^{-1}) : \mathcal{W}^q_- \to \mathcal{W}^q_-, & S_{p,-} &= M(\varphi^{-1}) : \mathcal{W}^p_{-,0} \to \mathcal{W}^p_{-,0}, \\ S_{*,p,+} &= T_{+,\varphi^{-1}} : \mathcal{W}^p_+ \to \mathcal{W}^p_+, & S_{*,q,+} &= T_{+,\psi^{-1}} : \mathcal{W}^q_{+,0} \to \mathcal{W}^q_{+,0}, \\ S_{*,q,-} &= T_{-,\psi} : \mathcal{W}^q_- \to \mathcal{W}^q_-, & S_{*,p,-} &= T_{-,\varphi} : \mathcal{W}^p_{-,0} \to \mathcal{W}^p_{-,0}, \end{aligned}$$

Then

$$\begin{aligned} S_{*,p,+}S_{p,+} &= I_{\mathcal{W}^p_+}, & S_{*,p,-}S_{p,-} &= I_{\mathcal{W}^p_{-,0}}, \\ S_{*,q,-}S_{q,-} &= I_{\mathcal{W}^q_-}, & S_{*,q,+}S_{q,+} &= I_{\mathcal{W}^q_{+,0}}. \end{aligned}$$

Also we have for $m \in \{p, q\}$ and $n \in \{p, q\}$, and $\rho \in \mathcal{W}^{n\times m}$ that

$$H_{+,\rho}S_{m,-} = S_{*,n,+}H_{+,\rho}, \quad H_{-,\rho}S_{m,+} = S_{*,n,-}H_{-,\rho}.$$

Finally for $\rho \in \mathcal{W}^{n\times m}_+$ we have $T_{+,\rho}S_{m,+} = S_{n,+}T_{+,\rho}$ and if $\rho \in \mathcal{W}^{n\times m}_-$ then $T_{-,\rho}S_{m,-} = S_{n,-}T_{-,\rho}$.

The following result is classical and is easy to prove using the inverse Fourier transform (see, e.g., [2, Section 2.3] or Sections XXII–XXIV in [6]).

Lemma 10.4. *Let $G : \mathcal{W}^q_- \to \mathcal{W}^p_+$, and assume that $GS_{q,-} = S_{*,p,+}G$. Then there exists a function $g \in \mathcal{W}^{p\times q}_+$ such that $G = H_{+,g}$. Similarly, if $H : \mathcal{W}^p_+ \to \mathcal{W}^q_-$ and $HS_{p,+} = S_{*,q,-}H$, then there exists a function $h \in \mathcal{W}^{q\times p}_-$ such that $H = H_{-,h}$.*

Appendix: Hankel and Wiener–Hopf integral operators

In this appendix, which consists of three subsections, we present a number of results that play an essential role in the proof of Theorem 1.2. In Subsection A.1 we recall the definition of a Hankel operator on $L^2(\mathbb{R}_+)$ and review some basic facts. In Subsection A.2 we present a theorem (partially new) characterising classical Hankel integral operators mapping $L^1(\mathbb{R}_+)^p$ into $L^1(\mathbb{R}_+)^q$. Two auxiliary results are presented in the final subsection.

A.1. Preliminaries about Hankel operators

We begin with some preliminaries about Hankel operators on $L^2(\mathbb{R}_+)$, mainly taken from or Section 1.8 in [17] or Section 9.1 in [2]. Throughout J is the flip over operator on $L^2(\mathbb{R})$ defined by $(Jf)(t) = f(-t)$. Furthermore, $\mathcal{F}$ denotes the Fourier transform on $L^2(\mathbb{R})$ defined by

$$(\mathcal{F}f)(\lambda) = \frac{1}{\sqrt{2\pi}} \int_{\mathbb{R}} e^{i\lambda t} f(t)\, dt.$$

It is well known (see, e.g., [2, Section 9.1, page 482]) that $\mathcal{F}$ is a unitary operator and

$$\mathcal{F}^* = \mathcal{F}^{-1} = J\mathcal{F} \quad \text{and} \quad J\mathcal{F} = \mathcal{F}J.$$

Given $\alpha \in L^\infty(\mathbb{R})$ we define the multiplier $m(\alpha)$ and the convolution operator $M(\alpha)$ defined by α to be the operators on $L^2(\mathbb{R})$ given by

$$(m(\alpha)f)(t) = \alpha(t)f(t), \quad f \in L^2(\mathbb{R}), \, t \in \mathbb{R}, \quad \text{and} \quad M(\alpha) = \mathcal{F}^{-1}m(\alpha)\mathcal{F}.$$

Given $f \in L^2(\mathbb{R})$ we have

$$\begin{aligned}
(M(\alpha)f)(t) &= \left(\mathcal{F}^{-1}m(\alpha)\mathcal{F}f\right)(t) = \left(J\mathcal{F}m(\alpha)\mathcal{F}f\right)(t) \\
&= \frac{1}{\sqrt{2\pi}} \int_{\mathbb{R}} e^{-its} \left(m(\alpha)\mathcal{F}f\right)(s)\, ds = \frac{1}{\sqrt{2\pi}} \int_{\mathbb{R}} e^{-its} \alpha(s)(\mathcal{F}f)(s)\, ds \\
&= \frac{1}{2\pi} \int_{\mathbb{R}} e^{-its}\alpha(s) \left(\int_{\mathbb{R}} e^{isr} f(r)\, dr \right) ds \\
&= \frac{1}{2\pi} \int_{\mathbb{R}} \int_{\mathbb{R}} e^{is(r-t)} \alpha(s) f(r)\, dr\, ds, \quad t \in \mathbb{R}.
\end{aligned} \tag{A.1}$$

By P and Q we denote the orthogonal projections on $L^2(\mathbb{R})$ of which the ranges are $L^2(\mathbb{R}_+)$ and $L^2(\mathbb{R}_-)$, respectively.

Definition A.1. Let $\alpha \in L^\infty(\mathbb{R})$. Then the *Hankel operator* defined by α is the operator on $L^2(\mathbb{R}_+)$ given by

$$H(\alpha) = PM(\alpha)J|_{L^2(\mathbb{R}_+)} : L^2(\mathbb{R}_+) \to L^2(\mathbb{R}_+).$$

The action of the Hankel operator $H(\alpha)$ on $f \in L^2(\mathbb{R}_+)$ is given by

$$\begin{aligned}
(H(\alpha)f)(t) &= (PM(\alpha)Jf)(t) = (M(\alpha)Jf)(t) \\
&= \frac{1}{2\pi} \int_{\mathbb{R}} \int_{\mathbb{R}} e^{is(r-t)} \alpha(s) f(-r)\, dr\, ds \\
&= \frac{1}{2\pi} \int_{\mathbb{R}} \int_{\mathbb{R}} e^{-is(t+r)} \alpha(s) f(r)\, dr\, ds \\
&= \frac{1}{2\pi} \int_{\mathbb{R}} \int_0^\infty e^{-is(t+r)} \alpha(s) f(r)\, dr\, ds, \quad t \geq 0.
\end{aligned} \tag{A.2}$$

The following result provides a characterization of which operators on $L^2(\mathbb{R}_+)$ are Hankel operators; see [16, Exercise (a) on page 199–200].

Theorem A.2. *A bounded linear operator K on $L^2(\mathbb{R}_+)$ is a Hankel operator if and only if $V_\tau^* K = KV_\tau$ for all $\tau \geq 0$, where for each $\tau \geq 0$ the operator V_τ is the transition operator on $L^2(\mathbb{R}_+)$ defined by*

$$(V_\tau f)(t) = \begin{cases} f(t-\tau), & t \geq \tau, \\ 0, & 0 \leq t \leq \tau, \end{cases} \tag{A.3}$$

Remark A.3. When we worked on this paper we assumed that the above theorem, which is a natural analogue of the intertwining shift relation theorem for discrete Hankel operators, to be true and that we only had to find a reference. The latter turned out to be a bit difficult. Various Hankel operator experts told us "of course, the result is true." But no reference. We asked Vladimir Peller, and he mailed us how the result could be proved using the beautiful relations between H^2 on the disc and H^2 on the upper half-plane, but again no reference. What to do? Should we include Peller's proof? November last year Albrecht Bötcher solved the problem. He referred us to Nikolski's book [16] which appeared recently in Spring 2017 and contains the result as an exercise. Other references remain welcome.

Next we consider the special case when the defining function α is given by

$$\alpha(\lambda) = \int_{\mathbb{R}} e^{i\lambda s} a(s)\, ds, \quad \text{where } a \in L^1(\mathbb{R}_+). \tag{A.4}$$

Then

$$(H(\alpha)f)(t) = \int_0^\infty a(t+s) f(s)\, ds, \quad t \in \mathbb{R}_+,\ f \in L^2(\mathbb{R}_+). \tag{A.5}$$

In this case one calls $H(\alpha)$ the *classical Hankel integral operator* defined by a. To prove (A.5) we may without loss of generality assume that f belongs to $L^1(\mathbb{R}_+) \cap L^2(\mathbb{R}_+)$ and α is rational. In that case using (A.4) we have

$$\begin{aligned}(H(\alpha)f)(t) &= \frac{1}{2\pi} \int_{\mathbb{R}} \left(\int_0^\infty e^{-is(t+r)} \alpha(s) f(r)\, dr \right) ds \\ &= \frac{1}{2\pi} \int_0^\infty \left(\int_{\mathbb{R}} e^{-is(t+r)} \alpha(s)\, ds \right) f(r)\, dr \\ &= \int_0^\infty a(t+r) f(r)\, dr, \quad t \geq 0.\end{aligned}$$

If α is given by (A.4), then α belongs to the Wiener algebra over $\mathbb{R}$ and thus $H(\alpha)$ also defines a bounded linear operator on $L^1(\mathbb{R}_+)$.

We shall also deal with Hankel operators defined by matrix-valued functions. Let α be a $q \times p$ matrix whose entries α_{ij}, $1 \leq i \leq q$, $1 \leq j \leq p$, are $L^\infty(\mathbb{R}_+)$ functions. Then $H(\alpha)$ will denote the Hankel operator from $L^2(\mathbb{R}_+)^p$ to $L^2(\mathbb{R}_+)^q$ defined by

$$H(\alpha) = \begin{bmatrix} H(\alpha_{11}) & \cdots & H(\alpha_{1p}) \\ \vdots & \dots & \vdots \\ H(\alpha_{q1}) & \cdots & H(\alpha_{qp}) \end{bmatrix}. \tag{A.6}$$

If the operators $H(\alpha_{ij})$, $1 \leq i \leq q$, $1 \leq j \leq p$, are all classical Hankel integral operators, then we call $H(\alpha)$ a classical Hankel integral operator too.

A.2. Classical Hankel integral operators on L^1 spaces

The main theorem of this section allows us to identify the classical Hankel integral operators among all operators from $L^1(\mathbb{R}_+)^p$ to $L^1(\mathbb{R}_+)^q$. We begin with some preliminaries about related Sobolev spaces.

Let n be a positive integer. By $\mathrm{SB}(\mathbb{R}_+)^n$ we denote the Sobolev space consisting of all functions $\varphi \in L^1(\mathbb{R}_+)^n$ such that φ is absolutely continuous on compact intervals of $\mathbb{R}_+$ and $\varphi' \in L^1(\mathbb{R}_+)^n$. Note that

$$\varphi \in \mathrm{SB}(\mathbb{R}_+)^n \quad \Longrightarrow \quad \varphi(t) = -\int_t^\infty \varphi'(s)\,ds, \quad t \geq 0. \tag{A.7}$$

The linear space $\mathrm{SB}(\mathbb{R}_+)^n$ is a Banach space with norm

$$\|\varphi\|_{\mathrm{SB}} = \|\varphi\|_{L^1} + \|\varphi'\|_{L^1}. \tag{A.8}$$

Furthermore, $\mathrm{SB}(\mathbb{R}_+)^n$ is continuously and densely embedded in $L^1(\mathbb{R}_+)^n$. More precisely, the map $j : \mathrm{SB}(\mathbb{R}_+)^n \to L^1(\mathbb{R}_+)^n$ defined by $j\varphi = \varphi$ is a continuous linear map which is one-to-one and has dense range. From (A.8) we see that j is a contraction. We are now ready to state and proof the main theorem of this section.

Theorem A.4. *An operator K from $L^1(\mathbb{R}_+)^p$ to $L^1(\mathbb{R}_+)^q$ is a classical Hankel integral operator if and only if the following two conditions are satisfied:*

(H1) *K maps $\mathrm{SB}(\mathbb{R}_+)^p$ boundedly into $\mathrm{SB}(\mathbb{R}_+)^q$;*

(H2) *there exists $k \in L^1(\mathbb{R}_+)^{q\times p}$ such that $(K\varphi)' + K\varphi' = k(\cdot)\varphi(0)$ for each $\varphi \in \mathrm{SB}(\mathbb{R}_+)^p$.*

Moreover, in that case the operator K is given by

$$(Kf)(t) = \int_0^\infty k(t+s)f(s)\,ds, \quad 0 \leq t < \infty, \tag{A.9}$$

where $k \in L^1(\mathbb{R}_+)^{q\times p}$ is the matrix function from (H2).

Proof. We split the proof into three parts. In the first part we show that the conditions (H1) and (H2) are necessary. The proof is taken from [5], and is given here for the sake of completeness. The second and third part concern the reverse implication which seems to be new. In the second part we assume that $p = q = 1$, and in the third part p and q are arbitrary positive integers.

Part 1. Let K on $L^1(\mathbb{R}_+)$ be a classical Hankel integral operator, and assume K is given by (A.9) with $k \in L^1(\mathbb{R}_+)$. Let $\varphi \in \mathrm{SB}(\mathbb{R}_+)$. Then

$$(K\varphi)(t) = \int_0^\infty k(t+s)\varphi(s)\,ds = \int_t^\infty k(s)\varphi(s-t)\,ds. \tag{A.10}$$

It follows that

$$\begin{aligned}\left(\frac{d}{dt}K\varphi\right)(t) &= -\int_t^\infty k(s)\varphi'(s-t)\,ds + k(t)\varphi(0)\\ &= -\int_0^\infty k(t+s)\varphi'(s)\,ds + k(t)\varphi(0) \qquad \text{(A.11)}\\ &= -\left(K\frac{d}{dt}\varphi\right)(t) + k(t)\varphi(0). \qquad \text{(A.12)}\end{aligned}$$

This proves (H2). From the first identity in (A.10) it follows that $K\varphi$ belongs to $L^1(\mathbb{R}_+)$. Since $\varphi' \in L^1(\mathbb{R}_+)$, we have $(K\varphi)' = K\varphi' + k\varphi(0) \in L^1(\mathbb{R}_+)$ and it follows that $K\varphi \in \mathrm{SB}(\mathbb{R}_+)$. We conclude that K maps $\mathrm{SB}(\mathbb{R}_+)$ into $\mathrm{SB}(\mathbb{R}_+)$. Furthermore, from (A.10) we see that

$$\|K\varphi\|_{L^1} \leq \|k\|_{L^1}\|\varphi\|_{L^1} \leq \|k\|_{L^1}\|\varphi\|_{\mathrm{SB}}.$$

From (A.11) (using $\varphi(0) = -\int_0^\infty \varphi'(s)\,ds$) it follows that

$$\begin{aligned}\|(K\varphi)'\|_{L^1} &\leq \|k\|_{L^1}\|\varphi'\|_{L^1} + \|k\|_{L^1}\|\varphi'\|_{L^1}\\ &\leq 2\|k\|_{L^1}\|\varphi'\|_{L^1} \leq 2\|k\|_{L^1}\|\varphi\|_{\mathrm{SB}}.\end{aligned}$$

Hence $\|K\varphi\|_{\mathrm{SB}} \leq 3\|k\|_{L^1}\|\varphi\|_{\mathrm{SB}}$. Thus $K|_{\mathrm{SB}(\mathbb{R}_+)}$ is a bounded operator on $\mathrm{SB}(\mathbb{R}_+)$, and item (H1) is proved.

PART 2. In this part $p = q = 1$, and we assume that items (H1) and (H2) are satisfied. Given k in item (H2), let H be the operator on $L^1(\mathbb{R}_+)$ defined by

$$(Hf)(t) = \int_0^\infty k(t+s)f(s)\,ds, \quad 0 \leq t < \infty.$$

Then H is a classical Hankel integral operator, and the first part of the proof tells us that $(H\varphi)' + H\varphi' = k(\cdot)\varphi(0)$ for each $\varphi \in \mathrm{SB}(\mathbb{R}_+)$. Now put $M = K - H$. Then M is an operator on $L^1(\mathbb{R}_+)$, and M maps $\mathrm{SB}(\mathbb{R}_+)$ into $\mathrm{SB}(\mathbb{R}_+)$. Furthermore, we have

$$(M\varphi)' = -M\varphi', \quad \varphi \in \mathrm{SB}(\mathbb{R}_+). \tag{A.13}$$

It suffices to prove that M is zero.

For $n = 0, 1, 2, \ldots$ let φ_n be the function on $\mathbb{R}_+$ defined by $\varphi_n(t) = t^n e^{-t}$, $0 \leq t < \infty$. Obviously, $\varphi_n \in \mathrm{SB}(\mathbb{R}_+)$. By induction we shall prove that $M\varphi_n$ is zero for each $n = 0, 1, 2, \ldots$. First we show that $M\varphi_0 = 0$. To do this note that $\varphi_0'(t) = -e^{-t} = -\varphi_0(t)$. Using (A.13) it follows that $\psi_0 := M\varphi_0$ satisfies

$$\psi_0' = (M\varphi_0)' = -M\varphi_0' = M\varphi_0 = \psi_0.$$

Thus ψ_0 satisfies the differential equation $\psi_0' = \psi_0$, and hence $\psi_0(t) = ce^t$ on $[0, \infty)$ for some $c \in \mathbb{C}$. On the other hand, $\psi_0 = M\varphi_0 \in \mathrm{SB}(\mathbb{R}_+) \subset L^1(\mathbb{R}_+)$. But then c must be zero, and we conclude that $M\varphi_0 = 0$.

Next, fix a positive integer $n \geq 1$, and assume $M\varphi_j = 0$ for $j = 0, \ldots, n-1$. Again we use (A.13). Since

$$\varphi_n'(t) = nt^{n-1}e_{-t} - t^n e^{-t} = n\varphi_{n-1} - \varphi_n,$$

we obtain

$$(M\varphi_n)' = -M\varphi_n' = nM\varphi_{n-1} + M\varphi_n.$$

But, by assumption, $M\varphi_{n-1} = 0$. Thus $(M\varphi_n)' = M\varphi_n$, and hence $\psi_n := M\varphi_n$ satisfies the differential equation $\psi_n' = \psi_n$. It follows that $\psi_n(t) = ce^t$ on $[0, \infty)$ for some $c \in \mathbb{C}$. On the other hand, $\psi_n = M\varphi_n \in \mathrm{SB}(\mathbb{R}_+) \subset L^1(\mathbb{R}_+)$. But then $c = 0$, and we conclude that $M\varphi_n = 0$.

By induction we obtain $M\varphi_j = 0$ for each $j = 0, 1, 2, \ldots$. But then $Mf = 0$ for any f of the form $f(t) = p(t)e^{-t}$, where p is a polynomial. The set of all these functions is dense in $L^1(\mathbb{R}_+)$. Since M is an operator on $L^1(\mathbb{R}_+)$, we conclude that $M = 0$.

PART 3. In this part we use the result of the previous part to prove the sufficiency of the conditions (H1) and (H2). Assume K from $L^1(\mathbb{R}_+)^p$ to $L^1(\mathbb{R}_+)^q$, and write K as a $q \times p$ operator matrix

$$K = \begin{bmatrix} K_{11} & \cdots & K_{1p} \\ \vdots & \dots & \vdots \\ K_{q1} & \cdots & K_{qp} \end{bmatrix}, \tag{A.14}$$

where K_{ij} is an operator on $L^1(\mathbb{R}_+)$ for $1 \le j \le p$ and $1 \le i \le q$. let

$$\tau_j : L^1(\mathbb{R}_+) \to L^1(\mathbb{R}_+)^p, \quad \tau_j f = [\delta_{j,k} f]_{k=1}^p \quad (f \in L^1(\mathbb{R}_+);$$

$$\pi_i : L^1(\mathbb{R}_+)^q \to L^1(\mathbb{R}_+), \quad \pi_j f = f_j, \quad (f = \begin{bmatrix} f_1 \\ \vdots \\ f_q \end{bmatrix} \in L^1(\mathbb{R}_+)^q).$$

Note that $K_{ij} = \pi_i K \tau_j$ for each i, j. Furthermore, we have

$$\tau_j \mathrm{SB}(\mathbb{R}_+) \subset \mathrm{SB}(\mathbb{R}_+)^p \quad \text{and} \quad \pi_j \mathrm{SB}(\mathbb{R}_+)^q \subset \mathrm{SB}(\mathbb{R}_+).$$

Now fix a pair i, j, $1 \le j \le p$ and $1 \le i \le q$. Then conditions (H1) and (H2) tell us that

(i) K_{ij} maps $\mathrm{SB}(\mathbb{R}_+)$ into $\mathrm{SB}(\mathbb{R}_+)$ and $K_{ij}|_{\mathrm{SB}(\mathbb{R}_+)}$ is a bounded operator from $\mathrm{SB}(\mathbb{R}_+)$ to $\mathrm{SB}(\mathbb{R}_+)$;
(ii) there exists $k_{ij} \in L^1(\mathbb{R}_+)$ such that $(K_{ij}\varphi)' + K_{ij}\varphi' = k_{ij}(\cdot)\varphi(0)$ for each $\varphi \in \mathrm{SB}(\mathbb{R}_+)$.

But then we can use the result of the second part of the proof which covers the case when $p = q = 1$. It follows that K_{ij} is a classical Hankel integral operator. Moreover, K_{ij} is given by

$$(K_{ij} f)(t) = \int_0^\infty k_{ij}(t+s) f(s)\, ds, \quad 0 \le t < \infty \quad \text{and} \quad f \in L^1(\mathbb{R}_+).$$

Here $k_{ij} \in L^1(\mathbb{R}_+)$ is the function appearing in item (ii) above. Recall that K is given by (A.14). Since the pair i, j is arbitrary, we see that K is a classical Hankel integral operator, and

$$(Kf)(t) = \int_0^\infty k(t+s) f(s)\, ds, \quad 0 \le t < \infty \quad \text{and} \quad f \in L^1(\mathbb{R}_+)^p,$$

where

$$k := \begin{bmatrix} k_{11} & \cdots & k_{1p} \\ \vdots & \dots & \vdots \\ k_{q1} & \cdots & k_{qp} \end{bmatrix} \in L^1(\mathbb{R}_+)^{q \times p}.$$

This completes the proof. □

The following corollary shows that if the operator K in Theorem A.4 is assumed to be a Hankel operator, i.e., $K = H(\alpha)$ for some $\alpha \in L^\infty(\mathbb{R})^{q \times p}$, then it suffices to verify (H1) to conclude that $H(\alpha)$ is a classical Hankel operator.

Corollary A.5. *Let $\alpha \in L^\infty(\mathbb{R})^{q\times p}$, and assume that $H(\alpha)$ maps $L^1(\mathbb{R}_+)^p$ into $L^1(\mathbb{R}_+)^q$. Furthermore, assume that $K = H(\alpha)$ satisfies condition* (H1) *in Theorem* A.4*, i.e., $H(\alpha)$ maps $\mathrm{SB}(\mathbb{R}_+)^p$ into $\mathrm{SB}(\mathbb{R}_+)^q$ and the operator $H(\alpha)|_{\mathrm{SB}(\mathbb{R}_+)^p}$ is a bounded operator from $\mathrm{SB}(\mathbb{R}_+)^p$ into $\mathrm{SB}(\mathbb{R}_+)^q$. Then $K = H(\alpha)$ also satisfies condition* (H2) *in Theorem* A.4*, and thus there exists $k \in L^1(\mathbb{R}_+)^{q\times p}$ such that*

$$\alpha(\lambda) = \int_{\mathbb{R}} e^{i\lambda s} k(s)\, ds, \quad \lambda \in \mathbb{R}. \tag{A.15}$$

In particular, $H(\alpha)$ is a classical Hankel integral operator.

Proof. We split the proof into two parts. In the first part we assume that $p = q = 1$. In the second part p and q are arbitrary positive integers, and we reduce the problem to the case considered in the first part.

PART 1. In this part we prove the theorem for the case when $p = q = 1$. Note that we assume that condition (H1) in Theorem A.4 is satisfied for $H(\alpha)$ in place of K. Take $\varphi \in \mathrm{SB}(\mathbb{R}_+)$. Since $H(\alpha)$ maps $\mathrm{SB}(\mathbb{R}_+)$ into $\mathrm{SB}(\mathbb{R}_+)$, the function $H(\alpha)\varphi$ also belongs to $\mathrm{SB}(\mathbb{R}_+)$, and hence $(H(\alpha)\varphi)'$ belongs to $L^1(\mathbb{R}_+)$. On the other hand, since $H(\alpha)$ maps $L^1(\mathbb{R}_+)$ into $L^1(\mathbb{R}_+)$ and $\varphi' \in L^1(\mathbb{R}_+)$, we also have $H(\alpha)\varphi' \in L^1(\mathbb{R}_+)$. Hence

$$\frac{d}{dt}(H(\alpha)\varphi) + H(\alpha)\frac{d}{dt}\varphi \in L^1(\mathbb{R}_+).$$

Using (A.2) we obtain

$$\begin{aligned}\frac{d}{dt}\Big(H(\alpha)\varphi(t)\Big) &= \frac{1}{2\pi}\int_{\mathbb{R}} (-is)e^{-its}\alpha(s)\left(\int_0^\infty e^{-isr}\varphi(r)\,dr\right) ds\\ &= \frac{1}{2\pi}\int_{\mathbb{R}} e^{-its}\alpha(s)\left(\int_0^\infty \left(\frac{d}{dr}e^{-isr}\right)\varphi(r)\,dr\right) ds,\end{aligned} \tag{A.16}$$

and

$$\Big(H(\alpha)\frac{d}{dt}\varphi\Big)(t) = \frac{1}{2\pi}\int_{\mathbb{R}} e^{-its}\alpha(s)\left(\int_0^\infty e^{-isr}\varphi'(r)\,dr\right) ds.$$

Now integration by parts yields

$$\begin{aligned}\int_0^\infty \left(\frac{d}{dr}e^{-isr}\right)\varphi(r)\,dr &= -\int_0^\infty e^{-isr}\varphi'(r)\,dr + \Big(e^{-isr}\varphi(r)\Big|_0^\infty\Big)\\ &= -\int_0^\infty e^{-isr}\varphi'(r)\,dr - \varphi(0).\end{aligned}$$

Hence

$$\int_0^\infty \left(\frac{d}{dr}e^{-isr}\right)\varphi(r)\,dr + \int_0^\infty e^{-isr}\varphi'(r) = -\varphi(0).$$

This implies that

$$-\frac{\varphi(0)}{2\pi}\int_{\mathbb{R}} e^{-its}\alpha(s)\,ds = \left(\frac{d}{dt}(H(\alpha)\varphi) + H(\alpha)\frac{d}{dt}\varphi\right)(t), \quad t \geq 0. \tag{A.17}$$

In case $\varphi(0) \neq 0$, it follows that (H2) holds for $K = H(\alpha)$ with $k \in L^1(\mathbb{R}_+)$ given by

$$k(t) := \frac{1}{\varphi(0)} \left(\frac{d}{dt}(H(\alpha)\varphi) + H(\alpha)\frac{d}{dt}\varphi \right)(t) = -\frac{1}{2\pi} \int_{\mathbb{R}} e^{-its}\alpha(s)\, ds,$$

which is independent of the choice of φ. On the other hand, in case $\varphi(0) = 0$, then (A.17) shows that (H2) still holds for this choice of k. We can thus use Theorem A.4 with $K = H(\alpha)$ to conclude that $H(\alpha)$ is a classical Hankel integral operator and α is defined by (A.15).

Part 2. In this part p and q are arbitrary positive integers. Since $\alpha \in L^\infty(\mathbb{R})^{q\times p}$, the function α is a $q \times p$ matrix function of which the (i,j)th entry $\alpha_{i,j}$ belongs to $L^\infty(\mathbb{R}_+)$. It follows that

$$K = H(\alpha) = \begin{bmatrix} H(\alpha_{11}) & \cdots & H(\alpha_{1p}) \\ \vdots & \cdots & \vdots \\ H(\alpha_{q1}) & \cdots & H(\alpha_{qp}) \end{bmatrix}.$$

Put $K_{ij} = H(\alpha_{ij})$, where $1 \leq j \leq p$ and $1 \leq i \leq q$. Now fix (i,j). Since K maps $L^1(\mathbb{R}_+)^p$ into $L^1(\mathbb{R}_+)^q$, the operator K_{ij} maps $L^1(\mathbb{R}_+)$ into $L^1(\mathbb{R}_+)$. Furthermore, since $K = H(\alpha)$ satisfies condition (H1) in Theorem A.4, the operator $K_{ij} = H(\alpha_{ij})$ satisfies condition (H1) in Theorem A.4 with $p = q = 1$. But then the result of the first part of the proof tells us that $K_{ij} = H(\alpha_{ij})$ satisfies condition (H2) in Theorem A.4 with $p = q = 1$. Thus, using (A.15), there exists $k_{ij} \in L^1(\mathbb{R}_+)$ such that

$$\alpha_{ij} = \int_{\mathbb{R}} e^{i\lambda s} k_{ij}(s)\, ds, \quad \lambda \in \mathbb{R}.$$

The latter holds for each $1 \leq j \leq p$ and $1 \leq i \leq q$. It follows that

$$\alpha(\lambda) = \int_{\mathbb{R}} e^{i\lambda s} k(s)\, ds, \text{ where } k = \begin{bmatrix} k_{11} & \cdots & k_{1p} \\ \vdots & \cdots & \vdots \\ k_{q1} & \cdots & k_{qp} \end{bmatrix} \in L^1(\mathbb{R}_+)^{q\times p}.$$

This completes the proof. □

A.3. Two auxiliary results

We present two lemmas concerning condition (H1) in Theorem A.4. We begin with some preliminaries. Let

$$M = I + H_{11}H_{12} + H_{21}H_{22}, \tag{A.18}$$

where

$$H_{11} : L^1(\mathbb{R}_+)^q \to L^1(\mathbb{R}_+)^p, \quad H_{12} : L^1(\mathbb{R}_+)^p \to L^1(\mathbb{R}_+)^q,$$

$$H_{21} : L^1(\mathbb{R}_+)^r \to L^1(\mathbb{R}_+)^p, \quad H_{22} : L^1(\mathbb{R}_+)^p \to L^1(\mathbb{R}_+)^r,$$

and we assume that H_{ij} is a classical Hankel integral operator for each $1 \leq i, j \leq 2$. We are interested in computing the inverse of M, assuming the inverse exists. Put $\widetilde{M} = I + \widetilde{H}_1 \widetilde{H}_2$, where

$$\widetilde{H}_1 = \begin{bmatrix} H_{11} & H_{21} \end{bmatrix} : \begin{bmatrix} L^1(\mathbb{R}_+)^q \\ L^1(\mathbb{R}_+)^r \end{bmatrix} \to L^1(\mathbb{R}_+)^p,$$

$$\widetilde{H}_2 = \begin{bmatrix} H_{12} \\ H_{22} \end{bmatrix} : L^1(\mathbb{R}_+)^p \to \begin{bmatrix} L^1(\mathbb{R}_+)^q \\ L^1(\mathbb{R}_+)^r \end{bmatrix}.$$

Note that the entries of $\widetilde{H}_1$ and $\widetilde{H}_2$ are classical Hankel integral operators, and

$$\widetilde{M} = I + \widetilde{H}_1 \widetilde{H}_2 = I + H_{11}H_{12} + H_{21}H_{22} = M.$$

It follows that M is invertible if and only if $\widetilde{M}$ is invertible, and in that case

$$\widetilde{M}^{-1} = M^{-1}. \tag{A.19}$$

Theorem 0.1 in [9] tells us how to compute $\widetilde{M}^{-1}$. This yields the following result.

Lemma A.6. *Assume M given by* (A.18) *is invertible. Then*

$$M^{-1} = I + K_1 + K_2 + K_3 + K_4, \tag{A.20}$$

where for each $j = 1, 2, 3, 4$ the operator K_j is a product of two classical Hankel integral operators. In particular, $M^{-1} = I + K$, where K is an operator on $L^1(\mathbb{R}_+)^p$ satisfying condition (H1) *in Theorem* A.4.

Proof. From Theorem 0.1 in [9] we know that

$$\widetilde{M}^{-1} = I + AB + CD, \tag{A.21}$$

where the operators A, B, C, D have the following operator matrix representation:

$$A = \begin{bmatrix} A_{11} & A_{12} \end{bmatrix}, \ B = \begin{bmatrix} B_{11} \\ B_{21} \end{bmatrix}, \ C = \begin{bmatrix} C_{11} & C_{12} \end{bmatrix}, \ D = \begin{bmatrix} D_{11} \\ D_{21} \end{bmatrix}.$$

and for each $i, j = 1, 2$ the entries A_{ij}, B_{ij}, C_{ij}, D_{ij} are classical Hankel integral operators. Using (A.19) and (A.21) it follows that

$$M^{-1} = \widetilde{M}^{-1} = I + \begin{bmatrix} A_{11} & A_{12} \end{bmatrix} \begin{bmatrix} B_{11} \\ B_{21} \end{bmatrix} + \begin{bmatrix} C_{11} & C_{12} \end{bmatrix} \begin{bmatrix} D_{11} \\ D_{21} \end{bmatrix}.$$

Thus (A.20) holds true with

$$K_1 = A_{11}B_{11}, \quad K_2 = A_{12}B_{21}, \quad K_3 = C_{11}D_{11}, \quad K_4 = C_{12}D_{22}.$$

Clearly for each $j = 1, 2, 3, 4$ the operator K_j is a product of two classical Hankel integral operators. Recall that for each classical Hankel integral operator H from $L^1(\mathbb{R}_+)^n$ to $L^1(\mathbb{R}_+)^n$ for some n and m we have H maps $\mathrm{SB}(\mathbb{R}_+)^n$ into $\mathrm{SB}(\mathbb{R}_+)^m$ and $H|_{\mathrm{SB}(\mathbb{R}_+)^n}$ is bounded as an operator from $\mathrm{SB}(\mathbb{R}_+)^n$ into $\mathrm{SB}(\mathbb{R}_+)^m$. It follows that the same is true if H is a sum or a product of classical Hankel integral operators. But then condition (H1) in Theorem A.4 is satisfied for $K = K_1 + K_2 + K_3 + K_4$. □

Lemma A.7. *Let $\tau \in L^1(\mathbb{R}_-)^{q\times p}$, and let T be the Wiener–Hopf integral operator mapping $L^1(\mathbb{R}_+)^p$ into $L^1(\mathbb{R}_+)^q$ defined by*

$$(Tf)(t) = \int_t^\infty \tau(t-s)f(s)\,ds, \quad 0 \le t < \infty \quad (f \in L^1(\mathbb{R}_+)^p). \tag{A.22}$$

Then T maps $\mathrm{SB}(\mathbb{R}_+)^p$ into $\mathrm{SB}(\mathbb{R}_+)^q$, and $T|_{\mathrm{SB}(\mathbb{R}_+)^p}$ is a bounded linear operator from $\mathrm{SB}(\mathbb{R}_+)^p$ into $\mathrm{SB}(\mathbb{R}_+)^q$.

Proof. We split the proof into two parts. In the first part we prove the lemma for the case when $p = q = 1$. In the second p and q are arbitrary positive integers, and we reduce the problem to the case considered in the first part.

PART 1. In this part we prove the lemma for the case when $p = q = 1$. To do this take $\varphi \in \mathrm{SB}(\mathbb{R}_+)$. Then

$$\begin{aligned}(T\varphi)(t) &= -\int_t^\infty \tau(t-s)\left(\int_s^\infty \varphi'(r)\,dr\right) ds\\ &= -\int_t^\infty \left(\int_t^r \tau(t-s)\,ds\right)\varphi'(r)\,dr, \quad 0 \le t < \infty.\end{aligned}$$

Put

$$\rho(r-t) := \int_0^{r-t} \tau(-u)\,du = \int_t^r \tau(t-s)\,ds, \quad 0 \le t \le r < \infty.$$

Note that $\rho(0) = 0$. Furthermore,

$$(T\varphi)(t) = -\int_t^\infty \rho(r-t)\varphi'(r)\,dr, \quad \text{and} \quad \rho'(t) = \tau(-t) \quad (0 \le t < \infty). \tag{A.23}$$

Using (A.22) with $f = \varphi$ we see that $\psi := T\varphi$ belongs to $L^1(\mathbb{R}_+)$. Furthermore, from the first identity in (A.23) it follows that ψ is absolutely continuous on compact intervals of $\mathbb{R}_+$. Using the Leibnitz rule and the second identity in (A.23), we obtain

$$\begin{aligned}\psi'(t) &= -\frac{d}{dt}\int_t^\infty \rho(r-t)\varphi'(r)\,dr\\ &= -\int_t^\infty \frac{\partial}{\partial t}\rho(r-t)\varphi'(r)\,dr + \rho(t-t)\varphi'(t)\\ &= -\int_t^\infty \tau(t-r)\varphi'(r)\,dr.\end{aligned} \tag{A.24}$$

Since $\tau \in L^1(\mathbb{R}_-)$ and $\varphi' \in L^1(\mathbb{R}_+)$, it follows that $\psi' \in L^1(\mathbb{R}_+)$. We conclude that ψ belongs to $\mathrm{SB}(\mathbb{R}_+)$.

It remains to show that $T|_{\mathrm{SB}(\mathbb{R}_+)}$ is bounded on $\mathrm{SB}(\mathbb{R}_+)$. As before let $\varphi \in L^1(\mathbb{R}_+)$, and let $\psi = T\varphi$. By $\|T\|$ we denote the norm of T as an operator on $L^1(\mathbb{R}_+)$. From the definition of T in (A.22) and using (A.8), we see that

$$\|\psi\|_{L^1} \le \|T\|\|\varphi\|_{L^1} \le \|T\|\|\varphi\|_{\mathrm{SB}}$$

On the other hand from (A.24) and using (A.8) we obtain

$$\|\psi'\|_{L^1} \leq \|T\|\|\varphi'\|_{L^1} \leq \|T\|\|\varphi\|_{\mathrm{SB}}.$$

Together these inequalities show (using (A.8)) that $\|\psi\|_{\mathrm{SB}} \leq \|T\|\|\varphi\|_{\mathrm{SB}}$. Thus $\|T\|_{\mathrm{SB}(\mathbb{R}_+)} \leq \|T\|$. This proves the lemma for the case when $p = q = 1$.

PART 2. In this part p and q are arbitrary positive integers. Since $\tau \in L^1(\mathbb{R}_-)^{q\times p}$, the function τ is a $q \times p$ matrix function of which the (i,j)th entry τ_{ij} belongs to $L^1(\mathbb{R}_-)$. It follows that

$$T = \begin{bmatrix} T_{11} & \cdots & T_{1p} \\ \vdots & \cdots & \vdots \\ T_{q1} & \cdots & T_{qp} \end{bmatrix}, \tag{A.25}$$

where for $1 \leq j \leq p$ and $1 \leq i \leq q$ the operator T_{ij} is the Wiener–Hopf integral operator on $L^1(\mathbb{R}_+)$ given by

$$(T_{ij})f(t) = \int_t^\infty \tau_{ij}(t-s)f(s)\,ds, \quad 0 \leq t < \infty \quad (f \in L^1(\mathbb{R}_+)).$$

From the first part of the proof we know that for each i, j the operator T_{ij} maps $\mathrm{SB}(\mathbb{R}_+)$ into $\mathrm{SB}(\mathbb{R}_+)$, and $T|_{\mathrm{SB}(\mathbb{R}_+)}$ is a bounded linear operator from $\mathrm{SB}(\mathbb{R}_+)$ into $\mathrm{SB}(\mathbb{R}_+)$. Now recall that T is given by (A.25). It follows that T maps $\mathrm{SB}(\mathbb{R}_+)^p$ into $\mathrm{SB}(\mathbb{R}_+)^q$, and $T|_{\mathrm{SB}(\mathbb{R}_+)^p}$ is a bounded linear operator from $\mathrm{SB}(\mathbb{R}_+)^p$ into $\mathrm{SB}(\mathbb{R}_+)^q$, which completes the proof. □

Acknowledgement. This work is based on the research supported in part by the National Research Foundation of South Africa. Any opinion, finding and conclusion or recommendation expressed in this material is that of the authors and the NRF does not accept any liability in this regard.

References

[1] H. Bart, I. Gohberg, M.A. Kaashoek, and A.C.M. Ran, *Factorization of matrix and operator functions: the state space method*, Oper. Theory Adv. Appl. **178**, Birkhäuser Verlag, Basel, 2008.

[2] A. Böttcher and B. Silbermann, *Analysis of Toeplitz operators*, Springer Verlag, 2006.

[3] N. Dunford and J.T. Schwartz, *Linear Operators. Part I: General Theory*, Interscience Publishers, INC., New York, 1964.

[4] R.L. Ellis and I. Gohberg, Orthogonal systems related to infinite Hankel matrices, *J. Funct. Analysis* **109** (1992), 155–198.

[5] R.L. Ellis and I. Gohberg, *Orthogonal systems and convolution operators*, Oper. Theory Adv. Appl. **140**, Birkhäuser Verlag, Basel, 2003.

[6] I. Gohberg, S. Goldberg, and M.A. Kaashoek, *Classes of Linear Operators II*, Birkhäuser Verlag, Basel, 1993.

[7] I. Gohberg and G. Heinig, On matrix valued integral operators on a finite interval with matrix kernels that depend on the difference of arguments, *Rev. Roumaine Math. Pures Appl.* **20** (1975), 55–73, (in Russian).

[8] I. Gohberg, M.A. Kaashoek, and H.J. Woerdeman, The band method for positive and contractive extension problems, *J. Operator Theory* **22** (1989), 109–155.

[9] G.J. Groenewald and M.A. Kaashoek, A Gohberg–Heinig type inversion formula involving Hankel operators, in: *Interpolation, Schur functions and moment problems*, Oper. Theory Adv. Appl. **165**, Birkhäuser Verlag, Basel, 2005, pp. 291–302.

[10] S. ter Horst, M.A. Kaashoek, and F. van Schagen, The discrete twofold Ellis–Gohberg inverse problem, *J. Math. Anal. Appl.* **452** (2017), 846–870.

[11] M.A. Kaashoek and L. Lerer, The band method and inverse problems for orthogonal matrix functions of Szegő–Kreĭn type, *Indag. Math.* **23** (2012), 900–920.

[12] M.A. Kaashoek and F. van Schagen, Inverting structured operators related to Toeplitz plus Hankel operators, in: *Advances in Structured Operator Theory and Related Areas. The Leonid Lerer Anniversary Volume*, Oper. Theory Adv. Appl. **237**, Birkhäuser Verlag, Basel, 2012, pp. 161–187.

[13] M.A. Kaashoek and F. van Schagen, Ellis–Gohberg identities for certain orthogonal functions II: Algebraic setting and asymmetric versions, *Math. Proc. R. Ir. Acad.*, **113A** (2) (2013), 107–130.

[14] M.A. Kaashoek and F. van Schagen, The inverse problem for Ellis–Gohberg orthogonal matrix functions, *Integral Equ. Oper. Theory* **80** (2014), 527–555.

[15] M.A. Kaashoek and F. van Schagen, The Ellis–Gohberg inverse problem for matrix-valued Wiener functions on the line, *Oper. Matrices* **10** (2016), 1009–1042.

[16] N. Nikolski, *Matrices et opérateurs de Toeplitz*, Calvage et Mounet, Paris, 2017.

[17] V. Peller, *Hankel operators and their applications*, Springer-Verlag, New York, Inc., 2003.

[18] C.E. Rickart, *General theory of Banach algebras*, Robert E. Kriger Publ. Co, New York, 1960.

[19] S. Roch and B. Silbermann, Toeplitz-like operators, quasicommutator ideals, numerical analysis I, *Math. Nachr.* **120** (1985), 141–173.

[20] S. Roch and B. Silbermann, Toeplitz and Hankel algebras – axiomatic and asymptotic aspects, *Operator Theory, Operator Algebras, and Matrix Theory*, Oper. Theory Adv. Appl. **267**, Birkhäuser Verlag, Basel, 2018, pp. 285–315.

S. ter Horst
Department of Mathematics
Unit for BMI, North West University
Potchefstroom, 2531 South Africa
e-mail: `Sanne.TerHorst@nwu.ac.za`

M.A. Kaashoek and F. van Schagen
Department of Mathematics, Vrije Universiteit Amsterdam
De Boelelaan 1081a
1081 HV Amsterdam, The Netherlands
e-mail: `m.a.kaashoek@vu.nl`
`f.van.schagen@vu.nl`

Operator Theory:
Advances and Applications, Vol. 272, 213–234

Abstract Interpolation Problem and Some Applications. II: Coefficient Matrices

A. Kheifets

Dedicated to Joe Ball on occasion of his 70th birthday

Abstract. The main content of this paper is Lectures 5 and 6 that continue lecture notes [20]. Content of Lectures 1–4 of [20] is reviewed for the reader's convenience in Sections 1–4, respectively. It is shown in Lecture 5 how residual parts of the minimal unitary extensions, that correspond to solutions of the problem, yield some boundary properties of the coefficient matrix-function. These results generalize the classical Nevanlinna–Adamjan–Arov–Kreĭn theorem. Lecture 6 discusses how further properties of the coefficient matrices follow from denseness of certain sets in the associated function model spaces. The structure of the dense set reflects the structure of the problem data.

Mathematics Subject Classification (2010). 47A20, 47A57, 30E05.

Keywords. Isometry; minimal unitary extension; residual part; de Branges–Rovnyak function space; dense set; coefficient matrix.

1. Abstract Interpolation Problem

1.1. Data of the problem

Data of the Abstract Interpolation Problem consists of the following components: a vector space X (without any topology), a positive-semidefinite sesquilinear form D on X, linear operators T_1 and T_2 on X, separable Hilbert spaces E_1 and E_2, linear mappings M_1 and M_2 from X to E_1 and E_2 respectively. The data pieces are connected by the following identity

$$D(T_2x, T_2y) - D(T_1x, T_1y) = \langle M_1x, M_1y\rangle_{E_1} - \langle M_2x, M_2y\rangle_{E_2}. \tag{1.1}$$

Let $\mathbb{D}$ be the open unit disc, $|\zeta| < 1$, and let $\mathbb{T}$ be the unit circle, $|\zeta| = 1$. Let $w(\zeta) : E_1 \to E_2$ be a contraction for every $\zeta \in \mathbb{D}$ and assume that $w(\zeta)$ is analytic in variable ζ. Functions of this type are called the Schur class (operator-valued) functions.

1.2. de Branges–Rovnyak function space

Let $L^2(E_2 \oplus E_1)$ be the space of vector functions on the unit circle $\mathbb{T}$ that are square summable against the Lebesgue measure. Space L^w is defined as the range of $L^2(E_2 \oplus E_1)$ under

$$\begin{bmatrix} I_{E_2} & w(t) \\ w(t)^* & I_{E_1} \end{bmatrix}^{1/2} \tag{1.2}$$

endowed with the range norm. The de Branges–Rovnyak space H^w is defined as a subspace of L^w that consists of functions

$$f = \begin{bmatrix} f_2 \\ f_1 \end{bmatrix} \in L^w, \quad f_2 \in H^2_+(E_2), \quad f_1 \in H^2_-(E_1).$$

1.3. Setting of the problem

A Schur class function $w : E_1 \to E_2$ is said to be a solution of the AIP with data (1.1), if there exists a linear mapping $F : X \to H^w$ such that for all $x \in X$

$$\text{i)} \qquad \|Fx\|^2_{H^w} \leq D(x,x); \tag{1.3}$$

$$\text{ii)} \qquad tFT_2x - FT_1x = \begin{bmatrix} I_{E_2} & w(t) \\ w(t)^* & I_{E_1} \end{bmatrix} \begin{bmatrix} -M_2x \\ M_1x \end{bmatrix}, \text{ a.e. } t \in \mathbb{T}. \tag{1.4}$$

One can write Fx as a vector of two components

$$Fx = \begin{bmatrix} F_+x \\ F_-x \end{bmatrix}$$

which are E_2- and E_1-valued, respectively. Then conditions $Fx \in H^w$ and (i) read as

$$\begin{aligned} &\text{(a)} \qquad F_+x \in H^2_+(E_2), \\ &\text{(b)} \qquad F_-x \in H^2_-(E_1), \\ &\text{(c)} \qquad \|Fx\|^2_{L^w} \leq D(x,x). \end{aligned}$$

Sometimes we will call the pair (w, F) a solution of the Abstract Interpolation Problem.

1.4. Special case

The following additional assumption on operators T_1 and T_2 is met in many concrete problems: the operators

$$(\zeta T_2 - T_1)^{-1} \text{ and } (T_2 - \overline{\zeta} T_1)^{-1} \tag{1.5}$$

exist for all $\zeta \in \mathbb{D}$ except for a discrete set. In this case condition (ii) can be written as explicit formulae for F_+ and F_-

$$(F^w_+x)(\zeta) = (w(\zeta)M_1 - M_2)(\zeta T_2 - T_1)^{-1}x\ , \tag{1.6}$$

$$(F^w_-x)(\zeta) = \overline{\zeta}(M_1 - w(\zeta)^*M_2)(T_2 - \overline{\zeta}T_1)^{-1}x\ . \tag{1.7}$$

From here one can see that under assumptions (1.5), for every solution w there exists only one F that satisfies (ii).

References to this section are [12, 13, 15, 24].

2. Examples

2.1. The Nevanlinna–Pick interpolation problem

In this section we recall several classical problems of analysis that can be included in the AIP scheme.

Problem 2.1. Let $\zeta_1, \ldots, \zeta_n, \ldots$ be a finite or infinite sequence of points in the unit disk $\mathbb{D}$; let $w_1, \ldots, w_n, \ldots$ be a sequence of complex numbers. One is interested in describing all the Schur class functions w such that

$$w(\zeta_k) = w_k. \tag{2.1}$$

The well-known solvability criterion for this problem is:

$$\left[\frac{1-\overline{w}_k w_j}{1-\overline{\zeta}_k \zeta_j}\right]_{k,j=1}^{n} \geq 0, \quad \text{for all} \quad n \geq 1. \tag{2.2}$$

We specify data of the Abstract Interpolation Problem (1.1) as follows: the space X consists of all sequences

$$x = \begin{bmatrix} x_1 \\ \vdots \\ x_n \\ \vdots \end{bmatrix} \tag{2.3}$$

that have a finite number of nonzero components;

$$D(x,y) = \sum_{k,j} \overline{y}_k \, \frac{1-\overline{w}_k w_j}{1-\overline{\zeta}_k \zeta_j} \, x_j \,, \quad x, y \in X;$$

$$T_1 = \begin{bmatrix} \zeta_1 & & & \\ & \ddots & & \\ & & \zeta_n & \\ & & & \ddots \end{bmatrix}, \quad T_2 = I_X; \tag{2.4}$$

$$E_1 = E_2 = \mathbb{C}^1; \quad M_1 x = \sum_j x_j, \quad M_2 x = \sum_j w_j x_j.$$

The operators T_1 and T_2 meet the special case assumption (1.5). Therefore, for every solution w there is only one corresponding mapping F^w that can be written in form (1.6), (1.7). It can be further explicitly computed as

$$(F^w_+x)(\zeta) = \sum\nolimits_j \frac{w(\zeta) - w_j}{\zeta - \zeta_j} x_j,$$

$$(F^w_-x)(\zeta) = \overline{\zeta} \sum\nolimits_j \frac{1 - \overline{w(\zeta)} w_j}{1 - \overline{\zeta}\zeta_j} x_j.$$

Since w has non-tangential boundary values, the function $F^w x$ extends to the boundary of the unit disk. Since for $|t| = 1$ we have $\overline{t} = 1/t$, $(F^w x)(t)$ further simplifies as follows:

$$(F^w x)(t) = \begin{bmatrix} 1 & w(t) \\ w(t)^* & 1 \end{bmatrix} \begin{bmatrix} -\sum_j \frac{w_j x_j}{t - \zeta_j} \\ \sum_j \frac{x_j}{t - \zeta_j} \end{bmatrix}, \qquad |t| = 1.$$

Theorem 2.2. *The solution set of the Nevanlinna–Pick problem coincides with the solution set of the Abstract Interpolation Problem with data* (2.3)–(2.4). *Moreover, for data of this type, inequality* (1.3) *turns into equality*

$$\|F^w x\|^2_{H^w} = D(x, x)$$

for every solution w and for every $x \in X$.

This example can be viewed as a special case of the one in the next subsection.

2.2. The Sarason problem

Problem 2.3. Let H^2_+ be the Hardy space of the unit disk. Let θ be an inner function, $K_\theta = H^2_+ \ominus \theta H^2_+$, $T^*_\theta x = P_+\overline{t}x$ $(x \in K_\theta)$, W^* be a contractive operator on K_θ that commutes with T^*_θ : $W^* T^*_\theta = T^*_\theta W^*$. Find all the Schur class functions w such that

$$W^* x = P_+ \overline{w} x \ .$$

We specify here the data of the Abstract Interpolation problem as follows: $X = K_\theta$;

$$\begin{gathered} D(x, x) = \|x\|^2_{K_\theta} - \|W^* x\|^2_{K_\theta}, x \in X; \\ T_1 = I_{K_\theta}, \quad T_2 = T^*_\theta; \\ E_1 = E_2 = \mathbb{C}^1, \quad M_1 x = (W^* x)(0), \quad M_2 x = x(0), \end{gathered} \tag{2.5}$$

where the latter notation stands for the value of an H^2_+ function at 0.

The operators T_1 and T_2 meet the special case assumption (1.5). Therefore, for every solution w there is only one corresponding mapping F^w that can be explicitly computed as

$$F^w x = \begin{bmatrix} 1 & w \\ \overline{w} & 1 \end{bmatrix} \begin{bmatrix} x \\ -W^* x \end{bmatrix},$$

(when variable is on $\mathbb{T}$).

Theorem 2.4. *The solution set of the Sarason Problem* 2.3 *coincides with the solution set of the Abstract Interpolation Problem with data* (2.5). *Moreover, for data* *of this type, inequality* (1.3) *turns into equality*

$$\|F^w x\|^2_{H^w} = D(x,x)$$

for every solution w and for every $x \in X$.

Remark 2.5. Let $\zeta_1, \dots, \zeta_n, \dots$ be a finite or infinite sequence of points in the unit disk $\mathbb{D}$; let $w_1, \dots, w_n, \dots$ be a sequence of complex numbers such that (2.2) holds. Let θ be the Blaschke product with zeros ζ_k, if the latter satisfy the Blaschke condition and $\theta = 0$ otherwise. Note that

$$\frac{1}{1 - t\overline{\zeta}_k} \in K_\theta.$$

We define

$$W^* \frac{1}{1 - t\overline{\zeta}_k} = \frac{\overline{w}_k}{1 - t\overline{\zeta}_k}.$$

W^* extends by linearity to a dense set in K_θ and further, due to (2.2), to a contraction on the whole K_θ. The set of solutions w of the Sarason Problem 2.3 with this θ and this W coincides with the set of solutions of the Nevanlinna–Pick Problem 2.1. However, the data of the Abstract Interpolation Problem in (2.5) differ from the ones in (2.3)–(2.4). Moreover, the coefficient matrices S in the description formula (4.16) for the solution sets are different and the associated universal colligations (4.5)–(4.7) $\mathcal{A}_0$ are non-equivalent.

References to this problem are [13, 23].

2.3. The boundary interpolation problem

Definition 2.6. A Schur class function w defined on the unit disk $\mathbb{D}$ is said to have an angular derivative in the sense of Carathéodory at a point $t_0 \in \mathbb{T}$ if there exists a nontangential unimodular limit

$$w_0 = \lim_{\zeta \to t_0} w(\zeta), \quad |w_0| = 1,$$

and there exists a nontangential limit

$$w_0' = \lim_{\zeta \to t_0} \frac{w(\zeta) - w_0}{\zeta - t_0}.$$

Theorem 2.7 (Carathéodory–Julia). *A Schur class function $w(\zeta)$ has an angular derivative at $t_0 \in \mathbb{T}$ if and only if*

$$D_{w,t_0} \overset{\text{def}}{=} \liminf_{\zeta \to t_0} \frac{1 - |w(\zeta)|^2}{1 - |\zeta|^2} < \infty \tag{2.6}$$

(here $|\zeta| < 1$, $\zeta \to t_0$ in an arbitrary way). In this case

$$w_0' = D_{w,t_0} \cdot \frac{w_0}{t_0} \qquad \text{and} \qquad \frac{1 - |w(\zeta)|^2}{1 - |\zeta|^2} \to D_{w,t_0}$$

as ζ goes to t_0 nontangentially. Note that $D_{w,t_0} \geq 0$ and $D_{w,t_0} = 0$ if and only if $w(\zeta)$ is a constant of modulus 1.

Theorem 2.8. *A Schur class function w has an angular derivative in the sense of Carathéodory at a point $t_0 \in \mathbb{T}$ if and only if there exists a unimodular constant w_0 such that*

$$\left|\frac{w(t)-w_0}{t-t_0}\right|^2 + \frac{1-|w(t)|^2}{|t-t_0|^2} \in L^1 \tag{2.7}$$

against the Lebesgue measure $m(dt)$ on $\mathbb{T}$. In this case

$$\int_{\mathbb{T}} \left(\left|\frac{w(t)-w_0}{t-t_0}\right|^2 + \frac{1-|w(t)|^2}{|t-t_0|^2}\right) m(dt) = D_{w,t_0},$$

where D_{w,t_0} is the same as in (2.6)*. In particular,* (2.7) *implies that*

$$\frac{w-w_0}{t-t_0} \in H_+^2.$$

Problem 2.9. Let t_0 be a point on the unit circle $\mathbb{T}$, let w_0 be a complex number, $|w_0| = 1$, and let $0 \leq D < \infty$ be a given non-negative number. One wants to describe all the Schur class functions w such that

$$w(\zeta) \to w_0 \quad \text{as} \quad \zeta \to t_0$$

nontangentially, and

$$D_{w,t_0} \leq D.$$

We specify here the data of the Abstract Interpolation Problem as follows:

$$\begin{aligned} X = \mathbb{C}^1, \quad D(x,x) &= \overline{x}Dx, \\ T_1 x = t_0 x, \quad T_2 x &= x, \\ E_1 = E_2 &= \mathbb{C}^1, \\ M_1 x = x, \quad M_2 x &= w_0 x. \end{aligned} \tag{2.8}$$

The operators T_1 and T_2 meet the special case assumption (1.5). Therefore, for every solution w there is only one corresponding mapping F^w that can be explicitly computed as

$$(F^w x)(t) = \begin{bmatrix} 1 & w(t) \\ \overline{w(t)} & 1 \end{bmatrix} \begin{bmatrix} -\frac{w_0}{t-t_0} \\ \frac{1}{t-t_0} \end{bmatrix} x, \quad |t| = 1.$$

Direct computation shows that

$$\|F^w x\|_{H^w}^2 = \overline{x} D_{w,t_0} x.$$

Theorem 2.10. *The solution set of the Boundary Interpolation Problem coincides with the solution set of the Abstract Interpolation Problem with data specified in* (2.8).

There is indeed inequality in (1.3) for some solutions w of the Abstract Interpolation Problem with data specified in (2.8). We will discuss this in more detail in Section 5.

References to this problem are [18, 26]. For a higher-order analogue of Theorem 2.7 and related boundary interpolation Problem 2.9 see [7, 8].

3. Solutions of the Abstract Interpolation Problem

3.1. Isometry defined by the data

We say that two vectors x_1 and x_2 in X are D equivalent if

$$D(x_1, y) = D(x_2, y), \quad \forall y \in X. \tag{3.1}$$

We consider the vector space of equivalence classes $\{[x],\ x \in X\}$. We define the inner product between two equivalence classes as

$$\langle [x], [y] \rangle \stackrel{\text{def}}{=} D(x, y) \,. \tag{3.2}$$

After completion we get a Hilbert space that will be denoted by H_0. We rewrite identity (1.1) as

$$D(T_2x, T_2y) + \langle M_1x, M_1y \rangle_{E_1} = D(T_1x, T_1y) + \langle M_2x, M_2y \rangle_{E_2},$$

or, using definition (3.2), as

$$\langle [T_2x], [T_2y] \rangle + \langle M_1x, M_1y \rangle_{E_1} = \langle [T_1x], [T_1y] \rangle + \langle M_2x, M_2y \rangle_{E_2}. \tag{3.3}$$

We set

$$d_V \stackrel{\text{def}}{=} \operatorname{Clos}\left\{ \begin{bmatrix} [T_1x] \\ M_1x \end{bmatrix}, \ x \in X \right\} \subseteq H_0 \oplus E_1 \tag{3.4}$$

and

$$\Delta_V \stackrel{\text{def}}{=} \operatorname{Clos}\left\{ \begin{bmatrix} [T_2x] \\ M_2x \end{bmatrix}, \ x \in X \right\} \subseteq H_0 \oplus E_2. \tag{3.5}$$

We define a mapping $V : d_V \to \Delta_V$ by the formula

$$V : \begin{bmatrix} [T_1x] \\ M_1x \end{bmatrix} \stackrel{\text{def}}{\to} \begin{bmatrix} [T_2x] \\ M_2x \end{bmatrix} \,. \tag{3.6}$$

In view of (3.3), V is an isometry.

Remark 3.1. An arbitrary isometry V from $H_0 \oplus E_1$ to $H_0 \oplus E_2$ may appear here under appropriate choice of the data in (1.1).

3.2. Unitary colligations, characteristic functions, Fourier representations

Let $H,\ E_1,\ E_2$ be separable Hilbert spaces. A unitary mapping $\mathcal{A}$ of $H \oplus E_1$ onto $H \oplus E_2$

$$\mathcal{A} : H \oplus E_1 \to H \oplus E_2 \tag{3.7}$$

is said to be a *unitary colligation*. The space H is called the *state space* of the colligation, E_1 is called the *input space*, and E_2 is called the *output space*. Both E_1 and E_2 are called *exterior spaces*. Sometimes it is convenient to write the colligation $\mathcal{A}$ as a block matrix:

$$\mathcal{A} = \begin{bmatrix} A & B \\ C & D \end{bmatrix} : \begin{bmatrix} H \\ E_1 \end{bmatrix} \to \begin{bmatrix} H \\ E_2 \end{bmatrix}. \tag{3.8}$$

The *characteristic function* of the unitary colligation is defined as

$$w(\zeta) = D + \zeta C(I_H - \zeta A)^{-1} B. \tag{3.9}$$

It is well defined on $\mathbb{D}$ analytic contractive operator-valued function from E_1 to E_2. The *Fourier representation* of the space H associated with the colligation $\mathcal{A}$ is defined as

$$(\mathcal{G}h)(\zeta) = \begin{bmatrix} (\mathcal{G}_+h)(\zeta) \\ (\mathcal{G}_-h)(\zeta) \end{bmatrix} = \begin{bmatrix} C(I_H - \zeta A)^{-1}h \\ \bar{\zeta}B^*(I_H - \bar{\zeta}A^*)^{-1}h \end{bmatrix}, \quad h \in H, \quad \zeta \in \mathbb{D}. \tag{3.10}$$

$\mathcal{G}$ maps space H onto the de Barnges–Rovnyak space H^w associated with the characteristic function w (see Section 1.2).

Definition 3.2. We define the residual subspace $H_{\text{res}} \subseteq H$ of the colligation $\mathcal{A}$ as the maximal subspace of H that reduces $\mathcal{A}$ (that is invariant for $\mathcal{A}$ and $\mathcal{A}^*$). Equivalently H_{res} can be defined as the maximal subspace of H that is invariant for A and A^*, and $C|_{H_{\text{res}}} = 0$, $B^*|_{H_{\text{res}}} = 0$. The simple part of the space H is defined as $H_{\text{simp}} = H \ominus H_{\text{res}}$. A unitary colligation $\mathcal{A}$ is said to be simple with respect to the exterior spaces E_1 and E_2 if H_{res} is trivial.

The following fact will be of crucial importance to us in Lecture 5.

Theorem 3.3. *Let $\mathcal{G}$ be the mapping defined in* (3.10)*. Then $\mathcal{G}$ maps H_{simp} onto H^w unitarily and $\mathcal{G}$ vanishes on H_{res}.*

3.3. Unitary extensions of the isometry V and solutions of the problem

Definition 3.4. We say that a unitary colligation $\mathcal{A}$ of form (3.8) is a unitary extension of the isometry V, defined in (3.6), if $H_0 \subseteq H$ and

$$\mathcal{A}|d_V = V. \tag{3.11}$$

An extension $\mathcal{A}$ of V is said to be *minimal* if it does not have a nontrivial residual subspace in $H \ominus H_0$. Note that minimal extension $\mathcal{A}$ may have a residual subspace in H, though.

Theorem 3.5. *A Schur class function $w : E_1 \to E_2$ and a mapping $F : X \to H^w$ solve Abstract Interpolation Problem* (1.3)–(1.4) *if and only if there exists a unitary colligation $\mathcal{A}$ of form* (3.8) *that minimally extends the isometry V (defined in* (3.6) *from the problem data) such that w is the characteristic function of $\mathcal{A}$*

$$w(\zeta) = D + \zeta C(I_H - \zeta A)^{-1}B \tag{3.12}$$

and

$$Fx = \mathcal{G}[x], \quad \forall x \in X, \tag{3.13}$$

where $[x]$ is the D-equivalence class of x (defined in (3.1)*) and $\mathcal{G}$ is the Fourier representation of the colligation $\mathcal{A}$ (defined in* (3.10)*).*

References to this section are [12, 13, 15, 24].

4. Parametric description of the solutions of the Abstract Interpolation Problem

4.1. Structure of minimal unitary extensions of V

Let V be an isometric colligation:

$$V : d_V \to \Delta_V \ , \quad d_V \subseteq H_0 \oplus E_1 \ , \quad \Delta_V \subseteq H_0 \oplus E_2 \ .$$

Let $\mathcal{A}$ be a minimal unitary extension of V:

$$\mathcal{A} : H \oplus E_1 \to H \oplus E_2 \ , \quad H \supseteq H_0 \ , \quad \mathcal{A}|d_V = V \ .$$

Let $d_V^{\perp}$ and $\Delta_V^{\perp}$ be the orthogonal complements of d_V in $H_0 \oplus E_1$ and Δ_V in $H_0 \oplus E_2$, respectively. Let

$$H_1 = H \ominus H_0. \tag{4.1}$$

Then the orthogonal complement of d_V in $H \oplus E_1$ is $H_1 \oplus d_V^{\perp}$ and the orthogonal complement of Δ_V in $H \oplus E_2$ is $H_1 \oplus \Delta_V^{\perp}$. Since $\mathcal{A}$ is a unitary operator and it maps d_V onto Δ_V ($\mathcal{A}|d_V = V$), $\mathcal{A}$ has to map the orthogonal complement onto the orthogonal complement, i.e., $H_1 \oplus d_V^{\perp}$ onto $H_1 \oplus \Delta_V^{\perp}$. Denote the restriction of $\mathcal{A}$ onto $H_1 \oplus d_V^{\perp}$ by $\mathcal{A}_1$. Thus, $\mathcal{A}_1$ is a unitary colligation,

$$\mathcal{A}_1 : H_1 \oplus d_V^{\perp} \to H_1 \oplus \Delta_V^{\perp}.$$

Since $\mathcal{A}$ is a minimal extension of V, $\mathcal{A}_1$ is a simple colligation with respect to $d_V^{\perp}$ and $\Delta_V^{\perp}$. Thus, the parameter of a minimal unitary extension $\mathcal{A}$ is an arbitrary simple unitary colligation $\mathcal{A}_1$ with input space $d_V^{\perp}$ and output space $\Delta_V^{\perp}$.

Let N_1 be an auxiliary copy of $d_V^{\perp}$, that is we assume that there exists a unitary mapping u_1 from $d_V^{\perp}$ onto N_1

$$u_1 : d_V^{\perp} \to N_1. \tag{4.2}$$

Also let N_2 be a copy of $\Delta_V^{\perp}$, that is there exists a unitary mapping u_2 from $\Delta_V^{\perp}$ onto N_2

$$u_2 : \Delta_V^{\perp} \to N_2. \tag{4.3}$$

In what follows it will be convenient to consider simple unitary colligations $\mathcal{A}_1$ with input space N_1 and output space N_2 (instead of $d_V^{\perp}$ and $\Delta_V^{\perp}$)

$$\mathcal{A}_1 : H_1 \oplus N_1 \to H_1 \oplus N_2. \tag{4.4}$$

4.2. Universal extension of V

Here we define a unitary colligation $\mathcal{A}_0$ that extends V in a different way:

$$\mathcal{A}_0 : H_0 \oplus E_1 \oplus N_2 \to H_0 \oplus E_2 \oplus N_1 \tag{4.5}$$

with

$$\mathcal{A}_0|d_V = V, \quad (d_V \subseteq H_0 \oplus E_1), \tag{4.6}$$

$$\mathcal{A}_0|d_V^{\perp} = u_1, \quad \mathcal{A}_0|N_2 = u_2^*, \tag{4.7}$$

where u_1 maps unitarily $d_V^{\perp}$ onto N_1 and u_2 maps unitarily $\Delta_V^{\perp}$ onto N_2. $\mathcal{A}_0$ is uniquely defined by V and the identification maps u_1 and u_2. Note that mappings

u_1 and u_2 can be chosen arbitrarily. We will call this choice a *normalization* of the universal colligation $\mathcal{A}_0$.

Similarly to (3.8), we write colligation $\mathcal{A}_0$ as a block matrix:

$$\mathcal{A}_0 = \begin{bmatrix} A_0 & B_0 \\ C_0 & D_0 \end{bmatrix} : \begin{bmatrix} H_0 \\ E_1 \oplus N_2 \end{bmatrix} \to \begin{bmatrix} H_0 \\ E_2 \oplus N_1 \end{bmatrix}. \tag{4.8}$$

We also introduce the characteristic function of $\mathcal{A}_0$

$$S(\zeta) = D_0 + \zeta C_0 (I_{H_0} - \zeta A_0)^{-1} B_0. \tag{4.9}$$

It is an analytic on $\mathbb{D}$ contractive operator-valued function from $E_1 \oplus N_2$ to $E_2 \oplus N_1$. Note that S depends on the data of the problem and on normalization (4.7) of $\mathcal{A}_0$. According to the structure of the input an the output spaces, we further break S into blocks

$$S(\zeta) = \begin{bmatrix} s_0(\zeta) & s_2(\zeta) \\ s_1(\zeta) & s(\zeta) \end{bmatrix} : \begin{bmatrix} E_1 \\ N_2 \end{bmatrix} \to \begin{bmatrix} E_2 \\ N_1 \end{bmatrix}. \tag{4.10}$$

The special structure (4.6), (4.7) of $\mathcal{A}_0$ forces

$$s(0) = 0. \tag{4.11}$$

We consider the *Fourier representation* of the space H_0 associated with the colligation $\mathcal{A}_0$

$$(\mathcal{G}_0 h_0)(\zeta) = \begin{bmatrix} (\mathcal{G}_{0+} h_0)(\zeta) \\ (\mathcal{G}_{0-} h_0)(\zeta) \end{bmatrix} = \begin{bmatrix} C_0 (I_{H_0} - \zeta A_0)^{-1} h_0 \\ \bar{\zeta} B_0^* (I_{H_0} - \bar{\zeta} A_0^*)^{-1} h_0 \end{bmatrix}, \tag{4.12}$$

$h_0 \in H_0, \zeta \in \mathbb{D}$. $\mathcal{G}_0$ maps space H_0 onto the de Branges–Rovnyak space H^S associated with the characteristic function S.

4.3. Description of solutions

Given unitary colligation $\mathcal{A}_0$ of form (4.5) and unitary colligation $\mathcal{A}_1$ of form (4.4), there is a procedure (called *feedback coupling*) that produces a colligation $\mathcal{A}$ of the form (3.7) with

$$H = H_0 \oplus H_1.$$

We do not discuss here the feedback coupling. A detailed explanation of the procedure is given, for instance, in [20], Lecture 4, Section 3, page 373. Here we just state some consequences of this procedure.

Theorem 4.1. *If $\mathcal{A}_0$ is defined in terms of V as in* (4.6)–(4.7) *and $\mathcal{A}_1$ is an arbitrary simple unitary colligation of form* (4.4), *then the feedback coupling $\mathcal{A}$ is a minimal unitary extension of V in the sense of Definition* 3.4. *Moreover, all minimal unitary extensions of V arise in this way.*

Theorem 4.2. *Given unitary colligation $\mathcal{A}_0$ of the form* (4.5) *and unitary colligation $\mathcal{A}_1$ of the form* (4.4). *Let unitary colligation $\mathcal{A}$ of form* (3.7) *be the feedback coupling of $\mathcal{A}_0$ and $\mathcal{A}_1$. Then*

$$w = s_0 + s_2 \omega (I_{N_1} - s\omega)^{-1} s_1, \tag{4.13}$$

where w is the characteristic function of $\mathcal{A}$, ω is the characteristic function of $\mathcal{A}_1$, and S is the characteristic function of $\mathcal{A}_0$ (see (4.10)*);*

$$\mathcal{G}\begin{bmatrix} h_1 \\ h_0 \end{bmatrix} = \begin{bmatrix} \psi\omega & I_{E_2} & 0 & 0 \\ 0 & 0 & \varphi^*\omega^* & I_{E_1} \end{bmatrix} \mathcal{G}_0 h_0 + \begin{bmatrix} \psi & 0 \\ 0 & \varphi^* \end{bmatrix} \mathcal{G}_1 h_1 \ , \tag{4.14}$$

$h_0 \in H_0, h_1 \in H_1$, where $\mathcal{G}$, $\mathcal{G}_1$ and $\mathcal{G}_0$ are Fourier representations of $\mathcal{A}$, $\mathcal{A}_1$ and $\mathcal{A}_0$, respectively, as defined in (3.10), (4.12),

$$\varphi = (I_{N_1} - s\omega)^{-1} s_1, \quad \psi = s_2 (I_{N_2} - \omega s)^{-1}. \tag{4.15}$$

Combining Theorems 3.5 and 4.2, we get a description of all solutions (w, F) of the Abstract Interpolation Problem (1.3)–(1.4).

Theorem 4.3. *Let V be the isometry defined by the data of the problem as in* (3.4)–(3.6)*. Let N_1 and N_2 be the spaces* (4.2), (4.3) *and let $\mathcal{A}_0$ be the unitary colligation defined in* (4.5)–(4.7)*. Let S be the characteristic function* (4.10) *of $\mathcal{A}_0$ and $\mathcal{G}_0$ be the Fourier representation* (4.12) *of H_0. Then the solution set (w, F) of the Abstract Interpolation Problem* (1.3)–(1.4) *is described as follows*

$$w = s_0 + s_2 \omega (I_{N_1} - s\omega)^{-1} s_1, \tag{4.16}$$

where ω is an arbitrary Schur class function from N_1 to N_2, S is the characteristic function (4.10) *of $\mathcal{A}_0$;*

$$Fx = \begin{bmatrix} \psi\omega & I_{E_2} & 0 & 0 \\ 0 & 0 & \varphi^*\omega^* & I_{E_1} \end{bmatrix} \mathcal{G}_0[x], \quad x \in X, \tag{4.17}$$

where $[x]$ is the D-equivalence class of x defined in (3.1),

$$\varphi = (I_{N_1} - s\omega)^{-1} s_1, \quad \psi = s_2 (I_{N_2} - \omega s)^{-1}, \tag{4.18}$$

ω, s, s_1, s_2 are the same as above. All functions in formula (4.17) *are considered on $\mathbb{T}$.*

References to this section are [3, 12, 13, 15, 24].

5. Lecture 5: Inequality $\|Fx\|^2 \leq D(x,x)$, residual parts of minimal unitary extensions and the Nevanlinna–Adamjan–Arov–Kreĭn type theorems

This lecture is focused on the inequality $\|Fx\|^2 \leq D(x,x)$ in the setting of the Abstract Interpolation Problem (AIP) (1.3), (1.4). The main goals are

- to explain the inequality in terms of the corresponding minimal unitary extension (3.11) $\mathcal{A}$ of the isometry V (3.4)–(3.6). For more details, see [13, 15];
- to give a formula for the quantitative characteristic of how the inequality is far from the equality.

After that we apply the latter formula to the case when the equality $\|Fx\|^2 = D(x,x)$ is known a priori (like in Problem 2.1 and Problem 2.3). This in turn yields certain boundary properties of the coefficient matrix $S(\zeta)$, defined in (4.10). In particular, this leads to generalizations of a classical Nevanlinna–Adamjan–Arov–Kreĭn theorem [1, 2]: for general semi-determinate Nehari problem [15, 16, 21] and for general Commutant Lifting problem [5, 6].

5.1. The equality $\|Fx\|^2 = D(x, x)$ and simplicity of the corresponding minimal unitary extension

Let a Schur class operator function $w : E_1 \to E_2$ and a mapping $F : X \to H^w$ be a solution of the AIP (1.3), (1.4). Let $V : d_V \to \Delta_V$, $d_V \subseteq H_0 \oplus E_1$, $\Delta_V \subseteq H_0 \oplus E_2$ be the isometry (3.4)–(3.6) associated to AIP data (1.1). Then, by Theorem 3.11, w is the characteristic function of a unitary colligation $\mathcal{A}$ of the form (3.8) that minimally extends the isometry V and

$$Fx = \mathcal{G}[x], \quad \forall x \in X, \tag{5.1}$$

where $[x]$ is the D-equivalence class of x (defined in (3.1)) and $\mathcal{G}$ is the Fourier representation of the colligation $\mathcal{A}$ (defined in (3.10)). By Theorem 3.3, $\mathcal{G}$ maps $H_{\text{simp}} \subseteq H$ onto H^w unitarily and $\mathcal{G}$ vanishes on $H_{\text{res}} \subseteq H$, see also Definition 3.2.

From here one can see what the equality

$$\|Fx\|^2_{H^w} = D(x, x), \ \forall x \in X \tag{5.2}$$

means: in view of (5.1) and definition (3.1)–(3.2) of H_0, equality (5.2) is the same as

$$\|\mathcal{G}[x]\|^2_{H^w} = \|[x]\|^2_{H_0} \ .$$

Since the lineal $\{[x], \ x \in X\}$ is dense in H_0, this means that $\mathcal{G}$ is isometric on H_0, i.e., $H_0 \subseteq H_{\text{simp}}$. The latter is equivalent to the inclusion $H_{\text{res}} \subseteq H \ominus H_0$. Since extension $\mathcal{A}$ is minimal, this is possible if and only if $H_{\text{res}} = \{0\}$. Thus, we arrive at the following

Proposition 5.1. $\|Fx\|^2_{H^w} = D(x, x)$, $\forall x \in X$ *if and only if the corresponding minimal unitary extension* $\mathcal{A}$ *of the isometry* V *is simple.*

Hence, a strict inequality in (5.2) may occur for some $x \in X$ if and only if the corresponding minimal extension $\mathcal{A}$ is non-simple, i.e., $H \supseteq H_{\text{res}} \neq \{0\}$. In this case $\mathcal{A}|H_{\text{res}}$ is a unitary operator on H_{res}.

5.2. Residual part of a minimal unitary extension and its spectral function

The theme of this section is the unitary operator $\mathcal{A}|_{H_{\text{res}}}$. By Theorem 4.1, every minimal unitary extension $\mathcal{A} : H \oplus E_1 \to H \oplus E_2$ of the isometry V is the feedback coupling of the universal unitary colligation $\mathcal{A}_0$, defined in (4.6)–(4.7), and a simple unitary colligation $\mathcal{A}_1$ of form (4.4):

$$\mathcal{A}_0 : H_0 \oplus E_1 \oplus N_2 \to H_0 \oplus E_2 \oplus N_1,$$
$$\mathcal{A}_1 : H_1 \oplus N_1 \to H_1 \oplus N_2,$$

where $H = H_0 \oplus H_1$. It follows from the feedback coupling procedure that if colligations $\mathcal{A}_0$ and $\mathcal{A}_1$ are not simple, then their residual parts are contained in the residual part of their coupling $\mathcal{A}$. However, the residual part of the latter colligation may be properly larger.

Here we are interested in the piece of the residual part of $\mathcal{A}$ that results from the feedback coupling procedure, but not from non-simplicity of the coupled colligations. Let $\mathcal{A}_0 : H_0 \oplus E_1 \oplus N_2 \to H_0 \oplus E_2 \oplus N_1$ and $\mathcal{A}_1 : H_1 \oplus N_1 \to H_1 \oplus N_2$ be simple unitary colligations. Let $\mathcal{A} : H \oplus E_1 \to H \oplus E_2$ be their feedback coupling, where $H = H_0 \oplus H_1$. Let $H = H_{\text{simp}} \oplus H_{\text{res}}$ be the simple and residual parts of the colligation $\mathcal{A}$, respectively. Let $U = \mathcal{A}|H_{\text{res}}$.

Definition 5.2. Let U be a unitary operator on a separable Hilbert space $\mathcal{H}$. Let N be an auxiliary Hilbert space and let $\Gamma : N \to \mathcal{H}$ be a linear operator from N into $\mathcal{H}$ such that $\Gamma(N)$ is a *cyclic subspace* for the operator U, i.e., the closed linear span of $\{U^k\Gamma(N)\}_{k\in\mathbb{Z}}$ coincides with $\mathcal{H}$. The operator function $a(\zeta) : N \to N$,

$$a(\zeta) \stackrel{\text{def}}{=} \frac{1}{2}\Gamma^* \frac{1_{\mathcal{H}} + \zeta U}{1_{\mathcal{H}} - \zeta U}\,\Gamma$$

is called a *spectral function* of the operator U. Clearly, a is analytic in $\mathbb{D}$.

Since U is a unitary operator, $a(\zeta) + a(\zeta)^* \geq 0$. Therefore, $a(\zeta)$ admits a Riesz–Herglotz representation

$$a(\zeta) = \frac{1}{2}\int_{\mathbb{T}} \frac{t+\zeta}{t-\zeta}\,\sigma(dt),$$

where $\sigma(dt)$ is an operator-valued measure ($\sigma(dt) : N \to N$) on the unit circle $\mathbb{T}$. The measure $\sigma(dt)$ is called a *spectral measure* of the operator U.

Remark 5.3. Different choices of the auxiliary space N and the operator Γ lead to different spectral functions. However, any of them defines the unitary operator U uniquely up to a unitary equivalence.

The next theorem gives a formula for the spectral function of the unitary operator $U = \mathcal{A}|H_{\text{res}}$ that is the residual part of the feedback coupling considered above in this section. Results of this type in the context of the cascade coupling go back to Yu.L. Šmulian [28] and were inspired by M. Livšits and M. Brodskii [9]. It was realized, in particular, as an obstacle for solving the Hilbert space invariant subspace problem by means of factorization of the characteristic function. In the same context it was investigated by B. Sz.-Nagy and C. Foiaş ([27], irreducible factorizations), and by L. de Branges ([10], overlapping subspaces), see also [4, 29].

Theorem 5.4 ([13, 15]). *Let unitary colligation $\mathcal{A} : H \oplus E_1 \to H \oplus E_2$ be the feedback coupling of simple unitary colligations $\mathcal{A}_0 : H_0 \oplus E_1 \oplus N_2 \to H_0 \oplus E_2 \oplus N_1$ and $\mathcal{A}_1 : H_1 \oplus N_1 \to H_1 \oplus N_2$, where $H = H_0 \oplus H_1$. Let*

$$S(\zeta) = \begin{bmatrix} s_0(\zeta) & s_2(\zeta) \\ s_1(\zeta) & s(\zeta) \end{bmatrix} : \begin{bmatrix} E_1 \\ N_2 \end{bmatrix} \to \begin{bmatrix} E_2 \\ N_1 \end{bmatrix}$$

be the characteristic function of the colligation $\mathcal{A}_0$ *and* $\omega(\zeta)$ *be the characteristic function of the colligation* $\mathcal{A}_1$. *Let* $H_{\mathrm{res}} \subseteq H$ *be the residual subspace of* $\mathcal{A}$. *Let*

$$U = \mathcal{A}|_{H_{\mathrm{res}}}$$

and let $N \overset{\mathrm{def}}{=} N_2 \oplus N_1$. *We define* $\Gamma : N \to H_{\mathrm{res}}$ *as*

$$\Gamma\left(\begin{bmatrix} n_2 \\ n_1 \end{bmatrix}\right) \overset{\mathrm{def}}{=} P_{H_{\mathrm{res}}}(P_{H_0}\mathcal{A}_0^*(0_{H_0} \oplus 0_{E_2} \oplus (\omega(0)^* n_2 + n_1)) \oplus P_{H_1}\mathcal{A}_1^*(0_{H_1} \oplus n_2)),$$

where P_{H_0}, P_{H_1} *are the orthogonal projections onto the corresponding subspaces,* $P_{H_{\mathrm{res}}}$ *is the orthogonal projection onto* H_{res}

$$P_{H_{\mathrm{res}}} = I_H - \mathcal{G}^*\mathcal{G},$$

where $\mathcal{G}$ *is defined in* (4.14). *Then* $\Gamma(N)$ *is a cyclic subspace for the operator* U *and the corresponding spectral function* $a_\omega(\zeta) : \begin{bmatrix} N_2 \\ N_1 \end{bmatrix} \to \begin{bmatrix} N_2 \\ N_1 \end{bmatrix}$ *is given by the formula*

$$a_\omega(\zeta) = \frac{1}{2}\begin{bmatrix} 0 & -\omega(0) \\ \omega(0)^* & 0 \end{bmatrix} + \overset{\circ}{a}_\omega(\zeta) \tag{5.3}$$

$$-\frac{1}{2}\int_{\mathbb{T}} \frac{t+\zeta}{t-\zeta} \begin{bmatrix} \psi^* & \omega\varphi \\ \omega^*\psi^* & \varphi \end{bmatrix} \begin{bmatrix} \mathbf{1}_{E_2} & w \\ w^* & \mathbf{1}_{E_1} \end{bmatrix}^{[-1]} \begin{bmatrix} \psi & \psi\omega \\ \varphi^*\omega^* & \varphi^* \end{bmatrix} m(dt),$$

where

$$\overset{\circ}{a}_\omega = \frac{1}{2}\,\frac{\mathbf{1}_{N_2\oplus N_1} + \begin{bmatrix} 0 & \omega \\ s & 0 \end{bmatrix}}{\mathbf{1}_{N_2\oplus N_1} - \begin{bmatrix} 0 & \omega \\ s & 0 \end{bmatrix}} = \begin{bmatrix} \overset{\circ}{\psi} - \frac{1}{2} & \omega\overset{\circ}{\varphi} \\ s\overset{\circ}{\psi} & \overset{\circ}{\varphi} - \frac{1}{2} \end{bmatrix},$$

$$\overset{\circ}{\varphi} = (\mathbf{1}_{N_1} - s\omega)^{-1}, \quad \overset{\circ}{\psi} = (\mathbf{1}_{N_2} - \omega s)^{-1}, \tag{5.4}$$

$$\varphi = \overset{\circ}{\varphi} s_1 = (\mathbf{1}_{N_1} - s\omega)^{-1} s_1, \quad \psi = s_2\overset{\circ}{\psi} = s_2(\mathbf{1}_{N_2} - \omega s)^{-1},$$

w *is the characteristic function of the colligation* $\mathcal{A}$

$$w = s_0 + s_2\omega(\mathbf{1} - s\omega)^{-1} s_1, \tag{5.5}$$

$m(dt)$ *is the normalized Lebesgue measure on the unit circle* $\mathbb{T}$.

Note that the spectral function a_ω of the feedback coupling depends on the characteristic functions of the coupled colligations only.

The real part of the spectral function $a_\omega(\zeta)$ can be expressed as

$$a_\omega(\zeta) + a_\omega(\zeta)^* = \overset{\circ}{a}_\omega(\zeta) + \overset{\circ}{a}_\omega(\zeta)^* \tag{5.6}$$

$$-\int_{\mathbb{T}} \frac{1-|\zeta|^2}{|t-\zeta|^2} \begin{bmatrix} \psi^* & \omega\varphi \\ \omega^*\psi^* & \varphi \end{bmatrix} \begin{bmatrix} \mathbf{1}_{E_2} & w \\ w^* & \mathbf{1}_{E_1} \end{bmatrix}^{[-1]} \begin{bmatrix} \psi & \psi\omega \\ \varphi^*\omega^* & \varphi^* \end{bmatrix} m(dt).$$

We will also need the following re-expression of $\overset{\circ}{a}_\omega$

$$\overset{\circ}{a}_\omega = \begin{bmatrix} \frac{1}{2}\ \frac{\mathbf{1}_{N_2}+\omega s}{\mathbf{1}_{N_1}-\omega s} & \omega\overset{\circ}{\varphi} \\ s\overset{\circ}{\psi} & \frac{1}{2}\ \frac{\mathbf{1}_{N_1}+s\omega}{\mathbf{1}_{N_1}-s\omega} \end{bmatrix} \tag{5.7}$$

with the same notations as in (5.4). Since

$$\overset{\circ}{a}_w + \overset{\circ}{a}{}_w^* \geq 0,$$

there exists an operator measure $\overset{\circ}{\sigma}(dt) : N \to N$ such that

$$\overset{\circ}{a}_\omega(\zeta) = \frac{1}{2}\begin{bmatrix} 0 & \omega(0) \\ -\omega(0)^* & 0 \end{bmatrix} + \frac{1}{2}\int_{\mathbb{T}} \frac{t+\zeta}{t-\zeta}\,\overset{\circ}{\sigma}(dt). \tag{5.8}$$

5.3. Property $\|Fx\|^2 = D(x,x)$ yields boundary properties of the coefficient matrix S and the parameter ω

In this section we show how formula (5.6) for the spectral function of the residual part yields boundary properties of the coefficient matrix-function

$$S = \begin{bmatrix} s_0 & s_2 \\ s_1 & s \end{bmatrix}.$$

Let w be a solution of the AIP and let $F^\omega : X \to H^w$ (we label it with parameter ω in view of formula (4.14)) be the corresponding mapping. Assume that the equality

$$\|F^\omega x\|^2_{H^w} = D(x,x) \tag{5.9}$$

holds for all $x \in X$. By Proposition 5.1, this means that the corresponding extension $\mathcal{A}$ is simple, i.e., the residual part is trivial. But then the spectral function of the residual part must be equal to zero. Therefore, formula (5.6) along with assumption (5.9) yield

$$\overset{\circ}{a}_\omega(\zeta) + \overset{\circ}{a}_\omega(\zeta)^* = \int_{\mathbb{T}} \frac{1-|\zeta|^2}{|t-\zeta|^2} \begin{bmatrix} \psi^* & \omega\varphi \\ \omega^*\psi^* & \varphi \end{bmatrix} \begin{bmatrix} \mathbf{1}_{E_2} & w \\ \omega^*\psi^* & \varphi \end{bmatrix}^{[-1]} \begin{bmatrix} \psi & \psi\omega \\ \varphi^*\omega^* & \varphi^* \end{bmatrix} m(dt). \tag{5.10}$$

The latter is equivalent to the following two properties:

1. $\overset{\circ}{\sigma}_\omega(dt)$ is absolutely continuous with respect to the Lebesgue measure;
2. $\overset{\circ}{a}_w(t) + \overset{\circ}{a}_\omega(t)^* = \begin{bmatrix} \psi^* & \omega\varphi \\ \omega^*\psi^* & \varphi \end{bmatrix} \begin{bmatrix} \mathbf{1}_{E_2} & w \\ w^* & \mathbf{1}_{E_1} \end{bmatrix}^{[-1]} \begin{bmatrix} \psi & \psi\omega \\ \varphi^*\omega^* & \varphi^* \end{bmatrix},$

almost everywhere on $\mathbb{T}$. Thus, these two properties are equivalent to

$$\|F^\omega x\|^2_{H^w} = D(x,x), \ \forall\, x \in X.$$

Problem 2.1 and Problem 2.3 possess the property

$$\|F^\omega x\|^2_{H^w} = D(x,x), \ \forall\, x \in X$$

for every solution w and every parameter ω that produces this w via formula (5.5) (actually, for these examples formula (5.5) gives a one to one correspondence between ω and w). Therefore, properties 1 and 2 above hold for every parameter ω in those examples.

Property 2 itself is equivalent to vanishing of the absolutely continuous part of the measure $\sigma_\omega(dt)$ that corresponds to $a_\omega(\zeta)$. It is the case in Problem 2.9 that $\overset{\circ}{\sigma}_\omega(dt)$ can be supported by a single point for every parameter ω. Hence, $\sigma_\omega(dt)$ has trivial absolutely continuous part for every parameter ω in this example. Thus, we have Property 2 for every parameter ω in Problem 2.9. Property 1 holds for some parameters and does not hold for the others in this problem. Analysis shows that in this problem the residual part is nontrivial if and only if

$$\lim_{\zeta\to t_0} \omega(\zeta) = \overline{s(t_0)},$$

where s is the right-bottom entry of the coefficient matrix S, and

$$\lim_{\zeta\to t_0} \frac{1-|\omega(\zeta)|^2}{1-|\zeta|^2} < \infty.$$

Note that $|s(t_0)| = 1$ always in this problem.

We reformulate the above Properties 1 and 2 using formulas (5.6) and (5.7) of this Lecture. Property 1 is equivalent to the following property

1′. Measures $\overset{\circ}{\sigma}{}^1_w(dt)$ and $\overset{\circ}{\sigma}{}^2_\omega(dt)$ are absolutely continuous with respect to the Lebesgue measures, where

$$\frac{\mathbf{1}_{N_2}+\omega(\zeta)s(\zeta)}{\mathbf{1}_{N_2}-\omega(\zeta)s(\zeta)} = \int_{\mathbb{T}} \frac{t+\zeta}{t-\zeta}\,\overset{\circ}{\sigma}{}^2_\omega(dt) \quad\text{and}\quad \frac{\mathbf{1}_{N_1}+s(\zeta)\omega(\zeta)}{\mathbf{1}_{N_1}-s(\zeta)\omega(\zeta)} = \int_{\mathbb{T}} \frac{t+\zeta}{t-\zeta}\,\overset{\circ}{\sigma}{}^2_\omega(dt).$$

A special case of Property 2, when parameter $\omega = 0$, reads as follows

2′. $$\begin{bmatrix} \mathbf{1}_{N_2} & s^* \\ s & \mathbf{1}_{N_1} \end{bmatrix} = \begin{bmatrix} s_2^* & 0 \\ 0 & s_1 \end{bmatrix} \begin{bmatrix} \mathbf{1}_{E_2} & s_0 \\ s_0^* & \mathbf{1}_{E_1} \end{bmatrix}^{[-1]} \begin{bmatrix} s_2 & 0 \\ 0 & s_1^* \end{bmatrix}$$

almost everywhere on $\mathbb{T}$.

It was shown in [15, 16] (under the assumptions that $\dim E_1 < \infty$ and $\dim E_2 < \infty$) that converse is also true, i.e., Property **2′** implies **2** for every parameter ω. It was also shown in [15, 16] under the same assumptions ($\dim E_1 < \infty$ and $\dim E_2 < \infty$) that Property **2′** is in turn equivalent to this property

2″. $\operatorname{rank}\,(\mathbf{1}_{E_1,\oplus N_2} - S^*S) = \operatorname{rank}\,(\mathbf{1}_{E_1} - s_0^* s_0) - \dim N_1$, a.e. on $\mathbb{T}$.

This result contains, in particular, the classical Nevanlinna–Adamjan–Arov–Kreĭn theorem: if $\dim N_1 = \dim E_1$ (i.e., the problem is completely indeterminate), then

$$\mathbf{1} - S^*S = 0$$

a.e. on $\mathbb{T}$. That is, the coefficient matrix S in inner.

6. Lecture 6: Properties of the coefficient matrices via dense sets in the function model spaces

Let S be the characteristic function (4.10) of the universal colligation $\mathcal{A}_0$ (4.8)

$$S(\zeta) = \begin{bmatrix} s_0(\zeta) & s_2(\zeta) \\ s_1(\zeta) & s(\zeta) \end{bmatrix} : \begin{bmatrix} E_1 \\ N_2 \end{bmatrix} \to \begin{bmatrix} E_2 \\ N_1 \end{bmatrix}.$$

It serves as the coefficient matrix in the parametrization formula (4.16). Since S is a Schur class operator-function, there exists the de Branges–Rovnyak function space H^S associated to it (similar to the definition in Section 1.2). We discuss here an approach that employs some dense sets in the space H^S. This approach was applied to the Sarason problem in [17], to the Nehari problem in [19, 21] and to the general Commutant Lifting problem in [5, 6].

6.1. The data of AIP suggest a dense set in H^S

Let $\mathcal{G}^0$ be the Fourier representation of the colligation $\mathcal{A}_0$ defined in (4.12). $\mathcal{G}^0$ maps the space H^0 onto the de Branges–Rovnyak function space H^S contractively. We define mapping $F^S : X \to H^S$ as

$$F^S x \stackrel{\text{def}}{=} \mathcal{G}^0[x], \tag{6.1}$$

where $[x]$ is defined in (3.1), (3.2). Observe that the lineal $\{F^S x,\ x \in X\}$ is dense in H^S, since the lineal $\{[x],\ x \in X\}$ is dense in H^0.

It follows from definition (4.5)–(4.7) of the colligation $\mathcal{A}_0$ and definition (4.12) of the Fourier representation $\mathcal{G}_0$ that F^S possess the properties similar to (1.3) and (1.4):

i) $$\|F^S x\|^2_{H^S} \le D(x,x), \quad x \in X;$$

ii) $$tF^S T_2 x - F^S T_1 x = \begin{bmatrix} \mathbf{1}_{E_2 \oplus N_1} & S \\ S^* & \mathbf{1}_{E_1 \oplus N_2} \end{bmatrix} \begin{bmatrix} -M_2 x \\ 0 \\ M_1 x \\ 0 \end{bmatrix},$$

a.e. on $\mathbb{T}$, where the first zero stands for the zero vector of the space N_1 and the second zero stands for the zero vector of the space N_2.

Under the Special Case assumptions (1.5), property ii) can be re-expressed as follows:

$$\begin{aligned} (F^S_+ x)(\zeta) &= \left(S(\zeta) \begin{bmatrix} M_1 \\ 0 \end{bmatrix} - \begin{bmatrix} M_2 \\ 0 \end{bmatrix} \right) (\zeta T_2 - T_1)^{-1} x \\ (F^S_- x)(\zeta) &= \bar{\zeta} \left(\begin{bmatrix} M_1 \\ 0 \end{bmatrix} - S(\zeta)^* \begin{bmatrix} M_2 \\ 0 \end{bmatrix} \right) (T_2 - \bar{\zeta} T_1)^{-1} x, \quad |\zeta| < 1. \end{aligned} \tag{6.2}$$

Here

$$\begin{bmatrix} M_1 \\ 0 \end{bmatrix} : X \to \begin{bmatrix} E_1 \\ N_2 \end{bmatrix}, \qquad \begin{bmatrix} M_2 \\ 0 \end{bmatrix} : X \to \begin{bmatrix} E_2 \\ N_1 \end{bmatrix}.$$

The structure of the dense set $\{F^S x, x \in X\}$ in H^S carries all the information about properties of the matrix-function S. It will be demonstrated in the next sections how this approach works for Sarason problem.

6.2. Coefficient matrix of the Sarason problem and associated function model space

In Section 2.2 we considered the scalar-valued case of the Sarason problem. We consider more general case now. Let θ be an inner operator-function, $\theta(\zeta) : E_2' \to E_2$, $\theta^*\theta = \mathbf{1}_{E_2'}$, a.e. on $\mathbb{T}$, E_2' and E_2 be separable Hilbert spaces. Let $K_\theta = H_+^2(E_2) \ominus \theta H_+^2(E_2')$, where $H_+^2(E)$ stands for the vector Hardy space with coefficients in the space E. Let $T_\theta^* x = P_+ \bar{t} x$, $x \in K_\theta$. Let W^* be a contractive operator, $W^* : K_\theta \to H_+^2(E_1)$, where E_1 is a separable Hilbert space, such that $W^* T_\theta^* = P_+ \bar{t} W^*$. Consider the following interpolation problem: find all the Schur class functions $w(\zeta) : E_1 \to E_2$ that satisfy

$$W^* x = P_+ w^* x, \; x \in K_\theta \; .$$

The associated AIP data are:

$$X = K_\theta, \; T_1 = I_X, \; T_2 = T_\theta^*,$$

E_1 and E_2 are the spaces introduced above, $M_1 x = (W^* x)(0)$, $M_2 x = x(0)$, where the latter notations stands for the value at 0 of an $H_+^2(E_1)$ and an $H_+^2(E_2)$ function, respectively,

$$D(x, x) = \langle (I - WW^*)x, x \rangle_{K_\theta} .$$

The Fourier representation F^w is unique for every solution w and can be expressed as

$$F^w x = \begin{bmatrix} \mathbf{1}_{E_2} & w \\ w^* & \mathbf{1}_{E_1} \end{bmatrix} \begin{bmatrix} x \\ -W^* x \end{bmatrix}.$$

The Fourier representation F^S of the colligation $\mathcal{A}_0$ (see (6.2)) can also be expressed as

$$F^S x = \begin{bmatrix} \mathbf{1}_{E_2 \oplus N_1} & S \\ S^* & \mathbf{1}_{E_1 \oplus N_2} \end{bmatrix} \begin{bmatrix} x \\ 0 \\ -W^* x \\ 0 \end{bmatrix}, \quad x \in K_\theta \; .$$

Since s_0 is also a solution (corresponding to the parameter $\omega = 0$), we can write the latter expression as

$$F^S x = \begin{bmatrix} \mathbf{1}_{E_2 \oplus N_1} & S \\ S^* & \mathbf{1}_{E_1 \oplus N_2} \end{bmatrix} \begin{bmatrix} x \\ 0 \\ -P_+ s_0^* x \\ 0 \end{bmatrix}, \quad x \in K_\theta . \tag{6.3}$$

As it was observed in Section 6.1, the set

$$\{F^S x, \ x \in K_\theta\} \subseteq H^S$$

is dense in H^S.

6.3. Properties of the coefficient matrices of the Sarason problem

Observe first, that the bottom entry of the vector $F^S x$ (i.e., the second component of the vector $F^S_- x$) in (6.3) is $s_2^* x$, and it is an $H^2_-(N_2)$ function, since $F^S x \in H^S$. Thus,

$$s_2^* x \in H^2_-(N_2), \quad \forall x \in K_\theta.$$

In other words,

$$\langle s_2^* x, \ h_+ \rangle_{L^2(N_2)} = 0$$

for all $h_+ \in H^2_+(N_2)$ and $x \in K_\theta$. Equivalently,

$$\langle x, \ s_2 h_+ \rangle_{L^2(E_2)} = 0$$

for all $h_+ \in H^2_+(N_2)$ and $x \in K_\theta$. Hence, $s_2 h_+ \in \theta H^2_+(E_2')$ for all $h_+ \in H^2_+(N_2)$. This means that

$$s_2 = \theta \tilde{s}_2$$

for a Schur class function $\tilde{s}_2 : N_2 \to E_2'$. The next theorem shows how the denseness property works.

Theorem 6.1 ([14]). *Let the* $S = \begin{bmatrix} s_0 & \theta \tilde{s}_2 \\ s_1 & s \end{bmatrix}$ *be the coefficient matrix of the Sarason problem. Then* s_1 *is an outer function (i.e., the lineal* $s_1 H^2_+(E_1)$ *is dense in* $H^2_+(N_1)$*) and* $\tilde{s}_2$ *is a* $*$*-outer function (i.e., the lineal* $\tilde{s}_2^* H^2_-(E_2')$ *is dense in* $H^2_-(N_2)$*).*

Sketch of the proof. To prove that s_1 is an outer function we need to check that the assumption

$$P_+ s_1^* h_+ = 0, \quad h_+ \in H^2_+(N_1) \tag{6.4}$$

implies $h_+ = 0$. Consider the vector

$$\begin{bmatrix} 1 & S \\ S^* & 1 \end{bmatrix} \begin{bmatrix} \begin{bmatrix} 0 \\ h_+ \end{bmatrix} \\ -P_+ S^* \begin{bmatrix} 0 \\ h_+ \end{bmatrix} \end{bmatrix} = \begin{bmatrix} 1 & S \\ S^* & 1 \end{bmatrix} \begin{bmatrix} 0 \\ h_+ \\ -P_+ s_1^* h_+ \\ -P_+ s^* h_+ \end{bmatrix}. \tag{6.5}$$

It belongs to H^S for every $h_+ \in H^2_+(N_1)$. Assumption (6.4) makes it orthogonal to the lineal (6.3). Since (6.3) is dense in H^S, the orthogonality forces (6.5) to be zero. The latter in turn yields $h_+ = 0$. □

We return now to Problem 2.3, i.e., to the scalar Sarason problem ($E_1 = E_2 = E_2' = \mathbb{C}^1$). Assume also that the problem is indeterminate (i.e., permits a non-unique solution, i.e., $N_1 = N_2 = \mathbb{C}^1$). In this case S is an inner matrix-function (see Lecture 5) and it can be normalized so that $s_1 = \tilde{s}_2 = a$ is an outer function. The following criterion is a consequence of the denseness of the lineal (6.3) in H^S.

Theorem 6.2 ([17, 19], also [6]). *A* 2×2 *inner matrix function*

$$S(\zeta) = \begin{bmatrix} s_0 & s_2 \\ s_1 & s \end{bmatrix} = \begin{bmatrix} s_0 & \theta a \\ a & s \end{bmatrix},$$

(*where* a *is an outer function,* $s(0) = 0$) *is the coefficient matrix of an indeterminate Sarason Problem* 2.3 *if and only if*

$$\begin{bmatrix} P_-\bar{s}_1 \\ \bar{s} \end{bmatrix} \in \operatorname{clos}\left\{ \begin{bmatrix} P_-\bar{s}_0 x \\ \bar{s}_2 x \end{bmatrix}, \ x \in K_\theta \right\}, \tag{6.6}$$

where the closure is understood in the L^2 *sense.*

Remark 6.3. Property (6.6) is equivalent to the following one

$$\inf_{x \in K_\theta} \|P_- S^* \left(\begin{bmatrix} 0 \\ 1 \end{bmatrix} - \begin{bmatrix} x \\ 0 \end{bmatrix} \right) \|^2_{H^2_-} = 0,$$

equivalently,

$$\inf_{x \in K_\theta} \left\langle SP_-S^* \left(\begin{bmatrix} 0 \\ 1 \end{bmatrix} - \begin{bmatrix} x \\ 0 \end{bmatrix} \right), \left(\begin{bmatrix} 0 \\ 1 \end{bmatrix} - \begin{bmatrix} x \\ 0 \end{bmatrix} \right) \right\rangle_{L^2} = 0. \tag{6.7}$$

Further simplification of property (6.7) is unknown. A discussion on this matter was given in [22]. An application to uniqueness of the inverse scattering for CMV matrices was given in [11, 25].

References

[1] V.M. Adamjan, D.Z. Arov, M.G. Kreĭn, *Infinite Hankel matrices and generalized problem of Carathéodory–Fejér and I. Schur.*, Funkt. Anal. i Prilozhen. **2:4** (1968) 1–17, Russian. English Transl., Funct. Analysis and Appl. **2** (1968) 269–281.

[2] V.M. Adamjan, D.Z. Arov, M.G. Kreĭn, *Some function theoretic problems connected with the theory of spectral measures of isometric operators*, in: Lecture Notes in Mathematics, Springer-Verlag, **1043** (1984) 160–163; or **1573** (1994) 183–185.

[3] D.Z. Arov and L.Z. Grossman, *Scattering matrices in the theory of extensions of isometric operators*, Dokl. Akad. Nauk SSSR 270:1 (1983), 17–20. Russian; English translation in Soviet Math. Dokl. 27:3 (1983), 518–521.

[4] J.A. Ball, *Factorization and model theory for contraction operators with unitary part*, Mem. Amer. Math. Soc. **13** (1978), no. 198

[5] J. Ball, A. Kheifets, *The Inverse Commutant Lifting Problem. I: Coordinate-Free Formalism*, Integral Equations and Operator Theory **70** (2011) 17–62.

[6] J. Ball, A. Kheifets, *The Inverse Commutant Lifting Problem. II: Hellinger Functional-Model Spaces*, Complex Analysis and Operator Theory **7** (2013) 873–907.

[7] V. Bolotnikov, A. Kheifets, *A higher order analogue of the Carathéodory–Julia theorem*, Journal of Functional Analysis **237** 1 (2006) 350–371

[8] V. Bolotnikov, A. Kheifets, *A higher order Carathéodory–Julia theorem and related boundary interpolation problem*, Operator Theory: Advances and Applications **179** (2008) 63–102,

[9] M.S. Brodskii, M.S. Livšic, *Spectral analysis of non-selfadjoint operators and intermediate systems*, (Russian) Uspehi Mat. Nauk (N.S.) **13** (1958) no. 1(79), 3–85. Engl. Transl.: Amer. Math. Soc. Transl. (2) **13** (1960) 265–346.

[10] L. de Branges, *Factorization and invariant subspaces*, J. Math. Anal. Appl. **29** (1970) 163–200.

[11] L. Golinskii, A. Kheifets, P. Yuditskii, *Scattering theory for CMV matrices: uniqueness, Helson–Szegő and strong Szegő theorems*, Integral Equations Operator Theory **69** (2011) no. 4, 479–508.

[12] V.E. Katsnel'son, A.Y. Kheifets, and P.M. Yuditskii, *Abstract Interpolation Problem and Theory of Extensions of Isometric Operators*, in Operators in function spaces and problems in function theory **146** (1987) 83–96 Naukova Dumka, Kiev (Russian); English translation in Oper. Theory Adv. Appl. **95** (1997) 283–298 Birkhäuser, Basel.

[13] A.Ya. Kheifets, *Parseval equality in abstract interpolation problem and coupling of open systems*, Teor. Funk., Funk. Anal. i ikh Prilozhen **49** (1988) 112–120, **50** (1988) 98–103, Russian. English transl.: J. Sov. Math. **49**, 4 (1990) 1114–1120, **49**, 6 (1990) 1307–1310.

[14] A.Ya. Kheifets, *Generalized bitangential Schur–Nevanlinna–Pick problem, related Parseval equality and scattering operator*, deposited in VINITI, 11.05.1989, No. 3108–B89 Dep., 1–60, 1989, Russian.

[15] A.Ya. Kheifets, *Scattering matrices and Parseval equality in Abstract Interpolation Problem*, PhD thesis, 1990, Russian.

[16] A.Ya. Kheifets, *Nevanlinna - Adamjan - Arov - Kreĭn theorem in semi-determinate case*, Teor. Funkt., Funkt. Anal. i ikh Prilozhen. **56** (1991) 128–137, Russian. Engl. transl.: Journal of Mathematical Sciences **76**, 4 (1995) 2542–2549.

[17] A.Ya. Kheifets, *Regularization of gamma-generating pairs and exposed points in* H^1, Journal of Functional Analysis **130** 2 (1995) 310–333.

[18] A.Ya. Kheifets, *Hamburger moment problem: Parseval equality and Arov-singularity*, Journal of Functional Analysis, **141** 2 (1996) 374–420

[19] A. Kheifets, *Nehari problem and exposed points of the unit ball in the Hardy space*, Proceedings of the Ashquelon conference, Israel, 1996, in Israel Mathematical Conference Proceedings, **11** (1997) 145–151

[20] A. Kheifets, *Abstract interpolation problem and some applications*, Lecture notes given in framework of Holomorphic Spaces semester, fall 1995, in: Holomorphic

Spaces (S. Axler, J. McCarthy, D. Sarason, editors), MSRI Publications, **33** (1998) 351–381, Berkeley, California.

[21] A. Kheifets, *Parameterization of solutions of the Nehari Problem and nonorthogonal dynamics*, Operator Theory: Advances and Applications, **115** (2000) 213–233, Birkhäuser Verlag, Basel.

[22] A. Kheifets, *Density of domain of the weighted Hilbert transform*, Byrnes, James S. (ed.), Twentieth century harmonic analysis – a celebration. Proceedings of the NATO Advanced Study Institute, Il Ciocco, Italy, July 2–15, 2000. Dordrecht: Kluwer Academic Publishers. NATO Sci. Ser. II, Math. Phys. Chem. **33**, 374–375 (2001). Zentrallblatt Review Zbl 0991.44501, https://zbmath.org/?q=an:0991.44501

[23] A.Ya. Kheifets, P.M. Yuditskii, *Interpolation of operator that commutes with compressed shift by Schur class functions*, Teor. Funk., Funk. Anal. i ikh Prilozhen, **40** (1983) 129–136, Russian

[24] A.Y. Kheifets and P.M. Yuditskii, *Analysis and extension of V.P. Potapov's approach to interpolation problems with applications to the generalized bi-tangential Schur–Nevanlinna–Pick problem and J-inner-outer factorization*, Oper. Theory Adv. Appl. **72** (1994) 133–161.

[25] A. Volberg, P. Yuditskii, *On the inverse scattering problem for Jacobi matrices with the spectrum on an interval, a finite system of intervals or a Cantor set of positive length*, Comm. Math. Phys. **226** (2002) no. 3 567–605.

[26] D. Sarason, Sub-Hardy Hilbert spaces in the unit disk, University of Arkansas Lecture Notes in the Mathematical Sciences, Wiley, New York, 1994.

[27] B. Sz.-Nagy, C. Foiaş, *Harmonic analysis of operators on Hilbert space*, Translated from the French and revised North-Holland Publishing Co., Amsterdam-London; American Elsevier Publishing Co., Inc., New York; Akademiai Kiado, Budapest 1970 xiii+389 pp.

[28] Yu.L. Šmulian, *Some points of the theory of operators with finite nonhermitian rank*, Matem. Sborn. **57** (99) (1962) 105–136, Russian.

[29] Ja.S. Švarcman, On invariant subspaces of a dissipative operator and the divisors of its characteristic function, Funkcional. Anal. i Priložen. **4** (1970), no. 4, 85–86. Engl. transl.: Functional Anal. Appl. **4** (1970), 342–343.

A. Kheifets
Department of Mathematical Sciences
University of Massachusetts Lowell
One University Avenue
Lowell, MA 01854, USA
e-mail: Alexander_Kheifets@uml.edu

Operator Theory:
Advances and Applications, Vol. 272, 235–247

Analytic Interpolation into the Tetrablock and a μ-synthesis Problem

Zinaida A. Lykova, N.J. Young and Amos E. Ajibo

To Joe Ball in esteem and friendship

Abstract. We give a solvability criterion for a special case of the μ-synthesis problem. That is, we prove the necessity and sufficiency of a condition for the existence of an analytic 2×2 matrix-valued function on the disc subject to a bound on the structured singular value and satisfying a finite set of interpolation conditions. To do this we prove a realization theorem for analytic functions from the disc to the tetrablock. We also obtain a solvability criterion for the problem of analytic interpolation from the disc to the tetrablock.

Mathematics Subject Classification (2010). 32F45, 30E05, 93B36, 93B50.

Keywords. Tetrablock, interpolation, matricial Nevanlinna–Pick problem, μ-synthesis problem.

1. Introduction

The application of operators on Hilbert space to complex analysis has enjoyed about a century of fruitful development since G. Pick gave it a spectacular impulse in [26]. For the second half of the century Joseph Ball been a prolific contributor to the theory. One important strand in his research has been to broaden the range of application of operator-theoretic methods to ever new areas of function theory. In this paper we also take a small step in extending established methods to a new type of domain.

Many of the operator-theoretic methods developed over the first half-century work most smoothly for problems involving analytic functions from the open unit disc $\mathbb{D}$ into certain very special domains, such as Cartan domains. Such is the case for the methods originated by D. Sarason [27] and V.M. Adamyan, D.Z. Arov and

The first and second authors were partially supported by the UK Engineering and Physical Sciences Research Council grants EP/N03242X/1. The third author was supported by the Government of Nigeria.

M.G. Kreĭn [1]. Ball has been prominent from an early stage both in introducing new approaches to the classical problems and in extending the class of domains to which the methods apply. As examples of new approaches, witness a series of joint papers with J. W. Helton [10] in the 1980s on the application of a novel Lax–Beurling theorem in H^2 spaces over Kreĭn spaces, and a little later, on the exploitation of J-inner-outer factorization of operator-valued analytic functions on the disc. More recently Ball has been active in the extension of operator-theoretic methods to multivariable, time-varying, nonlinear and noncommutative variants of classical problems [5, 6, 8]. It remains a worthwhile goal to explore the reach of operator-theoretic methods in function theory, and in this paper we apply them to the *tetrablock*, the domain in $\mathbb{C}^3$ defined by

$$\mathcal{E} = \{(x_1, x_2, x_3) \in \mathbb{C}^3 : 1 - x_1 z - x_2 w + x_3 zw \neq 0 \text{ for all } z, w \in \overline{\mathbb{D}}\}. \tag{1.1}$$

Operator-theoretic techniques can be applied to many problems in analysis. One that is relevant to this paper is the problem of analytic interpolation:

Given distinct points $\lambda_1, \dots, \lambda_N$ in a domain D in some $\mathbb{C}^d$ and target points $w_1, \dots, w_N$ in a subset E of some Banach space X, determine whether there exists an analytic map $F : D \to E$ such that $F(\lambda_j) = w_j$ for $j = 1, \dots, N$.

We shall call such a problem a *finite interpolation problem for* $\mathrm{Hol}(D, E)$, the latter symbol denoting the set of holomorphic maps from D to E.

The exemplar for solutions of finite interpolation problems of this type is the above-mentioned theorem of Pick [26], which provides an elegant criterion for the existence of an interpolating function F in the case that $D = \mathbb{D}$ and $E = \overline{\mathbb{D}}$. Among the many papers in the literature stemming from Pick's theorem, we mention a relevant paper of Ball and Bolotnikov [4], which extends Pick's theorem (and related results) to the case that the target set E has the form

$$E = \{z \in \mathbb{C}^d : \|P(z)\| \leq 1\} \tag{1.2}$$

where P is a matrix-valued polynomial in d variables and $\|\cdot\|$ is the operator norm with respect to the Euclidean norms. It seems to be difficult to extend established operator-theoretic methods much beyond this class of target sets while preserving the concreteness of Pick's original criterion, although there are extensions that make use of more abstract notions [5, 17].

In this paper we extend methods from [3, 12] to study the analytic interpolation problem in which the domain D is $\mathbb{D}$ and the target set E is the closure $\overline{\mathcal{E}}$ of the tetrablock in $\mathbb{C}^3$. The closed tetrablock, being the closure of $\mathcal{E}$, is given by

$$\overline{\mathcal{E}} = \{(x_1, x_2, x_3) \in \mathbb{C}^3 : 1 - x_1 z - x_2 w + x_3 zw \neq 0 \text{ for all } z, w \in \mathbb{D}\}. \tag{1.3}$$

$\overline{\mathcal{E}}$ is not of the form (1.2), and so its function theory is not accessible to the results of [4]. $\overline{\mathcal{E}}$ is the closure of a domain which is both inhomogeneous and non-convex, and so it is to be expected that its function theory should be more complicated in some respects than that of the classical domains. We derive a criterion for solvability (Theorem 4.1) of the finite interpolation problem for $\mathrm{Hol}(\mathbb{D}, \overline{\mathcal{E}})$, which while concrete and potentially capable of numerical checking, is not verifiable in rational arithmetic (unlike Pick's criterion). The original motivation for the study of the tetrablock was its connection with a special case of the μ-synthesis problem, which arises in the theory of robust control. H^∞ control is another topic on which Ball has published widely (for example, [7, 8]). We explain the connection between the tetrablock and a case of the μ-synthesis problem in Section 5 below.

In addition to a criterion for interpolation we also prove two further results in the function theory associated with $\mathcal{E}$. The first is a realization formula for analytic maps from $\mathbb{D}$ to $\overline{\mathcal{E}}$ (Theorem 3.2) and the second is a solvability criterion for a certain interpolation problem for matrix-valued analytic functions on $\mathbb{D}$, which is a case of the μ-synthesis problem (Theorem 5.3).

2. The tetrablock

The complex geometry, function theory and operator theory of $\mathcal{E}$ have attracted much attention over the past 10 years, for example [11, 12, 19–21, 24, 25, 28]. Although $\mathcal{E}$ was first studied because of its application to a problem in control theory, it has turned out to be interesting to specialists in several complex variables and operator theory.

We shall require some elementary properties of the tetrablock. The following function plays an important role [2].

Definition 2.1. The rational function Ψ is defined for $(z, x_1, x_2, x_3) \in \mathbb{C}^4$ such that $x_2 z \neq 1$ by

$$\Psi(z, x_1, x_2, x_3) = \frac{x_3 z - x_1}{x_2 z - 1}.$$

Ψ is analytic on the complement of the variety $x_2 z = 1$ in $\mathbb{C}^4$. Note that, for $x \in \mathbb{C}^3$ such that $x_1 x_2 = x_3$, the function $\Psi(\cdot, x)$ is constant and equal to x_1.

Numerous characterizations of the closed tetrablock are given in [2, Theorem 2.4]. Here are four of them.

Proposition 2.2. *Let $x = (x_1, x_2, x_3) \in \mathbb{C}^3$. The following are equivalent.*

1. *$x \in \overline{\mathcal{E}}$;*
2. *$|\Psi(z, x)| \leq 1$ for all $z \in \mathbb{D}$ and if $x_1 x_2 = x_3$ then, in addition, $|x_2| \leq 1$;*
3. *$|x_2 - \overline{x_1} x_3| + |x_1 x_2 - x_3| \leq 1 - |x_1|^2$ and if $x_1 x_2 = x_3$ then, in addition, $|x_2| \leq 1$;*
4. *$|x_1 - \overline{x_2} x_3| + |x_1 x_2 - x_3| \leq 1 - |x_2|^2$ and if $x_1 x_2 = x_3$ then, in addition, $|x_1| \leq 1$;*
5. *there is a 2×2 matrix $A = [a_{ij}]_{i,j=1}^2$ such that $\|A\| \leq 1$ and $x = (a_{11}, a_{22}, \det A)$.*

We denote by $\mathcal{A}(\mathcal{E})$ the algebra of continuous functions on $\overline{\mathcal{E}}$ that are analytic on $\mathcal{E}$. By [2, Theorem 2.9], $\overline{\mathcal{E}}$ is polynomially convex, and so the maximal ideal space of $\mathcal{A}(\mathcal{E})$ is $\overline{\mathcal{E}}$. Hence the Shilov boundary of $\mathcal{A}(\mathcal{E})$ is a subset of $\overline{\mathcal{E}}$, called the distinguished boundary of $\mathcal{E}$ and denoted by $b\mathcal{E}$. The following alternative descriptions of $b\mathcal{E}$ (among others) are given in [2, Theorem 7.1].

Proposition 2.3. *Let $x = (x_1, x_2, x_3) \in \mathbb{C}^3$. The following are equivalent.*

1. $x \in b\mathcal{E}$;
2. $x \in \overline{\mathcal{E}}$ *and* $|x_3| = 1$;
3. $x_1 = \overline{x_2} x_3$, $|x_3| = 1$ *and* $|x_2| \le 1$.

By [2, Corollary 7.2], $b\mathcal{E}$ is homeomorphic to $\overline{\mathbb{D}} \times \mathbb{T}$. We denote the unit circle by $\mathbb{T}$.

Definition 2.4. An *$\mathcal{E}$-inner function* is an analytic function $\varphi : \mathbb{D} \to \overline{\mathcal{E}}$ such that the radial limit

$$\lim_{r \to 1-} \varphi(r\lambda) \text{ exists and belongs to } b\mathcal{E} \tag{2.1}$$

for almost all $\lambda \in \mathbb{T}$ with respect to Lebesgue measure.

By Fatou's Theorem, the radial limit (2.1) exists for almost all $\lambda \in \mathbb{T}$ with respect to Lebesgue measure. Note that, for an $\mathcal{E}$-inner function $\varphi = (\varphi_1, \varphi_2, \varphi_3) : \mathbb{D} \to \overline{\mathcal{E}}$, φ_3 is an inner function on $\mathbb{D}$ in the classical sense.

A finite interpolation problem for $\mathrm{Hol}(\mathbb{D}, \overline{\mathcal{E}})$ has a solution if and only if it has a rational $\mathcal{E}$-inner solution – see [12, Theorem 8.1].

3. Realization formulae and the tetrablock

A *realization formula* for a class of functions is an expression for a general function in the class in terms of operators on Hilbert space. We shall give a realization formula for the class $\mathrm{Hol}(\mathbb{D}, \overline{\mathcal{E}})$.

The *Schur class* of operator-valued or matricial functions (of a given type) is the set of analytic operator- or matrix-valued functions F on $\mathbb{D}$ bounded by 1 in norm, that is, satisfying

$$||F(\lambda)|| \le 1 \qquad \text{for all } \lambda \in \mathbb{D},$$

where $||\cdot||$ denotes the operator norm. In particular, the space of analytic 2×2 matrix functions F on $\mathbb{D}$ such that $||F|| \le 1$ for all $\lambda \in \mathbb{D}$ is called the 2×2 *Schur class* and is denoted by $\mathcal{S}^{2\times 2}$.

The best-known realization theorem is for the Schur class (see for example [23, Chapter VI.3]). If, for some Hilbert spaces H, U and Y,

$$\begin{bmatrix} A & B \\ C & D \end{bmatrix} : H \oplus U \to H \oplus Y \tag{3.1}$$

is a contractive operator, then, for any $z \in \mathbb{D}$,

$$||D + Cz(1 - Az)^{-1}B|| \le 1.$$

Conversely, any function in the Schur class (of functions from $\mathbb{D}$ to the space of bounded linear operators from U to Y) has such a representation, in which the block operator matrix (3.1) is a unitary operator from $H \oplus U$ to $H \oplus Y$.

It will be convenient to use some standard engineering notation. If H, U and Y are Hilbert spaces and

$$A : H \to H, \qquad B : U \to H,$$
$$C : H \to Y, \qquad D : U \to Y$$

are bounded linear operators, then we define

$$\left[\begin{array}{c|c} A & B \\ \hline C & D \end{array}\right]$$

to be the operator-valued function

$$z \mapsto D + Cz(1 - Az)^{-1}B : U \to Y$$

defined for all $z \in \mathbb{C}$ such that $1 - Az$ is invertible.

The following result [12, Theorem 7.1] relates $\operatorname{Hol}(\mathbb{D}, \overline{\mathcal{E}})$ to $\mathcal{S}^{2\times 2}$.

Proposition 3.1. *Let $x \in \operatorname{Hol}(\mathbb{D}, \overline{\mathcal{E}})$. There exists a unique function*

$$F = [F_{ij}]_1^2 \in \mathcal{S}^{2\times 2}$$

such that

$$x = (F_{11}, F_{22}, \det F), \tag{3.2}$$

where $(\det F)(z)$ *denotes the determinant of the matrix* $\det F(z)$ *for* $z \in \mathbb{D}$, *and*

$$|F_{12}| = |F_{21}| \text{ a.e. on } \mathbb{T},\ F_{21} \text{ is either 0 or outer, and } F_{21}(0) \geq 0.$$

Moreover, for all $\mu, \lambda \in \mathbb{D}$ and all $w, z \in \mathbb{C}$ such that

$$1 - F_{22}(\mu)w \neq 0 \text{ and } 1 - F_{22}(\lambda)z \neq 0,$$

$$\begin{aligned} 1 - \overline{\Psi(w, x(\mu))}\Psi(z, x(\lambda)) = (1 - \overline{w}z)\overline{\gamma(\mu, w)}\gamma(\lambda, z) \\ + \eta(\mu, w)^*(I - F(\mu)^*F(\lambda))\eta(\lambda, z), \end{aligned} \tag{3.3}$$

where

$$\gamma(\lambda, z) := (1 - F_{22}(\lambda)z)^{-1}F_{21}(\lambda) \text{ and } \eta(\lambda, z) := \begin{bmatrix} 1 \\ z\gamma(\lambda, z) \end{bmatrix}. \tag{3.4}$$

Conversely, if $F \in \mathcal{S}^{2\times 2}$ then

$$(F_{11}, F_{22}, \det F) \in \operatorname{Hol}(\mathbb{D}, \overline{\mathcal{E}}).$$

Here is an outline of the construction of F for a given $x \in \operatorname{Hol}(\mathbb{D}, \overline{\mathcal{E}})$. Certainly $|x_1(\lambda)|,\ |x_2(\lambda)| \leq 1$ for all $\lambda \in \mathbb{D}$. If $x_1x_2 = x_3$, then we may simply define F by

$$F = \begin{bmatrix} x_1 & 0 \\ 0 & x_2 \end{bmatrix}.$$

In the case that $x_1x_2 \neq x_3$, the H^∞ function $x_1x_2 - x_3$ is nonzero and so it has an inner-outer factorization which can be written in the form

$$x_1x_2 - x_3 = \varphi e^C,$$

where φ is inner, e^C is outer and $\mathrm{e}^C(0) \geq 0$. Let F be defined by

$$F = \begin{bmatrix} x_1 & \varphi e^{\frac{1}{2}C} \\ e^{\frac{1}{2}C} & x_2 \end{bmatrix}.$$

Then clearly

$$\det F = x_1x_2 - \varphi \mathrm{e}^C = x_1x_2 - x_1x_2 + x_3 = x_3$$

and

$$|F_{12}| = \mathrm{e}^{\mathrm{Re}\,\frac{1}{2}C} = |F_{21}| \text{ a.e. on } \mathbb{T},\ F_{21} \text{ is outer, and } F_{21}(0) \geq 0.$$

The crux of the proof in [12] is to show that F is in the Schur class.

We may combine Proposition 3.1 with the classical realization formula to obtain a realization for $\mathrm{Hol}(\mathbb{D}, \overline{\mathcal{E}})$.

Theorem 3.2. *A function*

$$x = (x_1, x_2, x_3) : \mathbb{D} \to \mathbb{C}^3$$

maps $\mathbb{D}$ analytically into $\overline{\mathcal{E}}$ if and only if there exist a Hilbert space H and a unitary operator

$$\begin{bmatrix} A & B \\ C & D \end{bmatrix} : H \oplus \mathbb{C}^2 \to H \oplus \mathbb{C}^2 \tag{3.5}$$

such that, for all $\lambda \in \mathbb{D}$,

$$x_1(\lambda) = \left[\begin{array}{c|c} A & B_1 \\ \hline C_1 & D_{11} \end{array}\right](\lambda) = D_{11} + C_1\lambda(1 - A\lambda)^{-1}B_1, \tag{3.6}$$

$$x_2(\lambda) = \left[\begin{array}{c|c} A & B_2 \\ \hline C_2 & D_{22} \end{array}\right](\lambda) = D_{22} + C_2\lambda(1 - A\lambda)^{-1}B_2, \tag{3.7}$$

and

$$x_3(\lambda) = \det \left[\begin{array}{c|c} A & B \\ \hline C & D \end{array}\right](\lambda) = \det[D + C\lambda(1 - A\lambda)^{-1}B], \tag{3.8}$$

where

$$B = [B_1 \quad B_2] : \mathbb{C}^2 \to H, \quad C = \begin{bmatrix} C_1 \\ C_2 \end{bmatrix} : H \to \mathbb{C}^2 \text{ and } D = \left[D_{ij}\right]_{i,j=1}^2.$$

Proof. Let $x : \mathbb{D} \to \overline{\mathcal{E}}$ be analytic. By Proposition 3.1 there exists F in the Schur class such that

$$x = (F_{11}, F_{22}, \det F).$$

By the realization theorem for the Schur class there exist a Hilbert space H and a unitary operator $\begin{bmatrix} A & B \\ C & D \end{bmatrix}$ on $H \oplus \mathbb{C}^2$ such that, for all $\lambda \in \mathbb{D}$,

$$\begin{aligned} F(\lambda) &= \left[\begin{array}{c|c} A & B \\ \hline C & D \end{array}\right](\lambda) \\ &= D + C\lambda(1 - A\lambda)^{-1}B \\ &= \begin{bmatrix} D_{11} & D_{12} \\ D_{21} & D_{22} \end{bmatrix} + \begin{bmatrix} C_1 \\ C_2 \end{bmatrix} \lambda(1 - A\lambda)^{-1}[B_1\ B_2]. \end{aligned}$$

Thus

$$F_{11} = \left[\begin{array}{c|c} A & B_1 \\ \hline C_1 & D_{11} \end{array}\right],$$

$$F_{22} = \left[\begin{array}{c|c} A & B_2 \\ \hline C_2 & D_{22} \end{array}\right],$$

$$\det F = \det \left[\begin{array}{c|c} \mathrm{A} & \mathrm{B} \\ \hline \mathrm{C} & \mathrm{D} \end{array}\right],$$

and so equations (3.6), (3.7) and (3.8) hold.

Conversely, let a Hilbert space H and operators A, B, C and D satisfy displayed formulae (3.5) to (3.8). Then x is clearly analytic in $\mathbb{D}$. We must show that $x(\mathbb{D}) \subset \overline{\mathcal{E}}$. Let

$$\chi = \left[\begin{array}{c|c} A & B \\ \hline C & D \end{array}\right] = [\chi_{ij}],$$

so that $\|\chi\|_\infty \leq 1$ by the realization theorem for the Schur class. We have

$$\chi_{jj} = \left[\begin{array}{c|c} A & B_j \\ \hline C_j & D_{jj} \end{array}\right], \qquad j = 1, 2,$$

and so

$$x_1 = \chi_{11}, \qquad x_2 = \chi_{22} \qquad \text{and} \qquad x_3 = \det \chi.$$

Hence, for any $\lambda \in \mathbb{D}$,

$$x(\lambda) = (\chi_{11}(\lambda), \chi_{22}(\lambda), \det \chi(\lambda))$$

where $\chi(\lambda)$ is a 2×2 matrix of norm at most 1. Hence, by Proposition 2.2, for any $\lambda \in \mathbb{D}$, $x(\lambda) \in \overline{\mathcal{E}}$. Thus $x(\mathbb{D}) \subset \overline{\mathcal{E}}$. □

4. The finite interpolation problem for $\mathrm{Hol}(\mathbb{D}, \overline{\mathcal{E}})$

In this section we prove the main theorem of the paper, a criterion for the solvability of the problem in the title of the section.

By *matricial Nevanlinna–Pick data* we mean a finite set $\lambda_1, \ldots, \lambda_n$ of distinct points in $\mathbb{D}$, where $n \in \mathbb{N}$, and an equal number of "target matrices" $W_1, \ldots, W_n$, of type (say) $m \times p$. We write these data

$$\lambda_k \mapsto W_k, \qquad k = 1, \ldots, n. \tag{4.1}$$

We say that these data are *solvable* if there exists a function F in the Schur class such that $F(\lambda_k) = W_k$, $k = 1, \ldots, n$. The *Nevanlinna–Pick problem* is to ascertain whether prescribed data are solvable. By the classical theorem of Pick, or more precisely its extension to matricial data (for example, [9]), the Nevanlinna–Pick problem with data (4.1) is solvable if and only if the "Pick matrix"

$$\left[\frac{I - W_k^* W_\ell}{1 - \overline{\lambda}_k \lambda_\ell}\right]_{k,\ell=1}^n$$

is positive.

The following is our criterion for the solvability of an interpolation problem for $\mathrm{Hol}(\mathbb{D},\overline{\mathcal{E}})$. It relates the problem to a family of classical matricial Nevanlinna–Pick problems.

Theorem 4.1. *Let $\lambda_1,\dots,\lambda_n$ be distinct points in $\mathbb{D}$ and let $(x_1^k,x_2^k,x_3^k)\in\mathcal{E}$ for $k=1,\dots,n$. The following statements are equivalent.*

(1) *There exists an analytic function $x:\mathbb{D}\to\overline{\mathcal{E}}$ such that*

$$x(\lambda_k)=(x_1^k,x_2^k,x_3^k),\quad k=1,\dots,n; \tag{4.2}$$

(2) *there exist $b_k,c_k\in\mathbb{C}$ such that*

$$b_kc_k=x_1^kx_2^k-x_3^k,\qquad k=1,\dots,n, \tag{4.3}$$

and the Nevanlinna–Pick problem with data

$$\lambda_k\mapsto\begin{bmatrix} x_1^k & b_k\\ c_k & x_2^k\end{bmatrix},\qquad k=1,\dots,n, \tag{4.4}$$

is solvable.

Proof. (1)⇒(2) Suppose there is an analytic function $x:\mathbb{D}\to\overline{\mathcal{E}}$ such that equation (4.2) holds. By Proposition 3.1, there is a function F in the 2×2 Schur class, that is, that $\|F\|_\infty\le 1$ such that

$$x=(F_{11},\ F_{22},\ \det F). \tag{4.5}$$

Let $b_k=F_{12}(\lambda_k)$ and $c_k=F_{21}(\lambda_k)$, $k=1,\dots,n$. Then

$$F(\lambda_k)=\begin{bmatrix} x_1(\lambda_k) & F_{12}(\lambda_k)\\ F_{21}(\lambda_k) & x_2(\lambda_k)\end{bmatrix}=\begin{bmatrix} x_1^k & b_k\\ c_k & x_2^k\end{bmatrix},$$

and so

$$x_3^k=x_3(\lambda_k)=x_1^kx_2^k-b_kc_k.$$

Thus

$$b_kc_k=x_1^kx_2^k-x_3^k,\quad k=1,\dots,n.$$

Hence the equations (4.3) are satisfied, and for this choice of b_k,c_k the matricial Nevanlinna–Pick problem with the data (4.4) is solvable by F. Condition (2) of the theorem is satisfied.

(2)⇒(1) Suppose that b_k, c_k exist such that the equations (4.3) hold and the Nevanlinna–Pick problem with data (4.4) is solvable by a function $F\in\mathcal{S}^{2\times 2}$. Let

$$x=(F_{11},F_{22},\det F).$$

By Proposition 2.2, $x\in\mathrm{Hol}(\mathbb{D},\overline{\mathcal{E}})$.

Since conditions (4.4) hold, for $k=1,\dots,n$,

$$\begin{aligned} x_1(\lambda_k)&=F_{11}(\lambda_k)=x_1^k,\\ x_2(\lambda_k)&=F_{22}(\lambda_k)=x_2^k,\\ x_3(\lambda_k)&=\det F(\lambda_k)=x_1(\lambda_k)x_2(\lambda_k)-b_kc_k=x_3^k. \end{aligned}$$

Thus condition (1) of the theorem holds. □

As a consequence we obtain a solvability criterion for a finite interpolation problem for $\mathrm{Hol}(\mathbb{D},\overline{\mathcal{E}})$ in terms of the positivity of one of a family of matrices.

Corollary 4.2. *Let $\lambda_1,\dots,\lambda_n$ be distinct points in $\mathbb{D}$ and let $(x_1^k,x_2^k,x_3^k)\in\mathcal{E}$ such that $x_1^k x_2^k\neq x_3^k$ for $k=1,\dots,n$. The following three statements are equivalent.*

(1) *There exists a function $x\in\mathrm{Hol}(\mathbb{D},\overline{\mathcal{E}})$ such that*

$$x(\lambda_k)=(x_1^k,x_2^k,x_3^k)\quad \text{for } k=1,\dots,n;$$

(2) *there exist $b_1,\dots,b_n$ and $c_1,\dots,c_n\in\mathbb{C}$ such that $b_k c_k = x_1^k x_2^k - x_3^k$ for $k=1,\dots,n$ and*

$$\left[\frac{I-\begin{bmatrix} x_1^k & b_k \\ c_k & x_2^k\end{bmatrix}^*\begin{bmatrix} x_1^\ell & b_\ell \\ c_\ell & x_2^\ell\end{bmatrix}}{1-\overline{\lambda_k}\lambda_\ell}\right]_{k,\ell=1}^{n}\geq 0. \tag{4.6}$$

(3) *there exists a rational $\mathcal{E}$-inner function $x:\mathbb{D}\to\overline{\mathcal{E}}$ such that*

$$x(\lambda_k)=(x_1^k,x_2^k,x_3^k),\quad k=1,\dots,n. \tag{4.7}$$

The statement (1)⇔(2) follows immediately from Theorem 4.1 and Pick's theorem. The statement that (1)⇔(3) is contained in [12, Theorem 8.1].

5. The structured singular value and the tetrablock

The original motivation for the study of the tetablock in [2] was the wish to throw light on the "problem of μ-synthesis", which arises in the theory of robust control [13, 15, 16, 18]. Operator theorists were greatly intrigued to learn around 1980 that some of their favourite theorems, such as those of Pick and Nehari, played a significant role in some problems of engineering design. It was immediately a challenge to extend those theorems to provide an analysis of some of the more subtle optimization problems posed by engineers. Up to now operator theorists have had limited success in meeting this challenge, and so it remains relevant to analyse test cases of these optimization problems.

The symbol μ is used to denote the *structured singular value* of a matrix relative to a space of linear transformations – see [14, 18] for full definitions. The structured singular value is a cost function which generalises the operator norm and is designed to reflect structural information about modelling uncertainty. The usual operator norm of a matrix and the spectral radius of a square matrix are both instances of μ. Accordingly, two special cases of the μ-synthesis problem are the classical Nevanlinna–Pick problem and its spectral variant, in which the operator norm is replaced by the spectral radius. The tetrablock is associated with a third special case of μ, which we denote by μ_{Diag}.

Definition 5.1. Diag denotes the space of diagonal 2×2 matrices over $\mathbb{C}$. For any 2×2 matrix A,

$$\mu_{\mathrm{Diag}}(A) \stackrel{\mathrm{def}}{=} (\inf\{\|X\| : X \in \mathrm{Diag},\ 1 - AX \text{ is singular}\})^{-1}.$$

In the event that $1 - AX$ is nonsingular for every $X \in \mathrm{Diag}$ we define $\mu_{\mathrm{Diag}}(A)$ to be 0.

The tetrablock is connected to μ_{Diag} by the simple fact [2, Theorem 9.1] that a 2×2 matrix $A = [a_{ij}]$ satisfies $\mu_{\mathrm{Diag}}(A) < 1$ if and only if $(a_{11}, a_{22}, \det A) \in \mathcal{E}$.

The μ_{Diag}*-synthesis problem* is the following special case of μ-synthesis.

Given distinct points $\lambda_1, \dots, \lambda_n$ in $\mathbb{D}$ and 2×2 matrices $W_1, \dots, W_n$, where $n \geq 1$, find conditions for the existence of an analytic 2×2 matrix-valued function F on $\mathbb{D}$ such that

$$F(\lambda_k) = W_k, \qquad k = 1, 2, \dots, n,$$

and

$$\mu_{\mathrm{Diag}}(F(\lambda)) \leq 1 \quad \text{for all } \lambda \in \mathbb{D}.$$

There is a Matlab package [22] for the numerical solution of this problem, but as yet no very efficient algorithm and little supporting theory. The problem can be reduced to the finite interpolation problem for $\mathrm{Hol}(\mathbb{D}, \overline{\mathcal{E}})$, as is shown in the following result, which is [2, Theorem 9.2].

Proposition 5.2. *Let $\lambda_1, \dots, \lambda_n$ be distinct points in $\mathbb{D}$ and let*

$$W_k = \left[w_{ij}^k\right]_{i,j=1}^2 \qquad \text{for } k = 1, \dots, n,$$

be 2×2 matrices such that $w_{12}^k w_{21}^k \neq 0$ for $k = 1, \dots, n$. The following statements are equivalent.

1. *There exists an analytic function $F : \mathbb{D} \to \mathbb{C}^{2\times 2}$ such that*

$$F(\lambda_k) = W_k \quad \text{for } k = 1, \dots, n$$

and

$$\sup_{\lambda \in \mathbb{D}} \mu_{\mathrm{Diag}}(F(\lambda)) \leq 1; \tag{5.1}$$

2. *there exists an analytic function $\varphi \in \mathrm{Hol}(\mathbb{D}, \overline{\mathcal{E}})$ such that*

$$\varphi(\lambda_k) = (w_{11}^k, w_{22}^k, \det W_k) \quad \text{for } k = 1, \dots, n. \tag{5.2}$$

On combining Proposition 5.2 with Theorem 4.1 we obtain a criterion for the solvability of a μ_{Diag}-synthesis problem.

Theorem 5.3. *Let $\lambda_1, \dots, \lambda_n$ be distinct points in $\mathbb{D}$ and let*

$$W_k = \left[w_{ij}^k\right]_{i,j=1}^2 \qquad \text{for } k = 1, \dots, n$$

be 2×2 matrices such that $w_{12}^k w_{21}^k \neq 0$ for $k = 1, \dots, n$. The following statements are equivalent.

1. *There exists an analytic function $F : \mathbb{D} \to \mathbb{C}^{2\times 2}$ such that*

$$F(\lambda_k) = W_k \quad \text{for } k = 1, \dots, n$$

and

$$\sup_{\lambda \in \mathbb{D}} \mu_{\mathrm{Diag}}(F(\lambda)) \leq 1; \tag{5.3}$$

2. *there exist* $b_1, \dots, b_n$ *and* $c_1, \dots, c_n \in \mathbb{C}$ *such that*

$$b_k c_k = w_{11}^k w_{22}^k - \det W_k \quad \textit{for } k = 1, \dots, n$$

and

$$\left[\frac{I - \begin{bmatrix} w_{11}^k & b_k \\ c_k & w_{22}^k \end{bmatrix}^* \begin{bmatrix} w_{11}^\ell & b_\ell \\ c_\ell & w_{22}^\ell \end{bmatrix}}{1 - \overline{\lambda_k} \lambda_\ell} \right]_{k,\ell=1}^n \geq 0. \tag{5.4}$$

Proof. By Theorem 5.2, since $w_{12}^k w_{21}^k \neq 0$ for $k = 1, \dots, n$, condition (1) is equivalent to the existence of $\varphi \in \mathrm{Hol}(\mathbb{D}, \overline{\mathcal{E}})$ such that

$$\varphi(\lambda_k) = (w_{11}^k, w_{22}^k, \det W_k) \quad \text{for } k = 1, \dots, n.$$

Hence, by Theorem 4.1, the matricial Nevanlinna–Pick problem with data

$$\lambda_k \mapsto \begin{bmatrix} w_{11}^k & b_k \\ c_k & w_{22}^k \end{bmatrix}, \qquad k = 1, \dots, n,$$

is solvable for some $b_k, c_k \in \mathbb{C}$ satisfying

$$b_k c_k = w_{11}^k w_{22}^k - \det W_k \quad \text{for } k = 1, \dots, n.$$

By the matricial version of Pick's theorem, this matricial Nevanlinna–Pick problem is solvable if and only if the Pick condition (5.4) is satisfied. □

The proof suggests a way to construct solutions of a μ_{Diag}-synthesis problem. Suppose we are given $\lambda_1, \dots, \lambda_n, W_1, \dots, W_n$ such that condition (2) of Theorem 5.3 is satisfied, and we can somehow find suitable numbers $b_1, \dots, b_n, c_1, \dots, c_n$ to make the Pick matrix positive. This is of course a nonconvex problem, and we do not know of a good algorithm to find the b_k and c_k. Once they are found, however, there are various ways of constructing functions χ in the 2×2 Schur class such that

$$\chi(\lambda_k) = \begin{bmatrix} w_{11}^k & b_k \\ c_k & w_{22}^k \end{bmatrix}, \qquad k = 1, \dots, n,$$

(see, for example, [9]). The function

$$\varphi = (\chi_{11}, \chi_{22}, \det \chi)$$

is then an analytic function from $\mathbb{D}$ to $\overline{\mathcal{E}}$ such that

$$\varphi(\lambda_k) = (w_{11}^k, w_{22}^k, \det W_k), \qquad k = 1, \dots, n,$$

and the outline following Proposition 3.1 shows how to use φ to construct an analytic 2×2 matrix-valued function F in $\mathbb{D}$ such that $F(\lambda_k) = W_k$ and $\mu_{\mathrm{Diag}}(F(\lambda)) \leq 1$ for all $\lambda \in \mathbb{D}$.

References

[1] V.M. Adamyan, D.Z. Arov and M.G. Kreĭn, Analytic properties of Schmidt pairs of a Hankel operator and generalized Schur–Takagi problem, *Mat. Sbornik* **86** (1971) 33–73.

[2] A.A. Abouhajar, M.C. White and N.J. Young, A Schwarz lemma for a domain related to μ-synthesis, *J. Geom. Anal.*, **17** (4) (2007) 717–750.

[3] J. Agler and N.J. Young, The two-point spectral Nevanlinna–Pick problem, *Integral Equ. Oper. Theory* **37** (2000) 375–385.

[4] J.A. Ball and V. Bolotnikov, Nevanlinna–Pick interpolation for Schur–Agler class functions on domains with matrix polynomial defining function in $\mathbb{C}^n$, *New York J. Math.* **11** (2005) 247–290.

[5] J.A. Ball and M.D. Guerra Huamán, Test functions, Schur–Agler classes and transfer-function realizations: the matrix-valued setting, *Complex Anal. Oper. Theory* **7** (2003) 529–575.

[6] J.A. Ball, G. Marx and V. Vinnikov, Interpolation and transfer-function realization for the noncommutative Schur–Agler class, *Operator theory in different settings and related applications*, pp. 23–116, Oper. Theory Adv. Appl., **OT 263**, Birkhäuser/Springer, Cham, 2018.

[7] J.A. Ball and S. ter Horst, Robust control, multidimensional systems and multivariable Nevanlinna–Pick interpolation, In: Topics in Operator Theory, pp. 13–88, **OT 203** Birkhäuser, Basel, 2010.

[8] J.A. Ball and S. ter Horst, Multivariable operator-valued Nevanlinna–Pick interpolation: a survey, In: Operator Algebras, Operator Theory and Applications, pp. 1–73, **OT 195** Birkhäuser, Basel, 2010.

[9] J.A. Ball, I.C. Gohberg and L. Rodman, *Interpolation of Rational Matrix Functions*, Operator Theory: Advances and Applications **45**, Birkhäuser, Basel, 1990.

[10] J.A. Ball and J.W. Helton, Interpolation problems of Pick–Nevanlinna and Loewner types for meromorphic matrix functions: parametrization of the set of all solutions. *Integral Equ. Oper. Theory* **9** (1986) 155–203.

[11] T. Bhattacharyya, The tetrablock as a spectral set, *Indiana Univ. Math. J.* **63** (6) (2014) 1601–1629.

[12] D.C. Brown, Z.A. Lykova and N.J. Young, A rich structure related to the construction of holomorphic matrix functions, *J. Funct. Anal.* **272** (2017) 1704–1754.

[13] J.C. Doyle, Analysis of feedback systems with structured uncertainties. *Proc. IEE-D* **129** (1982) 242–250.

[14] J.C. Doyle and A. Packard, The complex structured singular value, *Automatica J. IFAC*, **29** (1993) 71–109.

[15] J.C. Doyle, Structured uncertainty in control system design, *24th IEEE Conference on Decision and Control*, **24** (1985) 260–265.

[16] J.C. Doyle and G. Stein, Multivariable feedback design: concepts for a classical/modern synthesis, *IEEE Transactions on Automatic Control*, **26** (1981) 4–16.

[17] M.A. Dritschel and S. McCullough, Test functions, kernels, realizations and interpolation, in *Operator Theory, Structured Matrices and Dilations: T. Constantinescu*

Memorial Volume (ed. M. Bakonyi, A. Gheondea, M. Putinar and J. Rovnyak), pp. 153–179, Theta Series in Advanced Mathematics, Bucharest, 2007.

[18] G.E. Dullerud and F.G. Paganini, *A Course in Robust Control Theory: A Convex Approach*, Springer, 2000.

[19] A. Edigarian, L. Kosiński and W. Zwonek, The Lempert theorem and the tetrablock, *J. Geom. Anal.* **23** (4) (2013) 1818–1831.

[20] M. Jarnicki and P. Pflug, *Invariant Distances and Metrics in Complex Analysis*, 2nd Extended Edition, De Gruyter, Berlin, 2013.

[21] L. Kosiński and W. Zwonek, Nevanlinna–Pick problem and uniqueness of left inverses in convex domains, symmetrized bidisc and tetrablock, *J. Geom. Anal.* **26** (2016) 1863–1890.

[22] *Matlab μ-Analysis and Synthesis Toolbox*, The Math Works Inc., Natick, Massachusetts (http://www.mathworks.com/products/muanalysis/).

[23] B. Sz-Nagy and C. Foiaş, *Harmonic Analysis of Operators on Hilbert Space*, Akadémiai Kiadó, Budapest, 1968.

[24] N. Nikolov, P.J. Thomas and M. Trybula, Gromov (non)hyperbolicity of certain domains in $\mathbb{C}^2$, *Forum Math.* **28** (2016) 783–794.

[25] S. Pal, The failure of rational dilation on the tetrablock, *J. Funct. Anal.* **269** (2015) 1903–1924.

[26] G. Pick, Über die Beschränkungen analytischer Funktionen, welche durch vorgegebene Funktionswerte bewirkt werden, *Math. Ann.* **77** (1916) 7–23.

[27] D. Sarason, Generalised interpolation in H^∞, *Trans. Amer. Math. Soc.* **127** (2) (1967) 179–203.

[28] N.J. Young, The automorphism group of the tetrablock, *J. London Math. Soc.* (2) **77** (2008) 757–770.

Zinaida A. Lykova and Amos E. Ajibo
School of Mathematics, Statistics and Physics
Newcastle University
Newcastle upon Tyne NE1 7RU, UK

N.J. Young
School of Mathematics, Statistics and Physics
Newcastle University
Newcastle upon Tyne NE1 7RU, UK

and

School of Mathematics
Leeds University
Leeds LS2 9JT, UK

Operator Theory:
Advances and Applications, Vol. 272, 249–305

A Gleason Solution Model for Row Contractions

Robert T.W. Martin and Andriamanankasina Ramanantoanina

Dedicated to, and inspired by research of, Joseph A. Ball

Abstract. In the de Branges–Rovnyak functional model for contractions on Hilbert space, any completely non-coisometric (CNC) contraction is represented as the adjoint of the restriction of the backward shift to a de Branges–Rovnyak space, $\mathscr{H}(b)$, associated to a contractive analytic operator-valued function, b, on the open unit disk.

We extend this model to a large class of CNC contractions of several copies of a Hilbert space into itself (including all CNC row contractions with commuting component operators). Namely, we completely characterize the set of all CNC row contractions, T, which are unitarily equivalent to an extremal Gleason solution for a de Branges–Rovnyak space, $\mathscr{H}(b_T)$, contractively contained in a vector-valued Drury–Arveson space of analytic functions on the open unit ball in several complex dimensions. Here, a Gleason solution is the appropriate several-variable analogue of the adjoint of the restricted backward shift and the characteristic function, b_T, belongs to the several-variable Schur class of contractive multipliers between vector-valued Drury–Arveson spaces. The characteristic function, b_T, is a unitary invariant, and we further characterize a natural sub-class of CNC row contractions for which it is a complete unitary invariant.

1. Introduction

The de Branges–Rovnyak and Sz.-Nagy–Foiaş functional models are two widely-used and powerful approaches to the representation theory of contractions on Hilbert space [20, 21, 50]. These two constructions provide equivalent models for completely non-unitary (CNU) contractions [7, 11, 43, 48]. In this paper we focus on the de Branges–Rovnyak model and its several-variable extension to the setting of contractions from several copies of a Hilbert space into itself.

The first author acknowledges support of NRF CPRR Grants 90551 and 105837.

A linear contraction, $T : \mathcal{H} \to \mathcal{J}$, between Hilbert spaces, $\mathcal{H}, \mathcal{J}$, is called *completely non-coisometric* (CNC) if it has no co-isometric restriction to a non-trivial subspace. In the de Branges–Rovnyak model for CNC contractions on a (single) Hilbert space, the model operator acts on the (unique) reproducing kernel Hilbert space (RKHS), $\mathcal{H}(k^b)$, associated to an operator-valued contractive analytic function b on the open unit disk, $\mathbb{D}$, in the complex plane [7, 11]. Here the positive, sesqui-analytic de Branges–Rovnyak kernel k^b is

$$k^b(z,w) := \frac{I - b(z)b(w)^*}{1 - zw^*}.$$

In the above $z^* := \overline{z}$ denotes complex conjugation. The RKHS $\mathscr{H}(b) := \mathcal{H}(k^b)$ is called the de Branges–Rovnyak space of b, and in the case where $b \equiv 0$ we recover the Szegő kernel for the classical Hardy space of analytic functions in the unit disk.

Recall that the *shift*, $S : H^2(\mathbb{D}) \to H^2(\mathbb{D})$ is the canonical isometry of multiplication by z and its adjoint, the *backward shift*, S^*, acts as the difference quotient:

$$(S^*h)(z) = \frac{h(z) - h(0)}{z}; \qquad h \in H^2(\mathbb{D}).$$

The shift plays a central role in the classical theory of Hardy spaces [25, 29, 42]. If T is any CNC contraction, there is a (essentially unique) contractive, operator-valued analytic function, b_T, on the unit disk, $b_T(z) \in \mathscr{L}(\mathcal{H}, \mathcal{K})$ (*i.e.*, a member of the operator-valued *Schur class*), so that T is unitarily equivalent to X where $X^* := S^*|_{\mathscr{H}(b_T)}$ is the restriction of the backward shift of the vector-valued Hardy space $H^2(\mathbb{D}) \otimes \mathcal{K}$ to the de Branges–Rovnyak space $\mathscr{H}(b_T)$ [7, 11]. (Any de Branges–Rovnyak space $\mathscr{H}(b)$ associated to a contractive, operator-valued analytic function, b, on the disk is always contractively contained in vector-valued Hardy space and is always co-invariant for the shift [48].) This provides a natural model for CNC contractions as adjoints of restrictions of the backward shift to de Branges–Rovnyak reproducing kernel Hilbert spaces, and this is the model we extend to several variables in this paper.

A canonical several-variable extension of the Hardy space of the disk is the Drury–Arveson space, H^2_d, the unique RKHS of analytic functions on the open unit ball $\mathbb{B}^d := (\mathbb{C}^d)_1$ corresponding to the several-variable Szegő kernel. (If Y is a Banach space, let $(Y)_1$ denote the open unit ball and let $[Y]_1$ denote the closed unit ball.) The Schur classes of contractive, operator-valued functions on the disk are promoted to the multi-variable Schur classes, $\mathscr{S}_d(\mathcal{J}, \mathcal{K})$, of contractive, operator-valued multipliers between vector-valued Drury–Arveson spaces $H^2_d \otimes \mathcal{J}$ and $H^2_d \otimes \mathcal{K}$ (see Subsection 2.1), and the appropriate analogue of the adjoint of the restricted backward shift in this several-variable setting is a contractive solution to the *Gleason problem* in $\mathscr{H}(b)$: A contraction $X = (X_1, \dots, X_d) : \mathscr{H}(b) \otimes \mathbb{C}^d \to \mathscr{H}(b)$ is called a *Gleason solution* (or a solution to the Gleason problem) for $\mathscr{H}(b)$ if its adjoint acts as the multi-variable difference quotient:

$$(zX^*f)(z) := z_1(X_1^*f)(z) + \cdots + z_d(X_d^*f)(z) = f(z) - f(0); \qquad \forall f \in \mathscr{H}(b).$$

The concepts of *contractive* and extremal contractive (or more simply, *extremal*) Gleason solution are further norm/positivity constraints and we will review these basic definitions in the upcoming Subsection 2.4, see Definition 2.5.

Given any contraction between Hilbert spaces, $T : \mathcal{H} \to \mathcal{K}$, recall that the defect operator, D_T of T is defined as

$$D_T := \sqrt{I - T^*T}.$$

We say that a *row contraction, i.e.*, a contraction from several copies of a Hilbert space into itself, $T = (T_1, \dots, T_d) : \mathcal{H} \otimes \mathbb{C}^d \to \mathcal{H}$, obeys the *commutative CNC condition*, and we write: T is CCNC if

$$\mathcal{H} = \bigvee_{z \in \mathbb{B}^d} (I - Tz^*)^{-1} \mathrm{Ran}\,(D_{T^*}).$$

Here, and throughout, $\bigvee$ denotes closed linear span. We will prove that any CCNC row contraction T is automatically CNC (Corollary 3.4), and that any *d-contraction* (a row contraction with d mutually commuting component operators) is CNC if and only if it is CCNC. One of the main results of this paper is the extension of the de Branges–Rovnyak model for CNC contractions to the class of all CCNC row contractions:

Theorem 1 (Theorem 5.14). *A row contraction $T : \mathcal{H} \otimes \mathbb{C}^d \to \mathcal{H}$ is CCNC (obeys the commutative CNC condition) if and only if it is unitarily equivalent to an extremal (contractive) Gleason solution in a multi-variable de Branges–Rovnyak space $\mathscr{H}(b)$ for a Schur-class multiplier $b \in \mathscr{S}_d(\mathcal{J}, \mathcal{K})$.*

If T is unitarily equivalent to an extremal Gleason solution X^b for $\mathscr{H}(b)$, then the characteristic function $b := b_T$ is a unitary invariant for T: If T_1, T_2 are unitarily equivalent CCNC row contractions, then b_{T_1} coincides weakly with b_{T_2}. One can choose $b_T \in \mathscr{S}_d(\mathrm{Ran}\,(D_T), \mathrm{Ran}\,(D_{T^*}))$.

Remark 1.1. In the above, two row contractions $T^{(k)} : \mathcal{H}_k \otimes \mathbb{C}^d \to \mathcal{H}_k$, $k = 1, 2$ are said to be unitarily equivalent, denoted $T^{(1)} \simeq T^{(2)}$, if there is an onto isometry $U : \mathcal{H}_1 \to \mathcal{H}_2$ so that

$$UT^{(1)} = T^{(2)}(U \otimes I_d).$$

Equivalently, $UT_j^{(1)}U^* = T_j^{(2)}; \quad 1 \leq j \leq d$. Two operator-valued Schur class functions $b_k \in \mathscr{S}_d(\mathcal{H}_k, \mathcal{J}_k)$ are said to *coincide weakly* if there is an onto isometry $U : \mathcal{J}_1 \to \mathcal{J}_2$ so that

$$Ub_1(z)b_1(w)^*U^* = b_2(z)b_2(w)^*; \qquad \forall\ z, w \in \mathbb{B}^d.$$

See Definition 4.11 and Definition 5.11 for the definitions of the Schur-class characteristic functions of a CNC row partial isometry, and an arbitrary CNC row contraction, respectively.

Remark 1.2. As we will show in Subsection 2.16, the above theorem is an easy consequence of the colligation and transfer-function realization theory of Ball–Bolotnikov–Fang [6, 9].

In [30, 31], the concept of a *quasi-extreme multiplier* for Drury–Arveson space was introduced. This concept is a several-variable extension of a 'Szegő approximation property', the salient idea being that this property is equivalent to being an extreme point of the Schur class in the classical, single-variable, scalar-valued setting (see Section 6 and Definition 6.3). Recently, it has been shown that quasi-extreme implies extreme in the scalar-valued, several-variable setting as well [33]. If b is a quasi-extreme Schur multiplier, then $\mathscr{H}(b)$ has a unique contractive (and extremal) Gleason solution (see Theorem 6.2). We say that a CCNC row contraction T is *quasi-extreme* (QE) if its characteristic function b_T is a quasi-extreme multiplier. Our second main result characterizes the class of all QE row contractions:

Theorem 2 (Theorem 6.9, Theorem 6.10). *A row contraction $T : \mathcal{H} \otimes \mathbb{C}^d \to \mathcal{H}$ is QE if and only if it is CCNC and obeys the QE condition:*

$$\operatorname{Ker}(T)^{\perp} \subseteq \bigvee_{z \in \mathbb{B}^d} z^*(I - Tz^*)^{-1}\operatorname{Ran}(D_{T^*}).$$

This happens if and only if T is unitarily equivalent to the (unique) extremal Gleason solution in a de Branges–Rovnyak space $\mathscr{H}(b)$ for a quasi-extreme $b \in \mathscr{S}_d(\mathcal{J}, \mathcal{K})$. The QE characteristic function $b_T := b$ of a QE row contraction T is a complete unitary invariant: two QE row contractions T_1, T_2 are unitarily equivalent if and only if their characteristic functions coincide weakly.

In summary, given the following strict hierarchy of classes of row contractions on Hilbert space,

$$CNU \supsetneqq CNC \supsetneqq CCNC \supsetneqq QE,$$

this paper constructs a commutative de Branges–Rovnyak model for the CCNC and QE classes.

Previous work on the representation theory and functional models for row contractions include [7, 10, 15–17, 45, 46]. The theory of Popescu [45, 46] for CNC row contractions constructs a Sz.-Nagy–Foiaş-type model by studying the structure of the space of the minimal row isometric dilation of the row contraction, and defines a non-commutative 'characteristic function' (*i.e.*, an element of the free Schur class, see [8]) which is a complete unitary invariant. This theory is extended to arbitrary CNU row contractions by Ball–Vinnikov in [15].

The papers [7, 16, 17] of Ball–Bolotnikov and Bhattacharyya–Eschmeier–Sarkar study the model theory of d-contractions (row contractions with d mutually commuting component operators). In [16, 17], the classical Sz.-Nagy–Foiaş model and the definition of the Sz.-Nagy–Foiaş characteristic function is extended to CNC d-contractions, and [7] recovers this theory as a special case of the general model developed in [15] for arbitrary CNU row contractions. In this model theory for d-contractions, the characteristic function is an element of the (operator-valued, several-variable) Schur classes, and this characteristic function is a complete unitary invariant for CNC d-contractions [17, Theorem 3.6], [7, Theorem 5.2]. In particular, [7, Theorem 5.7], proves that T is a CNC d-contraction if and only if

T is unitarily equivalent to a contractive Gleason solution, X^T acting on $\mathscr{H}(b_T)$, where b_T is the characteristic function of T. In this case, since T has commuting component operators, the results [10, Theorem 3.5, Theorem 3.6] imply that $\mathscr{H}(b^T) \subset H^2_d \otimes \mathcal{K}$ is co-invariant for the (vector-valued) Arveson d-shift, $S \otimes I_{\mathcal{K}}$, that $(X^T)^* = (S^* \otimes I_{\mathcal{K}})|_{\mathscr{H}(b^T)}$, and that X^T is the unique contractive Gleason solution for $\mathscr{H}(b^T)$ which has commuting component operators (*i.e.*, which is a d-contraction).

More generally, the seminal work of Ball–Bolotnikov–Fang on transfer-function realization theory for the multi-variable Schur class provides a commutative de Branges–Rovnyak model for arbitrary CCNC row contractions [6, 7, 9, 10] (see Subsection 2.16). This series of papers clearly demonstrates that the appropriate multi-variable generalization of the notion of restriction of the backward shift to a de Branges–Rovynak space is (the adjoint of) a contractive Gleason solution, and that contractive Gleason solutions should play the role of the model operator in a several-variable de Branges–Rovnyak model. Moreover, as we will see in the upcoming Subsection 2.16, Theorem 1 follows directly from the transfer-function realization theory of [9], and this provides a (commutative) multi-variable de Branges–Rovnyak model for arbitrary CCNC row contractions.

1.3. Overview

We will provide an alternate approach to Theorem 1 which is inspired by classical work of Kreĭn and Livšic [27, 35–39]. M.S. Livšic developed a functional model and the notion of a characteristic function for arbitrary CNU partial isometries with equal defect indices in [38, 39]. Similarly, in a series of papers, M.G. Kreĭn constructed a general functional model for unbounded symmetric operators with equal defect indices [35–37] (the reference [27] provides an overview of this theory). Recall here, that the classes of partial isometries and symmetric linear transformations on Hilbert space are equivalent under a natural bijection, the *Cayley Transform*, a fractional linear transformation of the closed complex upper half-plane onto the closed unit disk (minus the point $\{1\}$) [1, Chapter VII]. In [2, 24, 40], several key aspects of these approaches were combined, and slightly generalized with the introduction of the concept of a *model map* for a symmetric linear transformation or partial isometry with equal defect indices [2, Section 2.1], [24, Section 3], [40, Definition 5.1].

In Section 4 we extend this approach to several-variables and apply it to develop the representation theory of CCNC row partial isometries. Along the way we will obtain several results of independent interest. Corollary 4.15 proves that any extremal Gleason solution for a multivariable de Branges–Rovnyak space $\mathscr{H}(b)$ with $b(0) = 0$ acts as multiplication by the independent variables on its initial space (the projection onto the initial space prevents this from being a d-contraction in general). Our definition, Definition 4.11, of the (operator-valued) Schur-class characteristic function, b_V, of any CCNC row partial isometry, V, is a direct multi-variable analogue of Livšic's original definition from [39]. In Section 5 we study operator-Möbius transformations (Frostman shifts) of Schur-class functions, and we show that if T is any CCNC row contraction with canonical decomposition $T = V - C$ where $V = TP_{\mathrm{Ker}(D_T)}$ is a CCNC row partial isometry, that the characteristic function b_V, of V, is the Frostman shift of b_T which vanishes at 0.

Our Livšic-type characteristic function, b_T, is equivalent to the Sz.-Nagy–Foiaş-type characteristic function of T as constructed for CNC d-contractions in [7, 16, 17] (see Proposition 5.15). Theorem 1 shows that the map $T \mapsto b_T$ is a surjection of CCNC row contractions onto (weak equivalence classes of) multivariable operator-valued Schur-class functions, and that T is in the inverse image of $b \in \mathscr{S}_d(\mathcal{J}, \mathcal{K})$ if and only if it is unitarily equivalent to some extremal Gleason solution for $\mathscr{H}(b)$. The characteristic function is known to be a complete unitary invariant for the class of commutative CCNC row contractions (CNC d-contractions) [7, 16, 17]. The characteristic function is still a unitary invariant for the general class of all CCNC row contractions (see Theorem 5.14), but it is not a complete unitary invariant (see Subsection 4.17). Theorem 6.9 shows that the characteristic function is also a complete unitary invariant for the smaller sub-class of QE row contractions (which are again generally non-commuting).

2. Preliminaries and background

2.1. Vector-valued RKHS

We will be using the theory of vector-valued reproducing kernel Hilbert spaces throughout this paper, as presented in, *e.g.*, [44]. Recall the following basic facts from RKHS theory:

Given a set X, and an auxiliary Hilbert space $\mathcal{H}$, an $\mathcal{H}$-valued RKHS $\mathcal{K}$ on X is a Hilbert space of $\mathcal{H}$-valued functions on X so that for any $x \in X$ the linear point evaluation maps $K_x^* \in \mathscr{L}(\mathcal{K}, \mathcal{H})$ defined by

$$K_x^* F = F(x) \in \mathcal{H}; \qquad F \in \mathcal{K},$$

are bounded. We write $K_x := (K_x^*)^* \in \mathscr{L}(\mathcal{H}, \mathcal{K})$ for the Hilbert space adjoint. The operator-valued function $K : X \times X \to \mathscr{L}(\mathcal{H})$:

$$K(x, y) := K_x^* K_y \in \mathscr{L}(\mathcal{H}); \qquad x, y \in X,$$

is called the *reproducing kernel* of $\mathcal{K}$. One usually writes $\mathcal{K} = \mathcal{H}(K)$. The reproducing kernel K of any vector-valued RKHS on X is a *positive kernel function* on X: A function $K : X \times X \to \mathscr{L}(\mathcal{H})$ is an operator-valued positive kernel function on X if for any finite set of distinct points $\{x_k\}_{k=1}^N \subset X$, the matrix

$$[K(x_i, x_j)] \in \mathscr{L}(\mathcal{H}) \otimes \mathbb{C}^{N \times N},$$

is positive semi-definite. The vector-valued extension of the theory of RKHS developed by Aronszajn and Moore (see, *e.g.*, [44]) shows that there is a bijection between positive $\mathscr{L}(\mathcal{H})$-valued kernel functions on $X \times X$ and RKHS of $\mathcal{H}$-valued functions on X. Namely, given any positive kernel K on X there is a unique RKHS $\mathcal{K}$ on X so that K is its reproducing kernel, $\mathcal{K} = \mathcal{H}(K)$.

Any two RKHS $\mathcal{H}(k)$, $\mathcal{H}(K)$ with $\mathscr{L}(\mathcal{J})$ and $\mathscr{L}(\mathcal{K})$-valued positive kernel functions k, K on some set X, respectively, can be naturally equipped with a *multiplier space*:

$$\mathrm{Mult}(\mathcal{H}(k), \mathcal{H}(K)) := \{F : X \to \mathscr{L}(\mathcal{J}, \mathcal{K}) |\ Fh \in \mathcal{H}(K)\ \forall h \in \mathcal{H}(k)\}.$$

Here, and throughout $\mathcal{H}, \mathcal{J}, \mathcal{K}$ denote separable (or finite-dimensional) Hilbert spaces. That is, $\mathrm{Mult}(\mathcal{H}(k), \mathcal{H}(K))$ is the space of all operator-valued functions which multiply elements of $\mathcal{H}(k)$ into $\mathcal{H}(K)$. Viewing multipliers, F, (elements of the multiplier space) as linear maps, M_F, from $\mathcal{H}(k)$ into $\mathcal{H}(K)$, standard functional analysis arguments show that any multiplier is a bounded linear map, $\mathrm{Mult}(\mathcal{H}(k), \mathcal{H}(K)) \subset \mathscr{L}(\mathcal{H}(k), \mathcal{H}(K))$, and $\mathrm{Mult}(\mathcal{H}(k), \mathcal{H}(K))$ is closed in the weak operator topology. In the particular case where $\mathcal{H}(k) = \mathcal{H}(K)$,

$$\mathrm{Mult}(\mathcal{H}(K)) := \mathrm{Mult}(\mathcal{H}(K), \mathcal{H}(K)$$

is a unital WOT-closed algebra of bounded linear operators on $\mathcal{H}(K)$, the *multiplier algebra*.

We work in the setting of vector-valued Drury–Arveson space $H^2_d \otimes \mathcal{H}$, where $\mathcal{H}$ is finite-dimensional or separable. This is the vector-valued reproducing kernel Hilbert space $\mathcal{H}(k)$ of $\mathcal{H}$-valued functions on the ball $\mathbb{B}^d := (\mathbb{C}^d)_1$ corresponding to the several-variable, operator-valued Szegő kernel:

$$k(z, w) := \frac{1}{1 - zw^*} I_{\mathcal{H}}; \qquad z, w \in \mathbb{B}^d.$$

Here, $zw^* := (w, z)_{\mathbb{C}^d}$, all inner products are assumed to be conjugate linear in the first argument. The Drury–Arveson space is arguably the canonical several-variable generalization of the classical Hardy space $H^2_1 = H^2(\mathbb{D})$, at least from an operator-theoretic viewpoint [4, 5, 49].

We will use the notation $H^\infty_d \otimes \mathscr{L}(\mathcal{J}, \mathcal{K}) := \mathrm{Mult}(H^2_d \otimes \mathcal{J}, H^2_d \otimes \mathcal{K})$ (the multiplier spaces are the closure of this algebraic tensor product in the weak operator topology). The *Schur class* is the closed unit ball of the multiplier space: $\mathscr{S}_d(\mathcal{J}, \mathcal{K}) := [H^\infty_d \otimes \mathscr{L}(\mathcal{J}, \mathcal{K})]_1$. In the single variable ($d = 1$) and scalar-valued ($\mathcal{H} = \mathbb{C}$) setting we recover the classical Hardy space $H^2(\mathbb{D})$ and algebra $H^\infty(\mathbb{D})$ of analytic functions on the disk which embed isometrically into L^2, L^∞ of the unit circle $\mathbb{T}$, respectively, by taking non-tangential boundary limits [29].

As in the single-variable case, given any Schur-class function $b \in \mathscr{S}_d(\mathcal{J}, \mathcal{K})$, one can construct a positive kernel function k^b on $\mathbb{B}^d \times \mathbb{B}^d$, the de Branges–Rovnyak kernel:

$$k^b(z, w) := \frac{I - b(z)b(w)^*}{1 - zw^*} \in \mathscr{L}(\mathcal{K}); \qquad z, w \in \mathbb{B}^d,$$

and the corresponding RKHS $\mathscr{H}(b) := \mathcal{H}(k^b)$ is called the de Branges–Rovnyak space associated to b. It is straightforward to show that a bounded analytic (operator-valued) function b on $\mathbb{B}^d$ belongs to the Schur class if and only if the above formula defines a positive kernel on $\mathbb{B}^d$ [13, Theorem 2.1], and we will use this fact frequently in the sequel.

It is easy to check that $k - k^b$ (with k the Szegő kernel of vector-valued Drury–Arveson space) is a positive kernel function so that standard vector-valued RKHS theory implies that $\mathscr{H}(b)$ is contractively contained in $H^2_d \otimes \mathcal{K}$ [44]. In the single-variable setting, any de Branges–Rovnyak space $\mathscr{H}(b)$ is co-invariant for the shift [48]. The natural several-variable generalization of the shift is the Arveson *d-shift*, $S : H^2_d \otimes \mathbb{C}^d \to H^2_d$, a row partial isometry on H^2_d, also denoted by $S = (S_1, \ldots, S_d)$ [4]. The component operators of S mutually commute (S is a d-contraction) and act as multiplication by the independent variables on H^2_d:

$$(S\mathbf{F})(z) = z_1 F_1(z) + \cdots + z_d F_d(z); \qquad \mathbf{F} = (F_1, \ldots, F_d) \in H^2_d \otimes \mathbb{C}^d.$$

In contrast to the classical case, multi-variable de Branges–Rovnyak spaces are generally not co-invariant for the component operators of the Arveson d-shift [10]. As described in the upcoming Subsection 2.4, the appropriate several-variable analogue of the adjoint of the restricted backward shift will be a *Gleason solution* for $\mathscr{H}(b)$, and (extremal) Gleason solutions will play the role of the model operator in our commutative model for CCNC row contractions.

2.2. Herglotz spaces

It will be convenient to define a second reproducing kernel Hilbert space $\mathscr{H}^+(H_b)$ associated to suitable 'square' $b \in \mathscr{S}_d(\mathcal{H})$.

In general we will say that a contraction $T \in \mathscr{L}(\mathcal{J}, \mathcal{K})$ is *pure* or *purely contractive* if $\|Tg\| < \|g\|$ for all g in $\mathcal{J}$, and that T is *strict* or *strictly contractive* if $\|T\| < 1$. A similar argument to [50, Proposition 2.1, Chapter V] (combined with a simple argument using automorphisms of the ball $\mathbb{B}^d$) shows that any $b \in \mathscr{S}_d(\mathcal{H}, \mathcal{J})$ decomposes as $b = b_0 + b_1$ on $\mathcal{H} = \mathcal{H}_0 \oplus \mathcal{H}_1$ where $b_0(z) := b(z)|_{\mathcal{H}_0}$ is purely contractive and b_1 is a constant isometry on $\mathbb{B}^d$ from $\mathcal{H}_1$ onto its range in $\mathcal{J}$. We say a Schur-class $b \in \mathscr{S}_d(\mathcal{J}, \mathcal{K})$ is *purely contractive* or *strictly contractive* if $b(z)$ is a pure or strict contraction, respectively, for all $z \in \mathbb{B}^d$. As discussed in [31, Section 1.8], b is strictly contractive if and only if $b(0)$ is a strict contraction. (This follows from the Schwarz Lemma for contractive analytic functions on $\mathbb{B}^d$ combined with an operator-Möbius transformation argument, see Lemma 5.2).

We say that $b \in \mathscr{S}_d(\mathcal{H})$ is *non-unital* if $I - b(z)$ is invertible for all $z \in \mathbb{B}^d$. Any strictly contractive $b \in \mathscr{S}_d(\mathcal{H})$ is certainly non-unital. The *Herglotz–Schur* class, $\mathscr{S}^+_d(\mathcal{H})$, is the set of all $\mathscr{L}(\mathcal{H})$-valued analytic functions on $\mathbb{B}^d$ such that the *Herglotz kernel*,

$$K^H(z, w) := \frac{1}{2} \frac{H(z) + H(w)^*}{1 - zw^*} \in \mathscr{L}(\mathcal{H}); \qquad z, w \in \mathbb{B}^d,$$

is an operator-valued positive kernel function. Any Herglotz–Schur function necessarily has positive real part. In particular, if $b \in \mathscr{S}_d(\mathcal{H})$ is square and non-unital, and

$$H_b(z) := (I - b(z))^{-1}(I + b(z)),$$

then,

$$\begin{aligned}
K^b(z,w) &:= K^{H_b}(z,w) \\
&= \frac{1}{2}(1-zw^*)^{-1}\left((I-b(z))^{-1}(I+b(z)) + (I+b(w)^*)(I-b(w)^*)^{-1}\right) \\
&= \frac{1}{2}k(z,w)(I-b(z))^{-1}\big((I+b(z))(I-b(w)^*) \\
&\qquad + (I-b(z))(I+b(w)^*)\big)(I-b(w)^*)^{-1} \\
&= \frac{1}{2}k(z,w)(I-b(z))^{-1}\big(I+b(z)-b(w)^*-b(z)b(w)^* \\
&\qquad + I - b(z) + b(w)^* - b(z)b(w)^*\big)(I-b(w)^*)^{-1} \\
&= (I-b(z))^{-1}k^b(z,w)(I-b(w)^*)^{-1},
\end{aligned}$$

is a positive kernel so that $H_b \in \mathscr{S}_d^+(\mathcal{H})$. As described in [31, Section 1.8], the maps

$$b \mapsto H_b := (I-b)^{-1}(I+b); \qquad \text{and} \qquad H \mapsto b_H := (H+I)^{-1}(H-I),$$

are compositional inverses and define bijections between $\mathscr{S}_d^+(\mathcal{H})$ and non-unital elements of $\mathscr{S}_d(\mathcal{H})$. If $H = H_b \in \mathscr{S}_d^+(\mathcal{H})$ we call $\mathscr{H}^+(H_b) := \mathcal{H}(K^b)$, the *Herglotz space* of b.

By standard vector-valued RKHS theory, the above relationship between the de Branges–Rovnyak and Herglotz kernels for non-unital $b \in \mathscr{S}_d(\mathcal{H})$ implies that there is a unitary multiplier $U_b : \mathscr{H}(b) \to \mathscr{H}^+(H_b)$.

Lemma 2.3. *The map $U_b : \mathscr{H}(b) \to \mathscr{H}^+(H_b)$ defined by multiplication by*

$$U_b(z) := (I-b(z))^{-1}, \tag{2.1}$$

is an onto isometry. The action of U_b on point evaluation kernels is

$$U_b k_z^b = K_z^b(I-b(z)^*).$$

It will be useful to consider the natural row partial isometry $V^b : \mathscr{H}^+(H_b) \otimes \mathbb{C}^d \to \mathscr{H}^+(H_b)$ on the Herglotz space of any square $b \in \mathscr{S}_d(\mathcal{H})$ defined by

$$z^*K_z^b h := \begin{bmatrix} \overline{z}_1 K_z^b h \\ \vdots \\ \overline{z}_d K_z^b h \end{bmatrix} \overset{V^b}{\mapsto} (K_z^b - K_0^b)h; \qquad h \in \mathcal{H}. \tag{2.2}$$

Verifying that this defines a partial isometry with

$$\operatorname{Ker}(V^b)^\perp = \bigvee_{z\in\mathbb{B}^d} z^*K_z^b\mathcal{H}, \qquad \text{and} \qquad \operatorname{Ran}\left(V^b\right) = \bigvee_{z\in\mathbb{B}^d} (K_z^b - K_0^b)\mathcal{H},$$

is a straightforward computation using the Herglotz kernel [31, Section 2].

2.4. Gleason solutions

As discussed previously, the de Branges–Rovnyak spaces $\mathscr{H}(b)$ for arbitrary $b \in \mathscr{S}_d(\mathcal{J}, \mathcal{K})$ are generally not co-invariant for the component operators S_j of the Arveson d-shift [10]. Instead, the appropriate replacement for the 'adjoint of the restricted backward shift' in the several-variable theory is a contractive solution to the *Gleason problem* [3, 6, 7, 10, 26]:

Definition 2.5. Let $b \in \mathscr{S}_d(\mathcal{J}, \mathcal{K})$ be a Schur-class function. A row contraction $X \in \mathscr{L}(\mathscr{H}(b) \otimes \mathbb{C}^d, \mathscr{H}(b))$ solves the *Gleason problem* in $\mathscr{H}(b)$ if

$$z(X^* f)(z) := z_1(X_1^* f)(z) + \cdots + z_d(X_d^* f)(z) = f(z) - f(0); \quad \forall f \in \mathscr{H}(b). \tag{2.3}$$

We say that a *Gleason solution* X is *contractive* if

$$XX^* \leq I - k_0^b(k_0^b)^*,$$

and we say that X is *extremal* if equality holds in the above.

It is easy to check that for any row contraction X on $\mathscr{H}(b)$, the Gleason solution condition (2.3) above is equivalent to:

$$(I - Xz^*)^{-1} k_0^b = k_z^b; \qquad \forall\, z \in \mathbb{B}^d, \tag{2.4}$$

and this property will also be used frequently in the sequel. Given $z \in \mathbb{B}^d$, and any Hilbert space $\mathcal{H}$, we will often view z as a strict contraction from $\mathcal{H} \otimes \mathbb{C}^d$ into $\mathcal{H}$: Define $z^* : \mathcal{H} \to \mathcal{H} \otimes \mathbb{C}^d$ by $z^* h := (\overline{z}_1 h, \dots, \overline{z}_d h)^T \in \mathcal{H} \otimes \mathbb{C}^d$ and for any $\mathbf{h} \in \mathcal{H} \otimes \mathbb{C}^d$, the adjoint map $z := (z^*)^* \in \mathscr{L}(\mathcal{H} \otimes \mathbb{C}^d, \mathcal{H})$ obeys:

$$z\mathbf{h} = z \begin{bmatrix} h_1 \\ \vdots \\ h_d \end{bmatrix} = z_1 h_1 + \cdots + z_d h_d, \qquad \text{and } \|z\|^2 = zz^* := (z, z)_{\mathbb{C}^d} < 1.$$

Remark 2.6. If X is any extremal Gleason solution for $\mathscr{H}(b)$ where $b \in \mathscr{S}_d(\mathcal{J}, \mathcal{K})$ obeys $b(0) = 0$, then

$$XX^* = I - k_0^b(k_0^b)^*,$$

is a projection so that X is a (row) partial isometry on $\mathscr{H}(b)$. In general, $P_0 := I - k_0^b k^b(0,0)^{-1}(k_0^b)^*$ is the projection onto the subspace of all functions $f \in \mathscr{H}(b)$ such that $f(0) = 0$, and

$$k^b(0,0) = I - b(0)b(0)^*,$$

so that an extremal Gleason solution X for $\mathscr{H}(b)$ is a row partial isometry if and only if $b(0) = 0$.

In the case where $d = 1$, the unique solution to equation (2.3) is the adjoint of the restriction of the backward shift S^* to $\mathscr{H}(b)$, so that adjoints of Gleason solutions are natural analogues of the restricted backward shift in the several-variable setting. Many references define a Gleason solution for $\mathscr{H}(b)$ as the adjoint of our definition above, we prefer to view it as a row contraction from several copies of a Hilbert space into itself. Contractive solutions X to the Gleason problem in $\mathscr{H}(b)$ always exist, although they are in general non-unique [6, 31]. Also note that the component operators of a contractive Gleason solution X for $\mathscr{H}(b)$ are generally non-commuting [10]. In fact, the existence of a commuting contractive Gleason solution X for $\mathscr{H}(b)$ is equivalent to co-invariance of $\mathscr{H}(b)$ with respect to the component operators of the Arveson d-shift, and in this case $X = (S^*|_{\mathscr{H}(b)})^*$ is a contractive commuting Gleason solution for $\mathscr{H}(b)$ [10, Theorem 3.5]. This happens, for example, if $b \in \mathscr{S}_d(\mathcal{J}, \mathcal{K})$ is an *inner multiplier, i.e.*, multiplication by b, $M_b : H^2_d \otimes \mathcal{J} \to H^2_d \otimes \mathcal{K}$ is a partial isometry so that $\mathscr{H}(b) \subset H^2_d \otimes \mathcal{K}$ is a co-invariant *model subspace* of vector-valued Drury–Arveson space [4, 16, 28, 41] (see Remark 2.15 below).

Similarly we define contractive Gleason solutions for any $b \in \mathscr{S}_d(\mathcal{J}, \mathcal{K})$:

Definition 2.7. A linear map $\mathbf{b} \in \mathscr{L}(\mathcal{J}, \mathscr{H}(b) \otimes \mathbb{C}^d)$,

$$\mathbf{b} = \begin{bmatrix} b_1 \\ \vdots \\ b_d \end{bmatrix}, \quad b_j \in \mathscr{L}(\mathcal{J}, \mathscr{H}(b)), \quad 1 \leq j \leq d,$$

is a solution to the Gleason problem for $b \in \mathscr{S}_d(\mathcal{J}, \mathcal{K})$ provided that

$$b(z) - b(0) = z \cdot \mathbf{b}(z) := \sum_{j=1}^{d} z_j b_j(z).$$

We say that $\mathbf{b}$ is a *contractive* Gleason solution for b if

$$\mathbf{b}^*\mathbf{b} \leq I - b(0)^* b(0),$$

and an *extremal* Gleason solution for b if equality holds in the above.

There is a surjection $\mathbf{b} \mapsto X(\mathbf{b})$ of contractive Gleason solutions for $b \in \mathscr{S}_d(\mathcal{J}, \mathcal{K})$ onto contractive Gleason solutions for $\mathscr{H}(b)$ given by the formula

$$X(\mathbf{b})^* k^b_z := z^* k^b_z - \mathbf{b} b(z)^*; \qquad z \in \mathbb{B}^d. \tag{2.5}$$

This surjection is injective if and only if $\bigcap_{z \in \mathbb{B}^d} \mathrm{Ker}(b(z)) = \{0\}$, and it preserves extremal Gleason solutions. This follows as in [31, Section 5] (which considers the case $\mathcal{J} = \mathcal{K}$).

If $b \in \mathscr{S}_d(\mathcal{H})$ is square and non-unital, then there is a bijection between contractive extensions D of the row partial isometry V^b defined on the Herglotz space $\mathscr{H}^+(H_b)$ (*i.e.*, $D : \mathscr{H}^+(H_b) \otimes \mathbb{C}^d \to \mathscr{H}^+(H_b)$, and $D(V^b)^* V^b = V^b$) and contractive Gleason solutions $\mathbf{b} = \mathbf{b}[D]$ given by

$$\mathbf{b}[D] := (U^*_b \otimes I_d) D^* K^b_0 (I - b(0)), \tag{2.6}$$

see [31, Section 5]. If $X = X(\mathbf{b}[D])$, we will simply write $X = X[D]$.

To apply the above parametrizaton of Gleason solutions to arbitrary Schur-class multipliers, it will be useful to define the *square extension* of any $b \in \mathscr{S}_d(\mathcal{J}, \mathcal{K})$: Any such b coincides with an element $b \in \mathscr{S}_d(\mathcal{J}', \mathcal{K}')$ where $\mathcal{J}' \subseteq \mathcal{K}'$ or $\mathcal{K}' \subseteq \mathcal{J}'$. Given any $b \in \mathscr{S}_d(\mathcal{J}, \mathcal{K})$, and assuming $\mathcal{J} \subseteq \mathcal{K}$ or $\mathcal{K} \subseteq \mathcal{J}$, the *square extension*, $[b]$, of b is

$$[b] := \begin{cases} \begin{bmatrix} b & 0_{\mathcal{K} \ominus \mathcal{J}, \mathcal{K}} \end{bmatrix} \in \mathscr{S}_d(\mathcal{K}); & \mathcal{J} \subset \mathcal{K} \\ \begin{bmatrix} b \\ 0_{\mathcal{J}, \mathcal{J} \ominus \mathcal{K}} \end{bmatrix} \in \mathscr{S}_d(\mathcal{J}); & \mathcal{K} \subset \mathcal{J} \end{cases}.$$

Remark 2.8.

1. If $J \subseteq \mathcal{K}$, then it follows that $\mathscr{H}(b) = \mathscr{H}([b])$. In this case it follows that $\mathbf{b}$ is a contractive Gleason solution for b if and only if there is a contractive Gleason solution $[\mathbf{b}]$ for $[b]$ so that $\mathbf{b} = [\mathbf{b}]|_{\mathcal{J}}$. Moreover if $[\mathbf{b}]$ is extremal, so is $\mathbf{b} = [\mathbf{b}]|_{\mathcal{J}}$.
2. If $\mathcal{K} \subseteq \mathcal{J}$ then $\mathscr{H}([b]) = \mathscr{H}(b) \oplus H^2_d \otimes (\mathcal{J} \ominus \mathcal{K})$. In this case $\mathbf{b}$ is a contractive Gleason solution for b if and only if $[\mathbf{b}] := \mathbf{b} \oplus \mathbf{0}$ is a contractive Gleason solution for b. Here $\mathbf{0} : \mathcal{J} \ominus \mathcal{K} \to H^2_d \otimes (\mathcal{J} \ominus \mathcal{K})$ sends every vector to $0 \in H^2_d \otimes (\mathcal{J} \ominus \mathcal{K})$. Again this $[\mathbf{b}]$ is extremal if and only if $\mathbf{b} = \begin{bmatrix} I_{\mathscr{H}(b)} \otimes I_d, & 0 \end{bmatrix} [\mathbf{b}]$ is extremal.

The statement of Theorem 5.14 implicitly claims that any extremal Gleason solution X for $\mathscr{H}(b)$, $b \in \mathscr{S}_d(\mathcal{J}, \mathcal{K})$, is a CCNC row contraction. This is easily established using the extremal condition:

Lemma 2.9. *Let $b \in \mathscr{S}_d(\mathcal{J}, \mathcal{K})$ be a multi-variable Schur-class function. Any extremal Gleason solution, X, for $\mathscr{H}(b)$ is a CCNC row contraction.*

Proof. Since X is extremal,

$$XX^* = I - k_0^b (k_0^b)^*,$$

and since $D^2_{X^*} = I - XX^*$, it follows that $\operatorname{Ran}(D_{X^*})$ is the range of the projection

$$k_0^b k^b(0,0)^{-1} (k_0^b)^*.$$

Since X is a contractive Gleason solution, it follows that

$$\begin{aligned} \bigvee_{z \in \mathbb{B}^d} (I - Xz^*)^{-1} \operatorname{Ran}(D_{X^*}) &= \bigvee (I - Xz^*)^{-1} k_0^b \mathcal{K} \\ &= \bigvee_{z \in \mathbb{B}^d} k_z^b \mathcal{K} = \mathscr{H}(b), \qquad \text{(by equation (2.4))} \end{aligned}$$

so that X is CCNC. □

2.10. Extremal Gleason solutions

Since extremal Gleason solutions for $\mathscr{H}(b)$, $b \in \mathscr{S}_d(\mathcal{J}, \mathcal{K})$ will provide a model for arbitrary CCNC row contractions, it is natural to ask whether extremal Gleason solutions always exist. The results below show that there are large classes of $b \in \mathscr{S}_d(\mathcal{J}, \mathcal{K})$, and hence $\mathscr{H}(b)$, which admit extremal Gleason solutions. (Recall that if $\mathbf{b}$ is an extremal Gleason solution for $b \in \mathscr{S}_d(\mathcal{J}, \mathcal{K})$, then $X(\mathbf{b})$ is an extremal Gleason solution for $\mathscr{H}(b)$.) We expect that any Schur class $b \in \mathscr{S}_d(\mathcal{J}, \mathcal{K})$ admits an extremal Gleason solution, but we do not have a proof of this in full generality.

Lemma 2.11. *Given $b \in \mathscr{S}_d(\mathcal{H})$, and a row contractive $D \supseteq V^b$, $\mathbf{b}[D]$ is extremal if and only if D is a co-isometry.*

Recall that $\mathbf{b}[D]$ is extremal implies that $X[D]$ is extremal. In particular, if V^b is itself a co-isometry, then $\mathscr{H}(b)$ has a unique contractive Gleason solution, and this solution is extremal. This property (of V^b being co-isometric) is equivalent to b being a quasi-extreme Schur multiplier, as defined in [30, 31]. See [30, 31] and Section 6.1 for examples of quasi-extreme multipliers, equivalent conditions, and details.

Proof. If D is a co-isometric extension of V^b, it follows readily from Formula (2.6), that $\mathbf{b}[D]$ is extremal.

Conversely, extremality of $\mathbf{b}[D]$ implies:

$$(K_0^b)^* DD^* K_0^b = K^b(0,0) = (K_0^b)^* K_0^b.$$

Then, for any $z \in \mathbb{B}^d$,

$$\begin{aligned}(K_z^b)^* DD^* K_0^b &= (DD^*(K_z^b - K_0^b))^* K_0^b + (K_0^b)^* DD^* K_0^b \\ &= (K_z^b - K_0^b)^* K_0^b + (K_0^b)^* K_0^b = (K_z^b)^* K_0^b.\end{aligned}$$

This proves that $DD^* K_0^b = K_0^b$ so that

$$\begin{aligned}DD^* K_z^b &= DD^*(K_z^b - K_0^b) + DD^* K_0^b \\ &= K_z^b - K_0^b + K_0^b = K_z^b,\end{aligned}$$

and $DD^* = I$. □

Remark 2.12. By the above lemma, $b \in \mathscr{S}_d(\mathcal{H})$ (and hence $\mathscr{H}(b)$) has extremal Gleason solutions if and only if $\dim\left(\mathrm{Ker}(V^b)\right) \geq \dim\left(\mathrm{Ran}\left(V^b\right)^{\perp}\right)$ (so that V^b has co-isometric extensions). The proofs of [31, Proposition 4.12, Proposition 4.15] can be adapted to show this is the case for large classes of $b \in \mathscr{S}_d(\mathcal{H})$:

Proposition 2.13. *If* $b \in \mathscr{S}_d(\mathcal{H})$ *is such that* $\dim(\mathscr{H}(b)) = \infty$ *and* $\dim(\mathcal{H}) < \infty$*, then* $\dim\left(\mathrm{Ker}(V^b)\right) = \infty$ *and* $\dim\left(\mathrm{Ran}\left(V^b\right)^{\perp}\right) \leq \dim(\mathcal{H})$.

Proof. Set $V := V^b$ and calculate,

$$(V_k^* V_j^* - V_j^* V_k^*) K_z^b = (\overline{z}_j V_k^* - \overline{z}_k V_j^* + V_k^* V_j^* - V_j^* V_k^*) K_0^b; \qquad z \in \mathbb{B}^d.$$

Since $\dim(\mathcal{H}) < \infty$ and $K_0^b \in \mathscr{L}(\mathcal{H}, \mathscr{H}^+(H_b))$, the above commutators have finite rank. It is also easy to check that $\mathrm{Ran}\,(V)^{\perp}$ consists of $\mathcal{H}$-valued constant functions so that $\dim\left(\mathrm{Ran}\,(V)^{\perp}\right) \leq \dim(\mathcal{H}) < \infty$, by assumption. Since we assume that $\mathscr{H}(b)$, and hence $\mathscr{H}^+(H_b)$ are infinite-dimensional, the V_k cannot all have finite rank by the definition of $V = V^b$. It follows that the image of V under the quotient $*$-homomorphism, π, onto the Calkin algebra (quotient by the compact operators) is a non-zero row co-isometry with commuting component operators. If $\mathrm{Ker}(V^b)$ is finite-dimensional, then $\pi(V)$ is a row isometry (in fact Cuntz unitary) with commuting component operators, and this is readily checked to be impossible. □

Similarly,

Proposition 2.14. *Let* $\mathbf{b} = (b_1, \ldots, b_d)^T$ *be any contractive Gleason solution for* $b \in \mathscr{S}_d(\mathcal{H})$*,* $d \geq 2$*. Then* $\dim\left(\mathrm{Ker}(V^b)\right) \geq \frac{d(d-1)}{2} \cdot \dim\left(\mathscr{H}(b) \ominus \bigvee_{j=1}^d b_j \mathcal{H}\right)$.

Proof. This is easily established using the proof of [31, Proposition 4.4]: Let $X = X(\mathbf{b})$ be a contractive Gleason solution for $\mathscr{H}(b)$, where $\mathbf{b}$ is any contractive Gleason solution for b (see Equation (2.5)). Then $\mathbf{b} = (b_1, \dots, b_d)^T \in \mathscr{H}(b) \otimes \mathbb{C}^d$, $d \geq 2$. Choose any $f \in \mathscr{H}(b)$ orthogonal to the linear span of the b_j, $1 \leq j \leq d$. It follows that

$$\begin{aligned} \langle k_z^b, (X_j f) \rangle &= \langle \overline{z}_j k_z^b - b_j b(z)^*, f \rangle \\ &= z_j f(z)_{\mathcal{H}}. \end{aligned}$$

This proves that $S_j f \in \mathscr{H}(b)$ for any $1 \leq j \leq d$. This in turn implies that $h_j := (I - b)^{-1} S_j f \in \mathscr{H}^+(H_b)$ and if we define $\mathbf{h} := -h_j \otimes e_i + h_i \otimes e_j \in \mathscr{H}^+(H_b) \otimes \mathbb{C}^d$, where $\{e_k\}$ is the standard basis of $\mathbb{C}^d$ and $i \neq j$, we then have that

$$\begin{aligned} \langle z^* K_z^b, \mathbf{h} \rangle &= -z_i h_j(z) + z_j h_i(z) \\ &= (-z_i z_j + z_j z_i)(I - b(z))^{-1} f(z) = 0. \end{aligned}$$

It follows that for any distinct pair (i, j); $1 \leq i, j \leq d$ and any $f \perp \bigvee_{j=1}^d \mathrm{Ran}\,(b_j)$, we can construct an element $\mathbf{h}^{(i,j)} \in \mathrm{Ker}(V^b)$. □

As discussed in Remark 2.8, if the square extension $[b]$, of any $b \in \mathscr{S}_d(\mathcal{J}, \mathcal{K})$ has extremal Gleason solutions, so does b. The above two propositions now imply that there are large classes of $b \in \mathscr{S}_d(\mathcal{J}, \mathcal{K})$ (and hence $\mathscr{H}(b)$) which admit extremal Gleason solutions.

Remark 2.15. If $b \in \mathscr{S}_d(\mathcal{J}, \mathcal{K})$ is an inner multiplier ($M_b : H_d^2 \otimes \mathcal{J} \to H_d^2 \otimes \mathcal{K}$ is a partial isometry), then $\mathscr{H}(b) = \mathrm{Ran}\,(M_b)^\perp$ is co-invariant for the component operators of the (vector-valued) Arveson d-shift, $S \otimes I_{\mathcal{K}}$, and

$$X_j^* := S_j^* \otimes I_{\mathcal{K}}|_{\mathscr{H}(b)},$$

defines an extremal Gleason solution (with commuting component operators). Indeed,

$$\begin{aligned} I - XX^* &= P^\perp_{\mathrm{Ran}(M_b)} (I - SS^*) P^\perp_{\mathrm{Ran}(M_b)} \\ &= (I - M_b M_b^*) k_0 k_0^* (I - M_b M_b^*) \\ &= k_0^b (k_0^b)^*. \end{aligned}$$

2.16. de Branges–Rovnyak model via transfer-function theory

In this section we will show that Theorem 5.14 (*i.e.*, Theorem 1 from the Introduction) is a straightforward consequence of the multi-variable transfer function realization theory of [9].

As part of [9, Theorem 1.3], J.A. Ball, V. Bolotnikov and Q. Fang proved that a $\mathscr{L}(\mathcal{J}, \mathcal{K})$-valued function b on $\mathbb{B}^d$ belongs to the multi-variable, operator-valued Schur class $\mathscr{S}_d(\mathcal{J}, \mathcal{K})$, if and only if there is an auxiliary Hilbert space, $\mathcal{H}$, and a contractive *colligation*, U:

$$U := \begin{bmatrix} A & B \\ C & D \end{bmatrix} =: \begin{bmatrix} A_1 & B_1 \\ \vdots & \vdots \\ A_d & B_d \\ C & D \end{bmatrix} : \begin{bmatrix} \mathcal{H} \\ \mathcal{J} \end{bmatrix} \to \begin{bmatrix} \mathcal{H} \otimes \mathbb{C}^d \\ \mathcal{K} \end{bmatrix},$$

so that b has the *transfer function realization*:

$$b(z) = D + C(I - zA)^{-1} zB.$$

Moreover in [9, Section 3], the authors construct a canonical *functional model realization* of $b \in \mathscr{S}_d(\mathcal{J}, \mathcal{K})$, providing a natural extension of the classical transfer function theory of de Branges and Rovnyak [20, 21]: If $A := X^*$ is the adjoint of any contractive Gleason solution for $\mathscr{H}(b)$, $C := (k_0^b)^*$, and $D := b(0)$, then there exists a (essentially unique [9, Corollary 2.9]) contractive Gleason solution $\mathbf{b} =: B$ for b so that

$$U_{dBR} := \begin{bmatrix} A & B \\ C & D \end{bmatrix} = \begin{bmatrix} X^* & \mathbf{b} \\ (k_0^b)^* & b(0) \end{bmatrix} : \begin{bmatrix} \mathscr{H}(b) \\ \mathcal{J} \end{bmatrix} \to \begin{bmatrix} \mathscr{H}(b) \otimes \mathbb{C}^d \\ \mathcal{K} \end{bmatrix},$$

is a contractive colligation providing a transfer-function realization for b. In fact, [32, Theorem 7.13] proves that $A^* = X = X(\mathbf{b})$ is the contractive Gleason solution for $\mathscr{H}(b)$ corresponding to the contractive Gleason solution $B = \mathbf{b}$ for b as in Subsection 2.4.

Remark 2.17. In this paper we focus on extremal Gleason solutions $X = X(\mathbf{b})$ as these will provide a model for arbitrary CCNC row contractions. It is easy to check that $\mathbf{b}$ (and hence $X(\mathbf{b})$) are extremal if and only if the corresponding canonical functional model colligation

$$U_{dBR} = \begin{bmatrix} X(\mathbf{b})^* & \mathbf{b} \\ (k_0^b)^* & b(0) \end{bmatrix},$$

is an isometry [31, Remark 4.18].

As described in [9, Section 3], any canonical functional model realization for $b \in \mathscr{S}_d(\mathcal{J}, \mathcal{K})$ is *weakly co-isometric* and *observable*. A colligation U, as above is weakly co-isometric if it is co-isometric on a certain subspace [9]. By [9, Proposition 1.5], U is weakly co-isometric if and only if:

$$k^b(z, w) = C^*(I - A^* z)^{-1}(I - w^* A)^{-1} C.$$

This characterization and Formula (2.4) show that any canonical functional model colligation U_{dBR} is always weakly-coisometric.

A contractive colligation U is said to be *observable* if

$$C(I - Az)^{-1} x = 0 \qquad \forall\ z \in \mathbb{B}^d$$

if and only if $x \equiv 0$ [9, Section 2]. Equivalently,

$$\bigvee_{z \in \mathbb{B}^d} (I - z^* A^*)^{-1} C^* \mathcal{K} = \mathcal{H}. \tag{2.7}$$

In the case where $U = U_{dBR}$ is a functional model colligation, $A = X^*$ is a contractive Gleason solution and $C = (k_0^b)^*$, so that, again by Formula (2.4),

$$\bigvee_{z \in \mathbb{B}^d} (I - z^* A^*)^{-1} C^* \mathcal{K} = \bigvee_{z \in \mathbb{B}^d} k_z^b \mathcal{K} = \mathscr{H}(b).$$

This shows that U_{dBR} is observable.

If T is any CCNC row contraction on $\mathcal{H}$, consider the Julia operator:

$$U_T := \begin{bmatrix} T^* & D_T \\ D_{T^*} & -T \end{bmatrix} : \begin{bmatrix} \mathcal{H} \\ \mathrm{Ran}\,(D_T) \end{bmatrix} \to \begin{bmatrix} \mathcal{H} \otimes \mathbb{C}^d \\ \mathrm{Ran}\,(D_{T^*}) \end{bmatrix}.$$

It is easy to verify that U_T is an onto isometry, so that in particular, it is a weakly co-isometric colligation. Since T is CCNC,

$$\mathcal{H} = \bigvee_{z \in \mathbb{B}^d} (I - Tz^*)^{-1} \mathrm{Ran}\,(D_{T^*}),$$

proving that U_T is also observable (by Formula 2.7).

Let $b_T \in \mathscr{S}_d(\mathrm{Ran}\,(D_T), \mathrm{Ran}\,(D_{T^*}))$ be the transfer-function corresponding to U_T. Since U_T is observable and weakly co-isometric, [9, Theorem 3.4] implies that U_T is unitarily equivalent to a canonical functional model colligation, U_{dBR}, for b_T. In particular, T is unitarily equivalent to X, a contractive Gleason solution for $\mathscr{H}(b_T)$. Finally, since U_T, and hence U_{dBR} is an isometry, it follows as in Remark 2.17 that X is necessarily an extremal Gleason solution, and this completes a proof of Theorem 5.14.

3. Completely non-coisometric row contractions

In this section we characterize completely non-coisometric (CNC) row contractions and motivate the definition of CCNC row contractions (contractions obeying the commutative CNC condition). As in the classical setting a row contraction $T : \mathcal{H} \otimes \mathbb{C}^d \to \mathcal{H}$ is CNC if there is no non-trivial co-invariant subspace $\mathcal{H}' \subset \mathcal{H}$ (co-invariant for each component operator T_k of T, $1 \le k \le d$) so that $T^*|_{\mathcal{H}'}$ is an isometry of $\mathcal{H}'$ into $\mathcal{H}' \otimes \mathbb{C}^d$. As observed in [7], a row contraction $T : \mathcal{H} \otimes \mathbb{C}^d \to \mathcal{H}$ is CNC if and only if $\bigvee_{\alpha \in \mathbb{F}^d} T^\alpha \mathrm{Ran}\,(D_{T^*}) = \mathcal{H}$, where, as before, the defect operator of any contraction $T : \mathcal{J} \to \mathcal{K}$ is $D_T := \sqrt{I_{\mathcal{J}} - T^*T}$, and $\mathbb{F}^d$ denotes the free monoid on d generators. For convenience, we will provide a proof based on methods ultimately due to Kreĭn.

In the above, recall that the free semigroup (or monoid), $\mathbb{F}^d$, on $d \in \mathbb{N}$ letters, is the multiplicative unital semigroup of all finite products or *words* in the d letters $\{1, \dots, d\}$. That is, given words $\alpha := i_1 \dots i_n$, $\beta := j_1 \dots j_m$, $i_k, j_l \in \{1, \dots, d\}$; $1 \le k \le n$, $1 \le l \le m$, their product $\alpha\beta$ is defined by concatenation:

$$\alpha\beta = i_1 \dots i_n j_1 \dots j_m,$$

and the unit is the empty word, $\emptyset$, containing no letters. Given $\alpha = i_1 \cdots i_n$, we use the standard notation $|\alpha| = n$ for the length of the word α. Let $\mathbb{N}^d$ be the unital additive semigroup or monoid of d-tuples of non-negative integers. By the universality property of the free unital semigroup $\mathbb{F}^d$, there is a unital semigroup epimorphism $\lambda : (\mathbb{F}^d, \cdot) \to (\mathbb{N}^d, +)$, the *letter counting map* which sends a given word $\alpha = i_1 \cdots i_n \in \mathbb{F}^d$ to $\mathbf{n} = (n_1, \dots, n_d) \in \mathbb{N}^d$ where n_k is the number of times the letter k appears in the word α. Given any $\alpha = i_1 \cdots i_k \in \mathbb{F}^d$, and any row contraction $T = (T_1, \dots, T_d)$, we use the standard notation

$$T^\alpha := T_{\alpha_1} \cdots T_{\alpha_k},$$

and for any $\mathbf{n} \in \mathbb{N}^d$ we define the symmetrized monomial:

$$T^{\mathbf{n}} := \sum_{\alpha;\ \lambda(\alpha)=\mathbf{n}} T^\alpha.$$

Initially, we focus on the case of a (row) partial isometry, $V : \mathcal{H} \otimes \mathbb{C}^d \to \mathcal{H}$, and we define the *restricted range spaces*,

$$\mathcal{R}(V - z) := \mathrm{Ran}\,((V - z)V^*V); \qquad z \in \mathbb{C}^d,$$

as the range of $V - z$ restricted to the initial space of V. The orthogonal complement, $\mathcal{R}(V - z)^\perp$, will be called a z-deficiency space or the *z-defect space*. More generally, as in [34], consider the *non-commutative* (NC) *open unit ball*:

$$\mathbb{B}^d_{\mathbb{N}} := \coprod_{n=1}^{\infty} \mathbb{B}^d_n; \qquad \mathbb{B}^d_n := \left(\mathbb{C}^{n\times n} \otimes \mathbb{C}^d\right)_1.$$

Elements $Z \in \mathbb{B}^d_n$ are viewed as strict (row) contractions from $\mathbb{C}^n \otimes \mathbb{C}^d$ into $\mathbb{C}^n$:

$$Z =: (Z_1, \dots, Z_d); \qquad Z_k \in \mathbb{C}^{n\times n}.$$

In particular, $\mathbb{B}^d_1 \simeq \mathbb{B}^d = (\mathbb{C}^d)_1$ can be identified with the open unit ball of $\mathbb{C}^d$.

Definition 3.1. For any $Z \in \mathbb{B}^d_n$, $n \in \mathbb{N}$, let

$$\begin{aligned} \mathcal{R}(V - Z) &:= \mathrm{Ran}\,([(V \otimes I_n) - (I_{\mathcal{H}} \otimes Z)]\,(V^*V \otimes I_n)) \\ &= [I_{\mathcal{H}} \otimes I_n - (I_{\mathcal{H}} \otimes Z)(V^* \otimes I_n)]\,\mathrm{Ran}\,(V) \otimes \mathbb{C}^n \\ &=: (I - ZV^*)\mathrm{Ran}\,(V) \otimes \mathbb{C}^n, \\ \text{where } ZV^* &:= V_1^* \otimes Z_1 + \cdots + V_d^* \otimes Z_d \in (\mathscr{L}(\mathcal{H} \otimes \mathbb{C}^n))_1. \end{aligned} \tag{3.1}$$

The main results of this section will be:

Theorem 3.2. *Let $V : \mathcal{H} \otimes \mathbb{C}^d \to \mathcal{H}$ be a row partial isometry. The subspace*

$$\mathcal{H}' := \{h \in \mathcal{H} |\ h \otimes \mathbb{C}^n \subseteq \mathcal{H}'_n \quad \forall\, n \in \mathbb{N}\}; \qquad \mathcal{H}'_n := \bigcap_{Z \in \mathbb{B}^d_n} \mathcal{R}(V - Z),$$

is the largest co-invariant subspace for V on which V^* acts isometrically.

In particular, V is CNC if and only if

$$\mathcal{H} = (\mathcal{H}')^{\perp} = \bigvee_{\lambda \in \mathbb{C}^n;\ Z \in \mathbb{B}^d_n} \lambda (I - VZ^*)^{-1} \mathrm{Ran}\,(V)^{\perp} \otimes \mathbb{C}^n = \bigvee_{\alpha \in \mathbb{F}^d} V^{\alpha} \mathrm{Ran}\,(V)^{\perp}.$$

In the above, as in Subsection 2.4, we view any $\lambda \in \mathbb{C}^n$ as a row operator, $\lambda : \mathcal{H} \otimes \mathbb{C}^n \to \mathcal{H}$. This theorem is a generalization of a characterization of CNU partial isometries due to Kreĭn [27, Chapter 1, Theorem 2.1]. Kreĭn's result is proven in the setting of unbounded symmetric operators. This can be restated in terms of partial isometries using the Cayley transform, a fractional linear transformation that implements a bijection between partial isometries and symmetric linear transformations [1].

In the special case where V is commutative, *i.e.*, a d-contraction, this yields:

Theorem 3.3. *Let $V : \mathcal{H} \otimes \mathbb{C}^d \to \mathcal{H}$ be a d-partial isometry. The subspace*

$$\mathcal{H}' := \bigcap_{z \in \mathbb{B}^d} \mathcal{R}\,(V - z),$$

is the largest co-invariant subspace for V on which V^* acts isometrically.

In particular, V is CNC if and only if

$$\mathcal{H} = (\mathcal{H}')^{\perp} = \bigvee_{z \in \mathbb{B}^d = \mathbb{B}^d_1} (I - Vz^*)^{-1} \mathrm{Ran}\,(V)^{\perp} = \bigvee_{\mathbf{n} \in \mathbb{N}^d} V^{\mathbf{n}} \mathrm{Ran}\,(V)^{\perp}.$$

A key corollary of these results is:

Corollary 3.4. *A row contraction $T : \mathcal{H} \otimes \mathbb{C}^d \to \mathcal{H}$ is CNC if and only if*

$$\mathcal{H} = \bigvee_{\alpha \in \mathbb{F}^d} T^{\alpha} \mathrm{Ran}\,(D_{T^*}).$$

A d-contraction $T : \mathcal{H} \otimes \mathbb{C}^d \to \mathcal{H}$ is CNC if and only if

$$\mathcal{H} = \bigvee_{\mathbf{n} \in \mathbb{N}^d} T^{\mathbf{n}} \mathrm{Ran}\,(D_{T^*}) = \bigvee_{z \in \mathbb{B}^d} (I - Tz^*)^{-1} \mathrm{Ran}\,(D_{T^*}).$$

In particular, if a (not necessarily commutative) row contraction T on $\mathcal{H}$ obeys

$$\mathcal{H} = \bigvee_{z \in \mathbb{B}^d} (I - Tz^*)^{-1} \mathrm{Ran}\,(D_{T^*}), \qquad \text{then } T \text{ is CNC.}$$

Motivated by the above corollary, we define:

Definition 3.5. A row contraction $T : \mathcal{H} \otimes \mathbb{C}^d \to \mathcal{H}$ obeys the *commutative CNC condition*, and we write: T is CCNC, if

$$\mathcal{H} = \bigvee_{z \in \mathbb{B}^d} (I - Tz^*)^{-1} \mathrm{Ran}\,(D_{T^*}).$$

Corollary 3.4 implies that any CCNC row contraction is CNC, and that any CNC d-contraction is CCNC.

Note, in particular, that if T is commutative, *i.e.*, T is a d-contraction, that

$$\bigvee_{\alpha \in \mathbb{F}^d} T^\alpha \mathrm{Ran}\,(D_{T^*}) = \bigvee_{\mathbf{n} \in \mathbb{N}^d} T^{\mathbf{n}} \mathrm{Ran}\,(D_{T^*}),$$

and this second set can be re-expressed as follows:

Lemma 3.6 ([30, Lemma 2.2]). *Given any row contraction T on $\mathcal{H}$,*

$$\bigvee_{z \in \mathbb{B}^d} (I - Tz^*)^{-1} = \bigvee_{\mathbf{n} \in \mathbb{N}^d} T^{\mathbf{n}}.$$

The next example below shows that not every CNC row contraction is CCNC:

Example 3.7 (The Arveson d-shift and the left free shift). The Arveson d-shift, S, on the Drury–Arveson space, H^2_d, is a commutative row partial isometry. In particular $D_{S^*} = k_0 k_0^*$ is a projection so that

$$\bigvee_{z \in \mathbb{B}^d} (I - Sz^*)^{-1} \mathrm{Ran}\,(D_{S^*}) = \bigvee_{z \in \mathbb{B}^d} k_z = H^2_d,$$

and S is CCNC. Observe that S is an extremal Gleason solution for the de Branges–Rovnyak space $\mathscr{H}(b) = H^2_d$ with $b \equiv 0 \in \mathscr{S}_d = \mathscr{S}_d(\mathbb{C})$.

For an example of a CNC row contraction which is not CCNC, consider the left free shift L on the full Fock space F^2_d. Recall here that the full Fock space over $\mathbb{C}^d$, F^2_d, is the direct sum of all tensor powers of $\mathbb{C}^d$:

$$F^2_d := \mathbb{C} \oplus \left(\mathbb{C}^d \otimes \mathbb{C}^d\right) \oplus \left(\mathbb{C}^d \otimes \mathbb{C}^d \otimes \mathbb{C}^d\right) \oplus \cdots = \bigoplus_{k=0}^{\infty} \left(\mathbb{C}^d\right)^{k \cdot \otimes}.$$

Fix an orthonormal basis $e_1, \ldots, e_d$ of $\mathbb{C}^d$. The left creation operators $L_1, \ldots, L_d$ are the operators which act as tensoring on the left by these basis vectors:

$$L_k f := e_k \otimes f; \qquad f \in F^2_d.$$

The left free shift is the row operator $L := (L_1, \ldots, L_d) : F^2_d \otimes \mathbb{C}^d \to F^2_d$. The left free shift, is, in fact, a row isometry: $L^* L = I_{F^2} \otimes I_d$. The orthogonal complement of the range of L is the vacuum vector 1 which spans the subspace $\mathbb{C} =: (\mathbb{C}^d)^{0 \cdot \otimes} \subset F^2_d$. A canonical orthonormal basis for F^2_d is then $\{e_\alpha\}_{\alpha \in \mathbb{F}^d}$ where $e_\alpha = L^\alpha 1$. It follows that L is CNC:

$$\bigvee_{\alpha \in F^2_d} L^\alpha \mathrm{Ran}\,(D_{L^*}) = \bigvee L^\alpha 1 = F^2_d.$$

However, L is not CCNC, since

$$\bigvee_{\mathbf{n} \in \mathbb{N}^d} L^{\mathbf{n}} \mathrm{Ran}\,(D_{L^*}) = \bigvee L^{\mathbf{n}} 1,$$

is equal to symmetric or Bosonic Fock space, a proper subspace of F^2_d (which is in fact canonically isomorphic to H^2_d, see, *e.g.*, [49, Section 4.5]).

It will be convenient to first establish several preliminary facts before proving Theorem 3.2 and its corollaries. Given a partial isometry $V : \mathcal{H}_1 \to \mathcal{H}_2$, and a contraction $T : \mathcal{K}_1 \to \mathcal{K}_2$ where $\mathcal{H}_k \subset \mathcal{K}_k$, we say that T extends V and write $V \subseteq T$ if $TV^*V = V$.

Lemma 3.8 ([31, Lemma 2.3]). *Let $V : \mathcal{H}_1 \to \mathcal{H}_2$ be a partial isometry and assume that $\mathcal{H}_k \subseteq \mathcal{J}_k$ for Hilbert spaces $\mathcal{H}_k, \mathcal{J}_k$, $k = 1, 2$. For any contraction $T : \mathcal{J}_1 \to \mathcal{J}_2$ the following are equivalent:*

(i) *T is a contractive extension of V, $V \subseteq T$.*
(ii) *T^* is a contractive extension of V^*, $V^* \subseteq T^*$.*
(iii) *There is a contraction $C : \mathcal{J}_1 \to \mathcal{J}_2$ such that $\mathrm{Ker}(C)^\perp \subseteq \mathcal{J}_1 \ominus \mathrm{Ker}(V)^\perp$, $\mathrm{Ran}\,(C) \subseteq \mathcal{J}_2 \ominus \mathrm{Ran}\,(V)$ and $T = V - C$.*

Lemma 3.9. *Let V be a row partial isometry on $\mathcal{H} \otimes \mathbb{C}^d$ and let $T \supseteq V$ be a row contractive extension to $\mathcal{K} \otimes \mathbb{C}^d$, $\mathcal{K} \supseteq \mathcal{H}$. Then for any $Z \in \mathbb{B}^d_n$,*

$$(I - TZ^*)^{-1} : \mathcal{K} \otimes \mathbb{C}^n \ominus (\mathrm{Ran}\,(V) \otimes \mathbb{C}^n) \to \mathcal{K} \otimes \mathbb{C}^n \ominus \mathcal{R}\,(V - Z), \quad \textit{is an isomorphism.}$$

Setting $T = V$, this shows $\mathcal{R}\,(V - Z)^\perp = (I - VZ^*)^{-1}\mathrm{Ran}\,(V)^\perp$. This holds, in particular, for $Z = z \in \mathbb{B}^d_1 = \mathbb{B}^d$ and $T \supseteq V$ acting on $\mathcal{H} \otimes \mathbb{C}^d$ (*e.g.*, $T = V$):

$$(I - Tz^*)^{-1} : \mathrm{Ran}\,(V)^\perp \to \mathcal{R}\,(V - z)^\perp \qquad \text{is an isomorphism.}$$

Proof. Since Z is strictly contractive, TZ^*, as defined above in equation (3.1), is a strict contraction so that $(I - TZ^*)^{-1}$ is well defined. Assume that $f \in \mathcal{K} \otimes \mathbb{C}^n \ominus \mathcal{R}\,(V - Z)$, and suppose $g \in \mathrm{Ran}\,(V) \otimes \mathbb{C}^n$ so that $g = (V \otimes I_n)G$ for some $G \in (\mathcal{H} \otimes \mathbb{C}^n) \otimes \mathbb{C}^d \ominus \mathrm{Ker}(V \otimes I_n)$. Then,

$$\begin{aligned} \langle (I - TZ^*)f, g\rangle &= \langle f, (I - ZT^*)(V \otimes I_n)G\rangle \\ &= \langle f, (I - ZV^*)(V \otimes I_n)G\rangle = 0. \end{aligned}$$

This proves that

$$(I - TZ^*)\left((\mathcal{K} \otimes \mathbb{C}^n) \ominus \mathcal{R}\,(V - Z)\right) \subseteq (\mathcal{K} \otimes \mathbb{C}^n) \ominus (\mathrm{Ran}\,(V) \otimes \mathbb{C}^n),$$

so that

$$(\mathcal{K} \otimes \mathbb{C}^n) \ominus \mathcal{R}\,(V - Z) \subseteq (I - TZ^*)^{-1}\left(\mathcal{K} \otimes \mathbb{C}^n \ominus \mathrm{Ran}\,(V) \otimes \mathbb{C}^n\right).$$

For the reverse inclusion suppose that $f \in \mathcal{K} \otimes \mathbb{C}^n \ominus \mathrm{Ran}\,(V \otimes I_n)$. Also assume that $g \in \mathcal{R}\,(V - Z)$, $g = (I - ZV^*)(V \otimes I_n)G$. Then,

$$\begin{aligned} \langle (I - TZ^*)^{-1}f, g\rangle &= \langle f, (I - ZT^*)^{-1}(I - ZV^*)VG\rangle \qquad (VG := V \otimes I_n G) \\ &= \langle f, (I - ZT^*)^{-1}(I - ZT^*)VG\rangle \\ &= \langle f, VG\rangle = 0. \end{aligned}$$

□

Lemma 3.10. *Suppose that $\mathscr{L} \subseteq \mathcal{H}$ is co-invariant for V, and that $V^*|_{\mathscr{L}}$ is an isometry. Then for any $Z \in \mathbb{B}^d_n$, $\mathscr{L} \otimes \mathbb{C}^n \subseteq \mathcal{R}\,(V - Z)$.*

Proof.

$$\begin{aligned}\mathcal{R}(V-Z) &\supseteq (I-ZV^*)V\otimes I_n(\mathscr{L}\otimes\mathbb{C}^n\otimes\mathbb{C}^d)\\ &= (I-ZV^*)\mathscr{L}\otimes\mathbb{C}^n; \qquad V \text{ is co-isometric, hence onto } \mathscr{L}\\ &= \mathscr{L}\otimes\mathbb{C}^n; \qquad (I-ZV^*) \text{ is invertible, hence onto.}\end{aligned}$$ □

Definition 3.11. For any $n\in\mathbb{N}$ define $\mathcal{H}'_n \subset \mathcal{H}\otimes\mathbb{C}^n$ by

$$\mathcal{H}'_n := \bigcap_{Z\in\mathbb{B}^d_n} \mathcal{R}(V-Z).$$

Also define $\mathcal{H}'\subseteq\mathcal{H}$ by

$$\begin{aligned}\mathcal{H}' &:= \{h\in\mathcal{H} |\ h\otimes\mathbb{C}^n \subseteq \mathcal{H}'_n,\ \forall n\in\mathbb{N}\}\\ &= \{h\in\mathcal{H} |\ \lambda^*h\in\mathcal{H}'_n\ \forall\lambda\in\mathbb{C}^n\ \forall n\in\mathbb{N}\}.\end{aligned}$$

In the above, as before, given any $\lambda\in\mathbb{C}^n$, we view $\lambda:\mathcal{H}\otimes\mathbb{C}^n\to\mathcal{H}$ as a row operator.

Lemma 3.12. *$\mathcal{H}'$ is a closed subspace, and*

$$\begin{aligned}(\mathcal{H}')^\perp &= \bigvee_{\substack{\lambda,\tau\in\mathbb{C}^n\\ Z\in\mathbb{B}^d_n;\ n\in\mathbb{N}}} \lambda(I-VZ^*)^{-1}\tau^*\mathrm{Ran}\,(V)^\perp\\ &= \bigvee_{\substack{\lambda\in\mathbb{C}^n\\ Z\in\mathbb{B}^d_n;\ n\in\mathbb{N}}} \lambda(I-VZ^*)^{-1}\mathrm{Ran}\,(V)^\perp\otimes\mathbb{C}^n.\end{aligned}$$

Proof. Let h_k be Cauchy in $\mathcal{H}'$ with limit $h\in\mathcal{H}$. Then for any fixed $n\in\mathbb{N}$ and $\tau\in\mathbb{C}^n$, τ^*h_k is Cauchy in $\mathcal{H}'\otimes\mathbb{C}^n$, and hence the limit τ^*h belongs to $\mathcal{H}'_n$, since $\mathcal{H}'_n$ is clearly closed. Since this holds for any n, we obtain that $h\in\mathcal{H}'$.

For the proof of the second statement suppose that $h\in\mathcal{H}$ is orthogonal to any vector of the form

$$\lambda(I-VZ^*)^{-1}\tau^*\mathrm{Ran}\,(V)^\perp: \qquad \lambda,\tau\in\mathbb{C}^n,\ Z\in\mathbb{B}^d_n.$$

This happens if and only if:

$$\begin{aligned}& h\perp\lambda(I-VZ^*)^{-1}\mathrm{Ran}\,(V)^\perp\otimes\mathbb{C}^n; \qquad \lambda\in\mathbb{C}^n,\ Z\in\mathbb{B}^d_n,\ n\in\mathbb{N}\\ \Leftrightarrow\ & h\otimes\mathbb{C}^n\perp \underbrace{(I-VZ^*)^{-1}\mathrm{Ran}\,(V)^\perp\otimes\mathbb{C}^n}_{=\mathcal{R}(V-Z)^\perp \text{ by Lemma 3.9}}; \qquad Z\in\mathbb{B}^d_n,\ n\in\mathbb{N}.\end{aligned}$$

This proves that h has this property if and only if $h\otimes\mathbb{C}^n\subseteq\mathcal{R}(V-Z)$ for any $Z\in\mathbb{B}^d_n$, *i.e.*, if and only if $h\otimes\mathbb{C}^n\subseteq\mathcal{H}'_n$ for any $n\in\mathbb{N}$ and therefore if and only if $h\in\mathcal{H}'$. □

Lemma 3.13. *If $\mathscr{L}\subseteq\mathcal{H}$ is co-invariant for the row partial isometry V and V^* is isometric on $\mathscr{L}$ then $\mathscr{L}\subseteq\mathcal{H}'$.*

Proof. By Lemma 3.10, $\mathscr{L} \otimes \mathbb{C}^n \subseteq \mathcal{H}'_n$ for any $n \in \mathbb{N}$. In particular, given any $l \in \mathscr{L}$, $\tau^* l$ belongs to $\mathcal{H}'_n$ for any $\tau \in \mathbb{C}^n$ and any $n \in \mathbb{N}$. Therefore $l \in \mathcal{H}'$ by definition and $\mathscr{L} \subseteq \mathcal{H}'$. □

The following technical fact provides a useful description of $(\mathcal{H}')^\perp$.

Lemma 3.14.
$$\bigvee_{\substack{\lambda,\tau\in\mathbb{C}^n \\ Z\in\mathbb{B}^d_n;\ n\in\mathbb{N}}} \lambda(I - ZV^*)^{-1}\tau^* = \bigvee_{\alpha\in\mathbb{F}^d} (V^*)^\alpha.$$

Proof. First note that anything in the left-hand side (LHS) of the above equation is a linear combination of products of the $V_1^*, \dots, V_d^*$, and so it follows that the left-hand side is contained in the right-hand side (RHS).

We will prove the converse inductively on the length, $N = |\alpha|$ of a word $\alpha \in \mathbb{F}^d$. First, taking $n = 1$, we have by Lemma 3.6 that $(V^*)^{\mathbf{n}}$ belongs to the LHS for all $\mathbf{n} \in \mathbb{N}^d$. In particular $V_k^* \in LHS$ for all $1 \le k \le d$ so that the inductive hypothesis holds for $N = 1$. Assume that the inductive hypothesis holds for all $\alpha \in \mathbb{F}^d$ of length less than or equal to $K \in \mathbb{N}$. That is, $|\alpha| \le K$ implies that $(V^*)^\alpha$ belongs to the LHS. To complete the induction step we need to prove that given any $\beta \in \mathbb{F}^d$ of length $K+1$ that $(V^*)^\beta$ belongs to the LHS.

Any such β can be written $\beta = j\alpha$ where $j \in \{1, \dots, d\}$ and $|\alpha| = K$. By the hypothesis $(V^*)^\alpha$ is the norm-limit of finite linear combinations of terms of the form
$$\lambda(I - ZV^*)^{-1}\tau^*,$$
where $\lambda, \tau \in \mathbb{C}^n$ and $Z \in \mathbb{B}^d_n$ for some $n \in \mathbb{N}$. It therefore suffices to prove that for any such term, and any fixed $1 \le j \le d$, we can find $W \in \mathbb{B}^d_m$ and $\Lambda, \Gamma \in \mathbb{C}^m$ so that
$$\Lambda(I - WV^*)^{-1}\Gamma^* = V_j^*\lambda(I - ZV^*)^{-1}\tau^*.$$
For simplicity fix $j = 1$. The other cases will follow from an analogous argument.

Choose $W = (W_1, \dots, W_d) \in \mathbb{B}^d_{2n}$ as follows:
$$W_1 := \begin{bmatrix} Z_1 & 0_n \\ rI_n & 0_n \end{bmatrix}; \qquad W_k := \begin{bmatrix} Z_k & 0 \\ 0 & 0 \end{bmatrix}; \quad 2 \le k \le d,$$
where if $\|Z\|^2 = s < 1$, choose $0 < r < 1$ small enough so that $1 > s(1 + r^2)$.

Then,
$$I - WW^* = \begin{bmatrix} I - ZZ^* & -rZ_1 \\ -rZ_1^* & I \end{bmatrix},$$
and by Schur complement theory [18, Appendix A.5.5], this is strictly positive if and only if $I - ZZ^* - r^2 Z_1 Z_1^* > 0$. By our choice of r, this is the case, and it follows that $W \in \mathbb{B}^d_{2n}$ is strictly contractive. Observe that
$$WV^* = \begin{bmatrix} ZV^* & 0 \\ rI_n \otimes V_1^* & 0 \end{bmatrix}; \quad \text{and} \quad (WV^*)^k = \begin{bmatrix} (ZV^*)^k & 0 \\ r(I_n \otimes V_1^*)(ZV^*)^{k-1} & 0 \end{bmatrix}; \quad k \ge 2.$$

It follows that

$$(I - WV^*)^{-1} = \begin{bmatrix} (I - ZV^*)^{-1} & 0 \\ r(I_n \otimes V_1^*)(I - ZV^*)^{-1} & I_n \otimes I_{\mathcal{H}} \end{bmatrix}.$$

Taking $\Lambda := (0_n, \lambda)$ and $\Gamma = (\tau, 0_n)$ then yields

$$\Lambda(I - WV^*)^{-1}\Gamma^* = rV_1^*\lambda(I - ZV^*)^{-1}\tau^*,$$

and the inductive step follows. □

We now have all the necessary ingredients to prove the main result of this section:

Proof of Theorem 3.2. By Lemma 3.10, any co-invariant subspace, $\mathscr{L}$, on which V^* acts isometrically is contained in $\mathcal{H}'$. It remains to show that $\mathcal{H}'$ is co-invariant for V, and that $V^*|_{\mathcal{H}'}$ is an isometry.

Let U be a co-isometric extension (*e.g.*, a Cuntz unitary dilation) of V on $\mathcal{K} \supseteq \mathcal{H}$. If $h \in \mathcal{H}'$, then $\tau^* h \in \mathcal{H}'_n$. Hence,

$$\tau^* h \perp \mathcal{K} \otimes \mathbb{C}^n \ominus \mathcal{R}(V - Z) = (I - UZ^*)^{-1}\left((\mathcal{K} \otimes \mathbb{C}^n) \ominus \mathrm{Ran}\,(V) \otimes \mathbb{C}^n\right).$$

This happens if and only if

$$(I - ZU^*)^{-1}\tau^* h \in \mathrm{Ran}\,(V) \otimes \mathbb{C}^n,$$

and this implies

$$\bigvee_{\substack{\tau,\lambda \in \mathbb{C}^n \\ Z \in \mathbb{B}^d_n}} \lambda(I - ZU^*)^{-1}\tau^* h \in \mathrm{Ran}\,(V).$$

By the last lemma this happens if and only if

$$\bigvee_{\alpha \in \mathbb{F}^d} (U^*)^\alpha h \in \mathrm{Ran}\,(V).$$

Given any $VF \in \mathrm{Ran}\,(V)$, it is easy to see that by definition

$$(I - ZU^*)\tau^* VF \in \mathcal{R}(V - Z).$$

Hence,

$$\begin{aligned} \mathcal{R}(V - Z) &\supseteq (I - ZU^*) \bigvee_{\substack{h \in \mathcal{H}' \\ \lambda \in \mathbb{C}^n; \alpha,\beta \in \mathbb{F}^d}} \tau^*(U^*)^\beta (U^*)^\alpha h \\ &\supseteq (I - ZU^*) \bigvee_{\substack{h \in \mathcal{H}';\ \alpha \in \mathbb{F}^d \\ \tau,\lambda,\kappa \in \mathbb{C}^n; W \in \mathbb{B}^d_n}} \tau^*\lambda(I - WU^*)^{-1}\kappa^*(U^*)^\alpha h, \quad \text{(by Lemma 3.14)} \\ &\supseteq \bigvee_{\substack{h \in \mathcal{H}'; \alpha \in \mathbb{F}^d \\ \kappa \in \mathbb{C}^n}} \kappa^*(U^*)^\alpha h, \end{aligned}$$

so that $(U^*)^\alpha h \in \mathcal{H}'$ for all $\alpha \in \mathbb{F}^d$ and all $h \in \mathcal{H}'$.

Given any $h \in \mathcal{H}' \subset \mathrm{Ran}\,(V)$, $h = VH$ where $V^*VH = H$ so that

$$V^*h = V^*VH = U^*VH = U^*h.$$

This proves that $\mathcal{H}'$ is co-invariant for V (*i.e.*, co-invariant for each V_k), and that V is co-isometric on $\mathcal{H}'$. □

Although Theorem 3.3 is an immediate consequence of Theorem 3.2 under the assumption that V is a d-partial isometry, the above proof can be modified to prove Theorem 3.3 directly.

Lemma 3.15. *Let $T : \mathcal{H}\otimes\mathbb{C}^d \to \mathcal{H}$ be a row contraction. Then $\mathcal{H}\otimes\mathbb{C}^d$ decomposes as $\mathcal{H}\otimes\mathbb{C}^d = \mathcal{H}_0 \oplus \mathcal{H}_1$ and $\mathcal{H} = \mathcal{H}_0' \oplus \mathcal{H}_1'$ where $V := TP_{\mathcal{H}_0}$ is a row partial isometry with initial space $\mathcal{H}_0 \subseteq \mathcal{H}\otimes\mathbb{C}^d$ and final space $\mathcal{H}_0' \subseteq \mathcal{H}$. The row contraction $C := -TP_{\mathcal{H}_1}$ is a pure row contraction,* i.e., $\|C\mathbf{h}\| < \|\mathbf{h}\|$ *for any* $\mathbf{h} \in \mathcal{H}_1 \otimes \mathbb{C}^d$, *with final space* $\mathrm{Ran}\,(C) \subseteq \mathcal{H}_1' = \mathrm{Ran}\,(V)^\perp$.

Proof. Let $\mathcal{H}_0 := \mathrm{Ran}\,(D_T)^\perp = \mathrm{Ker}(D_T)$, where recall $D_T = \sqrt{I - T^*T}$. It is clear that $V|_{\mathcal{H}_0}$ is an isometry, and can be extended to a row partial isometry on $\mathcal{H}\otimes\mathbb{C}^d$ with initial space $\mathcal{H}_0$. It follows that $C := V - T$ is a pure contraction and by Lemma 3.8 the initial space of C is contained in $\mathcal{H}_1 = \mathcal{H}_0^\perp$ and the final space of C is contained in $\mathcal{H}_1' = (\mathcal{H}_0')^\perp$. □

Definition 3.16. The above decomposition $T = V - C$ of any row contraction on $\mathcal{H}$ into a row partial isometry, V, on $\mathcal{H}\otimes\mathbb{C}^d$ and a pure row contraction C on $\mathcal{H}\otimes\mathbb{C}^d$ with $\mathrm{Ker}(C)^\perp \subseteq \mathrm{Ker}(V)$ and $\mathrm{Ran}\,(C) \subseteq \mathrm{Ran}\,(V)^\perp$, will be called the *isometric-pure decomposition* of T.

Proof of Corollary 3.4. By Lemma 3.15, it follows that T is CNC if and only if its partial isometric part V is CNC. Since $T \supseteq V$ is a row contractive extension of V on $\mathcal{H}$, Lemma 3.9, Lemma 3.14 and Theorem 3.2 imply that T is CNC if and only if

$$\begin{aligned}\mathcal{H} &= \bigvee_{\lambda\in\mathbb{C}^n;\ Z\in\mathbb{B}^d_n} \lambda(I - VZ^*)^{-1}\mathrm{Ran}\,(V)^\perp \otimes \mathbb{C}^n \\ &= \bigvee_{\lambda\in\mathbb{C}^n;\ Z\in\mathbb{B}^d_n} \lambda(I - TZ^*)^{-1}\mathrm{Ran}\,(D_{T^*}) \otimes \mathbb{C}^n \\ &= \bigvee_{\alpha\in\mathbb{F}^d} T^\alpha \mathrm{Ran}\,(D_{T^*}).\end{aligned}$$

If T is CCNC then it is clearly CNC, since in this case,

$$\bigvee_{\alpha\in\mathbb{F}^d} T^\alpha \mathrm{Ran}\,(D_{T^*}) \supseteq \bigvee_{\mathbf{n}\in\mathbb{N}^d} T^{\mathbf{n}} \mathrm{Ran}\,(D_{T^*}) = \mathcal{H}.$$ □

4. Model maps for CCNC row partial isometries

Let $V : \mathcal{H} \otimes \mathbb{C}^d \to \mathcal{H}$ be a CCNC row partial isometry. By definition,

$$\mathcal{H} = \bigvee_{z \in \mathbb{B}^d} \mathcal{R}(V - z)^{\perp} = \bigvee_{z \in \mathbb{B}^d} (I - Vz^*)^{-1} \mathrm{Ran}\,(V)^{\perp},$$

and V is CNC by the results of the previous section.

Definition 4.1. A *model triple*, $(\gamma, \mathcal{J}_\infty, \mathcal{J}_0)$ for a CCNC row partial isometry V on $\mathcal{H}$ consists of auxiliary Hilbert spaces $\mathcal{J}_\infty, \mathcal{J}_0$ of dimension $\mathrm{Ker}(V)$ and $\mathrm{Ran}\,(V)^{\perp}$, respectively, and a *model map*, γ, defined on $\mathbb{B}^d \cup \{\infty\}$:

$$\gamma : \begin{cases} \mathbb{B}^d \to \mathscr{L}(\mathcal{J}_0, \mathcal{R}(V - z)^{\perp}) \\ \{\infty\} \to \mathscr{L}(\mathcal{J}_\infty, \mathrm{Ker}(V)) \end{cases},$$

such that $\gamma(z)$ is a linear isomorphism for any $z \in \mathbb{B}^d$ and $\gamma(0), \gamma(\infty)$ are onto isometries.

We say that $(\Gamma, \mathcal{J}_\infty, \mathcal{J}_0)$ is an *analytic model triple* and that Γ is an *analytic model map* if $z \mapsto \Gamma(z)$ is anti-analytic for $z \in \mathbb{B}^d$.

Lemma 3.9 shows that any row contractive extension $T \supseteq V$ of a row partial isometry on $\mathcal{H}$ gives rise to an analytic model map: Let $\Gamma_T(0) : \mathcal{J}_0 \to \mathrm{Ran}\,(V)^{\perp}$, $\Gamma_T(\infty) : \mathcal{J}_\infty \to \mathrm{Ker}(V)$ be any two fixed onto isometries and then

$$\Gamma_T(z) := (I - Tz^*)^{-1} \Gamma_T(0); \qquad z \in \mathbb{B}^d,$$

defines an analytic model triple. Most simply, we are free to choose $T = V$.

Let $(\Gamma, \mathcal{J}_\infty, \mathcal{J}_0)$ be an analytic model triple for a CCNC row partial isometry V on $\mathcal{H}$. For any $h \in \mathcal{H}$, define

$$\hat{h}^{\Gamma}(z) := \Gamma(z)^* h,$$

an analytic $\mathcal{J}_0$-valued function on $\mathbb{B}^d$. (When it is clear from context, we will sometimes omit the superscript Γ.) Let $\hat{\mathcal{H}}^{\Gamma}$ be the vector space of all $\hat{h} = \hat{h}^{\Gamma}$ for $h \in \mathcal{H}$. We endow $\hat{\mathcal{H}}^{\Gamma}$ with an inner product:

$$\langle \hat{h}, \hat{g} \rangle_{\Gamma} := \langle h, g \rangle_{\mathcal{H}}.$$

Proposition 4.2. *The sesquilinear form $\langle \cdot, \cdot \rangle_{\Gamma}$ is an inner product on $\hat{\mathcal{H}}^{\Gamma}$, and $\hat{\mathcal{H}}^{\Gamma}$ is a Hilbert space with respect to this inner product.*

The Hilbert space $\hat{\mathcal{H}}^{\Gamma}$ is the reproducing kernel Hilbert space of $\mathcal{J}_0$-valued analytic functions on $\mathbb{B}^d$ with reproducing kernel

$$\hat{K}^{\Gamma}(z, w) := \Gamma(z)^* \Gamma(w) \in \mathscr{L}(\mathcal{J}_0), \qquad \hat{\mathcal{H}}^{\Gamma} = \mathcal{H}(\hat{K}^{\Gamma}).$$

The map $\hat{U}^{\Gamma} : \mathcal{H} \to \hat{\mathcal{H}}^{\Gamma}$ defined by $\hat{U}^{\Gamma} h = \hat{h}^{\Gamma}$ is an onto isometry and

$$\hat{K}_z^{\Gamma} g = \hat{U}^{\Gamma} \Gamma(z) g = \widehat{\Gamma(z) g}; \qquad g \in \mathcal{J}_0.$$

Proof. This is all pretty easy to verify. For simplicity we omit the superscript Γ. In order to show that $\langle \hat{h}, \hat{g} \rangle_\Gamma := \langle h, g \rangle_{\mathcal{H}}$ is an inner product, the only non-immediate property to check is that there are no non-zero vectors of zero length with respect to this sesquilinear form, or equivalently that $\hat{h}(z) = 0$ for all $z \in \mathbb{B}^d$ implies that $h = 0$. This is clear since $\hat{h}(z) = \Gamma(z)^* h$, and since V is CCNC,

$$\bigvee_{z \in \mathbb{B}^d} \operatorname{Ran}(\Gamma(z)) = \bigvee_{z \in \mathbb{B}^d} \mathcal{R}(V - z)^\perp = \mathcal{H},$$

so that $\bigcap_{z \in \mathbb{B}^d} \operatorname{Ker}(\Gamma(z)^*) = \{0\}$.

The map $\hat{U} : \mathcal{H} \to \hat{\mathcal{H}}$ is an onto isometry by definition of the inner product in $\hat{\mathcal{H}}$. For any $g \in \mathcal{J}_0$, compute

$$\begin{aligned} \langle \hat{U}\Gamma(z) g, \hat{h} \rangle_{\hat{\mathcal{H}}} &= \langle \Gamma(z) g, h \rangle_{\mathcal{H}} = \langle g, \Gamma(z)^* h \rangle_{\mathcal{J}_0} \\ &= \langle g, \hat{h}(z) \rangle_{\mathcal{J}_0} \quad =: \langle g, \hat{K}_z^* \hat{h} \rangle_{\mathcal{J}_0}. \end{aligned} \tag{4.1}$$

This proves simultaneously that $\hat{\mathcal{H}}$ is a RKHS of analytic $\mathcal{J}_0$-valued functions on $\mathbb{B}^d$ with point evaluation maps

$$\hat{K}_z = \hat{U}\Gamma(z) \in \mathscr{L}(\mathcal{J}_0, \hat{\mathcal{H}}),$$

and reproducing kernel

$$\hat{K}(z, w) = \hat{K}_z^* \hat{K}_w = \Gamma(z)^* \Gamma(w) \in \mathscr{L}(\mathcal{J}_0). \qquad \square$$

Proposition 4.3. *Let $V : \mathcal{H} \otimes \mathbb{C}^d \to \mathcal{H}$ be a CCNC row partial isometry with analytic model map Γ. The image, $\hat{V}^\Gamma := \hat{U}^\Gamma V (\hat{U}^\Gamma)^*$ of V under the unitary map $\hat{U}^\Gamma$ onto the model RKHS $\hat{\mathcal{H}}^\Gamma$ has initial and final spaces*

$$\begin{aligned} &\operatorname{Ker}(\hat{V}^\Gamma)^\perp \subseteq \{\hat{\mathbf{h}} \in \hat{\mathcal{H}}^\Gamma \otimes \mathbb{C}^d \mid z\hat{\mathbf{h}}(z) = \hat{g}(z) \text{ for some } \hat{g} \in \hat{\mathcal{H}}^\Gamma \text{ with } \|\hat{\mathbf{h}}\| = \|\hat{g}\|\}, \\ &\operatorname{Ran}\left(\hat{V}^\Gamma\right) = \{\hat{h} \in \hat{\mathcal{H}}^\Gamma \mid \hat{h}(0) = 0\}, \end{aligned}$$

and $\hat{V}^\Gamma$ acts as multiplication by $z = (z_1, \ldots, z_d)$ on its initial space: For any $\hat{\mathbf{h}} = (\hat{h}_1, \ldots, \hat{h}_d)^T \in \operatorname{Ker}(\hat{V}^\Gamma)^\perp$,

$$(\hat{V}^\Gamma \hat{\mathbf{h}})(z) = \left((\hat{V}_1^\Gamma, \ldots, \hat{V}_d^\Gamma) \begin{bmatrix} \hat{h}_1 \\ \vdots \\ \hat{h}_d \end{bmatrix} \right)(z) = z_1 \hat{h}_1(z) + \cdots + z_d \hat{h}_d(z) = z\hat{\mathbf{h}}(z).$$

Proof. Again, we omit the superscript Γ to simplify notation. First suppose that $\hat{\mathbf{h}} \in \operatorname{Ker}(\hat{V})^\perp$. Then,

$$\begin{aligned} \hat{V}\hat{\mathbf{h}} &= (\hat{U}V\mathbf{h})(z) = \Gamma(z)^*(V\mathbf{h}) = \Gamma(z)^* \left((V - z)\mathbf{h} + z\mathbf{h} \right) \\ &= \Gamma(z)^* z\mathbf{h} \quad \left(\text{since } (V - z)\mathbf{h} = (V - z)V^*V\mathbf{h} \in \mathcal{R}(V - z) = \operatorname{Ran}(\Gamma(z))^\perp\right) \\ &= z(\Gamma(z)^* \otimes I_d)\mathbf{h} = z\hat{\mathbf{h}}(z). \end{aligned}$$

This proves that $\hat{V}$ acts as multiplication by z on its initial space and that

$$\mathrm{Ker}(\hat{V})^\perp \subseteq \{\hat{\mathbf{h}} \in \hat{\mathcal{H}} \otimes \mathbb{C}^d |\ z\hat{\mathbf{h}}(z) \in \hat{\mathcal{H}} \text{ and } \|z\hat{\mathbf{h}}\| = \|\hat{\mathbf{h}}\|\},$$

since $\hat{V}$ is a partial isometry. The range statement is clear since $0 = \hat{h}(0) = \Gamma(0)^* h$ if and only if $h \in \mathrm{Ran}\,(V)$. □

Any model triple $(\gamma, \mathcal{J}_\infty, \mathcal{J}_0)$ for a CCNC row partial isometry V can be used to define a *characteristic function*, b_V^γ, on $\mathbb{B}^d$ as follows: First consider

$$D^\gamma(z) := \gamma(z)^*\gamma(0) = \hat{K}^\gamma(z, 0) \in \mathscr{L}(\mathcal{J}_0),$$
$$N^\gamma(z) := (\gamma(z)^* \otimes I_d)\gamma(\infty) \in \mathscr{L}(\mathcal{J}_\infty, \mathcal{J}_0 \otimes \mathbb{C}^d).$$

The next lemma below shows that $D^\gamma(z)$ is always a bounded, invertible operator on $\mathcal{J}_0$, so that we can define

$$b_V^\gamma(z) := D^\gamma(z)^{-1} z N^\gamma(z) \in \mathscr{L}(\mathcal{J}_\infty, \mathcal{J}_0). \tag{4.2}$$

The function b_V^γ is called a (representative) *characteristic function* of the CCNC row partial isometry V. Observe that if $\gamma = \Gamma$ is an analytic model map then b_V^Γ is analytic on $\mathbb{B}^d$.

Lemma 4.4. *Given any model triple $(\gamma, \mathcal{J}_\infty, \mathcal{J}_0)$ for a CCNC row partial isometry $V : \mathcal{H} \otimes \mathbb{C}^d \to \mathcal{H}$, the operator $D^\gamma(z) := \gamma(z)^*\gamma(0)$ is invertible for any $z \in \mathbb{B}^d$.*

Proof. Let $D(z) := D^\gamma(z) = \gamma(z)^*\gamma(0) \in \mathscr{L}(\mathcal{J}_0)$, we first show that $D(z)^*$ has trivial kernel. Suppose that $D(z)^* g = 0$. Then for any $h \in \mathcal{J}_0$,

$$\begin{aligned} 0 = \langle h, D(z)^* g\rangle_{\mathcal{J}_0} &= \langle h, \gamma(0)^*\gamma(z) g\rangle_{\mathcal{J}_0} \\ &= \langle \gamma(0) h, \gamma(z) g\rangle_{\mathcal{H}}, \end{aligned} \tag{4.3}$$

since we assume $D(z)^* g = 0$. By definition, $\gamma(z) : \mathcal{J}_0 \to \mathcal{R}\,(V - z)^\perp$ is an isomorphism (bounded, onto, invertible), and $\gamma(0)$ is an isometry onto $\mathrm{Ran}\,(V)^\perp$. By Lemma 3.9, $\gamma(z) g = (I - Vz^*)^{-1}\gamma(0) f$ for some non-zero $\gamma(0) f \in \mathrm{Ran}\,(V)^\perp$. Using that $\gamma(0) f \perp \mathrm{Ran}\,(V)$, and that $Vz^* = V_1 z_1^* + \cdots + V_d z_d^*$ is a strict contraction for $z \in \mathbb{B}^d$,

$$\begin{aligned} 0 &= \langle \gamma(0) f, \gamma(z) g\rangle_{\mathcal{H}} && \text{(Equation 4.3 with } h = f) \\ &= \langle \gamma(0) f, (I - Vz^*)^{-1}\gamma(0) f\rangle \\ &= \sum_{k=0}^{\infty} \langle \gamma(0) f, (Vz^*)^k \gamma(0) f\rangle && \text{(convergent geometric series)} \\ &= \langle \gamma(0) f, \gamma(0) f\rangle && ((Vz^*)^k\gamma(0) f \in \mathrm{Ran}\,(V) \text{ for } k \geq 1), \end{aligned}$$

a contradiction. In the case where $\mathcal{J}_0$ is finite-dimensional, this is enough to prove invertibility.

It remains to prove that $D(z)$ is bounded below. If not, then there is a sequence $(g_n) \subset \mathcal{J}_0$ such that $\|g_n\| = 1$ for all n, but $\|D(z) g_n\| \to 0$. Set $G_n := \gamma(0) g_n$, $\|G_n\| = 1$ since $\gamma(0)$ is an isometry. Again using that $\gamma(z)$:

$\mathcal{J}_0 \to \mathcal{R}(V-z)^\perp$, and $(I - Vz^*) : \mathcal{R}(V-z)^\perp \to \mathrm{Ran}(V)^\perp$ are isomorphisms, if $H_n \in \mathcal{R}(V-z)^\perp$ and $h_n \in \mathcal{J}_0$ are chosen so that $(I - Vz^*)H_n = G_n$ and $\gamma(z)h_n = H_n$, then the norms of the sequences (h_n), (H_n) are uniformly bounded above and below by strictly positive constants. Then,

$$\begin{aligned}\langle h_n, D(z)g_n\rangle &= \langle \gamma(z)h_n, \gamma(0)g_n\rangle = \langle H_n, G_n\rangle \\ &= \langle H_n, (I - Vz^*)H_n\rangle \\ &= \langle H_n, (I - zV^*Vz^*)H_n\rangle - \langle H_n, (V-z)V^*Vz^*H_n\rangle \\ &= \langle H_n, (I - zV^*Vz^*)H_n\rangle \quad (\text{Since } H_n \perp \mathcal{R}(V-z).) \\ &\geq (1 - |z|^2)\|H_n\|^2.\end{aligned}$$

Since there are constants $C, c > 0$ so that $\|h_n\| \leq C$, and $\|H_n\| \geq c > 0$, the limit of the left-hand side is zero while the right-hand side is bounded below by a strictly positive constant, a contradiction. □

Definition 4.5. Let b_1, b_2 be two analytic functions on $\mathbb{B}^d$ taking values in $\mathscr{L}(\mathcal{H}_1, \mathcal{J}_1)$ and $\mathscr{L}(\mathcal{H}_2, \mathcal{J}_2)$, respectively. We say that b_1, b_2 *coincide* if there are fixed unitary $R \in \mathscr{L}(\mathcal{J}_1, \mathcal{J}_2), Q \in \mathscr{L}(\mathcal{H}_1, \mathcal{H}_2)$ so that

$$Rb_1(z) = b_2(z)Q; \qquad z \in \mathbb{B}^d.$$

This clearly defines an equivalence relation on such functions, and we call the corresponding equivalence classes coincidence classes.

Lemma 4.6. *Let V be a CCNC row partial isometry. Let b_V^γ be a characteristic function for V defined using any model triple $(\gamma, \mathcal{J}_\infty, \mathcal{J}_0)$. The coincidence class of b_V^γ is invariant under the choice of model triple.*

It follows, in particular, that any characteristic function b_V^γ for V is analytic whether or not γ is an analytic model map, since b_V^γ coincides with b_V^Γ for any analytic model map Γ (and analytic model maps always exist).

Proof. Let $(\gamma, \mathcal{J}_\infty, \mathcal{J}_0), (\delta, \mathcal{K}_\infty, \mathcal{K}_0)$ be any two choices of model triples for V. Since both $\mathcal{K}_0, \mathcal{J}_0$ are isomorphic to $\mathrm{Ran}(V)^\perp$ and $\mathcal{K}_\infty, \mathcal{J}_\infty$ are isomorphic to $\mathrm{Ker}(V)$, we can define onto isometries $R = \gamma(0)^*\delta(0) \in \mathscr{L}(\mathcal{K}_0, \mathcal{J}_0)$ and $Q := \gamma(\infty)^*\delta(\infty) \in \mathscr{L}(\mathcal{K}_\infty, \mathcal{J}_\infty)$. Moreover $C_z := \gamma(z)^*(\delta(z)\delta(z)^*)^{-1}\delta(z)$ is a linear isomorphism (bounded and invertible) of $\mathcal{K}_0$ onto $\mathcal{J}_0$ for any $z \in \mathbb{B}^d$. As before,

$$\begin{aligned}D^\gamma(z) &:= \gamma(z)^*\gamma(0); \\ N^\gamma(z) &= (\gamma(z)^* \otimes I_d)\gamma(\infty), \\ b_V^\gamma(z) &:= D^\gamma(z)^{-1}zN^\gamma(z),\end{aligned}$$

and b_V^δ is defined analogously. It follows that:

$$\begin{aligned}Rb_V^\delta(z)Q &= RD^\delta(z)^{-1}C_z^{-1}C_z zN^\delta(z)Q \\ &= \left(C_z D^\delta(z)R^*\right)^{-1} z\left((C_z \otimes I_d)N^\delta(z)Q\right).\end{aligned}$$

In particular,

$$\begin{aligned}C_z D^\delta(z)R^* &= \gamma(z)^*(\delta(z)\delta(z)^*)^{-1}\delta(z)\delta(z)^*\delta(0)\delta(0)^*\gamma(0)\\ &= \gamma(z)^*\gamma(0)\\ &= D^\gamma(z),\end{aligned}$$

and similarly

$$(C_z \otimes I_d)N^\delta(z)Q = N^\gamma(z),$$

so that $Rb_V^\delta(z)Q = b^\gamma(z)$, and b_V^δ, b_V^γ belong to the same equivalence (coincidence) class. □

Theorem 4.7. *Let $\hat{\mathfrak{H}}^\Gamma$ be the abstract model RKHS on $\mathbb{B}^d$ defined using an analytic model triple $(\Gamma, \mathfrak{J}_0, \mathfrak{J}_\infty)$ for V. The reproducing kernel for $\hat{\mathfrak{H}}^\Gamma$ can be written:*

$$\begin{aligned}\hat{K}^\Gamma(z,w) &= \frac{D^\Gamma(z)D^\Gamma(w)^* - zN^\Gamma(z)N^\Gamma(w)^*w^*}{1 - zw^*}\\ &= D^\Gamma(z)\frac{I - b_V^\Gamma(z)b_V^\Gamma(w)^*}{1 - zw^*}D^\Gamma(w)^* \in \mathscr{L}(\mathfrak{J}_0).\end{aligned}$$

It follows that $b_V^\Gamma \in \mathscr{S}_d(\mathfrak{J}_\infty, \mathfrak{J}_0)$ is Schur-class, and that multiplication by $D^\Gamma(z)$ is an isometry, $M^\Gamma := M_{D^\Gamma}$, of $\mathscr{H}(b_V^\Gamma)$ onto $\hat{\mathfrak{H}}^\Gamma$.

Proof. As before, we omit the superscript Γ for the proof. Consider the projections:

$$P_0 := P_{\mathrm{Ran}(\hat{V})^\perp} = \hat{U}\Gamma(0)\Gamma(0)^*\hat{U}^*$$

and

$$P_\infty := P_{\mathrm{Ker}(\hat{V})} = (\hat{U} \otimes I_d)\Gamma(\infty)\Gamma(\infty)^*(\hat{U}^* \otimes I_d).$$

Then consider

$$\begin{aligned}\hat{K}_z^* z\hat{V}^*\hat{K}_w &= \left(\hat{K}_w^*\hat{V}(I - P_\infty)z^*\hat{K}_z^*\right)^*\\ &= \left(\hat{K}_w^* w(I - P_\infty)z^*\hat{K}_z\right)^*\\ &= \hat{K}_z^* z(I - P_\infty)w^*\hat{K}_w\\ &= zw^*\hat{K}(z,w) - \hat{K}_z^* zP_\infty w^*\hat{K}_w.\end{aligned}$$

In the above we used that $\hat{V}$ acts as multiplication by z on its initial space and $I - P_\infty = P_{\mathrm{Ker}(V)}^\perp$ is the projection onto this initial space.

The same expression can be evaluated differently:

$$\begin{aligned}\hat{K}_z^* z\hat{V}^*\hat{K}_w &= \hat{K}_z^* z\hat{V}^*(I - P_0)\hat{K}_w\\ &= \hat{K}_z^*\hat{V}\hat{V}^*(I - P_0)\hat{K}_w\\ &= \hat{K}_z^*(I - P_0)\hat{K}_w \qquad (I - P_0) = P_{\mathrm{Ran}(\hat{V})} = \hat{V}\hat{V}^*\\ &= \hat{K}(z,w) - \hat{K}_z^* P_0\hat{K}_w.\end{aligned}$$

Again, in the above we used that $\hat{V}$ acts as multiplication by z on its initial space, the range of $\hat{V}^*$.

Equating these two expressions and solving for $\hat{K}(z,w)$ yields:

$$\hat{K}(z,w) = \frac{\hat{K}_z^* P_0 \hat{K}_w - \hat{K}_z^* z P_\infty w^* \hat{K}_w}{1 - zw^*}.$$

Use that $\hat{K}_z = \hat{U}\Gamma(z)$, $P_0 = \hat{U}\Gamma(0)\Gamma(0)^*\hat{U}^*$, $P_\infty = (\hat{U} \otimes I_d)\Gamma(\infty)\Gamma(\infty)^*(\hat{U}^* \otimes I_d)$, and that

$$\hat{K}_z^* z P_\infty w^* \hat{K}_w = \Gamma(z)^* z \Gamma(\infty)\Gamma(\infty)^* w^* \Gamma(w)$$

to obtain

$$\hat{K}(z,w) = \frac{D(z)D(w)^* - zN(z)N(w)^* w^*}{1 - zw^*}; \qquad z, w \in \mathbb{B}^d.$$

In particular it follows that $b_V = D(z)^{-1} z N(z) \in \mathscr{S}_d(\mathfrak{J}_\infty, \mathfrak{J}_0)$ as claimed (by [13, Theorem 2.1]). □

Given an analytic model map Γ for a CCNC row partial isometry V on $\mathcal{H}$, let $U^\Gamma : \mathcal{H} \to \mathscr{H}(b_V^\Gamma)$ denote the canonical onto isometry $U^\Gamma := (M^\Gamma)^{-1}\hat{U}^\Gamma$, where, recall, $(M^\Gamma)^{-1} = M_{D^\Gamma}^{-1} = M_{(D^\Gamma)^{-1}} = (M^\Gamma)^*$, since $M^\Gamma = M_{D^\Gamma}$ is a unitary multiplier of $\mathscr{H}(b_V^\Gamma)$ onto $\hat{\mathcal{H}}^\Gamma$.

Theorem 4.8. *Let $V : \mathcal{H} \otimes \mathbb{C}^d \to \mathcal{H}$ be a CCNC row partial isometry with analytic model triple $(\Gamma, \mathfrak{J}_\infty, \mathfrak{J}_0)$. The image, $X^\Gamma := U^\Gamma V((U^\Gamma)^* \otimes I_d)$, of V under the corresponding canonical unitary is an extremal Gleason solution for $\mathscr{H}(b_V^\Gamma)$.*

If $\Gamma(z) = \Gamma_T(z) := (I - Tz^*)^{-1}\Gamma(0)$ is an analytic model map corresponding to a contractive extension $T \supseteq V$ on $\mathcal{H}$, then $X^{\Gamma_T} = X(\mathbf{b}^T)$ is the unique contractive Gleason solution corresponding to the extremal Gleason solution

$$\mathbf{b}^T := (U^{\Gamma_T} \otimes I_d)\Gamma_T(\infty), \qquad \textit{for } b_V^{\Gamma_T} \in \mathscr{S}_d(\mathfrak{J}_\infty, \mathfrak{J}_0).$$

Proof. (As before we will omit superscripts.) Since multiplication by $D(z)^{-1}$ is an isometric multiplier of $\hat{\mathcal{H}}$ onto $\mathscr{H}(b_V)$, $M_{D^{-1}} = M_D^*$, Proposition 4.3 implies that $X := M_D^* \hat{V} M_D$ acts as multiplication by z on its initial space, and

$$\mathrm{Ran}\,(X) = \{f \in \mathscr{H}(b_V) |\ f(0) = 0\}.$$

Hence, $XX^* = I - k_0^b(k_0^b)^*$ and

$$\begin{aligned} z(X^*f)(z) &= (XX^*f)(z) \\ &= f(z) - k^{b_V}(z,0)f(0) \\ &= f(z) - f(0), \end{aligned}$$

since $b_V^\Gamma(0) = 0$. This proves that X is a (contractive) extremal Gleason solution for $\mathscr{H}(b_V^\Gamma)$.

To see that $\mathbf{b} := (M_{D^{-1}}\hat{U} \otimes I_d)\Gamma(\infty)$ is a contractive extremal Gleason solution for b_V^Γ, let $b := b_V^\Gamma$ and calculate:

$$\begin{aligned}\mathbf{b}(z) &= (z^* k_z^b)^* \mathbf{b} \\ &= \left(z^* \hat{U}^* (M_{D^{-1}}^*) k_z^b\right)^* \Gamma(\infty) \\ &= \left(z^* \hat{U}^* \hat{K}_z (D(z)^{-1})^*\right)^* \Gamma(\infty) \\ &= \left(z^* \Gamma(z)(D(z)^{-1})^*\right)^* \Gamma(\infty) \\ &= D(z)^{-1} z (\Gamma(z)^* \otimes I_d)\Gamma(\infty) \\ &= D(z)^{-1} z N(z) = b(z).\end{aligned}$$

This proves that $\mathbf{b}^\Gamma =: \mathbf{b}$ is a Gleason solution (since $b_V^\Gamma(0) = 0$). Furthermore,

$$\mathbf{b}^*\mathbf{b} = \Gamma(\infty)^*\Gamma(\infty) = I_{\mathfrak{J}_\infty},$$

so that $\mathbf{b}^\Gamma = \mathbf{b}$ is contractive and extremal.

To see that $X = X^{\Gamma_T} = X(\mathbf{b}^{\Gamma_T})$, calculate the action of $X^* - w^*$ on point evaluation maps,

$$\begin{aligned}(X^* - w^*)k_w^b &= \left((M_D^{-1}\hat{U} \otimes I_d)(V^* - w^*)\hat{U}^* M_D\right) k_w^b \\ &= (M_D^{-1}\hat{U} \otimes I_d)(V^* - w^*)\Gamma(w)(D(w)^{-1})^*,\end{aligned}$$

and compare this to

$$\begin{aligned}\mathbf{b}^\Gamma b(w)^* &= (M_D^{-1}\hat{U} \otimes I_d)\Gamma(\infty)N(w)^* w^* (D(w)^{-1})^* \\ &= (M_D^{-1}\hat{U} \otimes I_d)\Gamma(\infty)\Gamma(\infty)^*(\Gamma(w) \otimes I_d) w^* (D(w)^{-1})^* \\ &= (M_D^{-1}\hat{U} \otimes I_d)P_{\mathrm{Ker}(V)} w^* \Gamma(w)(D(w)^{-1})^* \\ &= (M_D^{-1}\hat{U} \otimes I_d)(I - V^*V) w^* \Gamma(w)(D(w)^{-1})^*.\end{aligned}$$

Under the assumption that $\Gamma = \Gamma_T$ for a contractive extension $T \supseteq V$ on $\mathcal{H}$, recall that by Lemma 3.15, $T = V - C$, where C is a pure row contraction with $\mathrm{Ker}(C)^\perp \subseteq \mathrm{Ker}(V)$ and $\mathrm{Ran}\,(C) \subseteq \mathrm{Ran}\,(V)^\perp$. Applying that $\Gamma_T(z) = (I - Tz^*)^{-1}\Gamma_T(0)$,

$$\begin{aligned}\mathbf{b}b(w)^* &= (M_D^{-1}\hat{U} \otimes I_d)(I - V^*(T + C)) w^* \Gamma_T(w)(D(w)^{-1})^* \\ &= (M_D^{-1}\hat{U} \otimes I_d) w^* \Gamma_T(w)(D(w)^{-1})^* \\ &\quad - (M_D^{-1}\hat{U} \otimes I_d)V^*(Cw^*\Gamma_T(w) + \Gamma_T(w) - \Gamma_T(0))(D(w)^{-1})^* \\ &= (M_D^{-1}\hat{U} \otimes I_d)(w^* - V^*)\Gamma_T(w)(D(w)^{-1})^*,\end{aligned}$$

and it follows that $X^* k_w^b = w^* k_w^b - \mathbf{b}^{\Gamma_T} b_V^{\Gamma_T}(w)^*$, proving the claim. □

Remark 4.9. Lemma 4.6 shows that the coincidence class of any characteristic function b_V^γ of a CCNC row partial isometry V is invariant under the choice of model triple $(\gamma, \mathfrak{J}_\infty, \mathfrak{J}_0)$ and Theorem 4.7 shows that $b_V^\gamma \in \mathscr{S}_d(\mathfrak{J}_\infty, \mathfrak{J}_0)$ belongs to the Schur class. It will also be useful to define *weak coincidence* of Schur-class functions as in [17, Definition 2.4]:

Definition 4.10. The *support* of $b \in \mathscr{S}_d(\mathcal{H}, \mathcal{J})$ is

$$\mathrm{supp}(b) := \bigvee_{z \in \mathbb{B}^d} \mathrm{Ran}\,(b(z)^*) = \mathcal{H} \ominus \bigcap_{z \in \mathbb{B}^d} \mathrm{Ker}(b(z)).$$

Schur-class multipliers $b_1 \in \mathscr{S}_d(\mathcal{H}, \mathcal{J})$ and $b_2 \in \mathscr{S}_d(\mathcal{H}', \mathcal{J}')$ *coincide weakly* if $b_1' := b_1|_{\mathrm{supp}(b_1)}$ coincides with $b_2' := b_2|_{\mathrm{supp}(b_2)}$.

By [17, Lemma 2.5], $b_1 \in \mathscr{S}_d(\mathcal{H}, \mathcal{J})$ and $b_2 \in \mathscr{S}_d(\mathcal{H}', \mathcal{J}')$ coincide weakly if and only if there is an onto isometry $V : \mathcal{J} \to \mathcal{J}'$ so that

$$Vb_1(z)b_1(w)^*V^* = b_2(z)b_2(w)^*; \qquad z, w \in \mathbb{B}^d,$$

i.e., if and only if $\mathscr{H}(b_2) = \mathscr{H}(Vb_1)$.

Weak coincidence also defines an equivalence relation on Schur-class functions, and we define:

Definition 4.11. The *characteristic function*, b_V, of a CCNC row partial isometry V is the weak coincidence class of any Schur-class characteristic function $b_V^\gamma \in \mathscr{S}_d(\mathcal{J}_0, \mathcal{J}_\infty)$ constructed using any model triple $(\gamma, \mathcal{J}_0, \mathcal{J}_\infty)$ for V. We will often abuse terminology and simply say that any b_V^γ is the characteristic function, b_V, of V.

Note that the characteristic function, b_V, of any CCNC row partial isometry always vanishes at 0, $b_V(0) = 0$, and (as discussed in Subsection 2.2) this implies that b_V is strictly contractive on $\mathbb{B}^d$. In the single variable case when $d = 1$, the above definition of the characteristic function reduces to that of the *Livšic characteristic function* of the partial isometry V [2, 24, 39].

Remark 4.12. There is an alternative proof of Theorem 4.7 using the colligation or transfer function theory of [6, 9, 13]. Any model triple for a CCNC row partial isometry, V, provides a unitary colligation and transfer-function realization for b_V:

Lemma 4.13 ([47, Lemma 3.7]). *Let V be a CCNC row partial isometry on $\mathcal{H}$, and let $(\gamma, \mathcal{J}_\infty, \mathcal{J}_0)$ be any model triple for V. Then*

$$\Xi^\gamma := \begin{bmatrix} V^* & \gamma(\infty) \\ \gamma(0)^* & 0 \end{bmatrix} : \begin{bmatrix} \mathcal{H} \\ \mathcal{J}_\infty \end{bmatrix} \to \begin{bmatrix} \mathcal{H} \otimes \mathbb{C}^d \\ \mathcal{J}_0 \end{bmatrix},$$

is a unitary (onto isometry) colligation with transfer function equal to b_V^γ.

In [47, Theorem 3.14], the above lemma was applied to provide an alternate proof of Theorem 4.7.

Lemma 4.14. *Let $b \in \mathscr{S}_d(\mathcal{J}, \mathcal{K})$ be any Schur-class function such that $b(0) = 0$. The characteristic function, b_X, of any extremal Gleason solution, X, for $\mathscr{H}(b)$ coincides weakly with b.*

Proof. By Remark 2.6 and Lemma 2.9, any such X is a CCNC row partial isometry.

To calculate the characteristic function of X, use the analytic model map $\Gamma = \Gamma_X$, with $\Gamma(0) := k_0^b$, an isometry of $\mathcal{K}$ onto $\operatorname{Ran}(X)^\perp$ since $b(0) = 0$. Then,

$$\Gamma(z) = (I - Xz^*)^{-1} k_0^b = k_z^b, \tag{4.4}$$

and

$$D(z) = \Gamma(z)^* \Gamma(0) = (k_z^b)^* k_0^b = k^b(z, 0) = I_{\mathcal{K}}.$$

It then follows that $zN(z) = b_X(z)$, and by equation (4.4) and Theorem 4.7, the reproducing kernel for the model space $\widehat{\mathscr{H}(b)}$ is

$$\hat{K}(z, w) = \Gamma(z)^* \Gamma(w) = k^b(z, w) = k^{b_X}(z, w); \qquad z, w \in \mathbb{B}^d.$$

This proves that $\mathscr{H}(b_X^\Gamma) = \mathscr{H}(b)$ and [17, Lemma 2.5] implies that b_X^Γ coincides weakly with b. □

Corollary 4.15. *If X is an extremal Gleason solution for any Schur-class b such that $b(0) = 0$ then X acts as multiplication by $z = (z_1, \dots, z_d)$ on its initial space.*

Proof. If we take $\Gamma = \Gamma_X$ as our model map for X, the above example shows that $\widehat{\mathscr{H}(b)}^\Gamma = \mathscr{H}(b)$ and $X = \hat{X}$, so that X acts as multiplication by z on its initial space by Proposition 4.3. □

Remark 4.16. Theorem 4.7 and Lemma 4.14 show that the (weak coincidence class of any) characteristic function is a unitary invariant: If X, Y are two CCNC row partial isometries which are unitarily equivalent then one can choose model triples for X, Y so that they have the same characteristic function. Conversely if the characteristic functions of X, Y coincide weakly, $\mathscr{H}(b_X) = \mathscr{H}(Ub_Y)$ for a fixed unitary U, then both X, Y are unitarily equivalent to extremal Gleason solutions for $\mathscr{H}(b_X)$ (it is easy to see that multiplication by U is a constant unitary multiplier of $\mathscr{H}(b_X)$ onto $\mathscr{H}(Ub_X)$ taking extremal Gleason solutions onto extremal Gleason solutions).

Unless $d = 1$, the characteristic function of a CCNC row partial isometry is not a complete unitary invariant. Indeed, if $b \in \mathscr{S}_d(\mathcal{J}, \mathcal{K})$, $b(0) = 0$, has two extremal Gleason solutions X, X' which are not unitarily equivalent, the above results show that X, X' are two non-equivalent row partial isometries with the same characteristic function. The subsection below provides examples of such Schur-class functions.

4.17. The characteristic function is not a complete unitary invariant

Let $b \in \mathscr{S}_d(\mathcal{H})$ be any square Schur-class function with $b(0) = 0$ (for simplicity) and $\operatorname{supp}(b) = \mathcal{H}$. Recall that since $b(0)$ is a strict contraction, so is $b(z)$ for any $z \in \mathbb{B}^d$.

Lemma 4.18. *Suppose that $b \in \mathscr{S}_d(\mathcal{H})$ obeys $b(0) = 0$, and that $\mathscr{H}(b)$ has two extremal Gleason solutions X, Y which are unitarily equivalent, $WX = YW$, for some unitary $W \in \mathscr{L}(\mathscr{H}(b))$. Then W is a constant unitary multiplier: There is* *a constant unitary* $R \in \mathscr{L}(\mathcal{H})$ *so that* $Wk_z^b = k_z^b R$, $(WF)(z) = R^*F(z)$ *for all* $F \in \mathscr{H}(b)$ *and* $z \in \mathbb{B}^d$. *Moreover, given any* $z, w \in \mathbb{B}^d$, $Rb(z)b(w)^* = b(z)b(w)^*R$.

Proof. Suppose that X, Y are unitarily equivalent, $WX = YX$ for some unitary $W : \mathscr{H}(b) \to \mathscr{H}(b)$. Since both X, Y are extremal and $b(0) = 0$, they are both row partial isometries with the same range projection $XX^* = I - k_0^b(k_0^b)^* = YY^*$. It follows that the range of k_0^b is a reducing subspace for W so that there is a unitary $R \in \mathscr{L}(\mathcal{H})$ such that

$$Wk_0^b = k_0^b R.$$

Moreover, by Property (2.4), given any $z \in \mathbb{B}^d$,

$$\begin{aligned} Wk_z^b &= W(I - Xz^*)^{-1}k_0^b \\ &= (I - Yz^*)^{-1}Wk_0^b \\ &= (I - Yz^*)^{-1}k_0^b R = k_z^b R, \end{aligned}$$

so that W is multiplication by the fixed unitary operator $R^* \in \mathscr{L}(\mathcal{H})$. Since W is a unitary multiplier of $\mathscr{H}(b)$ onto itself, RKHS theory implies that

$$k^b(z, w) = R^*k^b(z, w)R; \qquad z, w \in \mathbb{B}^d,$$

and the final claim follows. □

Corollary 4.19. *If* $b \in \mathscr{S}_d(\mathbb{C})$ *satisfies the conditions:*

$$b(0) = 0 \quad \textit{and} \quad \dim\left(\mathrm{Ker}(V^b)\right) > \dim\left(\mathrm{Ran}\left(V^b\right)^\perp\right) \neq 0,$$

then $\mathscr{H}(b)$ *has two extremal Gleason solutions which are not unitarily equivalent.*

Proof. In this case $\mathcal{H} = \mathbb{C}$ the dimension of $\mathrm{Ran}\left(V^b\right)^\perp$ is either 0 or 1. By assumption $\dim\left(\mathrm{Ker}(V^b)\right) > 1 = \dim\left(\mathrm{Ran}\left(V^b\right)\right)$ so that we can define two co-isometric extensions D, d of V^b on $\mathscr{H}^+(H_b)$ so that

$$D^*K_0^b \perp d^*K_0^b.$$

As described in Subsection 2.10 and Subsection 2.4, the contractive Gleason solutions $X[D]$ and $X[d]$ are extremal.

Suppose that $X[D], X[d]$ are unitarily equivalent, $WX[d] = X[D]W$ for a unitary W on $\mathscr{H}(b)$. By the previous Lemma, W is a constant unitary multiplier by a unitary operator $R \in \mathscr{L}(\mathbb{C})$. That is, W is simply multiplication by a unimodular constant $R = \alpha \in \mathbb{T}$, $W = \alpha I_{\mathscr{H}(b)}$. In particular, $X[D] = WX[d]W^* = X[d]$ so that $\mathbf{b}[D]b(w)^* = \mathbf{b}[d]b(w)^*$ and $\mathbf{b}[D] = \mathbf{b}[d]$. Equivalently,

$$D^*K_0^b = d^*K_0^b, \qquad \text{a contradiction.}$$

□

Example 4.20. To construct examples of a Schur-class functions $b \in \mathscr{S}_d(\mathbb{C})$ which admit non-unitarily equivalent extremal Gleason solutions, let $b \in \mathscr{S}_d(\mathbb{C})$, $d \geq 2$ be any Schur-class function obeying $b(0) = 0$ and such that V^b is not a co-isometry. Such a Schur-class function is called non quasi-extreme (see Section 6.1), and it is not difficult to apply Theorem 6.2 below to show, for example, that if $b \in \mathscr{S}_d(\mathbb{C})$

is any Schur-class function and $0 < r < 1$, $rb \in \mathscr{S}_d(\mathbb{C})$ is non quasi-extreme. Moreover, $\mathscr{H}(rb)$ is infinite-dimensional (this is just H^2_d with a new norm since $\mathscr{H}(rb) = \mathrm{Ran}\left(D_{M^*_{rb}}\right)$ equipped with the norm that makes $D_{M^*_{rb}}$ a co-isometry onto its range as in [48]) so that rb satisfies the assumptions of Propostion 2.14, and hence those of Corollary 4.19.

Corollary 4.21. *The characteristic function, b_X, of a CCNC row partial isometry, X, is not a complete unitary invariant: There exist CCNC row partial isometries X_1, X_2 which are not unitarily equivalent and yet satisfy $b_{X_1} = b_{X_2}$.*

This is in contrast to the result of [17, Theorem 3.6], which shows that the Sz.-Nagy–Foiaş characteristic function of any CNC d-contraction (which is a CCNC row contraction by Corollary 3.4) is a complete unitary invariant for CNC d-contractions. We will also later prove in Proposition 5.15, that our characteristic function, b_T, of any CCNC row contraction coincides weakly with the Sz.-Nagy–Foiaş characteristic function of T.

Proof. Let $b \in \mathscr{S}_d = \mathscr{S}_d(\mathbb{C})$ be any Schur-class function satisfying the assumptions of Corollary 4.19 (as in the above example). Then there exist extremal Gleason solutions, X_1, X_2 for $\mathscr{H}(b)$ which are not unitarily equivalent. Lemma 4.14 implies that $b_{X_1} = b_{X_2}$ proving the claim. □

We conclude this section with a calculation that will motivate an approach to extending our commutative de Branges–Rovnyak model for CCNC row partial isometries to arbitrary CCNC row contractions:

4.22. Frostman shifts

Suppose that X is any extremal Gleason solution for $\mathscr{H}(b)$, where $b \in \mathscr{S}_d(\mathcal{J}, \mathcal{K})$ is purely contractive on the ball, but we do not assume that $b(0) = 0$. Recall that b is purely contractive on the ball if and only if for any $z \in \mathbb{B}^d$, $\|b(z)g\| < \|g\|$ for all $g \in \mathcal{J}$, *i.e.*, $b(z)$ is a pure contraction. It is further not difficult to show, as in [50, Proposition V.2.1], that if $b(z)$ is a pure contraction for $z \in \mathbb{B}^d$, then so is $b(z)^*$. In particular, both defect operators $D_{b(0)}, D_{b(0)^*}$ are injective and have dense ranges, *i.e.*, they are quasi-affinities. In the case where $b(0) \neq 0$, X is not a row partial isometry, but it is still CCNC by Lemma 2.9 so that we can consider the isometric-purely contractive decomposition $X = V - C$ of X, and we calculate the characteristic function, b_V, of the CCNC partial isometric part, V, of X.

By Lemma 3.15, $V^* = X^* P_{\mathrm{Ker}(D_{X^*})}$, and since X is extremal,

$$D^2_{X^*} = I - XX^* = k^b_0 (k^b_0)^*.$$

It follows that

$$\mathrm{Ker}(D_{X^*}) = \mathrm{Ker}(D^2_{X^*}) = \left(\bigvee k^b_0 \mathcal{K}\right)^{\perp},$$

so that if $b(0)$ is a strict contraction,

$$V^* = X^* P_{\mathrm{Ker}(D_{X^*})} = X^* \left(I - k^b_0 k^b(0,0)^{-1} (k^b_0)^*\right).$$

If $b(0)$ is a strict contraction, we can then define the onto isometry $\Gamma(0) : \mathcal{K} \to \mathrm{Ran}\,(V)^{\perp}$ by $\Gamma(0) := k_0^b D_{b(0)^*}^{-1}$. If $b(0)$ is not a strict contraction, one can define $\Gamma(0) : \mathrm{Ran}\,(D_{b(0)^*}) \to \mathrm{Ran}\,(V)^{\perp}$ by:

$$\Gamma(0)D_{b(0)^*}g := k_0^b g.$$

It is easy to check that this is a densely defined isometry with dense range in $\mathrm{Ran}\,(V)^{\perp}$, and this extends by continuity to an isometry of $\mathcal{K}$ onto $\mathrm{Ran}\,(V)^{\perp}$.

Similarly, since X is an extremal Gleason solution for $\mathscr{H}(b)$, as described in Section 2.4, there is an extremal Gleason solution $\mathbf{b}$ for b so that $X^* k_w^b = w^* k_w^b - \mathbf{b}b(w)^*$. Since $\mathbf{b}$ is extremal, $\mathbf{b}^*\mathbf{b} = I - b(0)^*b(0) = D_{b(0)}^2$, so that if $b(0)$ is a strict contraction, $\mathbf{b}D_{b(0)}^{-1}$ also defines an isometry.

Lemma 4.23. *Define* $\Gamma_0(\infty)$ *on* $\mathrm{Ran}\,(D_{b(0)})$ *by* $\Gamma_0(\infty)D_{b(0)}h = \mathbf{b}h$. *Then* $\Gamma_0(\infty)$ *extends by continuity to an isometry of* $\mathcal{J}$ *into* $\mathrm{Ker}(V)$.

If $b(0)$ is a strict contraction then $\Gamma_0(\infty) = \mathbf{b}D_{b(0)}^{-1}$ and the proof simplifies.

Proof. Showing that $\Gamma_0(\infty)$ extends to an isometry is straightforward. To show that $\Gamma_0(\infty)$ maps into $\mathrm{Ker}(V)$, use that $V^* = X^*(I - \Gamma(0)\Gamma(0)^*)$ since $\Gamma(0)$ is an isometry onto $\mathrm{Ran}\,(V)^{\perp}$, and observe that

$$k^b(w,0) = (k_w^b)^* k_0^b = (k_w^b)^*\Gamma(0)\Gamma(0)^* k_0^b.$$

It follows that for any $h \in \mathcal{K}$ and $w \in \mathbb{B}^d$, since $\Gamma(0)$ is onto $\bigvee k_0^b\mathcal{K}$, there is a $g \in \mathcal{K}$ so that $\Gamma(0)\Gamma(0)^* k_w^b h = k_0^b g$, or, equivalently,

$$k^b(0,w)h = (k_0^b)^*\Gamma(0)\Gamma(0)^* k_w^b h = k^b(0,0)g.$$

Then, calculating on kernel maps,

$$\begin{aligned} D_{b(0)}\Gamma_0(\infty)^* V^* k_w^b h &= \mathbf{b}^* X^*(I - \Gamma(0)\Gamma(0)^*)k_w^b h \\ &= \mathbf{b}^*(w^* k_w^b - \mathbf{b}b(w)^*)h - \mathbf{b}^* X^* k_0^b g \\ &= (b(w)^* - b(0)^* - (I - b(0)^*b(0))b(w)^*)\,h + D_{b(0)}^2 b(0)^* g \\ &= (-b(0)^* + b(0)^*b(0)b(w)^*)h + b(0)^* k^b(0,0)g \\ &= -b(0)^*h + b(0)^*b(0)b(w)^*h + b(0)^*k^b(0,w)h = 0. \qquad \square \end{aligned}$$

It follows that we can choose an isometry $\Gamma_1(\infty) : \mathcal{J}' \to \mathrm{Ker}(V) \ominus \mathrm{Ran}\,(\Gamma_0(\infty))$ so that

$$\Gamma(\infty) := \Gamma_0(\infty) \oplus \Gamma_1(\infty) : \mathcal{J} \oplus \mathcal{J}' \to \mathrm{Ker}(V),$$

is an onto isometry.

Lemma 4.24. *The range of* $\Gamma_1(\infty)$ *is in the kernel of* $z(k_z^b)^*$ *for any* $z \in \mathbb{B}^d$. *That is,* $z(\Gamma(z)^* \otimes I_d)\Gamma_1(\infty) = 0$.

Proof. For simplicity assume $b(0)$ is a strict contraction. The argument can be extended to the general case as before. If $\mathbf{h} \in \mathrm{Ran}\,(\Gamma_1(\infty))$ then $\mathbf{h} \in \mathrm{Ker}(V)$, where $V = (I - k_0^b k^b(0,0)^{-1}(k_0^b)^*)X$ and $\mathbf{h} \in \mathrm{Ker}(\mathbf{b}^*) = \mathrm{Ran}\,(\mathbf{b})^\perp$. Then calculate:

$$\begin{aligned} z(k_z^b)^*\mathbf{h} &= \left(X^* k_z^b + \mathbf{b}b(z)^*\right)^* \mathbf{h} \\ &= \left(X^* k_0^b k^b(0,0)^{-1}(k_0^b)^* k_z^b + V^* k_z^b\right)^* \mathbf{h} \\ &= \left(\mathbf{b}b(0)^* k^b(0,0)^{-1} k^b(0,z)\right)^* \mathbf{h} = 0. \end{aligned}$$

□

Since X is a contractive extension of the CCNC row partial isometry V, set

$$\Gamma(z) := (I - Xz^*)^{-1}\Gamma(0) = k_z^b D_{b(0)^*}^{-1}.$$

This defines an analytic model triple, $(\Gamma, \mathcal{J} \oplus \mathcal{J}', \mathcal{K})$ for V, and we now calculate the characteristic function of V using this model triple: First note that by the previous Lemma,

$$\begin{aligned} zN^\Gamma(z) &= z(\Gamma(z)^* \otimes I_d)\left(\Gamma_0(\infty) \oplus \Gamma_1(\infty)\right) \\ &= z(\Gamma(z)^* \otimes I_d)\Gamma_0(\infty) + 0_{\mathcal{J}',\mathcal{K}} =: zN_0^\Gamma(z) + 0_{\mathcal{J}',\mathcal{K}}, \end{aligned}$$

so that $b_V^\Gamma(z) = D^\Gamma(z)^{-1} zN^\Gamma(z)$ coincides weakly with

$$b^{\langle 0 \rangle}(z) := D^\Gamma(z)^{-1} zN_0^\Gamma(z) \in \mathscr{L}(\mathcal{J}, \mathcal{K}). \tag{4.5}$$

Even if b is only purely contractive on the ball, $D^\Gamma(z)$ is a bounded, invertible operator for $z \in \mathbb{B}^d$ by Lemma 4.4, and since $b^{\langle 0 \rangle}(0) = 0$, $b^{\langle 0 \rangle} \in \mathscr{S}_d(\mathcal{J}, \mathcal{K})$ is strictly contractive. Note that the construction of $b^{\langle 0 \rangle}$ is independent of the choice of extremal Gleason solution, X, for $\mathscr{H}(b)$.

In the case where $b(0)$ is a strict contraction, the formula for $b^{\langle 0 \rangle}$ can be computed more explicitly:

$$\begin{aligned} D^\Gamma(z) &= \Gamma(z)^*\Gamma(0) \\ &= \Gamma(0)^*(I - zX^*)^{-1}\Gamma(0) \\ &= D_{b(0)^*}^{-1}(k_0^b)^*(I - zX^*)^{-1} k_0^b D_{b(0)^*}^{-1} \\ &= D_{b(0)^*}^{-1}(k_z^b)^* k_0^b D_{b(0)^*}^{-1} \\ &= D_{b(0)^*}^{-1}(I_{\mathcal{J}_0} - b(z)b(0)^*) D_{b(0)^*}^{-1}. \end{aligned}$$

Similarly,

$$\begin{aligned} zN_0^\Gamma(z) &= z(\Gamma(z)^* \otimes I_d)\Gamma_0(\infty) \\ &= D_{b(0)^*}^{-1} z(k_z^b \otimes I_d)^* \mathbf{b} D_{b(0)}^{-1} \\ &= D_{b(0)^*}^{-1}(b(z) - b(0)) D_{b(0)}^{-1}, \end{aligned}$$

and finally,

$$\begin{aligned} b^{\langle 0 \rangle}(z) &= D^\Gamma(z)^{-1} zN_0^\Gamma(z) \\ &= D_{b(0)^*}\left(I_{\mathcal{K}} - b(z)b(0)^*\right)^{-1}\left(b(z) - b(0)\right) D_{b(0)}^{-1}. \end{aligned} \tag{4.6}$$

In summary, we have proven:

Corollary 4.25. *Let $b \in \mathscr{S}_d(\mathcal{J}, \mathcal{K})$ be a strictly contractive Schur-class function and let X be any extremal Gleason solution for $\mathscr{H}(b)$ with isometric-purely contractive decomposition $X = V - C$. Then the characteristic function b_V, of the partial isometric part, V, of X coincides weakly with $b^{\langle 0 \rangle} \in \mathscr{S}_d(\mathcal{J}, \mathcal{K})$.*

Let $b^{\langle 0 \rangle} = b_V^{\Gamma}|_{\mathcal{J}} =: \Phi_{b(0)} \circ b$, where for any strict contractions $\alpha, \beta \in \mathscr{L}(\mathcal{J}, \mathcal{K})$, we define

$$\Phi_\alpha(\beta) := D_{\alpha^*} \left(I_{\mathcal{K}} - \beta \alpha^* \right)^{-1} (\beta - \alpha) D_\alpha^{-1} \in \mathscr{L}(\mathcal{J}, \mathcal{K}). \tag{4.7}$$

As we will show in the following section, for any strict contraction $\alpha \in \mathscr{L}(\mathcal{J}, \mathcal{K})$, the operator-Möbius transformation $\Phi_\alpha : [\mathscr{L}(\mathcal{J}, \mathcal{K})]_1 \to [\mathscr{L}(\mathcal{J}, \mathcal{K})]_1$ is an automorphism (*i.e.*, a bijection) of the unit ball of $\mathscr{L}(\mathcal{J}, \mathcal{K})$, and $\Phi_\alpha : \mathscr{S}_d(\mathcal{J}, \mathcal{K}) \to \mathscr{S}_d(\mathcal{J}, \mathcal{K})$ maps the Schur class onto itself. Given any strict contraction $\alpha \in \mathscr{L}(\mathcal{J}, \mathcal{K})$ and $b \in \mathscr{S}_d(\mathcal{J}, \mathcal{K})$, $\Phi_\alpha \circ b$ is an operator-generalization of what is called a Frostman shift of the Schur-class function b in the classical case where b is a contractive analytic function on the disk [22, 23, 25]. (This definition will be extended suitably to pure contractions.) The above example shows that if X is any extremal Gleason solution for $\mathscr{H}(b)$ with partial isometric part V, then $b_V = b^{\langle 0 \rangle} = \Phi_{b(0)} \circ b$ is the Frostman shift of b vanishing at the origin.

5. Commutative de Branges–Rovnyak model for CCNC row contractions

The goal of this section is to show that if $T = V - C : \mathcal{H} \otimes \mathbb{C}^d \to \mathcal{H}$ is any CCNC row contraction with partial isometric part V and purely contractive part C, then T is unitarily equivalent to an extremal Gleason solution X for $\mathscr{H}(b)$, where the (purely contractive) Schur-class function $b = b_T$ is a certain Frostman shift of the characteristic function b_V of V. As in the case where $T = V$ is a row partial isometry, the characteristic function b_T, of T, will be a unitary invariant for CCNC row contractions.

5.1. Automorphisms of the unit ball of $\mathscr{L}(\mathcal{H}, \mathcal{K})$

Recall that $b \in \mathscr{S}_d(\mathcal{H}, \mathcal{K})$ is said to be strictly contractive if $b(z)$ is a strict contraction for all $z \in \mathbb{B}^d$ and purely contractive if $b(z)$ is a pure contraction for all $z \in \mathbb{B}^d$, *i.e.*, if $\|b(z)h\| < \|h\|$ for any $h \in \mathcal{H}$ and $z \in \mathbb{B}^d$. Further recall that b is strictly contractive if and only if $b(0)$ is a strict contraction so that, in particular, b_V is strictly contractive for any CCNC row partial isometry V (since $b_V(0) = 0$).

Consider the closed unit ball $[\mathscr{L}(\mathcal{H}, \mathcal{K})]_1$, where $\mathcal{H}, \mathcal{K}$ are Hilbert spaces. Assume that $\alpha \in [\mathscr{L}(\mathcal{H}, \mathcal{K})]_1$ is a pure contraction so that the defect operators D_α, D_{α^*} have closed, densely defined inverses. The *α-Frostman transformation*, Φ_α, and the inverse α-Frostman transformation, Φ_α^{-1} are the maps defined on the open unit ball $(\mathscr{L}(\mathcal{H}, \mathcal{K}))_1$ by

$$\Phi_\alpha(\beta) := D_{\alpha^*} \left(I_{\mathcal{K}} - \beta \alpha^* \right)^{-1} (\beta - \alpha) D_\alpha^{-1}, \tag{5.1}$$

and

$$\Phi_\alpha^{-1}(\beta) := \Phi_{-\alpha^*}(\beta^*)^* = D_{\alpha^*}^{-1}(\beta+\alpha)(I_{\mathcal{H}}+\alpha^*\beta)^{-1}D_\alpha.$$

To be precise, for any pure contraction α, $\Phi_\alpha(\beta)$ is initially defined on the dense domain $\mathrm{Ran}\,(D_\alpha)$, and $(\Phi_\alpha^{-1}(\beta))^*$ is defined on the dense domain $\mathrm{Ran}\,(D_{\alpha^*})$. The next lemma below shows that both of these densely defined transformations are contractions, and hence extend by continuity to the entire space. If α is actually a strict contraction, we extend these definitions of $\Phi_\alpha, \Phi_\alpha^{-1}$ to the closed unit ball.

Lemma 5.2. *Let* $\alpha \in [\mathscr{L}(\mathcal{H},\mathcal{K})]_1$.

(i) *If* α *is a pure contraction, the Frostman transformations* $\Phi_\alpha, \Phi_\alpha^{-1}$ *map* $(\mathscr{L}(\mathcal{H},\mathcal{K}))_1$ *into pure contractions in* $[\mathscr{L}(\mathcal{H},\mathcal{K})]_1$.

(ii) *If* α *is a strict contraction,* $\Phi_\alpha, \Phi_\alpha^{-1}$ *are compositional inverses and define bijections of* $[\mathscr{L}(\mathcal{H},\mathcal{K})]_1$ *onto itself.*

(iii) *Given any two strict contractions* β, γ, Φ_α *and* Φ_α^{-1} *obey the identities:*

$$I - \Phi_\alpha(\beta)\Phi_\alpha(\gamma)^* = D_{\alpha^*}(I-\beta\alpha^*)^{-1}(I-\beta\gamma^*)(I-\alpha\gamma^*)^{-1}D_{\alpha^*}, \text{ and,}$$
$$I - \Phi_\alpha^{-1}(\beta)\Phi_\alpha^{-1}(\gamma)^* = D_\alpha^*(I+\beta\alpha^*)^{-1}(I-\beta\gamma^*)(I+\alpha\gamma^*)^{-1}D_{\alpha^*}.$$

Proof. Assume that α is a pure contraction so that $D_\alpha := \sqrt{I-\alpha^*\alpha}$ has dense range. Given any $h = D_\alpha g \in \mathrm{Ran}\,(D_\alpha)$, and any strict contraction β, consider:

$$\begin{aligned}&\|h\|^2 - \|\Phi_\alpha(\beta)h\|^2\\ &\quad= \langle D_\alpha^2 g, g\rangle - \langle \underbrace{(\beta^*-\alpha^*)(I_{\mathcal{K}}-\alpha\beta^*)^{-1}D_{\alpha^*}^2(I_{\mathcal{K}}-\beta\alpha^*)^{-1}(\beta-\alpha)}_{\text{(A)}}g, g\rangle.\end{aligned}$$

Observe that

$$(I-\beta\alpha^*)^{-1}(\beta-\alpha) = \beta(I-\alpha^*\beta)^{-1}D_\alpha^2 - \alpha.$$

This identity is easily verified by multiplying both sides from the left by $(I-\beta\alpha^*)$. Substitute this identity into the expression (A) to obtain:

$$\begin{aligned}\text{(A)} &= D_\alpha^2(I-\beta^*\alpha)^{-1}\beta^*D_{\alpha^*}^2\beta(I-\alpha^*\beta)^{-1}D_\alpha^2 + \alpha^*D_{\alpha^*}^2\alpha\\ &\quad - D_\alpha^2(I-\beta^*\alpha)^{-1}\beta^*D_{\alpha^*}^2\alpha - \alpha^*D_{\alpha^*}^2\beta(I-\alpha^*\beta)^{-1}D_\alpha^2\\ &= D_\alpha^2(I-\beta^*\alpha)^{-1}\beta^*D_{\alpha^*}^2\beta(I-\alpha^*\beta)^{-1}D_\alpha^2 + D_\alpha^2\alpha^*\alpha\\ &\quad - D_\alpha^2(I-\beta^*\alpha)^{-1}\beta^*\alpha D_\alpha^2 - D_\alpha^2\alpha^*\beta(I-\alpha^*\beta)^{-1}D_\alpha^2.\end{aligned}$$

Applying the identity $D_\alpha^2 - D_\alpha^2\alpha^*\alpha = D_\alpha^4$,

$$\begin{aligned}D_\alpha^2 - \text{(A)} &= D_\alpha^4 - D_\alpha^2(I-\beta^*\alpha)^{-1}\beta^*D_{\alpha^*}^2\beta(I-\alpha^*\beta)^{-1}D_\alpha^2\\ &\quad + D_\alpha^2(I-\beta^*\alpha)^{-1}\beta^*\alpha D_\alpha^2 + D_\alpha^2\alpha^*\beta(I-\alpha^*\beta)^{-1}D_\alpha^2.\\ &= D_\alpha^2(I-\beta^*\alpha)^{-1}\left[(I-\beta^*\alpha)(I-\alpha^*\beta) - \beta^*D_{\alpha^*}^2\beta\right.\\ &\quad \left.+ \beta^*\alpha(I-\alpha^*\beta) + (I-\beta^*\alpha)\alpha^*\beta\right](I-\alpha^*\beta)^{-1}D_\alpha^2\\ &= D_\alpha^2(I-\beta^*\alpha)^{-1}(I-\beta^*\beta)(I-\alpha^*\beta)^{-1}D_\alpha^2.\end{aligned}$$

This proves that for any $h \in \mathrm{Ran}\,(D_\alpha)$,

$$\|h\|^2 - \|\Phi_\alpha(\beta)h\|^2 = \langle D_\alpha(I-\beta^*\alpha)^{-1}(I-\beta^*\beta)(I-\alpha^*\beta)^{-1}D_\alpha h, h\rangle > 0,$$

and it follows that $\Phi_\alpha(\beta)$ is a pure contraction (and it will be a strict contraction if α is strict). If α is a strict contraction then the above formulas make sense with β only a pure contraction, and $\Phi_\alpha(\beta)$ will be a pure contraction in this case.

Assuming that α, β are both strict contractions, the expression $\Phi_\alpha^{-1}(\Phi_\alpha(\beta))$ is well defined. It remains to calculate:

$$\Phi_\alpha^{-1}(\Phi_\alpha(\beta)) = D_{\alpha^*}^{-1}\underbrace{(\Phi_\alpha(\beta)+\alpha)}_{\text{(N)}}\cdot\underbrace{(I+\alpha^*\Phi_\alpha(\beta))^{-1}}_{\text{(D)}^{-1}}D_\alpha.$$

The denominator, (D), evaluates to:

$$\begin{aligned}\text{(D)} &= D_\alpha(I-\alpha^*\beta)^{-1}\left(I-\alpha^*\beta+\alpha^*\beta-\alpha^*\alpha\right)D_\alpha^{-1}\\ &= D_\alpha(I-\alpha^*\beta)^{-1}D_\alpha,\end{aligned}$$

while the numerator, (N), evaluates to

$$\begin{aligned}\text{(N)} &= D_{\alpha^*}(I-\beta\alpha^*)^{-1}\left(\beta-\alpha+(I-\beta\alpha^*)D_{\alpha^*}^{-1}\alpha D_\alpha\right)D_\alpha^{-1}\\ &= D_{\alpha^*}(I-\beta\alpha^*)^{-1}\beta D_\alpha.\end{aligned}$$

It follows that the full expression is

$$\Phi_\alpha^{-1}(\Phi_\alpha(\beta)) = (I-\beta\alpha^*)^{-1}\beta(I-\alpha^*\beta) = \beta.$$

The remaining assertions are similarly easy to verify. □

5.3. Frostman shifts of Schur functions

For any strictly contractive $b \in \mathscr{S}_d(\mathcal{J},\mathcal{K})$, it was proven in Subsection 4.22 (see Equation 4.6) that

$$b^{\langle 0\rangle}(z) := D_{b(0)^*}(I_{\mathcal{J}_0} - b(z)b(0)^*)^{-1}(b(z)-b(0))D_{b(0)}^{-1} = (\Phi_{b(0)}\circ b)(z),$$

belongs to $\mathscr{S}_d(\mathcal{J},\mathcal{K})$ and vanishes at 0. In particular, it follows that $b^{\langle 0\rangle}$ is strictly contractive on the ball (since it vanishes at 0).

Definition 5.4. Let $b \in \mathscr{S}_d(\mathcal{J},\mathcal{K})$ be a purely contractive Schur-class multiplier. The 0-*Frostman shift of* b is the strictly contractive Schur-class function $b^{\langle 0\rangle} \in \mathscr{S}_d(\mathcal{J},\mathcal{K})$ of Equation (4.5).

For any pure contraction $\alpha \in \mathscr{L}(\mathcal{J},\mathcal{K})$, the α-*Frostman shift of* b is the purely contractive (strictly contractive if α is a strict contraction) analytic operator-valued function $b^{\langle\alpha\rangle} := \Phi_\alpha^{-1}\circ b^{\langle 0\rangle}$.

Theorem 5.6 below will prove that the α-Frostman shift $b^{\langle\alpha\rangle}$ of any $b \in \mathscr{S}_d(\mathcal{J},\mathcal{K})$ also belongs to the Schur class.

Lemma 5.5. *For any purely contractive* $b \in \mathscr{S}_d(\mathcal{J},\mathcal{K})$, $b = b^{\langle b(0)\rangle} = (b^{\langle 0\rangle})^{\langle b(0)\rangle} = \Phi_{b(0)}^{-1}\circ b^{\langle 0\rangle}$.

Proof. If $b(0)$ is a strict contraction, this follows immediately from Lemma 5.2. If $b(0)$ is only a pure contraction, then consider

$$b^{\langle b(0)\rangle} = D_{b(0)^*}^{-1}\left(b^{\langle 0\rangle} + b(0)\right)\left(I + b(0)^* b^{\langle 0\rangle}\right)^{-1} D_{b(0)}.$$

Since $b^{\langle 0\rangle}$ is Schur-class and strictly contractive, Lemma 5.2 implies that for any pure contraction α, $b^{\langle\alpha\rangle} = \Phi_\alpha^{-1} \circ b^{\langle 0\rangle}$ is purely contractive. The formula above initially defines $b^{\langle b(0)\rangle}(z)^*$ on a dense domain, and Lemma 5.2 shows that this extends by continuity to a pure contraction on the entire space.

Recall the notation of Subsection 4.22, and equation (4.5), which defines $b^{\langle 0\rangle}(z) := D(z)^{-1} z N_0(z)$ with $D(z) = \Gamma(z)^*\Gamma(0)$, $zN_0(z) = z(\Gamma(z)^* \otimes I_d)\Gamma_0(\infty)$, and $\Gamma(z), \Gamma_0(\infty)$ are the analytic model maps defined in Subsection 4.22. Then consider

$$\begin{aligned}
D_{b(0)^*}(D(z) + zN_0(z)b(0)^*)D_{b(0)^*} &= k^b(z,0) + z\left(k_z^b \otimes I_d\right)^* \Gamma_0(\infty) D_{b(0)} b(0)^* \\
&= k^b(z,0) + z\left(k_z^b \otimes I_d\right)^* \mathbf{b} b(0)^* \\
&= I - b(0)b(0)^* = D_{b(0)^*}^2.
\end{aligned}$$

This proves that $D(z) + zN_0(z)b(0)^* = I$. Next consider the denominator term,

$$\begin{aligned}
I + b(0)^* b^{\langle 0\rangle} &= I + b(0)^* D(z)^{-1} z N_0(z) \\
&= I + b(0)^*(I - zN_0(z)b(0)^*)^{-1} z N_0(z) \\
&= I + (I - b(0)^* z N_0(z))^{-1} - I.
\end{aligned}$$

The full expression is then

$$\begin{aligned}
D_{b(0)^*} b^{\langle b(0)\rangle}(z) &= (D(z)^{-1} z N_0(z) + b(0))(I - b(0)^* z N_0(z)) D_{b(0)} \\
&= \left(D(z)^{-1} z N_0(z) - D(z)^{-1} z N_0(z) b(0)^* z N_0(z) + b(0) - b(0)b(0)^* z N_0(z)\right) D_{b(0)} \\
&= \left(D(z)^{-1} z N_0(z) - D(z)^{-1}(I - D(z)) z N_0(z) + b(0) - b(0)b(0)^* z N_0(z)\right) D_{b(0)} \\
&= b(0) D_{b(0)} + D_{b(0)^*}^2 z N_0(z) D_{b(0)} \\
&= D_{b(0)^*} b(0) + D_{b(0)^*} z \mathbf{b}(z) \\
&= D_{b(0)^*} b(z).
\end{aligned}$$

□

Theorem 5.6. *Given any purely contractive Schur function $b \in \mathscr{S}_d(\mathcal{H}, \mathcal{K})$, and any pure contraction $\alpha \in \mathscr{L}(\mathcal{H}, \mathcal{K})$, the α-Frostman shift, $b^{\langle\alpha\rangle} \in \mathscr{S}_d(\mathcal{H}, \mathcal{K})$ is also Schur-class and purely contractive,*

$$b^{\langle\alpha\rangle}(z) = D_{\alpha^*}^{-1}\left(b^{\langle 0\rangle}(z) + \alpha\right)\left(I + \alpha^* b^{\langle 0\rangle}(z)\right)^{-1} D_\alpha.$$

Multiplication by

$$M^{\langle\alpha\rangle}(z) := D_{\alpha^*}(I + b^{\langle 0\rangle}(z)\alpha^*)^{-1} = (I - b^{\langle\alpha\rangle}(z)\alpha^*)D_{\alpha^*}^{-1},$$

defines a unitary multipier, $M^{\langle\alpha\rangle}$, of $\mathscr{H}(b^{\langle 0\rangle})$ onto $\mathscr{H}(b^{\langle\alpha\rangle})$.

The unitary multiplier $M^{\langle\alpha\rangle}$ is a multivariable and operator analogue of a Crofoot multiplier or *Crofoot transform* [19, 25]. Since $M^{\langle\alpha\rangle}$ is a unitary multiplier, it follows, in particular, that $M^{\langle\alpha\rangle}(z)$ defines a bounded invertible operator for any $z \in \mathbb{B}^d$.

Proof. By the identities of Lemma 5.2,

$$\begin{aligned} I - b^{\langle\alpha\rangle}(z)b^{\langle\alpha\rangle}(w)^* &= I - \Phi_\alpha^{-1}(b^{\langle 0\rangle}(z))\Phi_\alpha^{-1}(b^{\langle 0\rangle}(w))^* \\ &= D_{\alpha^*}(I + b^{\langle 0\rangle}(z)\alpha^*)^{-1}\left(I - b^{\langle 0\rangle}(z)b^{\langle 0\rangle}(w)^*\right)(I + \alpha b^{\langle 0\rangle}(w)^*)^{-1}D_{\alpha^*}. \end{aligned}$$

Similarly, suppose α is a strict contraction (the argument can be modified to prove the case where α is pure), then

$$\begin{aligned} &I - b^{\langle 0\rangle}(z)b^{\langle 0\rangle}(w)^* \\ &\quad = D_{\alpha^*}(I - b^{\langle\alpha\rangle}(z))\alpha^*)^{-1}\left(I - b^{\langle\alpha\rangle}(z)b^{\langle\alpha\rangle}(w)^*\right)(I - \alpha b^{\langle\alpha\rangle}(w)^*)^{-1}D_{\alpha^*}. \end{aligned}$$

This proves simultaneously that $k^{\langle\alpha\rangle} := k^{b^{\langle\alpha\rangle}}$ is a positive kernel so that $b^{\langle\alpha\rangle} \in \mathscr{S}_d(\mathcal{H}, \mathcal{K}))$ by [13, Theorem 2.1], and that $M^{\langle\alpha\rangle}(z)$ as written above is a unitary multiplier of $\mathscr{H}(b^{\langle 0\rangle})$ onto $\mathscr{H}(b^{\langle\alpha\rangle})$. □

Theorem 5.7. *The map,*

$$\mathbf{b}^{\langle\alpha\rangle} \mapsto \mathbf{b}^{\langle 0\rangle} := (M^{\langle\alpha\rangle} \otimes I_d)^{-1}\mathbf{b}^{\langle\alpha\rangle}D_\alpha^{-1},$$

is a bijection from contractive Gleason solutions for $b^{\langle\alpha\rangle}$ onto contractive Gleason solutions for $b^{\langle 0\rangle}$ which preserves extremal solutions.

Proof. First, assume that α is a strict contraction. Let $\mathbf{b}^{\langle\alpha\rangle}$ be any contractive Gleason solution for $b^{\langle\alpha\rangle}$. Then,

$$\begin{aligned} z\mathbf{b}^{\langle 0\rangle}(z) &= z(M^{\langle\alpha\rangle}(z)^{-1} \otimes I_d)\mathbf{b}^{\langle\alpha\rangle}(z)D_\alpha^{-1} \\ &= M^{\langle\alpha\rangle}(z)^{-1}(b^{\langle\alpha\rangle}(z) - \alpha)D_\alpha^{-1} \\ &= D_{\alpha^*}(I - b^{\langle\alpha\rangle}(z)\alpha^*)^{-1}(b^{\langle\alpha\rangle}(z) - \alpha)D_\alpha^{-1} \\ &= b^{\langle 0\rangle}(z), \end{aligned}$$

and this equals $b^{\langle 0\rangle}(z) - b^{\langle 0\rangle}(0)$ since $b^{\langle 0\rangle}(0) = 0$. This shows that $\mathbf{b}^{\langle 0\rangle}$ is a Gleason solution, and

$$\begin{aligned} (\mathbf{b}^{\langle 0\rangle})^*\mathbf{b}^{\langle 0\rangle} &= D_\alpha^{-1}(\mathbf{b}^{\langle\alpha\rangle})^*\mathbf{b}^{\langle\alpha\rangle}D_\alpha^{-1} \\ &\le D_\alpha^{-1}(I - \alpha^*\alpha)D_\alpha^{-1} = I, \end{aligned}$$

so that $\mathbf{b}^{\langle 0\rangle}$ is a contractive Gleason solution which is extremal if $\mathbf{b}^{\langle\alpha\rangle}$ is. The converse follows similarly.

In the case where α is not a strict contraction, set $b := b^{\langle\alpha\rangle}$, so that $\alpha = b(0)$. Recall the notation of Subsection 4.22, and equation (4.5), which defines $b^{\langle 0\rangle}(z) := D(z)^{-1}zN_0(z)$ with $D(z) = \Gamma(z)^*\Gamma(0)$, $zN_0(z) = z(\Gamma(z)^* \otimes I_d)\Gamma_0(\infty)$, and $\Gamma(z), \Gamma_0(\infty)$ are the analytic model maps defined in Subsection 4.22. Namely,

$\Gamma(z)D_{\alpha^*} = k_0^b$, and $\Gamma_0(\infty)D_\alpha = \mathbf{b}$. As in the proof of Lemma 5.5, it follows that $D(z) + zN_0(z)\alpha^* = I$, so that

$$\begin{aligned} M^{\langle\alpha\rangle}(z) &= D_{\alpha^*}(I - b^{\langle 0\rangle}(z)\alpha^*)^{-1} \\ &= D_{\alpha^*}D(z)(D(z) + zN_0(z)\alpha^*)^{-1} = D_{\alpha^*}D(z). \end{aligned}$$

To prove that $\mathbf{b}^{\langle 0\rangle}$ is a contractive Gleason solution for $b^{\langle 0\rangle}$, first observe that

$$\begin{aligned} D_{\alpha^*}zN_0(z)D_\alpha &= D_{\alpha^*}\Gamma(z)^*z\Gamma_0(\infty)D_\alpha \\ &= (z^*k_z^b)^*\mathbf{b} = z\mathbf{b}(z) \\ &= b(z) - b(0) = b(z) - \alpha. \end{aligned}$$

It then follows that

$$\begin{aligned} & M^{\langle\alpha\rangle}(z)z\mathbf{b}^{\langle 0\rangle}(z)D_\alpha = z\mathbf{b}(z) \\ \Rightarrow\quad & D_{\alpha^*}D(z)z\mathbf{b}^{\langle 0\rangle}(z)D_\alpha = D_{\alpha^*}zN_0(z)D_\alpha \\ \Leftrightarrow\quad & D(z)z\mathbf{b}^{\langle 0\rangle}(z) = zN_0(z) \\ \Leftrightarrow\quad & z\mathbf{b}^{\langle 0\rangle}(z) = b^{\langle 0\rangle}(z). \end{aligned}$$

This proves that $\mathbf{b}^{\langle 0\rangle}$ is a Gleason solution for $b^{\langle 0\rangle}$ and contractivity follows as before. Again, the fact that $\mathbf{b} \mapsto \mathbf{b}^{\langle 0\rangle}$ is surjective follows similarly. $\square$

Proposition 5.8. *Let $b \in \mathscr{S}_d(\mathcal{H}, \mathcal{K})$ be a purely contractive Schur-class function, let $X = X(\mathbf{b})$ be an extremal Gleason solution for $\mathscr{H}(b)$, and let $\mathbf{b}$ be the corresponding contractive extremal Gleason solution for b: $X^*k_w^b = w^*k_w^b - \mathbf{b}b(w)^*$.*

If $\mathbf{b}^{\langle 0\rangle} = \left(M^{\langle b(0)\rangle} \otimes I_d\right)^* \mathbf{b}D_{b(0)}^{-1}$, $X^{\langle 0\rangle} := X(\mathbf{b}^{\langle 0\rangle})$ are the corresponding extremal Gleason solutions for $b^{\langle 0\rangle}$ and $\mathscr{H}(b^{\langle 0\rangle})$, and $X = V - C$ is the isometric-pure decomposition of X, then

$$V = M^{\langle b(0)\rangle}X^{\langle 0\rangle}(M^{\langle b(0)\rangle} \otimes I_d)^*.$$

Proof. As in Subsection 4.22, we have that $V^* = X^*P_{\mathrm{Ran}(X)}$, where

$$P_{\mathrm{Ran}(X)} = I - k_0^b k^b(0,0)^{-1}(k_0^b)^* = I - k_0^b D_{b(0)^*}^{-2}(k_0^b)^*.$$

Let $k^{\langle 0\rangle} = k^{b^{\langle 0\rangle}}$, we need to verify that $M^{\langle b(0)\rangle}X^{\langle 0\rangle}(M^{\langle b(0)\rangle} \otimes I_d)^* = V$. To prove this, check the action on point evaluation maps:

$$\begin{aligned} V^*k_w^b &= w^*k_w^b - \mathbf{b}b(w)^* + \mathbf{b}b(0)^*D_{b(0)^*}^{-2}(I - b(0)b(w)^*) \\ &= w^*k_w^b - \mathbf{b}D_{b(0)}^{-2}\left((I - b(0)^*b(0))b(w)^* - b(0)^*(I - b(0)b(w)^*)\right) \\ &= w^*k_w^b - \mathbf{b}D_{b(0)}^{-2}(b(w)^* - b(0)^*). \end{aligned}$$

Compare this to:

$$\begin{aligned}
&(M^{\langle b(0)\rangle} \otimes I_d)(X^{\langle 0\rangle})^*(M^{\langle b(0)\rangle})^* k_w^b \\
&= (M^{\langle b(0)\rangle} \otimes I_d)(X^{\langle 0\rangle})^* k_w^{b^{\langle 0\rangle}} M^{\langle b(0)\rangle}(w)^* \\
&= (M^{\langle b(0)\rangle} \otimes I_d)\left(w^* k_w^{\langle 0\rangle} - \mathbf{b}^{\langle 0\rangle} b^{\langle 0\rangle}(w)^*\right) M^{\langle b(0)\rangle}(w)^* \\
&= w^* k_w^b - \mathbf{b} D_{b(0)}^{-1} b^{\langle 0\rangle}(w)^* M^{\langle b(0)\rangle}(w)^* \\
&= w^* k_w^b - \mathbf{b} D_{b(0)}^{-2}(b(w)^* - b(0)^*).
\end{aligned}$$

□

5.9. Gleason solution model for CCNC row contractions

Let T be an arbitrary CCNC row contraction on $\mathcal{H}$ with partial isometric-purely contractive decomposition $T = V - C$.

Lemma 5.10. *Let V be a CCNC row partial isometry on $\mathcal{H}$, and let $(\gamma, \mathfrak{J}_\infty, \mathfrak{J}_0)$ be a model triple for V. Given any pure contraction $\delta \in [\mathscr{L}(\mathfrak{J}_\infty, \mathfrak{J}_0)]_1$, the map*

$$\delta \mapsto T_\delta := V - \gamma(0)\delta\gamma(\infty)^*,$$

is a bijection from pure contractions onto CCNC row contractions with partial isometric part V.

Proof. Since $\gamma(0) : \mathfrak{J}_0 \to \mathrm{Ran}\,(V)^\perp$ and $\gamma(\infty) : \mathfrak{J}_\infty \to \mathrm{Ker}(V)$ are onto isometries, it is clear that $\delta \mapsto -\gamma(0)\delta\gamma(\infty)^*$ maps pure contractions $\delta \in \mathscr{L}(\mathfrak{J}_\infty, \mathfrak{J}_0)$ onto all pure contractions in $\mathscr{L}(\mathrm{Ker}(V), \mathrm{Ran}\,(V)^\perp)$. It is also clear that T_δ is CCNC if and only if V is, and that V is the partial isometric part of T_δ. Conversely, given any *CCNC* row contraction T on $\mathcal{H}$ such that $T = V - C$, we have that $\delta := -\gamma(0)^* T\gamma(\infty) = \gamma(0)^* C\gamma(\infty)$. Since C is a pure row contraction, δ is a pure contraction and $T = T_\delta$. □

Definition 5.11. Let $T : \mathcal{H} \otimes \mathbb{C}^d \to \mathcal{H}$ be a CCNC row contraction with partial isometeric-purely contractive decomposition $T = V - C$. For any fixed model triple $(\gamma, \mathfrak{J}_\infty, \mathfrak{J}_0)$ of V, define

$$\delta_T^\gamma := -\gamma(0)^* T\gamma(\infty) = \gamma(0)^* C\gamma(\infty),$$

the *zero-point contraction* of T. The *characteristic function*, b_T, of T, is then any Schur-class function in the weak coincidence class of the δ_T^γ-Frostman shift of b_V^γ,

$$\begin{aligned}
b_T^\gamma &:= (b_V^\gamma)^{\langle \delta_T^\gamma \rangle} \in \mathscr{S}_d(\mathfrak{J}_\infty, \mathfrak{J}_0), \\
b_T^\gamma(z) &= D_{(\delta_T^\gamma)^*}^{-1} \left(b_V^\gamma(z) + \delta_T^\gamma\right)\left(I + (\delta_T^\gamma)^* b_V^\gamma(z)\right)^{-1} D_{\delta_T^\gamma}; \qquad z \in \mathbb{B}^d.
\end{aligned}$$

Since C is a pure row contraction, it follows that δ_T^γ is always a pure contraction, and that b_T^γ is purely contractive on the ball, by Lemma 5.2.

Lemma 5.12. *The coincidence class of b_T^γ is invariant under the choice of model triple $(\gamma, \mathfrak{J}_\infty, \mathfrak{J}_0)$.*

Proof. Let $(\gamma, \mathcal{J}_\infty, \mathcal{J}_0)$ and $(\varphi, \mathcal{K}_\infty, \mathcal{K}_0)$ be two model triples for V, $T = V - C$. By Lemma 4.6 we know that there are onto isometries $R : \mathcal{K}_\infty \to \mathcal{J}_\infty$ and $Q^* : \mathcal{K}_0 \to \mathcal{J}_0$ so that $\varphi(\infty) = \gamma(\infty)R$, $\varphi(0) = \gamma(0)Q^*$, and

$$b_V^\varphi = Q b_V^\gamma R.$$

Similarly,

$$\delta_T^\varphi = -\varphi(0)^* T \varphi(\infty) = Q \delta_T^\gamma R.$$

Finally, we calculate

$$\begin{aligned} b_T^\varphi &= (Q b_V^\gamma R)^{\langle Q\delta_T^\gamma R\rangle} \\ &= D^{-1}_{(Q\delta_T^\gamma R)^*} \left(Q(b_V^\gamma + \delta_T^\gamma)R\right)\left(I + R^*(\delta_T^\gamma)^* Q^* Q b_V^\gamma R\right)^{-1} D_{Q\delta_T^\gamma R} \\ &= Q D^{-1}_{(\delta_T^\gamma)^*} Q^* Q \left(b_V^\gamma + \delta_T^\gamma\right) R R^* \left(I + (\delta_T^\gamma)^* b_V^\gamma\right)^{-1} R R^* D_{\delta_T^\gamma} R \\ &= Q (b_V^\gamma)^{\langle \delta_T^\gamma\rangle} R = Q b_T^\gamma R. \end{aligned}$$

□

Lemma 5.13 (Extremal Gleason solutions). *Let $b \in \mathscr{S}_d(\mathcal{H}, \mathcal{K})$ be a Schur-class multiplier. If X is any extremal Gleason solution for $\mathscr{H}(b)$, then the characteristic function, b_X, of X, coincides weakly with b.*

Proof. Given any purely contractive $b \in \mathscr{S}_d(\mathcal{H}, \mathcal{K})$, Lemma 2.9 proved that any contractive extremal Gleason solution, X, for $\mathscr{H}(b)$ is a CCNC row contraction. Using the model constructed as in Subsection 4.22, we will now show that the characteristic function, b_X, of X, coincides weakly with b.

As discussed in Subsection 2.4, $X = X(\mathbf{b})$ for an extremal Gleason solution $\mathbf{b}$ for b, where $X(\mathbf{b})$ is given by Formula (2.5). Let $\mathbf{b}^{\langle 0\rangle} := (M^{\langle b(0)\rangle} \otimes I_d)^* \mathbf{b} D^{-1}_{b(0)}$ be the extremal Gleason solution for $b^{\langle 0\rangle}$ which corresponds uniquely to $\mathbf{b}$ as in Theorem 5.7, and let $X = V - C$ be the isometric-pure decomposition of X. Proposition 5.8 then implies that if $X^{\langle 0\rangle} := X(\mathbf{b}^{\langle 0\rangle})$ is the corresponding extremal Gleason solution for $\mathscr{H}(b^{\langle 0\rangle})$, that $V = M^{\langle b(0)\rangle} X^{\langle 0\rangle} (M^{\langle b(0)\rangle} \otimes I_d)^*$.

As in Subsection 4.22, it then follows that we can define an analytic model triple for V as follows. Let $\Gamma(0) := k_0^b k^b(0,0)^{-1/2} = k_0^b D^{-1}_{b(0)^*}$, $\Gamma(z) := (I - Xz^*)^{-1}\Gamma(0) = k_z^b D^{-1}_{b(0)^*}$, and $\Gamma(\infty) := \mathbf{b} D^{-1}_{b(0)} \oplus \Gamma(\infty)'$, where $\Gamma(\infty)' : \mathcal{K}' \to \mathrm{Ker}(V) \ominus \mathrm{Ran}\,(\mathbf{b})$ is an arbitrary onto isometry (that $\mathbf{b}$ maps into $\mathrm{Ker}(V)$ follows as in Lemma 4.23). As in Subsection 4.22, $(\Gamma, \mathcal{K} \oplus \mathcal{K}', \mathcal{K})$ is an analytic model triple for V, and we will use this triple to compute the characteristic function b_X^Γ. Using the relationship between V and $X^{\langle 0\rangle}$, it is easy to check that $b_V^\Gamma = b^{\langle 0\rangle} \oplus 0_{\mathcal{K}'}$ as in Subsection 4.22, and it remains to check that $\delta_X^\Gamma = b(0) \oplus 0_{\mathcal{K}'}$:

$$\begin{aligned} \delta_X^\Gamma &= -(k_0^b D^{-1}_{b(0)^*})^* X \Gamma(\infty) \\ &= -(X^* k_0^b D^{-1}_{b(0)^*})^* \mathbf{b} D^{-1}_{b(0)} \oplus \Gamma(\infty)' \\ &= D^{-1}_{b(0)^*} b(0) \mathbf{b}^* \left(\mathbf{b} D^{-1}_{b(0)} \oplus \Gamma(\infty)'\right) = b(0) \oplus 0_{\mathcal{K}'}. \end{aligned}$$

We conclude that $b_X^\Gamma := (b_V^\Gamma)^{\langle \delta_X^\Gamma\rangle}$ coincides weakly with $b = (b^{\langle 0\rangle})^{\langle b(0)\rangle}$. □

We are now sufficiently prepared to prove one of our main results:

Theorem 5.14. *A row contraction* $T : \mathcal{H} \otimes \mathbb{C}^d \to \mathcal{H}$ *is CCNC if and only if* T *is unitarily equivalent to an extremal Gleason solution* X^b *acting on a multi-variable de Branges–Rovnyak space* $\mathscr{H}(b)$ *for a purely contractive Schur-class* $b \in \mathscr{S}_d(\mathcal{J}, \mathcal{K})$.

If $T \simeq X^b$, *the characteristic function,* $b_T := b$ *is a unitary invariant: if two CCNC row contractions* T_1, T_2 *are unitarily equivalent, then their characteristic functions* b_{T_1}, b_{T_2} *coincide weakly. One can choose* $b_T \in \mathscr{S}_d(\mathrm{Ran}(D_T), \mathrm{Ran}(D_{T^*}))$.

In the above $\simeq$ denotes unitary equivalence. Recall that as shown in Subsection 4.17, the characteristic function of a CCNC row contraction is not a complete unitary invariant: there exist CCNC row contractions T_1, T_2 which have the same characteristic function but are not unitarily equivalent.

Proof. We have already proven that any extremal Gleason solution, X, for $\mathscr{H}(b)$, $b \in \mathscr{S}_d(\mathcal{J}, \mathcal{K})$ is a CCNC row contraction with characteristic function coinciding weakly with b in Lemma 2.9 and Lemma 5.13 above. Conversely, let T be a CCNC row contraction on $\mathcal{H}$ with isometric-contractive decomposition $T = V - C$.

Let $(\Gamma, \mathcal{J}_\infty, \mathcal{J}_0)$ be the analytic model triple $\Gamma = \Gamma_V$ for V. Recall that $\mathcal{J}_\infty \simeq \mathrm{Ker}(V) \simeq \mathrm{Ran}\,(D_T)$ and $\mathcal{J}_0 \simeq \mathrm{Ran}\,(V)^\perp \simeq \mathrm{Ran}\,(D_{T^*})$ ($\simeq$ means isomorphic as Hilbert spaces, *i.e.*, they have the same dimension). By Theorem 4.8, the unitary $U^\Gamma := (M^\Gamma)^{-1}\hat{U}^\Gamma : \mathcal{H} \to \mathscr{H}(b_V^\Gamma)$, where recall $M^\Gamma = M_{D^\Gamma}$, a unitary multiplier, is such that $X := U^\Gamma V (U^\Gamma)^* = X(\mathbf{b}^\Gamma)$ is an extremal Gleason solution for $\mathscr{H}(b_V^\Gamma)$ corresponding to the extremal Gleason solution $\mathbf{b}^\Gamma = U^\Gamma \Gamma(\infty)$ for b_V^Γ. It is clear that $(U^\Gamma\Gamma, \mathcal{J}_\infty, \mathcal{J}_0)$ is then an analytic model triple for X.

By Theorem 5.7, since $b_V^\Gamma = \Phi_{b_T^\Gamma(0)}(b_T^\Gamma) = (b_T^\Gamma)^{\langle 0 \rangle}$ is the 0-Frostman shift of b_T, there is a unique extremal Gleason solution $\mathbf{b}^T$ for $b_T = b_T^\Gamma$ so that

$$\mathbf{b}^\Gamma := (M^{\langle b_T(0)\rangle} \otimes I)^{-1} \mathbf{b}^T D_{b_T(0)}^{-1},$$

where, by definition, $b_T(0) = b_T^\Gamma(0) = \delta_T^\Gamma$. Proposition 5.8 then implies that if $X^T := X(\mathbf{b}^T)$ is the corresponding extremal Gleason solution for b_T with isometric-pure decomposition $X^T = V^T - C^T$, then we have that

$$\widetilde{X} := (M^{\langle b_T(0)\rangle})^* X^T (M^{\langle b_T(0)\rangle} \otimes I_d) = X - \widetilde{C},$$

where $(M^{\langle b_T(0)\rangle})^* V^T (M^{\langle b_T(0)\rangle} \otimes I_d) = X$.

Our goal is to prove that $T' = U^\Gamma T (U^\Gamma)^* \otimes I_d =: X - C'$ is equal to $\widetilde{X} = (M^{\langle b(0)\rangle})^* X^T M^{\langle b_T(0)\rangle} \otimes I_d = X - \widetilde{C}$ so that $T \simeq X^T$, *i.e.*, T is unitarily equivalent to the extremal Gleason solution X^T for $\mathscr{H}(b_T^\Gamma)$. By Lemma 5.10, it suffices to show that $\delta_{T'}^{U^\Gamma\Gamma} = \delta_{\widetilde{X}}^{U^\Gamma\Gamma}$. Let $b := b_T^\Gamma$, $b_V := b_V^\Gamma$, and calculate

$$\begin{aligned}\delta_{T'}^{U^\Gamma\Gamma} &= -\Gamma(0)^*(U^\Gamma)^* T' (U^\Gamma \otimes I_d)\Gamma(\infty) \\ &= -\Gamma(0)^* T \Gamma(\infty) = \delta_T^\Gamma = b_T^\Gamma(0).\end{aligned}$$

Similarly, since $X = X(\mathbf{b}^\Gamma)$ where $\mathbf{b}^\Gamma = U^\Gamma \Gamma(\infty)$,

$$\begin{aligned}
\delta_{\widetilde{X}}^{U^\Gamma \Gamma} &= -(U^\Gamma \Gamma(0))^* \widetilde{X} \mathbf{b}^\Gamma \\
&= -(M_{D^\Gamma}^* \hat{K}_0^\Gamma)^* (M^{\langle b(0) \rangle})^* X^T (M^{\langle b_T(0) \rangle} \otimes I_d)(M^{\langle b_T(0) \rangle} \otimes I)^{-1} \mathbf{b}^T D_{b_T(0)}^{-1} \\
&= -\left((M^{\langle b_T(0) \rangle})^{-1})^* k_0^{b_V} \right)^* X^T \mathbf{b}^T D_{b_T(0)}^{-1} \\
&= -\left((X^T)^* k_0^{b_T} (M^{\langle b_T(0) \rangle}(0)^{-1})^* \right)^* \mathbf{b}^T D_{b_T(0)}^{-1} \\
&= M^{\langle b_T(0) \rangle}(0)^{-1} b(0) (\mathbf{b}^T)^* \mathbf{b}^T D_{b_T(0)}^{-1} \\
&= D_{b_T(0)^*} (I - b(0) b(0)^*)^{-1} b(0) (I - b(0)^* b(0)) D_{b_T(0)}^{-1} \\
&= b_T(0) = \delta_T^\Gamma.
\end{aligned}$$

Finally, if T_1, T_2 are two unitarily equivalent row contractions, $T_2 = U T_1 U^*$ for a unitary $U : \mathcal{H}_1 \to \mathcal{H}_2$, and $T_k = V_k - C_k$, then given any analytic model triple $(\Gamma, \mathfrak{J}_\infty, \mathfrak{J}_0)$ for T_1, $(U\Gamma, \mathfrak{J}_\infty, \mathfrak{J}_0)$ is an analytic model triple for T_2, and $b_{T_1}^\Gamma = b_{T_2}^{U\Gamma}$. □

We conclude this section with the observation that our characteristic function, b_T, for any CCNC row contraction, T, coincides weakly with the Sz.-Nagy–Foiaş-type characteristic function of T, as defined for CNC d-contractions in [17]:

$$\Theta_T(z) := \left(-T + z D_{T^*} (I - zT^*)^{-1} D_T \right) |_{\mathrm{Ran}(D_T)} \in \mathscr{L}(\mathrm{Ran}\,(D_T), \mathrm{Ran}\,(D_{T^*})); \quad z \in \mathbb{B}^d.$$

Proposition 5.15. *Let T be a CCNC row contraction. The characteristic function, b_T, of T, coincides weakly with the Sz.-Nagy–Foiaş characteristic function, Θ_T.*

Proof. It suffices to show that given any purely contractive $b \in \mathscr{S}_d(\mathfrak{J}, \mathcal{K})$, and any contractive, extremal Gleason solution, X, for $\mathscr{H}(b)$ that $b_X =: b$ coincides weakly with Θ_X. Let $\mathbf{b}$ be a contractive extremal Gleason solution for b so that $X = X(\mathbf{b})$.

As in the proof of Lemma 5.13, consider the analytic model triple $(\Gamma, \mathcal{K} \oplus \mathcal{K}', \mathcal{K})$ where $\Gamma(0) = k_0^b D_{b(0)^*}^{-1}$, $\Gamma(z) := (I - Xz^*)^{-1} \Gamma(0)$, and $\Gamma(\infty) = \mathbf{b} D_{b(0)}^{-1} \oplus \Gamma(\infty)'$, where if $X = V - C$ is the isometric-pure decomposition of X, then $\Gamma(\infty)' : \mathcal{K}' \to \mathrm{Ker}(V) \ominus \mathrm{Ran}\,(\mathbf{b})$ is an onto isometry. Since $\Gamma(0), \Gamma(\infty)$ are onto isometries, Θ_X coincides with

$$-\Gamma(0)^* X \Gamma(\infty) + \Gamma(0)^* z D_{X^*} (I - zX^*)^{-1} D_X \Gamma(\infty),$$

and $\delta_X^\Gamma = -\Gamma(0)^* X \Gamma(\infty) = b(0) \oplus 0_{\mathcal{K}'}$ as in the proof of Lemma 5.13. Since X is extremal, one can verify (by uniqueness of the positive square root) that

$$D_{X^*} = \sqrt{k_0^b (k_0^b)^*} = k_0^b D_{b(0)^*}^{-1} (k_0^b)^*,$$

and it follows that

$$\Gamma(0)^* z D_{X^*}(I - zX^*)^{-1} D_X \Gamma(\infty) = D_{b(0)^*}^{-1}(k_0^b)^* k_0^b D_{b(0)^*}^{-1}(z^* k_z^b)^* \sqrt{I - X^* X}\Gamma(\infty)$$
$$= (z^* k_z^b)^* \sqrt{I - X^* X}\Gamma(\infty).$$

Since $\Gamma(\infty)$ is an isometry onto $\mathrm{Ker}(V)$, and $X^*X = V^*V + C^*C$, it follows that

$$\sqrt{I - X^* X}\Gamma(\infty) = \sqrt{P_{\mathrm{Ker}(V)} - P_{\mathrm{Ker}(V)} C^* C P_{\mathrm{Ker}(V)}}\Gamma(\infty).$$

Moreover, in Subsection 4.22, we calculated that

$$V^* = X^*(I - k_0^b D_{b(0)^*}^{-1}(k_0^b)^*),$$

and it follows from this that

$$C^* = \mathbf{b} D_{b(0)}^{-1} b(0)^* D_{b(0)^*}^{-1}(k_0^b)^*.$$

In particular, it follows that

$$(C^*C)^k = \mathbf{b} D_{b(0)}^{-1} \left(b(0)^* b(0)\right)^k D_{b(0)}^{-1} \mathbf{b}^*; \qquad k \in \mathbb{N},$$

and the functional calculus then implies that

$$\sqrt{I - X^* X}\Gamma(\infty) = \mathbf{b} D_{b(0)}^{-1} \sqrt{I - b(0)^* b(0)} D_{b(0)}^{-1} \mathbf{b}^* = \mathbf{b} D_{b(0)}^{-1} \mathbf{b}^*.$$

In conclusion we obtain

$$\Theta_X(z) \simeq b(0) \oplus 0_{\mathcal{K}'} + (z^* k_z^b)^* \mathbf{b} D_{b(0)}^{-1} \mathbf{b}^* \mathbf{b} D_{b(0)}^{-1} \oplus 0_{\mathcal{K}'}$$
$$= b(0) \oplus 0_{\mathcal{K}'} + (b(z) - b(0)) \oplus 0_{\mathcal{K}'} = b(z) \oplus 0_{\mathcal{K}'},$$

and Θ_X coincides weakly with $b = b_X$. □

6. QE row contractions

In this section we focus on the sub-class of quasi-extreme (QE) row contractions. This is the set of all CCNC row contractions, T, whose characteristic function b_T coincides weakly with a quasi-extreme Schur multiplier as defined and studied in [30, 31, 33]. We will see that the characteristic function is a complete unitary invariant for QE row contractions.

6.1. Quasi-extreme Schur multipliers

As discussed in the introduction, the concept of a *quasi-extreme* Schur-class multiplier was introduced in [30, 31], as a several-variable analogue of a 'Szegő approximation property' that is equivalent to being an extreme point of the Schur class in the single-variable, scalar-valued setting (see, *e.g.*, [33]).

In [31], the quasi-extreme property was defined for any non-unital and square $b \in \mathscr{S}_d(\mathcal{H})$ (recall from Subsection 2.2 that the non-unital assumption is needed to ensure that the corresponding Herglotz–Schur function H_b takes values in bounded operators), but we will require the extension of this property to arbitrary purely contractive and 'rectangular' $b \in \mathscr{S}_d(\mathcal{J}, \mathcal{K})$. For this purpose it will be useful to consider the square extension, $[b]$, of any $b \in \mathscr{S}_d(\mathcal{J}, \mathcal{K})$, as defined in Subsection 2.4.

Theorem 6.2. *Given any $b \in \mathscr{S}_d(\mathfrak{J},\mathfrak{K})$ such that $[b]$ is non-unital, the following are equivalent:*

(i) *b has a unique contractive Gleason solution and this solution is extremal.*
(ii) $\mathrm{supp}(b) = \mathfrak{J}$, *$\mathscr{H}(b)$ has a unique contractive Gleason solution, and this solution is extremal.*
(iii) *There is no non-zero $g \in \mathfrak{J}$ so that $bg \in \mathscr{H}(b)$.*
(iv) *There is no non-zero $\mathfrak{J}$-valued constant function $F \equiv g \in \mathscr{H}^+(H_{[b]})$, $g \in \mathfrak{J}$.*
(v) $K_0^{[b]}(I - b(0))\mathfrak{J} \subseteq \mathrm{Ran}\left(V^{[b]}\right)$.

Any Schur multiplier is said to be *quasi-extreme* if it obeys the assumptions and equivalent conditions of this theorem. If, for example, b is strictly contractive, then $[b]$ will be strictly contractive (and hence non-unital). For conditions (iv) and (v) of the above theorem we are assuming that either $\mathfrak{J} \subseteq \mathfrak{K}$ or $\mathfrak{K} \subseteq \mathfrak{J}$. There is no loss of generality with this assumption, since it is easy to see that $b \in \mathscr{S}_d(\mathfrak{J},\mathfrak{K})$ is quasi-extreme if and only if every member of its coincidence class is quasi-extreme. In the particular case where $\mathfrak{J} = \mathfrak{K}$ so that $b = [b]$, items (iv) and (v) reduce to:

(iv)′ $\mathscr{H}^+(H_b)$ contains no constant functions.
(v)′ V^b is a co-isometry,

see [31, Theorem 4.17]. Since the proof and proof techniques of Theorem 6.2 are very similar to those of [31, Section 4], we will not include it here. The equivalence of (iii) and (iv), for example, follows as in the proof of [31, Theorem 3.22]. An arbitrary purely contractive $b \in \mathscr{S}_d(\mathfrak{J},\mathfrak{K})$ may still not satisfy the assumptions of Theorem 6.2, *i.e.*, $[b]$ may not be non-unital, and so we define:

Definition 6.3. A purely contractive $b \in \mathscr{S}_d(\mathfrak{J},\mathfrak{K})$ is *quasi-extreme* if b has a unique contractive Gleason solution, and this solution is extremal.

In particular, the bijection between contractive Gleason solutions for b and (the strictly contractive) $b^{\langle 0 \rangle}$ of Lemma 5.7 implies:

Corollary 6.4. *A purely contractive $b \in \mathscr{S}_d(\mathfrak{J},\mathfrak{K})$ is quasi-extreme if and only if the α-Frostman shift $b^{\langle \alpha \rangle}$ is quasi-extreme for any pure contraction $\alpha \in [\mathscr{L}(\mathfrak{J},\mathfrak{K})]_1$.*

In particular, b is quasi-extreme if and only if the strictly contractive $b^{\langle 0 \rangle}$ is quasi-extreme (so that $b^{\langle 0 \rangle}$ obeys the equivalent properties of Theorem 6.2.)

Lemma 6.5. *If $b \in \mathscr{S}_d(\mathfrak{J},\mathfrak{K})$ is purely contractive and* $\mathrm{supp}(b) = \mathfrak{J}$, *then b is quasi-extreme if and only if $\mathscr{H}(b)$ has a unique contractive Gleason solution, and this solution is extremal.*

Proof. This follows from Formula (2.5), as in [31, Theorem 4.9, Theorem 4.4]. □

The next result will yield an abstract characterization of CCNC row contractions with quasi-extreme characteristic functions.

Theorem 6.6. *Let $b \in \mathscr{S}_d(\mathcal{J}, \mathcal{K})$ be a purely contractive Schur-class multiplier such that* $\mathrm{supp}(b) = \mathcal{J}$*. Then b is quasi-extreme if and only if there is an extremal Gleason solution, X, for $\mathscr{H}(b)$ so that*

$$\mathrm{Ker}(X)^\perp \subseteq \bigvee_{z\in\mathbb{B}^d} z^* k_z^b \mathcal{K} = \bigvee_{z\in B^d} z^*(I - Xz^*)^{-1}\mathrm{Ran}\,(D_{X^*})\,.$$

Proof. We will first prove that any purely contractive b has this property if and only if $b^{\langle 0\rangle}$ has this property. This will show that we can assume, without loss of generality, that b is strictly contractive so that the equivalent conditions of Theorem 6.2 apply.

Given a purely contractive $b \in \mathscr{S}_d(\mathcal{J}, \mathcal{K})$, with $\mathrm{supp}(b) = \mathcal{J}$, let $X = X(\mathbf{b})$ be a contractive and extremal Gleason solution for $\mathscr{H}(b)$. Recall that $X(\mathbf{b})$ is defined as in Formula (2.5), and that since $\mathrm{supp}(b) = \mathcal{J}$, $X(\mathbf{b})$ is extremal if and only if $\mathbf{b}$ is. Then

$$\mathrm{Ker}(X)^\perp = \overline{\mathrm{Ran}\,(X^*)} = \bigvee_{z\in\mathbb{B}^d} \left(z^* k_z^b - \mathbf{b}b(z)^*\right)\mathcal{K},$$

and it follows that b will have the desired property if and only if

$$\bigvee_{z\in\mathbb{B}^d} \mathbf{b}b(z)^*\mathcal{K} = \bigvee \mathbf{b}\mathcal{J} \subseteq \bigvee_{w\in\mathbb{B}^d} w^* k_w^b \mathcal{K}. \qquad (\mathrm{supp}(b) = \mathcal{J})$$

By the bijection between Gleason solutions for b and $b^{\langle 0\rangle}$, Lemma 5.7, b will have this property if and only if

$$\bigvee_{z\in\mathbb{B}^d} z^* k_z^{\langle 0\rangle}\mathcal{K} \supseteq (M^{\langle b(0)\rangle} \otimes I_d)^*\mathbf{b}\mathcal{J}; \qquad k^{\langle 0\rangle} := k^{b^{\langle 0\rangle}}$$

$$= \mathbf{b}^{\langle 0\rangle}\mathcal{J},$$

where $\mathbf{b}^{\langle 0\rangle} := (M^{\langle b(0)\rangle} \otimes I_d)^*\mathbf{b}D_{b(0)}^{-1}$ is a contractive and extremal Gleason solution for $b^{\langle 0\rangle}$. As above it follows that this happens if and only if $\mathrm{Ker}(X^{\langle 0\rangle})^\perp \subseteq \bigvee z^* k_z^{\langle 0\rangle}\mathcal{K}$, where $X^{\langle 0\rangle} := X(\mathbf{b}^{\langle 0\rangle})$, and b has the desired property if and only if its Frostman shift $b^{\langle 0\rangle}$ does. We can now assume without loss of generality that $b \in \mathscr{S}_d(\mathcal{J}, \mathcal{K})$ is strictly contractive and that $b(0) = 0$.

First suppose that b is quasi-extreme (QE) so that b has a unique contractive Gleason solution $\mathbf{b}$ which is extremal, by Theorem 6.2, and $X := X(\mathbf{b})$ is the unique contractive and extremal Gleason solution for $\mathscr{H}(b)$. As in the first part of the proof, it follows that b will have the desired property provided that $\mathbf{b}\mathcal{J} \subseteq \bigvee_{z\in\mathbb{B}^d} z^* k_z^b\mathcal{K}$. Assume without loss of generality that $\mathcal{J} \subseteq \mathcal{K}$ or $\mathcal{K} \subseteq \mathcal{J}$ and consider $a := [b]$, the (strictly contractive) square extension of b. As described in Subsection 2.4, any contractive Gleason solution for a is given by

$$\mathbf{a} = (U_a^* \otimes I_d)D^* K_0^a, \qquad (a(0) = 0)$$

where $D \supseteq V^a$ is a contractive extension of V^a on $\mathscr{H}^+(H_a)$, the Herglotz space of a, and $U_a : \mathscr{H}(a) \to \mathscr{H}^+(H_a)$ is the canonical unitary multiplier of Lemma

2.3. Choose $D = V^a$. In the first case where $\mathcal{J} \subset \mathcal{K}$, uniqueness of $\mathbf{b}$ implies that $\mathbf{b} = \mathbf{a}|_{\mathcal{J}}$ so that

$$\begin{aligned} \mathbf{b}\mathcal{J} &\subseteq (U_a^* \otimes I_d) V_a^* K_0^a \mathcal{K} \\ &\subseteq \bigvee_{z \in \mathbb{B}^d} z^* U_a^* K_z^a \mathcal{K} \\ &= \bigvee_{z \in \mathbb{B}^d} z^* k_z^b (I - a(z)^*)^{-1} \mathcal{K} = \bigvee z^* k_z^b \mathcal{K}. \end{aligned}$$

Similarly, in the second case where $\mathcal{K} \subseteq \mathcal{J}$,

$$\begin{aligned} \mathbf{b}\mathcal{J} &= \begin{bmatrix} I_{\mathscr{H}(b)} \otimes I_d, & 0 \end{bmatrix} \mathbf{a}\mathcal{J} \\ &\subseteq \begin{bmatrix} I, & 0 \end{bmatrix} \bigvee_{z \in \mathbb{B}^d} z^* k_z^a \mathcal{J} \\ &= \bigvee_{z \in \mathbb{B}^d} z^* \begin{bmatrix} I, & 0 \end{bmatrix} \begin{bmatrix} k_z^b & 0 \\ 0 & k_z \otimes I_{\mathcal{J} \ominus \mathcal{K}} \end{bmatrix} \begin{bmatrix} \mathcal{K} \\ \mathcal{J} \ominus \mathcal{K} \end{bmatrix} \\ &= \bigvee_{z \in \mathbb{B}^d} z^* k_z^b \mathcal{K}. \end{aligned}$$

Conversely, suppose that $\mathscr{H}(b)$ has an extremal Gleason solution, X, with the desired property. Then it follows, as above, that $X = X(\mathbf{b})$, where $\mathbf{b}$ is an extremal Gleason solution for b obeying $\mathbf{b}\mathcal{J} \subseteq \bigvee_{z \in \mathbb{B}^d} z^* k_z^b \mathcal{K}$. By Remark 2.8, setting $a = [b]$, we have that there is a contractive Gleason solution $\mathbf{a}$ for a such that either $\mathbf{b} = \mathbf{a}|_{\mathcal{J}}$ or $\mathbf{b} = \begin{bmatrix} I_{\mathscr{H}(b)} \otimes I_d, & 0 \end{bmatrix} \mathbf{a}$. Also, again by Subsection 2.4, there is a contractive extension $D \supseteq V^a$ so that $\mathbf{a} = \mathbf{a}[D]$.

Consider the first case where $\mathcal{J} \subseteq \mathcal{K}$. It follows that $\mathbf{b}$ has the form $\mathbf{b} = \mathbf{a}[D]|_{\mathcal{J}}$ so that

$$D^* K_0^a \mathcal{J} \subseteq \bigvee_{z \in \mathbb{B}^d} z^* K_z^a \mathcal{K} = \operatorname{Ker}(V^a)^\perp. \qquad \text{(Recall } b(0) = 0 = a(0).\text{)}$$

Since $D^* = (V^a)^* + C^*$ where $C^* : \operatorname{Ran}(V^a)^\perp \to \operatorname{Ker}(V^a)$ (by Lemma 3.8), it follows that

$$\begin{aligned} D^* K_0^a \mathcal{J} &= P_{\operatorname{Ker}(V^a)}^\perp D^* K_0^a \mathcal{J} \\ &= (V^a)^* K_0^a \mathcal{J}, \end{aligned}$$

is contained in $\operatorname{Ker}(V^a)^\perp$. Since we assume X and hence $\mathbf{b}$ are extremal,

$$0 = P_{\mathcal{J}} (K_0^a)^* (I - V^a (V^a)^*) K_0^a P_{\mathcal{J}},$$

and it follows that

$$K_0^a \mathcal{J} \subseteq \operatorname{Ran}(V^a),$$

so that b is quasi-extreme by Theorem 6.2.

In the second case where $\mathcal{K} \subseteq \mathcal{J}$, we have that $\mathbf{a} := \mathbf{b} \oplus \mathbf{0}$ is a contractive (and extremal) Gleason solution for $\mathscr{H}(a)$ so that there is a $D \supseteq V^a$ such that

$\mathbf{a} = \mathbf{a}[D]$. As before

$$\mathbf{a}[D]\mathcal{J} \subseteq \bigvee_{z\in\mathbb{B}^d} z^* k_z^a \mathcal{K}, \quad \text{and} \quad D^* K_0^a (I - b(0))\mathcal{J} \subseteq \mathrm{Ker}(V^a)^\perp.$$

Again, the same argument as above implies that b is QE. □

6.7. de Branges–Rovnyak model for quasi-extreme row contractions

Definition 6.8. A CCNC row contraction $T : \mathcal{H} \otimes \mathbb{C}^d \to \mathcal{H}$ with isometric-pure decomposition $T = V - C$ is said to be *quasi-extreme* (QE) if its characteristic function coincides weakly with a QE Schur multiplier.

We obtain a refined model for QE row contractions:

Theorem 6.9. *A row contraction $T : \mathcal{H} \otimes \mathbb{C}^d \to \mathcal{H}$ is QE if and only if T is unitarily equivalent to the (unique) contractive and extremal Gleason solution X in a multi-variable de Branges–Rovnyak space $\mathscr{H}(b)$ for a quasi-extreme and purely contractive Schur multiplier b.*

In particular, any QE row contraction, T, is unitarily equivalent to X^{b_T} where b_T is any characteristic function for T. The characteristic function, b_T, of T, is a complete unitary invariant: Any two QE row contractions T_1, T_2 are unitarily equivalent if and only if their characteristic functions coincide weakly.

Proof. This follows from Theorem 5.14 under the added assumption that the characteristic function of T is quasi-extreme. For the final statement simply note that if b_1, b_2 are quasi-extreme Schur functions that coincide weakly so that $\mathscr{H}(b_1) = \mathscr{H}(Ub_2)$ for some unitary U, it is easy to see that X^{b_1} is unitarily equivalent to X^{b_2} (via a constant unitary multiplier), where X^{b_1}, X^{b_2} are the unique, contractive, and extremal Gleason solutions for $\mathscr{H}(b_1)$ and $\mathscr{H}(b_2)$, respectively. □

We will conclude with an abstract characterization of the class of QE row contractions:

Theorem 6.10. *A row contraction $T : \mathcal{H} \otimes \mathbb{C}^d \to \mathcal{H}$ is QE if and only if*

$$\bigvee_{z\in\mathbb{B}^d} (I - Tz^*)^{-1}\mathrm{Ran}\,(D_{T^*}) = \mathcal{H}; \qquad T \textit{ is CCNC},$$

and

$$\mathrm{Ker}(T)^\perp \subseteq \bigvee_{z\in\mathbb{B}^d} z^*(I - Tz^*)^{-1}\mathrm{Ran}\,(D_{T^*}). \qquad (T \textit{ obeys the QE condition.})$$

Proof. Let T be a QE row contraction on $\mathcal{H}$. By Theorem 6.9, T is unitarily equivalent to the unique contractive and extremal Gleason solution, X^T for $\mathscr{H}(b_T)$. We can assume that $b_T = b_T|_{\mathrm{supp}(b_T)}$ so that b_T is QE by Theorem 6.9. By Theorem 6.6,

$$\mathrm{Ker}(X^T)^\perp \subseteq \bigvee_{z\in\mathbb{B}^d} z^*(I - X^T z^*)^{-1}\mathrm{Ran}\left(D_{(X^T)^*}\right),$$

and it follows that $T \simeq X^T$ also obeys the QE condition.

Conversely suppose that T is CCNC and T obeys the QE condition. Then $T \simeq X^T$, an extremal Gleason solution in $\mathscr{H}(b_T)$. Again we can assume that $b_T = b_T|_{\text{supp}(b_T)}$, and since T obeys the QE condition, so does X^T. Theorem 6.6 implies that b_T is quasi-extreme so that T is QE. □

Proposition 6.11. *If T is a QE row contraction on $\mathcal{H}$ with isometric-pure decomposition $T = V - C$, then its partial isometric part, V, is a QE row partial isometry.*

Proof. Since $\text{Ker}(T) \subseteq \text{Ker}(V)$ and $T \supseteq V$ is a QE contractive extension,

$$\text{Ker}(V)^\perp \subseteq \text{Ker}(T)^\perp \subseteq \bigvee_{z \in \mathbb{B}^d} z^*(I - Tz^*)^{-1}\text{Ran}\,(D_{T^*})$$

$$= \bigvee z^*(I - Vz^*)^{-1}\text{Ran}\,(D_{V^*}),$$

and V also obeys the QE condition. □

On the other hand,

Proposition 6.12. *Let V be a QE row partial isometry with model triple $(\gamma, \mathcal{J}, \mathcal{K})$. If $\delta \in \mathscr{L}(\mathcal{J}, \mathcal{K})$ is any pure contraction with $\text{Ker}(\delta)^\perp \subseteq \text{supp}(b_V^\gamma)$, then $T_\delta = V - \gamma(0)\delta\gamma(\infty)^*$ is a QE row contractive extension of V.*

Lemma 6.13. *Let $b \in \mathscr{S}_d(\mathcal{J}, \mathcal{K})$, $b(0) = 0$ be a Schur multiplier and $\delta \in [\mathscr{L}(\mathcal{J}, \mathcal{K})]_1$ be any pure contraction obeying $\text{Ker}(\delta)^\perp \subseteq \text{supp}(b)$. If $b' := b|_{\text{supp}(b)}$ and $\delta' := \delta|_{\text{supp}(b)}$, then $\text{supp}(b) = \text{supp}(b^{\langle\delta\rangle})$ and $b^{\langle\delta\rangle}|_{\text{supp}(b)} = (b')^{\langle\delta'\rangle}$.*

Proof. Let $\mathcal{H} := \text{supp}(b)$. Writing elements of $\mathcal{J} = \mathcal{H} \oplus (\mathcal{J} \ominus \mathcal{H})$ as two-component column vectors, let $\alpha := \delta|_{\mathcal{H}}$ and $a := b|_{\mathcal{H}}$. Recall that $b^\delta(z) = D_{\delta^*}^{-1}(b(z) + \delta)(I_{\mathcal{J}} + \delta^* b(z))^{-1} D_\delta$. Writing

$$\delta = \begin{bmatrix} \alpha, & 0_{\mathcal{J}\ominus\mathcal{H},\mathcal{K}} \end{bmatrix}, \quad \text{and} \quad b = \begin{bmatrix} a, & 0_{\mathcal{J}\ominus\mathcal{H},\mathcal{K}} \end{bmatrix},$$

it is easy to check that

$$b^{\langle\delta\rangle}(z) P_{\mathcal{J}\ominus\mathcal{H}} = D_{\delta^*}^{-1} \begin{bmatrix} a(z) + \alpha, & 0_{\mathcal{J}\ominus\mathcal{H},\mathcal{K}} \end{bmatrix} \begin{bmatrix} 0 & 0 \\ 0 & I_{\mathcal{J}\ominus\mathcal{H}} \end{bmatrix} = 0,$$

proving that $\text{supp}(b^{\langle\delta\rangle}) \subseteq \mathcal{H} = \text{supp}(b)$. The remaining assertions are similarly easy to verify. □

Proof. (of Proposition 6.12) If V is a QE row partial isometry, and $(\gamma, \mathcal{J}, \mathcal{K})$ is any model triple for V, then $b := (b_V^\gamma)|_{\text{supp}(b_V^\gamma)}$ is quasi-extreme. Given any pure contraction $\delta \in [\mathscr{L}(\mathcal{J}, \mathcal{K})]_1$, we can define, as in Lemma 5.10, the CCNC row contraction $T = T_\delta := V - \gamma(0)\delta\gamma(\infty)^*$, which, by definition, has the characteristic function $b_T^\gamma = (b_V^\gamma)^{\langle\delta\rangle}$. Under the assumption that $\text{Ker}(\delta)^\perp \subseteq \text{supp}(b_V^\gamma)$, the above lemma proves that b_T^γ coincides weakly with a Frostman shift of the quasi-extreme Schur-class function b, so that T is also QE. □

If, however, $\text{Ker}(\delta)^\perp$ is not contained in $\text{supp}(b_V^\gamma)$, T_δ can fail to be QE. That is, as the following simple example shows, there exist CCNC row contractions T with partial isometric part V such that V is QE but T is not.

Example 6.14. Let $b \in \mathscr{S}_d(\mathcal{H})$ be any purely contractive quasi-extreme multiplier and set

$$B := \begin{bmatrix} b & 0 \\ 0 & 0 \end{bmatrix} \in \mathscr{S}_d(\mathcal{H} \oplus \mathbb{C}), \qquad \delta := \begin{bmatrix} 0 & 0 \\ 0 & r \end{bmatrix}; \quad 0 < r < 1.$$

Then,

$$\begin{aligned} B^{\langle \delta \rangle}(z) &= D_{\delta^*}^{-1}(B(z) + \delta)(I + \delta^* B(z))^{-1} D_\delta \\ &= \begin{bmatrix} b(z) & 0 \\ 0 & r \end{bmatrix}, \end{aligned}$$

which cannot be quasi-extreme since $0 < r < 1$.

7. Outlook

Motivated by the characterization of CNC row contractions in Section 3, given any CNC row partial isometry, V, on $\mathcal{H}$, it is natural to extend our definition of model triple and model map to the non-commutative setting of non-commutative function theory [12, 34].

Namely, recall that the non-commutative (NC) open unit ball is the disjoint union $\mathbb{B}^d_{\mathbb{N}} := \coprod_{n=1}^\infty \mathbb{B}^d_n$, where

$$\mathbb{B}^d_n := \left(\mathbb{C}^{n \times n} \otimes \mathbb{C}_d\right)_1,$$

is viewed as the set of all strict row contractions (with d component operators) on $\mathbb{C}^n$, and $\mathbb{B}^d_1 \simeq \mathbb{B}^d$. A natural extension of our concept of model triple to the NC unit ball, $\mathbb{B}^d_{\mathbb{N}}$, would be a triple $(\gamma, \mathcal{J}_\infty, \mathcal{J}_0)$ consisting of two Hilbert spaces $\mathcal{J}_\infty \simeq \mathrm{Ker}(V)$, $\mathcal{J}_0 \simeq \mathrm{Ran}\,(V)^\perp$, and a map γ on $\mathbb{B}^d_{\mathbb{N}} \cup \{\infty\}$,

$$\gamma : \begin{cases} Z \in \mathbb{B}^d_n \mapsto \gamma(Z) \in \mathscr{L}(\mathcal{J}_0 \otimes \mathbb{C}^n, \mathcal{R}\,(V - Z)^\perp) \\ \{\infty\} \quad \mapsto \qquad \gamma(\infty) \in \mathscr{L}(\mathcal{J}_\infty, \mathrm{Ker}(V)) \end{cases},$$

where $\gamma(Z)$ is an isomorphism for each $Z \in \mathbb{B}^d_{\mathbb{N}}$ and $\gamma(0_n), \gamma(\infty)$ are onto isometries. We will call such a model map γ a non-commutative (NC) model map. In particular, as in Section 4, if $T \supseteq V$ is any contractive extension of V, and $\Gamma_T(0) : \mathcal{J}_0 \to \mathrm{Ran}\,(V)^\perp$, $\Gamma_T(\infty) : \mathcal{J}_\infty \to \mathrm{Ker}(V)$ are any fixed onto isometries, then

$$\Gamma_T(Z) := (I - TZ^*)^{-1}(\Gamma_T(0) \otimes I_n); \quad Z \in \mathbb{B}^d_n, \quad TZ^* := T_1 \otimes Z_1^* + \cdots + T_d \otimes Z_d^*,$$

defines an analytic NC model map for V (we expect $\Gamma_T(Z)$ will be anti-analytic in the sense of non-commutative function theory [34, Chapter 7]). Moreover, as in Section 4, for any analytic NC model map Γ, we expect that one can then define an abstract model space, $\hat{\mathcal{H}}^\Gamma$ with non-commutative reproducing kernel

$$\hat{K}^\Gamma(Z, W) = \Gamma(Z)^* \Gamma(W); \qquad Z, W \in \mathbb{B}^d_n,$$

and that this will be a non-commutative reproducing kernel Hilbert space (NC-RKHS) in the sense of [12, 14]. If this analogy continues to hold, it would be natural to use Γ to define a NC *characteristic function*, $B_T(Z)$, on $\mathbb{B}^d_{\mathbb{N}}$, and one would expect this to be an element of the free (left or right) Schur class of contractive NC multipliers between vector-valued Fock spaces over $\mathbb{C}^d$ [8, 32]. Ultimately, it would be interesting to investigate whether such an extended theory will yield an alternate approach to the NC de Branges–Rovnyak model for CNC row contractions as (adjoints of) the restriction of the adjoint of the left or right free shift on (vector-valued) full Fock space over $\mathbb{C}^d$ to the right or left non-commutative de Branges–Rovnyak spaces, $\mathscr{H}^L(B_T)$ or $\mathscr{H}^R(B_T)$ [8, 12, 14].

References

[1] N.I. Akhiezer and I.M. Glazman. *Theory of linear operators in Hilbert space.* Dover Publications, 1993.

[2] A. Aleman, R.T.W. Martin, and W.T. Ross. On a theorem of Livsic. *J. Funct. Anal.*, 264:999–1048, 2013.

[3] D. Alpay and H.T. Kaptanoğlu. Gleason's problem and homogeneous interpolation in Hardy and Dirichlet-type spaces of the ball. *J. Math. Anal. Appl.*, 276:654–672, 2002.

[4] W.B. Arveson. Subalgebras of C*-algebras III. *Acta Math.*, 181:159–228, 1998.

[5] W.B. Arveson. The curvature invariant of a Hilbert module over $\mathbb{C}[z_1, \dots, z_d]$. *J. Reine Angew. Math.*, 522:173–236, 2000.

[6] J.A. Ball and V. Bolotnikov. Canonical de Branges–Rovnyak model transfer-function realization for multivariable Schur-class functions. In *CRM Proceedings and Lecture Notes*, volume 51, pages 1–40, 2010.

[7] J.A. Ball and V. Bolotnikov. Canonical transfer-function realization for Schur multipliers on the Drury–Arveson space and models for commuting row contractions. *Indiana Univ. Math. J.*, 61:665–716, 2011.

[8] J.A. Ball, V. Bolotnikov, and Q. Fang. Schur-class multipliers on the Fock space: de Branges–Rovnyak reproducing kernel spaces and transfer-function realizations. In *Operator Theory, Structured Matrices, and Dilations, Theta Series Adv. Math., Tiberiu Constantinescu Memorial Volume*, volume 7, pages 101–130. 2007.

[9] J.A. Ball, V. Bolotnikov, and Q. Fang. Transfer-function realization for multipliers of the Arveson space. *J. Math. Anal. Appl.*, 333:68–92, 2007.

[10] J.A. Ball, V. Bolotnikov, and Q. Fang. Schur-class multipliers on the Arveson space: de Branges–Rovnyak reproducing kernel spaces and commutative transfer-function realizations. *J. Math. Anal. Appl.*, 341:519–539, 2008.

[11] J.A. Ball and T.L. Kriete. Operator-valued Nevanlinna–Pick kernels and the functional models for contraction operators. *Integral Equations Operator Theory*, 10:17–61, 1987.

[12] J.A. Ball, G. Marx, and V. Vinnikov. Noncommutative reproducing kernel Hilbert spaces. *J. Funct. Anal.*, 271:1844–1920, 2016.

[13] J.A. Ball, T.T. Trent, and V. Vinnikov. Interpolation and commutant lifting for multipliers on reproducing kernel Hilbert spaces. In *Operator Theory and Analysis, Oper. Theory Adv. Appl.*, volume 122, pages 89–138. Springer, 2001.

[14] J.A. Ball and V. Vinnikov. Formal reproducing kernel Hilbert spaces: the commutative and noncommutative settings. In *Reproducing kernel spaces and applications, Oper. Theory Adv. Appl.*, volume 143, pages 77–134. Birkhäuser Basel, 2003.

[15] J.A. Ball and V. Vinnikov. *Lax–Phillips scattering and conservative linear systems: A Cuntz-algebra multidimensional setting.* Number 837 in Mem. Amer. Math. Soc. 2005.

[16] T. Bhattacharyya, J. Eschmeier, and J. Sarkar. Characteristic function of a pure commuting contractive tuple. *Integral Equations Operator Theory*, 53:23–32, 2005.

[17] T. Bhattacharyya, J. Eschmeier, and J. Sarkar. On c.n.c. commuting contractive tuples. *Proc. Indian Acad. Sci. Math. Sci.*, 116:299–316, 2006.

[18] S. Boyd and L. Vandenberghe. *Convex optimization.* Cambridge University Press, 2004.

[19] R. Crofoot. Multipliers between invariant subspaces of the backward shift. *Pacific J. Math.*, 166:225–246, 1994.

[20] L. de Branges and J. Rovnyak. Appendix on square summable power series: Canonical models in quantum scattering theory. In *Perturbation theory and its applications in quantum mechanics*, pages 295–392. Wiley, 1966.

[21] L. deBranges and J. Rovnyak. *Square summable power series.* Courier Corporation, 2015.

[22] O. Frostman. *Potentiel d'équilibre et capacité des ensembles avec quelques applications à la théorie des fonctions.* PhD thesis, Lund University, 1935.

[23] O. Frostman. Sur les produits de Blaschke. *Kungl. Fysiografiska Sällskapets i Lund Förhandlingar* [*Proc. Roy. Physiog. Soc. Lund*], 12:169–182, 1942.

[24] S.R. Garcia, R.T.W. Martin, and W.T. Ross. Partial orders on partial isometries. *J. Operator Theory*, 75:409–442, 2016.

[25] S.R. Garcia, J. Mashreghi, and W.T. Ross. *Introduction to model spaces and their operators*, volume 148. Cambridge University Press, 2016.

[26] A.M. Gleason. Finitely generated ideals in Banach algebras. *J. Math. Mech.*, 13:125, 1964.

[27] M.I. Gorbachuk and V.I. Gorbachuk, editors. *M.G. Kreĭn's lectures on entire operators*, volume 97 of *Oper. Theory Adv. Appl.* Birkhäuser, 2012.

[28] D.C.V. Greene, S. Richter, and C. Sundberg. The structure of inner multipliers on spaces with complete Nevanlinna–Pick kernels. *J. Funct. Anal.*, 194:311–331, 2002.

[29] K. Hoffman. *Banach spaces of analytic functions.* Courier Corporation, 2007.

[30] M.T. Jury. Clark theory in the Drury–Arveson space. *J. Funct. Anal.*, 266:3855–3893, 2014.

[31] M.T. Jury and R.T.W. Martin. Aleksandrov–Clark theory for Drury–Arveson space. *Integral Equations Operator Theory*, Vol. 90, 2018.

[32] M.T. Jury and R.T.W. Martin. Noncommutative Clark measures for the free and abelian Toeplitz algebras. *J. Math. Anal. Appl.*, 456:1062–1100, 2017.

[33] M.T. Jury and R.T.W. Martin. Extremal multipliers of the Drury–Arveson space. *Proc. Amer. Math. Soc.*, 146:4293–4306, 2018.

[34] D.S. Kaliuzhnyi-Verbovetskyi and V. Vinnikov. *Foundations of free noncommutative function theory.* American Mathematical Society, 2014.

[35] M.G. Kreĭn. On Hermitian operators with deficiency indices one. *Dokl. Akad. Nauk SSSR*, 43:339–342, 1944.

[36] M.G. Kreĭn. On one remarkable class of Hermitian operators. *Dokl. Akad. Nauk SSSR*, 44:191–195, 1944.

[37] M.G. Kreĭn. The principal aspects of the theory of representation of Hermitian operators whose deficiency index is (m, m). *Ukrain. Mat. Zh.*, 2:3–66, 1949.

[38] M.S. Livšic. A class of linear operators on Hilbert space. *Mat. Sbornik*, 19:239–260, 1946.

[39] M.S. Livšic. On the theory of isometric operators with equal deficiency indices. *Dokl. Akad. Nauk SSSR (NS)*, 58:13–15, 1947.

[40] R.T.W. Martin. Extensions of symmetric operators I: The inner characteristic function case. *Concr. Oper.*, 2:53–97, 2015.

[41] S. McCullough and T.T. Trent. Invariant subspaces and Nevanlinna–Pick kernels. *J. Funct. Anal.*, 178:226–249, 2000.

[42] N.K. Nikolskii. *Treatise on the shift operator: spectral function theory.* Springer, 2012.

[43] N.K. Nikolskii and V.I. Vasyunin. Notes on two function models in the Bieberbach conjecture. In *The Bieberbach conjecture: Proceedings of the symposium on the occasion of the proof, Math. Surveys*, volume 21, pages 113–141. Amer. Math. Soc., 1986.

[44] V. Paulsen and M. Raghupathi. *An Introduction to the theory of reproducing kernel Hilbert spaces.* Cambridge Studies in Advanced Mathematics, 2016.

[45] G. Popescu. Characteristic functions for infinite sequences of noncommuting operators. *J. Oper. Theory*, 22:51–71, 1989.

[46] G. Popescu. Models for infinite sequences of noncommuting operators. *Acta Sci. Math.*, 53:355–368, 1989.

[47] A. Ramanantoanina. Gleason solutions and canonical models for row contractions. Master's thesis, University of Cape Town, 2017.

[48] D. Sarason. *Sub-Hardy Hilbert spaces in the unit disk.* John Wiley & Sons Inc., 1994.

[49] O. Shalit. Operator theory and function theory in Drury–Arveson space and its quotients. In *Handbook of Operator Theory*, pages 1125–1180. Springer, 2015.

[50] B. Sz.-Nagy and C. Foiaş. *Harmonic analysis of operators on Hilbert space.* Elsevier, 1970.

Robert T.W. Martin and Andriamanankasina Ramanantoanina
Department of Mathematics and Applied Mathematics
University of Cape Town, South Africa
e-mail: `rtwmartin@gmail.com`
`andriamanak@gmail.com`

Operator Theory:
Advances and Applications, Vol. 272, 307–328

On the Augmented Biot-JKD Equations with Pole-Residue Representation of the Dynamic Tortuosity

Miao-Jung Yvonne Ou and Hugo J. Woerdeman

Dedicated to our colleague and friend Joe Ball

Abstract. In this paper, we derive the augmented Biot-JKD equations, where the memory terms in the original Biot-JKD equations are dealt with by introducing auxiliary dependent variables. The evolution in time of these new variables are governed by ordinary differential equations whose coefficients can be rigorously computed from the JKD dynamic tortuosity function $T^D(\omega)$ by utilizing its Stieltjes function representation derived in [19], where an approach for computing the pole-residue representation of the JKD tortuosity is also proposed. The two numerical schemes presented in the current work for computing the poles and residues representation of $T^D(\omega)$ improve the previous scheme in the sense that they interpolate the function at infinite frequency and have much higher accuracy than the one proposed in [19].

Mathematics Subject Classification (2010). 35R30, 76S05, 41A21, 30E10.

Keywords. Biot-JKD equations, tortuosity, Stieltjes function, pole-residue approximation, two-sided residue interpolation.

1. Introduction

Poroelastic composites are two-phase composite materials consisting of elastic solid frames with fluid-saturated pore space. The study of poroelasticity plays an important role in biomechanics, seismology and geophysics due to the nature of objects of research in these fields, e.g., fluid saturated rocks, sea ice and cancellous bone. It is of great interest to model wave propagation in these materials. When the wave length is much larger than the scale of the microstructure of the composite,

---

The work of MYO is partially sponsored by NSF-DMS-1413039.
HJW is partially supported by Simons Foundation grant 355645.

homogenization theory can be applied to obtain the effective wave equations, in which the fluid and the solid coexist at every point in the region occupied by the poroelastic material. M.A. Biot derived the governing equations for wave propagation in linear poroelastic composite materials in [4] and [5]. The paper [4] deals with the low-frequency regime where the friction between the viscous pore fluid and the elastic solid can be assumed to be linear proportional with the difference between the effective pore fluid velocity and the effective solid velocity by a real number b, which is independent of frequency ω; this set of equations is referred to as the low-frequency Biot equation. When the frequency is higher than the critical frequency of the poroelastic material, b will be frequency-dependent; this is the subject of study in [5]. The exact form of b as a function of frequency was derived in [5] for pore space with its micro-geometry being circular tubes. A more general expression was derived in the seminal paper [14] by Johnson, Koplik and Dashen (JKD), where a causality argument was applied to derive the 'simplest' form of b as a function of frequency. This frequency dependence of b results in a time-convolution term in the time-domain poroelastic wave equations; the Laplace transform of the kernel in the time-convolution term is called the 'dynamic tortuosity' in the literature. The Biot-JKD equations refer to the Biot equations with b replaced by the JKD-tortuosity in equation (1) below.

In general, the dynamic tortuosity function $\mathbf{T}$ is a tensor, which is related to the symmetric, positive definite dynamic permeability tensor of the poroelastic material $\mathbf{K}(\omega)$ by $\mathbf{T}(\omega) = \frac{i\eta\phi}{\omega\rho_f}\mathbf{K}^{-1}(\omega)$. By the definition of dynamic tortuosity and dynamic permeability it is clear that their principal directions coincide. In the principal direction x_j, $j = 1, \ldots, 3$, of $\mathbf{K}$, the JKD tortuosity is

$$T_j^J(\omega) = \alpha_{\infty j}\left(1 - \frac{\eta\phi}{i\omega\alpha_{\infty j}\rho_f K_{0j}}\sqrt{1 - i\frac{4\alpha_{\infty j}^2 K_{0j}^2 \rho_f \omega}{\eta\Lambda_j^2\phi^2}}\right), \quad j = 1, 2, 3, \tag{1}$$

with the tunable geometry-dependent constant Λ_j, the dynamic viscosity of pore fluid $\eta = \rho_f \nu$, the porosity ϕ, the fluid density ρ_f, the static permeability K_{0j} and the infinite-frequency-tortuosity $\alpha_{\infty j}$; all of these parameters are positive real numbers. We refer to $T_j^J(\omega)$ as the JKD tortuosity function in the jth direction.

The time-domain low-frequency Biot's equations have been numerically solved by many authors. However, for high-frequency Biot equations such as Biot-JKD, the time convolution term remains a challenge for numerical simulation. Masson and Pride [16] defined a time convolution product to discretize the fractional derivative. Lu and Hanyga [15] developed a new method to calculate the shifted fractional derivative without storing and integrating the entire velocity histories. In recent years Blanc, Chiavassa and Lombard [7, 8] used an optimization procedure to approach the fractional derivative.

In this work, we will derive an equivalent system of the Biot-JKD equations without resorting to the fractional derivative technique. The advantage of this approach is that the new system of equations have the same structure as the low-frequency Biot equations but with more variables. Hence we refer to this system as the augmented Biot-JKD equations. A key step in this derivation is to utilize the Stieltjes function structure to compute from the given JKD tortuosity the coefficients of the additional terms.

The first-order formulation of the time-domain Biot equations consists of the strain-stress relations of the poroelastic materials and the equation of motions. The solid displacement $\mathbf{u}$, the pore fluid velocity relative to the solid $\boldsymbol{q}$ and the pore pressure p are the unknowns to be solved. In terms of the solid displacement $\mathbf{u}$, we define the following variables

$$\begin{aligned}
&\mathbf{v} := \partial_t \mathbf{u} \ \ \text{(solid velocity)},\\
&\mathbf{w} := \phi(\boldsymbol{U} - \mathbf{u}) \ \text{(fluid displacement relative to the solid)},\\
&\boldsymbol{q} := \partial_t \mathbf{w}, \ \zeta := -\nabla \cdot \mathbf{w},
\end{aligned}$$

where ϕ is the porosity. Here $\boldsymbol{U}$ is the averaged fluid velocity over a representative volume element. The spatial coordinates (x_1, x_2, x_3) are chosen to be aligned with the principal directions of the static permeability tensor $\boldsymbol{K}_0$ of the poroelastic material, which is known to be symmetric and positive definite.

When we let $\epsilon_{ij} := \frac{1}{2}(\frac{\partial u_i}{\partial x_j} + \frac{\partial u_j}{\partial x_i})$ be the linear strain of the solid part, then the stress-strain relation is given by [6]

$$\begin{bmatrix} \sigma_{11} \\ \sigma_{22} \\ \sigma_{33} \\ \sigma_{23} \\ \sigma_{13} \\ \sigma_{12} \\ p \end{bmatrix} = \begin{bmatrix} c^u_{11} & c^u_{12} & c^u_{13} & c^u_{14} & c^u_{15} & c^u_{16} & Ma_1 \\ c^u_{12} & c^u_{22} & c^u_{23} & c^u_{24} & c^u_{25} & c^u_{26} & Ma_2 \\ c^u_{13} & c^u_{23} & c^u_{33} & c^u_{34} & c^u_{35} & c^u_{36} & Ma_3 \\ c^u_{14} & c^u_{24} & c^u_{34} & c^u_{44} & c^u_{45} & c^u_{46} & Ma_4 \\ c^u_{15} & c^u_{25} & c^u_{35} & c^u_{45} & c^u_{55} & c^u_{56} & Ma_5 \\ c^u_{16} & c^u_{26} & c^u_{36} & c^u_{46} & c^u_{56} & c^u_{66} & Ma_6 \\ Ma_1 & Ma_2 & Ma_3 & Ma_4 & Ma_5 & Ma_6 & M \end{bmatrix} \begin{bmatrix} \epsilon_{11} \\ \epsilon_{22} \\ \epsilon_{33} \\ 2\epsilon_{23} \\ 2\epsilon_{13} \\ 2\epsilon_{12} \\ -\zeta \end{bmatrix},$$

where p is the pore pressure, c^u_{ij} are the elastic constants of the undrained frame, which are related to the elastic constants c_{ij} of the drained frame by $c^u_{ij} = c_{ij} + Ma_i a_j$, $i, j = 1, \ldots, 6$. In terms of the material bulk moduli κ_s and κ_f of the solid and the fluid, respectively, the fluid-solid coupling constants a_i and M are given by

$$\begin{aligned}
a_i &:= \begin{cases} 1 - \frac{1}{3\kappa_s}\sum_{k=1}^{3} c_{ik} & \text{for } i = 1, 2, 3, \\ -\frac{1}{3\kappa_s}\sum_{k=1}^{3} c_{ki} & \text{for } i = 4, 5, 6, \end{cases}\\
M &:= \frac{\kappa_s}{1 - \overline{\kappa}/\kappa_s - \phi(1 - \kappa_s/\kappa_f)},\\
\overline{\kappa} &:= \frac{c_{11} + c_{22} + c_{33} + 2c_{12} + 2c_{13} + 2c_{23}}{9}.
\end{aligned}$$

The six equations of motion are as follows

$$\sum_{k=1}^{3} \frac{\partial \sigma_{jk}}{\partial x_k} = \rho \frac{\partial v_j}{\partial t} + \rho_f \frac{\partial q_j}{\partial t}, \quad t > 0, \ j = 1, 2, 3, \tag{2}$$

$$-\frac{\partial p}{\partial x_j} = \rho_f \frac{\partial v_j}{\partial t} + \left(\frac{\rho_f}{\phi}\right) \check{\alpha}_j \star \frac{\partial q_j}{\partial t}, \quad t > 0, \ j = 1, 2, 3, \tag{3}$$

where $\star$ denotes the time-convolution operator, ρ_f and ρ_s are the density of the pore fluid and of the solid, respectively, $\rho := \rho_s(1-\phi) + \phi\rho_f$ and $\check{\alpha}_j$ is the inverse Laplace transform of the dynamic tortuosity $\alpha_j(\omega)$ with ω being the frequency. Here the Laplace transform of a function $f(t)$ is defined as

$$\hat{f}(\omega) := \mathcal{L}[f](s = -i\omega) := \int_0^\infty f(t)e^{-st}dt.$$

As said, we will use the notation $\check{g}(t) = \mathcal{L}^{-1}[g](t)$ for the inverse Laplace transform of a function $g(s)$. For instance,

$$\mathcal{L}^{-1}\left[\frac{1}{s-p}\right](t) = e^{pt}, \quad t > 0, \tag{4}$$

a basic fact we will use later; the converse relation $\mathcal{L}[e^{pt}](s) = \frac{1}{s-p}, \operatorname{Re} s > p$, is easily derived.

As a special case of the Biot-JKD equations, the low frequency Biot equation corresponds to

$$\hat{\alpha_j}(t) = \alpha_{\infty j}\delta(t) + \frac{\eta\phi}{K_{0j}\rho_f}H(t),$$

where $\delta(t)$ is the Dirac function, $H(t)$ the Heaviside function, η the dynamic viscosity of the pore fluid, and K_{0j} the static permeability in the x_j direction. This low-frequency tortuosity function corresponds to

$$\alpha_j(\omega) = \alpha_{\infty j} + \frac{\eta\phi/K_{0j}\rho_f}{-i\omega}.$$

In the Biot-JKD equation, we have $\alpha_j(\omega) = T_j^J(\omega)$.

According to Theorem 5.1 in [19], in the principal coordinates $\{x_j\}_{j=1}^3$ of the permeability tensor $\mathbf{K}$ and for ω such that $-\frac{i}{\omega} \in \mathbb{C} \setminus [0, \theta_1]$, the JKD dynamic tortuosity function has the following integral representation formula (IRF)

$$T_j^J(\omega) = a_j\left(\frac{i}{\omega}\right) + \int_0^{\theta_1} \frac{d\sigma_j(t)}{1 - i\omega t}, \quad a_j := \frac{\eta\phi}{\rho_f K_{0j}}, \quad j = 1, 2, 3, \tag{5}$$

where $0 < \theta_1 < \infty$ and the positive measure $d\sigma_j$ has a Dirac measure of strength $\alpha_{\infty j}$ sitting at $t = 0$; this is to take into account the asymptotic behavior of dynamic torturosity as frequency goes to ∞. This function is the analytic continuation of the usual dynamic tortuosity function in which $\omega \geq 0$. As a function of the new variable $s := -i\omega$, $\omega \in \mathbb{C}$, the singularities of (5) are included in the interval $(-\infty, -\frac{1}{\theta_1})$ and a simple pole sitting at $s = 0$. Therefore, if we define a new function for each $j = 1, 2, 3$,

$$D_j^J(s) := T_j^J(is) - \frac{a_j}{s} = \int_0^{\theta_1} \frac{d\sigma_j(t)}{1 + st},$$

then $D_j^J(s)$ is analytic in $\mathbb{C} \setminus (-\infty, -\frac{1}{\theta_1})$ on the s-plane. This type of functions are closely related to the well-known Stieltjes functions. The first approach we propose in this paper is based on the fact [12] that a Stieltjes function can be well approximated by its Padé approximant whose poles are all simple. The other approach proposed here for computing the pole-residue approximation of the dynamic tortuosity function is based on the result in [1].

We note that

$$D_j^J(s) = \alpha_{\infty j}\left(1 + \frac{\eta\phi}{s\alpha_{\infty j}\rho_f K_{0j}}\sqrt{1 + s\frac{4\alpha_{\infty j}^2 K_{0j}^2 \rho_f}{\eta\Lambda_j^2\phi^2}}\right) - \frac{a_j}{s} =: \alpha^J(s) - \frac{a_j}{s},$$

$j = 1, 2, 3$, is analytic away from the **branch cut** on $[0, C_1]$ along the real axis, where $C_1 := \frac{4\alpha_{\infty j}^2 K_{0j}^2}{\nu\phi^2\Lambda_j^2}$. Therefore,

$$D_j^J(s) = \int_0^{C_1} \frac{d\sigma_j^J(t)}{1+st} \approx \alpha_\infty + \sum_{k=1}^{M} \frac{r_k^j}{s - p_k^j} \text{ for } s \in \mathbb{C} \setminus (-\infty, -\frac{1}{C_1}], \quad (6)$$

with $r_k^j > 0$, $p_k^j < -\frac{1}{C_1} < 0$, $j = 1, 2, 3$, $k = 1, \ldots, M$, which can be computed from the dynamic permeability data $K_j(\omega)$ evaluated at M different frequencies in the frequency content of the initial waves. The special choice of $s = -i\omega$, $\omega \in \mathbb{R}$ in (6) provides a pole-residue approximation of $T_j^J(\omega)$, $j = 1, 2, 3$.

Applying Laplace transform to the convolution term in (3) with JKD tortuosity, i.e., $\alpha = \alpha^J$, (see, e.g., Theorem 9.2.7 in [10])

$$\begin{aligned}
\mathcal{L}[\check{\alpha}^J \star \frac{\partial q_j}{\partial t}](s) &= \alpha^J(s)(s\hat{q}_j) \\
&= \left(D^J(s) + \frac{a}{s}\right)(s\hat{q}_j) \\
&\approx \left(\alpha_\infty + \sum_{k=1}^{M} \frac{r_k}{s - p_k} + \frac{a}{s}\right)(s\hat{q}_j) \\
&= \alpha_\infty s\hat{q}_j + \sum_{k=1}^{M} \frac{r_k}{s - p_k}(s\hat{q}_j) + a\hat{q}_j \\
&= \alpha_\infty s\hat{q}_j + \left(a + \sum_{k=1}^{M} r_k\right)\hat{q}_j + \sum_{k=1}^{M} r_k p_k \frac{\hat{q}_j}{s - p_k}.
\end{aligned}$$

Notice that

$$s\hat{q}_j = \mathcal{L}\left[\partial_t q^j\right] + q_j(0).$$

Using (4), we have for $t > 0$

$$\begin{aligned}
\left(\check{\alpha}^J \star \frac{\partial q_j}{\partial t}\right)(\boldsymbol{x}, t) &:= \int_0^t \check{\alpha}^J(\tau)\frac{\partial q_j}{\partial t}(\boldsymbol{x}, t - \tau)d\tau \\
&\approx \alpha_\infty\left(\frac{\partial q_j}{\partial t} + \delta(0)q_j(0)\right) + \left(a + \sum_{k=1}^{M} r_k\right)q_j - \sum_{k=1}^{M} r_k(-p_k)e^{p_k t} \star q_j.
\end{aligned}$$

Applying a strategy similar to those in the literature [9], we define the auxiliary variables Θ_k, $k = 1, \ldots, M$, by

$$\Theta_k^{x_j}(\boldsymbol{x}, t) := (-p_k^j) e^{p_k^j t} \star q_j. \tag{7}$$

It can be easily checked that Θ_k, $k = 1, \ldots, M$, satisfies the following equation:

$$\partial_t \Theta_k^{x_j}(\boldsymbol{x}, t) = p_k^j \Theta_k^{x_j}(\boldsymbol{x}, t) - p_k^j q_j(\boldsymbol{x}, t).$$

For an anisotropic media, each principal direction x_j, $j = 1, 2, 3$, has a different tortuosity function α_j. We label the corresponding poles and residue as p_k^j and r_k^j and modify (7) accordingly. Replacing the convolution terms in (3) with the equations of $\Theta_k^{x_j}$, we obtain the following system that has no explicit memory terms:

$$\sum_{k=1}^{3} \frac{\partial \sigma_{jk}}{\partial x_k} = \rho \frac{\partial v_j}{\partial t} + \rho_f \frac{\partial q_j}{\partial t}, \quad t > 0, \tag{8}$$

$$\partial_t \Theta_k^{x_j}(\boldsymbol{x}, t) = p_k^j \Theta_k^{x_j}(\boldsymbol{x}, t) - p_k^j q_j(\boldsymbol{x}, t), \quad j = 1, 2, 3, \tag{9}$$

$$\begin{aligned} -\frac{\partial p}{\partial x_j} &= \rho_f \frac{\partial v_j}{\partial t} + \left(\frac{\rho_f \alpha_{\infty j}}{\phi} \right) \frac{\partial q_j}{\partial t} + \left(\frac{\eta}{K_{0j}} + \frac{\rho_f}{\phi} \sum_{k=1}^{M} r_k^j \right) q_j \\ &\quad - \left(\frac{\rho_f}{\phi} \right) \sum_{k=1}^{M} r_k^j \Theta_k^{x_j} + \delta(t) \frac{\rho_f \alpha_{\infty j}}{\phi} q_j(\boldsymbol{x}, 0), \; t > 0, \quad j = 1, 2, 3. \end{aligned} \tag{10}$$

We refer to this system as the augmented system of Biot-JKD equations in the principal directions of the permeability tensor $\mathbf{K}$.

In Section 2 we will explain two approaches how to compute r_k and p_k from the dynamic permeability data, and in Section 3 we will apply the two approaches on several numerical examples and compare their performance.

2. Numerical scheme for computing r_k and p_k

Since the function D^J results from subtracting the pole of T^J at $s = 0$, it has a removable singularity at $s = 0$ and is analytic away from its branch-cut located at $(-\infty, -1/C_1]$. Both approaches presented here are based on the fact that $D^J(s)$ is a Stieltjes function.

The problem to be solved is formulated as follows. Given the data of D^J at distinct values of $s = s_1, \ldots, s_M$, construct the pole-residue approximation of D^J such that

$$D^J(s) \approx D_{\text{est}}^J(s) := \alpha_\infty + \sum_{k=1}^{M} \frac{r_k}{s - p_k}, \tag{11}$$

where $r_k > 0$ and $p_k < 0$ for $k = 1, \ldots, M$, because of the positivity of the measure σ in the IRF.

2.1. Rational function approximation and partial fraction decomposition

The following approximation takes into account the asymptotic behavior of $\lim_{s\to\infty} D(s) = \alpha_\infty$ and hence can be considered as an improved version of the reconstruction approach for tortuosity in [19], which does not interpolate at infinity. Note that

$$\lim_{\omega\to 0^+} D(s=-i\omega) = \alpha_\infty + 2\left(\frac{\alpha_\infty}{\Lambda}\right)^2 \frac{K_0}{\phi}, \quad \lim_{\omega\to\infty} D(s=-i\omega) = \alpha_\infty.$$

By a theorem in [12], we know that the poles in the Padé approximant of $D^J(s)$ have to be contained in $(-\infty, -1/C_1]$ and are all simple with positive weight (residue). This implies that the constant term in the denominator in the Padé approximant can be normalized to one. According to the aforementioned theorem, if $(s, D^J(s))$ is an interpolation point with $\mathrm{Im}(s) \neq 0$, then $(\overline{s}, D^J(\overline{s}))$ must also be an interpolation point, where $\bar{\cdot}$ represents the complex conjugate. From the integral representation formula (IRF), we know that $D^J(\overline{s_k}) = \overline{D^J(s_k)}$.

Hence, the following approximation problem is considered: Given M data points $D^J(s_k = -i\omega_k) \in \mathbb{C}$, $k = 1, \ldots, M$, find

$$\boldsymbol{x} := (a_0, \ldots, a_{M-1}, b_1, \ldots, b_M)^t$$

such that

$$(S)\begin{cases} D^J(s_k) - \alpha_\infty = \dfrac{a_0 + a_1 s_k + \cdots + a_{M-1} s_k^{M-1}}{1 + b_1 s_k + \cdots + b_M s_k^M}, & k = 1, \ldots, M, \\ \overline{D^J(s_k)} - \alpha_\infty = \dfrac{a_0 + a_1 \overline{s_k} + \cdots + a_{M-1} \overline{s_k}^{M-1}}{1 + b_1 \overline{s_k} + \cdots + b_M \overline{s_k}^M}, & k = 1, \ldots, M, \end{cases}$$

where ω_k, $k = 1, \ldots, M$, are distinct positive numbers. We define

$$A := \begin{bmatrix} 1 & s_1 & s_1^2 & \cdots & s_1^{M-1} & -D^J(s_1)s_1 & -D^J(s_1)s_1^2 & \cdots & -D^J(s_1)s_1^M \\ 1 & s_2 & s_2^2 & \cdots & s_2^{M-1} & -D^J(s_2)s_2 & -D^J(s_2)s_2^2 & \cdots & -D^J(s_2)s_2^M \\ \vdots & \vdots & \vdots & & \vdots & \vdots & \vdots & & \vdots \\ 1 & s_M & s_M^2 & \cdots & s_M^{M-1} & -D^J(s_M)s_M & -D^J(s_M)s_M^2 & \cdots & -D^J(s_M)s_M^M \end{bmatrix},$$

$$\boldsymbol{d} := (D^J(s_1) - \alpha_\infty,\ D^J(s_2) - \alpha_\infty,\ \cdots\ D^J(s_M) - \alpha_\infty)^t \in \mathbb{C}^M,$$

$$\boldsymbol{x} := (a_0, \ldots, a_{M-1}, b_1, \ldots, b_M)^t \in \mathbb{R}^M.$$

Then (S) is equivalent to solving

$$\begin{bmatrix} \mathrm{Re}(A) \\ \mathrm{Im}(A) \end{bmatrix} \boldsymbol{x} = \begin{bmatrix} \mathrm{Re}(\boldsymbol{d}) \\ \mathrm{Im}(\boldsymbol{d}) \end{bmatrix}, \tag{12}$$

where Re() and Im() denote the entrywise real part and the imaginary part, respectively. After solving for $\boldsymbol{x}$, the poles and residues are then obtained by the partial fraction decomposition of the Padé approximant, i.e.,

$$\frac{a_0 + a_1 s + \cdots + a_M s^{M-1}}{1 + b_1 s + \cdots + b_M s^M} = \sum_{j=1}^{M} \frac{r_j}{s - p_j}.$$

2.2. Two-sided residue interpolation in the Stieltjes class

The second approach is based on the following theorem that can be considered as a special case of what is proved in [1]. The advantage of this method is that it explicitly identifies the poles p_k, $k = 1, \dots, M$ as the generalized eigenvalues of matrices constructed from the data. In addition, this approach allows for matrix-valued interpolation, which is useful when the data is given in terms of tensor values. We note that the interpolation problem below also appears in the recent paper [2], where the main focus is model reduction.

Let $\mathbb{C}^+ := \{z \in \mathbb{C} : \mathrm{Im}(z) > 0\}$. Given M interpolation data $(z_i, u_i, v_i) \in \mathbb{C}^+ \times \mathbb{C}^{p\times q} \times \mathbb{C}^{p\times q}$, we seek a $p \times p$ matrix-valued function $F(z)$ of the form

$$F(z) = \int_0^\infty \frac{d\mu(t)}{t-z}, \text{ with } \mu \text{ a positive } p\times p \text{ matrix-valued measure} \tag{13}$$

such that

$$F(z_i)u_i = v_i, \quad i = 1, \dots, M. \tag{14}$$

Theorem 2.1. *If there exists a solution $F(z)$ described as above, then the Hermitian matrices S_1 and S_2 defined via*

$$(S_1)_{ij} = \frac{u_i^* v_j - v_i^* u_j}{z_j - \overline{z_i}}, \quad (S_2)_{ij} := \frac{z_j u_i^* v_j - \overline{z_i} v_i^* u_j}{z_j - \overline{z_i}}, \quad i, j = 1, \dots, M,$$

are positive semidefinite. Conversely, if S_1 is positive definite and S_2 is positive semidefinite, then

$$F(z) := -C_+(zS_1 - S_1A - C_+^* C_-)^{-1}C_+^* = C_+(S_2 - zS_1)^{-1}C_+^*$$

is a solution to the interpolation problem. Here the asterisk denotes the conjugate transpose operator,

$$C_- := \begin{bmatrix} u_1 & \cdots & u_M \end{bmatrix}, C_+ := \begin{bmatrix} v_1 & \cdots & v_M \end{bmatrix}, A := \mathrm{diag}(z_i I_q)_{i=1}^M,$$

and I_q is the identity matrix of dimension q.

Proof. Suppose (13) and (14) are true. Then we have

$$u_i^* v_j - v_i^* u_j = u_i^*(F(z_j) - F(z_i)^*)u_j = (z_j - \overline{z_i})u_i^* \left(\int_0^\infty \frac{d\mu(t)}{(t-z_j)(t-\overline{z_i})} \right) u_j.$$

Thus

$$S_1 = \int_0^\infty \begin{bmatrix} \frac{u_1^*}{t-\overline{z_1}} \\ \vdots \\ \frac{u_M^*}{t-\overline{z_M}} \end{bmatrix} d\mu(t) \begin{bmatrix} \frac{u_1}{t-z_1} & \cdots & \frac{u_M}{t-z_M} \end{bmatrix} \geq 0,$$

$$S_2 = \int_0^\infty \begin{bmatrix} \frac{u_1^*}{t-\overline{z_1}} \\ \vdots \\ \frac{u_M^*}{t-\overline{z_M}} \end{bmatrix} t d\mu(t) \begin{bmatrix} \frac{u_1}{t-z_1} & \cdots & \frac{u_M}{t-z_M} \end{bmatrix} \geq 0.$$

Conversely, suppose $S_1 > 0$ and $S_2 \geq 0$. Notice that

$$A^* S_1 - S_1 A = C_+^* C_- - C_-^* C_+, \tag{15}$$
$$A^* S_2 - S_2 A = A^* C_+^* C_- - C_-^* C_+ A. \tag{16}$$

These equations uniquely determine S_1 and S_2 as the spectra of A and A^* do not overlap. Observe that if S_1 satisfies (15), then $S_2 := S_1 A + C_+^* C_-$ is the solution of (16). Therefore, we have

$$S_2 = S_1 A + C_+^* C_-.$$

Note that $S_2 - zS_1 = S_1^{\frac{1}{2}}(S_1^{-\frac{1}{2}} S_2 S_1^{-\frac{1}{2}} - z)S_1^{\frac{1}{2}}$. Since $S_1^{-\frac{1}{2}} S_2 S_1^{-\frac{1}{2}}$ has eigen values in $[0, \infty)$, $(S_1^{-\frac{1}{2}} S_2 S_1^{-\frac{1}{2}} - z)$ is invertible for $z \notin [0, \infty)$. Let (X, D) be the eigen decomposition such that

$$S_1^{-\frac{1}{2}} S_2 S_1^{-\frac{1}{2}} = XDX^* \text{ with } X = \begin{bmatrix} \boldsymbol{x}_1 & \cdots & \boldsymbol{x}_{qM} \end{bmatrix}, \, D = \operatorname{diag}(d_j)_{j=1}^{qM}.$$

Then we have for $z \notin [0, \infty)$

$$F(z) = \sum_{j=1}^{qM} \left(\frac{1}{d_j - z} \right) C_+ S_1^{-\frac{1}{2}} \boldsymbol{x}_j \boldsymbol{x}_j^* S_1^{-\frac{1}{2}} C_+^*,$$

and thus $F(z)$ has the required form with $d\mu$ being a atomic measure supported on $d_1, \ldots, d_{qM}$.

Furthermore, letting $\boldsymbol{e}_1, \ldots, \boldsymbol{e}_M$ be the standard basis vectors of $\mathbb{R}^M$, we have for $i = 1, \ldots, M$,

$$\begin{aligned}(z_i S_1 - S_1 A - C_+^* C_-)(\boldsymbol{e}_i \otimes I_q) &= S_1(z_i - I - A)(\boldsymbol{e}_i \otimes I_q) - C_+^* C_-(\boldsymbol{e}_i \otimes I_q) \\ &= 0 - C_+^* u_i = -C_+^* u_i.\end{aligned}$$

Thus

$$F(z_i)u_i = -C_+(z_i S_1 - S_1 A - C_+^* C_-)^{-1} C_+^* u_i = -C_+(-\boldsymbol{e}_i \otimes I_q) = v_i. \qquad \square$$

To apply this theorem to our problem, we first note that if we identify z in Theorem 2.1 with $-\frac{1}{s}$, then the IRF for $D_j^J(s)$ in (6), denoted by D^J for simplicity, can be written as

$$D^J(s) = (-z) \int_0^{\Theta_1} \frac{d\sigma^J}{t - z},$$

and

$$\begin{aligned}D^J(s) - \alpha_\infty &= (-z) \left[\int_0^{\Theta_1} \frac{d\sigma^J(t)}{t - z} - \frac{\alpha_\infty}{-z} \right] \\ &= (-z) \left[\int_0^{\Theta_1} \frac{d\sigma^J(t)}{t - z} - \int_0^{\Theta_1} \frac{\alpha_\infty \sigma(t)}{t - z} \right],\end{aligned} \tag{17}$$

where $\sigma(t)$ is a Dirac measure at $t = 0$. Since σ^J has a Dirac measure of strength α_∞, the function inside the parentheses in (17) is a Stieltjes function, which we denote by $F_{\text{new}}(z)$, i.e.,

$$D^J(s) - \alpha_\infty = (-z)F_{\text{new}}(z)$$

What we would like to harvest is the pole-residue approximation of $D(s)-\alpha_\infty$. To avoid truncation error, we rewrite all the formulas in Theorem 2.1 in terms of variable $s = -\frac{1}{z}$ as follows.

$$\begin{aligned} s_i &= -\frac{1}{z_i}, \quad u_i = \frac{1}{s_i}, \\ v_i &= D(s_i) - \alpha_\infty, \quad i = 1, \ldots, M, \\ (S_1)_{ij} &= \frac{-s_j D(s_j) + s_i^* D^*(s_i)}{s_i^* - s_j} + \frac{s_j \alpha_\infty - s_i^* \alpha_\infty}{s_i^* - s_j}, \\ (S_2)_{ij} &= \frac{-D(s_j) + D^*(s_i)}{s_j - s_i^*}. \end{aligned}$$

Consequently, we have the following representation for $D(s)$

$$D^J(s) \equiv \alpha_\infty + \frac{1}{s} F_{\text{new}}\left(-\frac{1}{s}\right) = \alpha_\infty + \sum_{j=1}^{qM} \left(\frac{1}{sd_j + 1}\right) C_+ S_1^{-\frac{1}{2}} \boldsymbol{x}_j \boldsymbol{x}_j^* S_1^{-\frac{1}{2}} C_+^*.$$

With the generalized eigenvalues $[\mathbf{V}, \mathbf{L}] := \text{eig}(S_2, S_1)$, where $\mathbf{V}$ is the matrix of generalized vectors and $\mathbf{L}$ the diagonal matrix of generalized eigenvalues such that

$$S_2 \mathbf{V} = S_1 \mathbf{V} \mathbf{L}, \tag{18}$$

we have

$$D^J(s) = \alpha_\infty + \sum_{k=1}^{qM} \frac{C_+ \mathbf{V}(:, k) \mathbf{V}(:, k)^* C_+^*}{1 + s\mathbf{L}(k, k)}.$$

Here we used Matlab notation: $\mathbf{V}(:, k)$ stands for the k'th column of $\mathbf{V}$.

3. Numerical examples

In this section, we apply both approaches in Section 2 to the examples of cancellous bone (S1) studied in [13], [11] and the epoxy-glass mixture (S2 and S3) and the sandstone (S4 and S5) examples studied in [8]. From prior results, it is known that the wider range the frequency is, the more ill-conditioned the corresponding matrices will be. We focus on the test case in [8], which applies the fractional derivate approach to deal with the memory term. In this case, time profile of the source term, denoted by $g(t)$ is a Ricker signal of central frequency $f_0 = 10^5\, s^{-1}$ and time-shift $t_0 = 1/f_0$, i.e.,

$$g(t) = \begin{cases} (2\pi^2 f_0^2 (t - t_0)^2 - 1) e^{-\pi^2 f_0^2 (t-t_0)^2}, & \text{if } 0 \leq t \leq 2t_0, \\ 0, & \text{otherwise;} \end{cases}$$

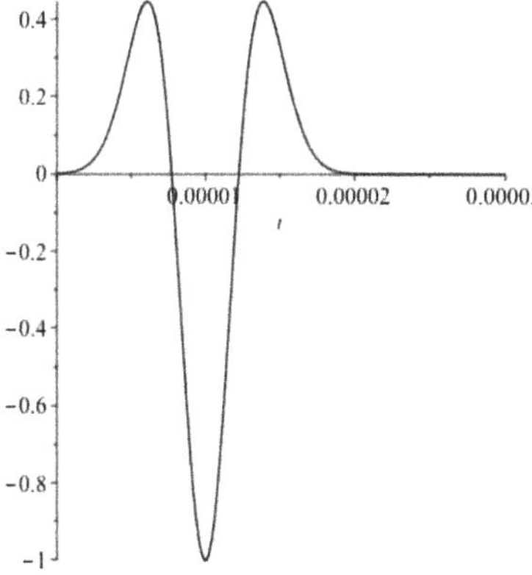

FIGURE 1. Ricker wavelet $g(t)$

See Figure 1. The spectrum content of $g(t)$ is visualized by its Fourier transform $\mathcal{F}\{g\}(\omega)$; see Figures 2 and 3. Since the real part and the imaginary part is symmetric and anti-symmetric with respect to $\omega = 0$, respectively, we only plot the $\omega \geq 0$ part of the graphs. Based on Figure 2 and Figure 3, we choose the frequency range in our numerical simulation to be from 10^{-3} Hz to 2×10^6 Hz.

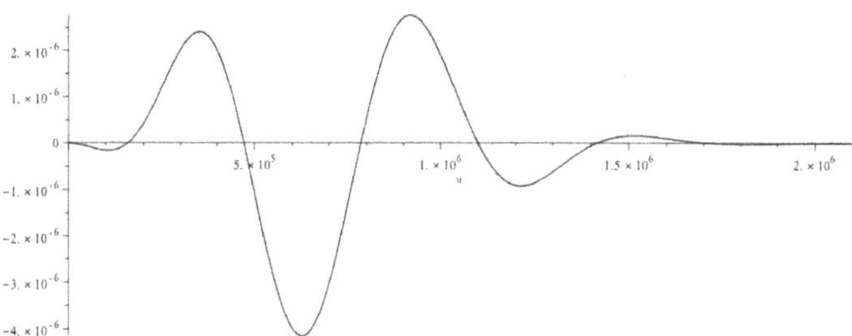

FIGURE 2. Spectral content of $g(t)$: real part of $\mathcal{F}\{g\}(\omega)$

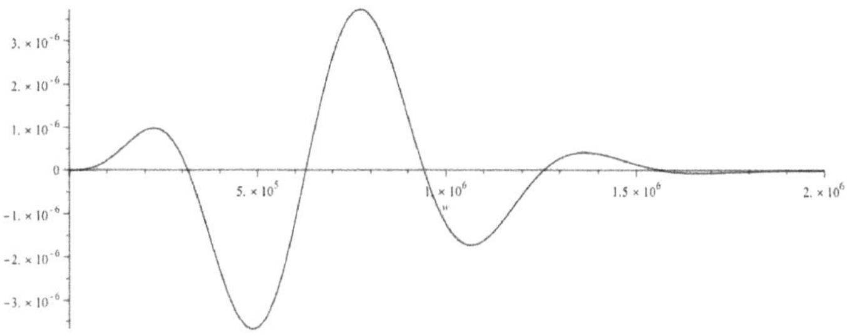

FIGURE 3. Spectral content of $g(t)$: imaginary part of $\mathcal{F}\{g\}(\omega)$

		S1	S2	S3	S4	S5
$\rho_f(Kg \cdot m^{-3})$	pore fluid density	1000	1040	1040	1040	1040
ϕ(unitless)	porosity	0.8	0.2	0.2	0.2	0.2
α_∞(unitless)	∞-frequency tortuosity	1.1	3.6	2.0	2.0	3.6
$K_0(m^2)$	static permeability	3e-8	1e-13	6e-13	6e-13	1e-13
$\nu(m^2 \cdot s^{-1})$	kinematic viscosity of pore fluid	1e-3/ρ_f	1e-3/ρ_f	1e-3/ρ_f	1e-3/ρ_f	1e-3/ρ_f
$\Lambda(m)$	structure constant	2.454e-5	3.790e-6	6.930e-6	2.190e-7	1.20e-7

TABLE 1. Biot-JKD parameters

We consider first equally spaced sample points. Similar to what was reported in [19], the relative error peaked near low frequency. This is due to fact that, in general, the function $D(s = -i\omega)$ varies the most near the lower end of ω. This observation leads to the alternative of log-distributed grid points, which in general perform better in terms of maximum relative errors but lead to more ill-conditioned matrices. For both the equally spaced and the log-spaced grids point, ill-conditioned matrices are involved. The ill-conditioning nature of the matrices A in Approach 1 and S_1, S_2 in Approach 2, together with the fact there is no obvious preconditioner available for these matrices, we resort to the multiprecision package *Advanpix* [17] for directly solving (12) and the subsequent partial fraction decomposition involved in Approach 1 and for solving the generalized eigenvalue problem (18). These real-valued poles and residues are then converted to double precision before we evaluate the relative errors

$$\text{rel\_err}(s) := \frac{|D^J(s) - D^J_{\text{est}}(s)|}{|D^J(s)|},$$

where the pole-residue approximation function D^J_{est} is defined in (11). We set the number of significant digits in *Advanpix* to be 90, which is much higher than the 15 decimal digits a 64-bit double-precision floating point format can represent.

The relative error rel_err with $M = 10$ for all the 5 media listed in Table 1 using Approach 2 is plotted in Figure 4, (a)–(e). The results of using equally space grids are in color blue while those by using log-distributed ones are in color red.

Among all the 5 media listed in Table 1, the cancellous bone S1 and the sandstones S4 and S5 are the most difficult one to approximate in the sense that it requires the largest M for achieving the same level of accuracy as for other media. The dynamic tortuosity functions of S1, S4 and S5 have large variation near low frequency and hence can be approximated much better when log-distributed grids are applied in the approximations. See Figure 11, (a) and (b).

In Tables 2 to 6, we list the condition numbers for both of the equally-spaced grid points and the log-distributed one. As can be seen, the condition numbers for matrices involved in Approach 1 with log-distributed grid points worsen very rapidly with the increase of M and the rescaling of volumes of A is not effective when compared with the equally spaced case. In Figures 4, (a)–(e), where the poles

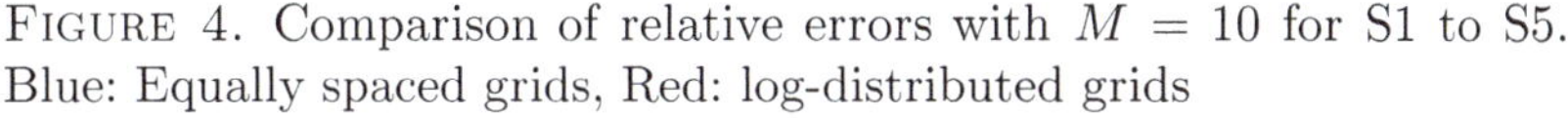

FIGURE 4. Comparison of relative errors with $M = 10$ for S1 to S5. Blue: Equally spaced grids, Red: log-distributed grids

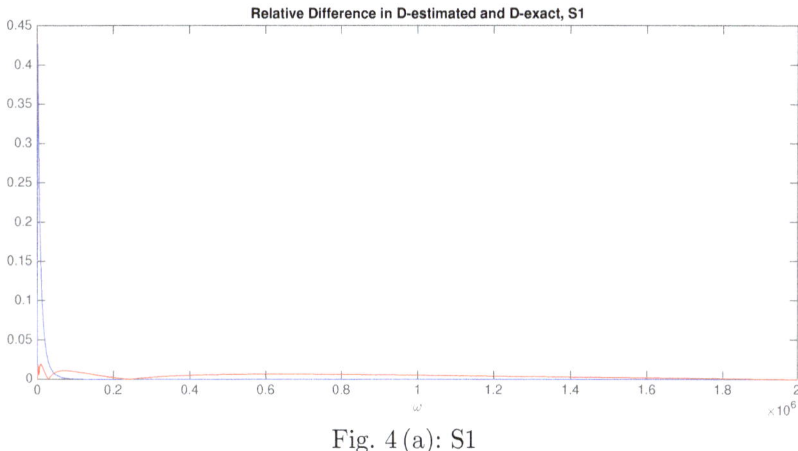

Fig. 4 (a): S1

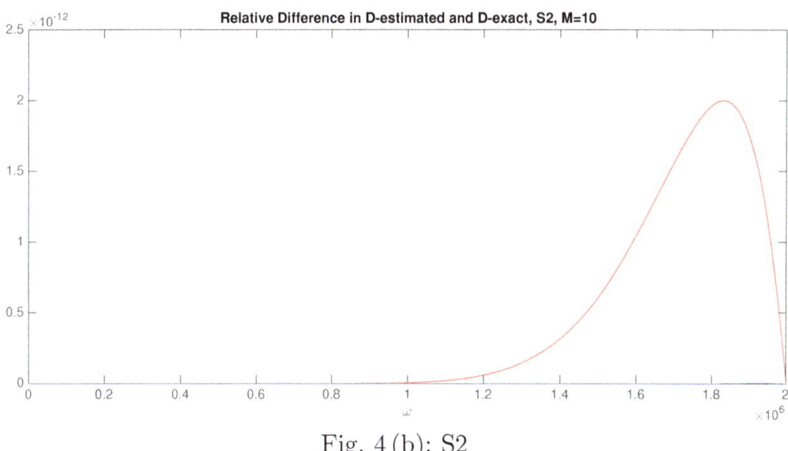

Fig. 4 (b): S2

and residues for $M = 14$ (computed with different combinations of methods) are plotted in log-log scale, we see that Approaches 1 and 2 give numerically identical results for all these 5 test media when the significant digits in the calculations are set much higher than the $\log_{10}$-scale of the condition numbers involved. The calculation is carried out by using 140 significant digits and it takes about 5 seconds with a single processor MacBook Pro.

In Figure 10, the relative error rel_err for approximations by using equally spaced grid and by log-distributed grids are presented. As can be seen from Figure 11 (a), the peak of error near the lower end of the frequency range is due to

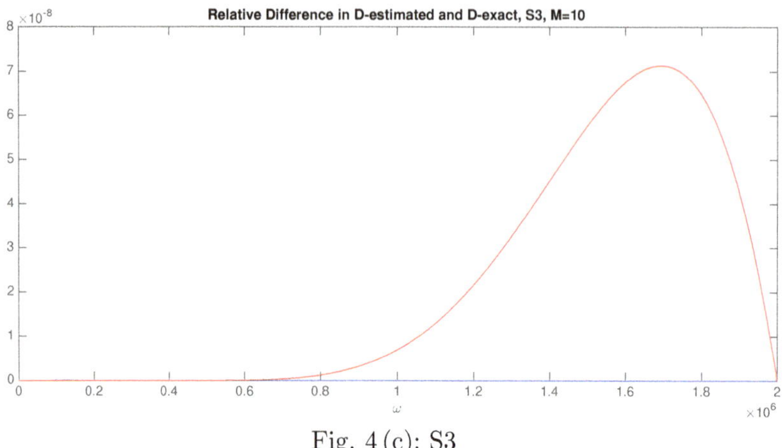

Fig. 4 (c): S3

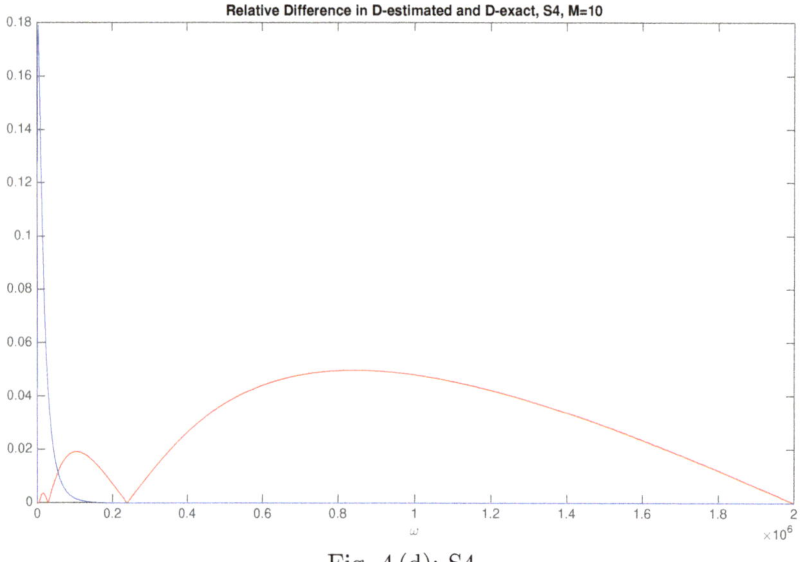

Fig. 4 (d): S4

the fact that the function being approximated needs more grid points there to resolve the variation. This is achieved by using the log-distributed grids. In Figure 11 (a) and (b), we plot D^J and its pole-residue approximation D^J_{est} to visualize the performance. Figure 11 (a) corresponds to the equally spaced grids while Figure 11 (b) to the log-distributed one. In both figures, these two functions are almost indiscernible except the imaginary parts in Figure 11 (a) near the lower end of frequency where rel_err peaks; both the colors black (imaginary part of D^J) and green (imaginary part of D^J_{est}) can be seen there.

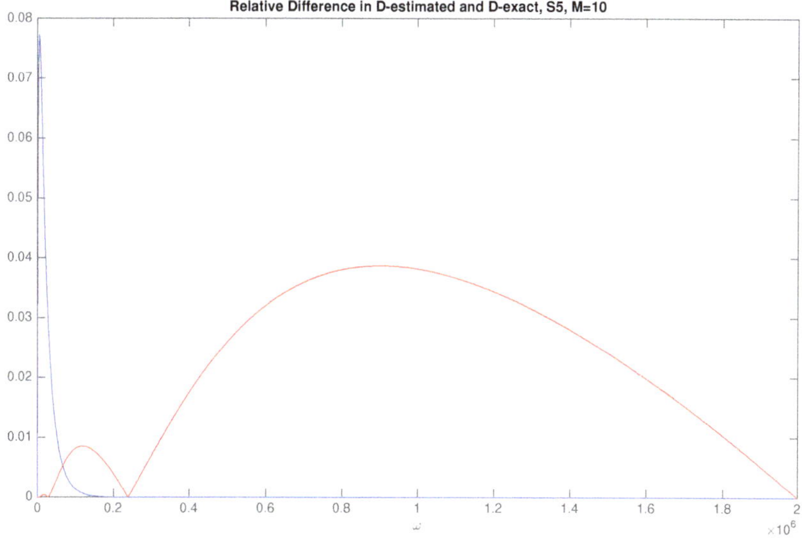

Fig. 4 (e): S5

	M=8	M=8	M=14	M=14
	Equally spaced	log-spaced	Equally spaced	log-spaced
A	48.9750	49.2328	86.7853	87.4283
B	10.0354	31.4021	15.2592	57.8602
S_1	13.3237	4.3014	23.7936	5.8767
S_2	9.9319	10.8471	30.3962	11.9522

TABLE 2. $\log_{10}$-Condition numbers of the matrices for material S1

	M=8	M=8	M=14	M=14
	Equally spaced	log-spaced	Equally spaced	log-spaced
A	50.3092	51.1536	88.1203	90.5197
B	20.1795	64.8793	31.3151	120.2955
S_1	15.1483	55.9887	29.1822	110.3533
S_2	15.1657	55.9766	29.1998	110.3456

TABLE 3. $\log_{10}$-Condition numbers of the matrices for material S2

	M=8	M=8	M=14	M=14
	Equally spaced	log-spaced	Equally spaced	log-spaced
A	49.9052	50.8287	87.7160	90.1957
B	16.5867	59.7475	24.6405	110.6680
S_1	12.3521	49.2816	24.0606	97.5660
S_2	12.5097	49.2655	24.2122	97.5514

TABLE 4. $\log_{10}$-Condition numbers of the matrices for material S3

	M=8	M=8	M=14	M=14
	Equally spaced	log-spaced	Equally spaced	log-spaced
A	50.8209	51.4408	88.6312	89.5011
B	11.8964	41.4130	16.7579	76.4563
S_1	11.5699	18.8133	22.0503	38.7220
S_2	14.6234	18.7833	25.0888	38.6911

TABLE 5. $\log_{10}$-Condition numbers of the matrices for material S4

	M=8	M=8	M=14	M=14
	Equally spaced	log-spaced	Equally spaced	log-spaced
A	51.0851	51.8269	88.8954	89.9913
B	12.2279	43.8165	17.1587	80.9135
S_1	11.3409	23.0730	21.8547	47.0251
S_2	13.8633	23.0477	24.3312	46.9971

TABLE 6. $\log_{10}$-Condition numbers of the matrices for material S5

4. Conclusions

In this paper, we utilize the Stieltjes function structure of the JKD dynamic tortuosity to derive an augmented system of Biot-JKD equations (8)–(10) that approximates the solution of the original Biot-JKD equations (2)–(3). Asymptotic behavior of the tortuosity function as $\omega \to \infty$ is enforced analytically before the numerical interpolation is carried out by Approach 1 and Approach 2. Due to the nature of the tortuosity functions of S1, S4 and S5, log-distributed interpolation points generally perform better than the equally distributed ones. We tested our approaches on 5 sets of poroelastic parameters obtaining from the existing literature and interpolated the JKD dynamic tortuosity equation to high accuracy through a frequency range that spans 9 orders of magnitude from 10^{-3} to 2×10^6.

The extremely ill-conditioned matrices are dealt with by using a multiprecision package *Advanpix* in which we set the significant digits of floating numbers to be 140. It turns out Approaches 1 and 2 give numerically identical results for all the test cases when the significant digits are set much higher than the $\log_{10}$-scale of the condition numbers involved. We think the exact link between these two approaches can be derived through the Barycentric forms for rational approximations [3], which in term provides an approach that can adapt the choice of grid points based on the data points so the Lebesgue constant is minimized [18]. This, as well as applying the results to matrix-valued data, will be explored in a later work.

Acknowledgment. We wish to thank Joe Ball, Daniel Alpay, Marc van Barel, Thanos Antoulas, Sanda Lefteriu, and Cosmin Ionita for their helpful suggestions. In addition, we thank the referee for the careful reading of our manuscript and the useful suggestions.

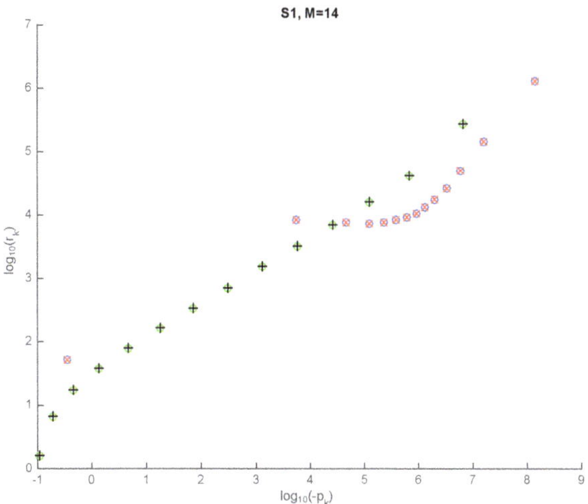

FIGURE 5. $(\log_{10}(-p_k), \log_{10}(r_k))$, $k = 1, \ldots, 14$. Blue circle: Approach 1 with Equally-spaced grids, Red x: Approach 2 with Equally-spaced grid, Green circle: Approach 1 with Log-spaced grids, Black +: Approach 2 with Log-spaced grids

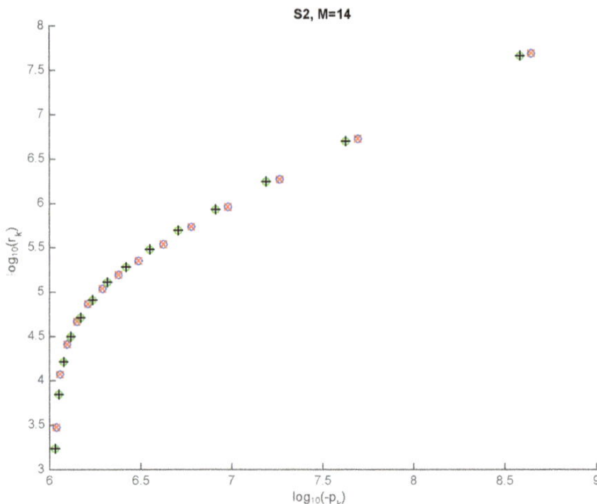

FIGURE 6. $(\log_{10}(-p_k), \log_{10}(r_k))$, $k = 1, \ldots, 14$. Blue circle: Approach 1 with Equally-spaced grids, Red x: Approach 2 with Equally-spaced grid, Green circle: Approach 1 with Log-spaced grids, Black +: Approach 2 with Log-spaced grids

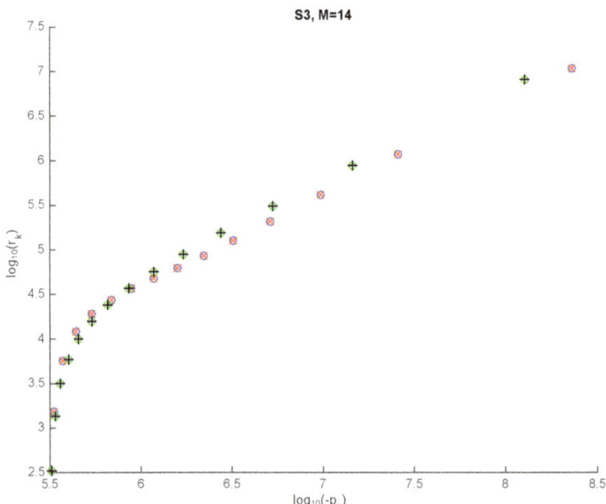

FIGURE 7. $(\log_{10}(-p_k), \log_{10}(r_k))$, $k = 1, \ldots, 14$. Blue circle: Approach 1 with Equally-spaced grids, Red x: Approach 2 with Equally-spaced grid, Green circle: Approach 1 with Log-spaced grids, Black +: Approach 2 with Log-spaced grids

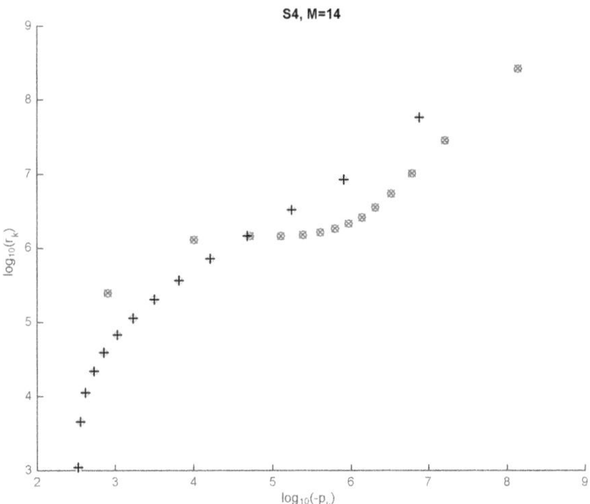

FIGURE 8. $(\log_{10}(-p_k), \log_{10}(r_k))$, $k = 1, \ldots, 14$. Blue circle: Approach 1 with Equally-spaced grids, Red x: Approach 2 with Equally-spaced grid, Green circle: Approach 1 with Log-spaced grids, Black +: Approach 2 with Log-spaced grids

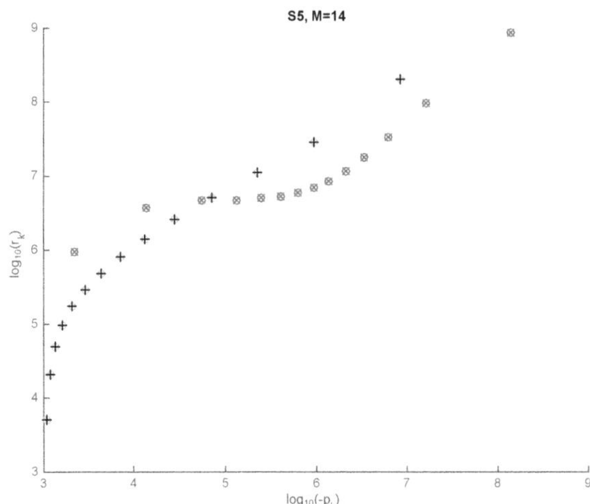

FIGURE 9. $(\log_{10}(-p_k), \log_{10}(r_k))$, $k = 1, \ldots, 14$. Blue circle: Approach 1 with Equally-spaced grids, Red x: Approach 2 with Equally-spaced grid, Green circle: Approach 1 with Log-spaced grids, Black +: Approach 2 with Log-spaced grids

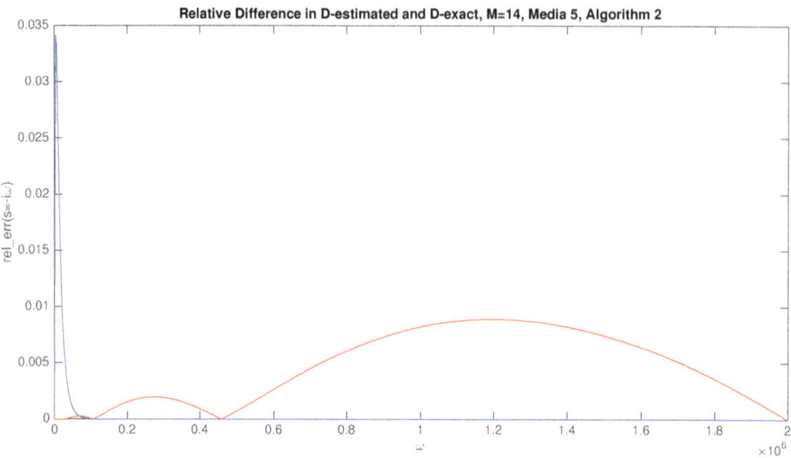

FIGURE 10. rel_err($s = -i\omega$),
Blue: equally spaced grids, Red: log-distributed grids

References

[1] D. Alpay, J.A. Ball, I. Gohberg, and L. Rodman. The two-sided residue interpolation in the Stieltjes class for matrix functions. *Linear Algebra and its Applications*, 208/209:485–521, 1994.

[2] A.C. Antoulas, S. Lefteriu, and A.C. Ionita. A tutorial introduction to the Loewner framework for model reduction. In *Model Reduction and Approximation*, volume 15 of *Comput. Sci. Eng.*, pages 335–376. SIAM, Philadelphia, PA, 2017.

[3] J.-P. Berrut and H.D. Mittelmann. Lebesgue constant minimizing linear rational interpolation of continuous functions over the interval. *Computers & Mathematics with Applications*, 33(6):77–86, 1997.

[4] M.A. Biot. Theory of propagation of elastic waves in a fluid-saturated porous solid. I. Low-frequency range. *The Journal of the Acoustical Society of America*, 28(2):168–178, 1956.

[5] M.A. Biot. Theory of propagation of elastic waves in a fluid-saturated porous solid. II. Higher frequency range. *The Journal of the Acoustical Society of America*, 28(2):179–191, 1956.

[6] M.A. Biot. Mechanics of deformation and acoustic propagation in porous media. *Journal of Applied Physics*, 33(4):1482–1498, 1962.

[7] E. Blanc, G. Chiavassa, and B. Lombard, A time-domain numerical modeling of two-dimensional wave propagation in porous media with frequency-dependent dynamic permeability, *The Journal of the Acoustical Society of America*, 134(6):4610–4623, 2013.

[8] E. Blanc, G. Chiavassa, and B. Lombard. Wave simulation in 2D heterogeneous transversely isotropic porous media with fractional attenuation: A Cartesian grid approach. *Journal of Computational Physics*, 275:118–142, 2014.

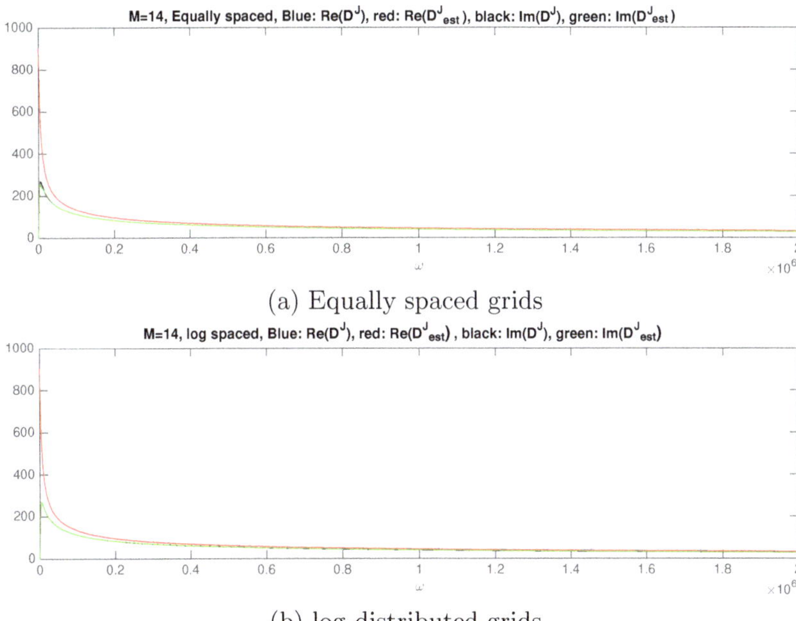

(a) Equally spaced grids

(b) log-distributed grids

FIGURE 11. Comparison of $D^J(s = -i\omega)$ and $D^J_{\text{est}}(s = -i\omega)$. Blue: $\text{Re}(D^J)$, Red: $\text{Re}(D^J_{\text{est}})$, Black: $\text{Im}(D^J)$, Green: $\text{Im}(D^J_{\text{est}})$

[9] J.M. Carcione. *Wave Fields in Real Media: Wave Propagation in Anisotropic, Anelastic and Porous Media.* Pergamon-Elsevier, Oxford, 2001.

[10] J.W. Dettman. *Applied Complex Variables.* Dover publications, 1965.

[11] M. Fellah, Z.E.A. Fellah, F.G. Mitri, E. Ogam, and C. Depollier. Transient ultrasound propagation in porous media using biot theory and fractional calculus: Application to human cancellous bone. *The Journal of the Acoustical Society of America*, 133(4):1867–1881, 2013.

[12] J.K. Gelfgren. Multipoint Padé approximants converging to functions of Stieltjes' type. In *Padé Approximation and its Applications, Amsterdam*, pages 197–207. Lecture Notes in Mathematics, vol 888. Springer, 1981.

[13] A. Hosokawa. Ultrasonic pulse waves in cancellous bone analyzed by finite-difference time-domain methods. *Ultrasonics*, 44:e227–e231, 2006.

[14] D.L. Johnson, J. Koplik, and R. Dashen. Theory of dynamic permeability and tortuosity in fluid-saturated porous media. *Journal of Fluid Mechanics*, 176(1):379–402, 1987.

[15] J.F. Lu and A. Hanyga, Wave field simulation for heterogeneous porous media with singular memory drag force, *J. Comput. Phys.* 208(2):651–674, 2005.

[16] Y.J. Masson and S.R. Pride, Finite-difference modeling of Biot's poroelastic equations across all frequencies, *Geophys.*, 75(2):N33–N44, 2010.

[17] Multiprecision Computing Toolbox for MATLAB 4.4.7.12736. Advanpix LLC., Yokohama, Japan, 2018.

[18] Y. Nakatsukasa, O. Sète, and L.N. Trefethen. The AAA algorithm for rational approximation. *SIAM Journal on Scientific Computing*, 40(3):A1494–A1522, 2018

[19] M.-J.Y. Ou. On reconstruction of dynamic permeability and tortuosity from data at distinct frequencies. *Inverse Problems*, 30(9):095002, 2014.

Miao-Jung Yvonne Ou
Department of Mathematical Sciences
University of Delaware
Newark, DE 19716, USA
e-mail: mou@udel.edu

Hugo J. Woerdeman
Department of Mathematics
Drexel University
3141 Chestnut Street
Philadelphia, PA 19104, USA
e-mail: hugo@math.drexel.edu

Operator Theory:
Advances and Applications, Vol. 272, 329–341

The Flow Equations of Resistive Electrical Networks

Arjan van der Schaft

Dedicated to Joe Ball at the occasion of his 70th birthday. It was a great pleasure to work with him on problems in nonlinear robust control theory [1, 2], *combining his nonlinear spectral factorization insights with concepts from Hamiltonian systems theory.*

Abstract. The analysis of the flow equations of electrical networks goes back to the classical work of Kirchhoff, but still poses interesting questions, of direct relevance to electrical power networks. A related open problem concerns the identifiability of the conductance parameters from the boundary behavior of the network.

Mathematics Subject Classification (2010). Primary 94C15, 05C21.

Keywords. Kirchhoff's problem, Laplacian matrix, Schur complement, identifiability.

1. Introduction

The theory of electrical engineering has triggered many interesting and elegant mathematical developments, which conversely turned out to be of great importance for engineering applications. Especially within systems and control theory one can witness a wide range of examples of this fruitful interplay; from network synthesis, impedance matching, uncertainty reduction, and robust control design, to large-scale network systems.

The present paper focusses on a classical, and in some sense very simple, example; namely the study of *linear resistive electrical networks*. The mathematical formulation of them, and the study of their properties, goes back to the 1847 paper [15] by Gustav Robert Kirchhoff (1824–1887), which laid the foundations for the theory of electrical networks, and at the same time can be seen as one of the early milestones in the development of (algebraic) graph theory. From a systems theory point of view resistive electrical networks constitute a beautiful example of an *interconnected* and *open*, although static (no dynamics), system. See [28] for stimulating reading, as well as [20] for related developments[1].

The paper is structured as follows. After recalling in Section 2 some basic notions of algebraic graph theory, Section 3 summarizes the graph-theoretic formulation of resistive networks. Section 4 starts with the description and solution of the classical problem posed by Kirchhoff concerning the existence and uniqueness of solutions of the flow equations of open resistive networks, and discusses some related problems. Although it is difficult to claim any originality in this very classical area (with the names of, e.g., Maxwell, Rayleigh, Thomson and Dirichlet attached to it), I believe there is still some originality in Propositions 4.3 and 4.4. Section 5 discusses within the same framework the identifiability of the conductance parameters of a resistive network from the knowledge of the map from boundary nodal voltages potentials to boundary nodal currents. It is explained how this problem is a discrete version of the inverse problem of Calderón [6], and conditions for identifiability are discussed. Finally, Section 6 indicates extensions to RLC-networks, and in particular the important load flow problem in electrical power networks. Section 7 contains the conclusions.

2. Preliminaries about graphs

First, let us recall from, e.g., [3, 14] the following standard notions regarding directed graphs. A *graph* is defined by a set of *nodes* (or vertices), and a set of *edges* (or links, branches), where the set of edges is identified with a subset of unordered pairs of nodes; $\{i, j\}$ denotes an edge between nodes i and j. We allow for multiple edges between nodes, but not for self-loops $\{i, i\}$. By endowing the edges with an orientation we obtain a *directed graph*, with edges corresponding to ordered pairs; ij denotes an edge from node i to node j. In the sequel, 'graph' will always mean 'directed graph'.

A directed graph with N nodes and M edges is specified by its $N \times M$ *incidence matrix*, denoted by $\mathcal{D}$. Every column of $\mathcal{D}$ corresponds to an edge of the graph, and contains exactly one 1 at the row corresponding to the tail node of the edge and one -1 at the row corresponding to its head node, while the other elements are 0. In particular, $\mathbb{1}^T\mathcal{D} = 0$ where $\mathbb{1}$ is the vector of all ones. Furthermore, $\ker \mathcal{D}^T = \operatorname{span} \mathbb{1}$ if and only if the graph is *connected* (any node can be reached from any other vertex by a sequence of, – undirected –, edges). In general, the dimension of $\ker \mathcal{D}^T$ is equal to the number of connected components. Throughout this paper we will adopt[2]

Assumption 2.1. *The graph is connected.*

---

[1]While writing the present paper I came across the recent paper [13], surveying, rather complementarily to the present paper, the interplay between algebraic graph theory and electrical networks from various angles.

[2]Without loss of generality since otherwise the analysis can be done for every connected component of the graph.

Corresponding to a directed graph we define, in the spirit of general k-complexes, the following vector spaces; see [21]. The *node space* Λ_0 of the directed graph is defined as the set of all functions from the node set $\{1, \ldots, N\}$ to $\mathbb{R}$. Obviously Λ_0 can be identified with $\mathbb{R}^N$. The dual space of Λ_0 is denoted by Λ^0. Furthermore, the *edge space* Λ_1 is defined as the linear space of functions from the edge set $\{1, \ldots, M\}$ to $\mathbb{R}$, with dual space denoted by Λ^1. Both spaces can be identified with $\mathbb{R}^M$. It follows that the incidence matrix $\mathcal{D}$ defines a linear map (denoted by the same symbol) $\mathcal{D} : \Lambda_1 \to \Lambda_0$ with adjoint map $\mathcal{D}^T : \Lambda^0 \to \Lambda^1$. In the context of the present paper the space Λ_1 corresponds to the *currents* through the edges, and the dual space Λ^1 to the space of *voltages* across the edges. Furthermore, as we will see later on, Λ_0 denotes the space of *nodal currents* (entering the nodes from the environment, or from electrical devices located at the nodes), and Λ^0 the space of *voltage potentials* at the nodes.

Finally, we note that it is straightforward to extend the network models described in this paper, as well as their dynamical versions (see, e.g., [21]), to more general cases. Indeed, for any linear space $\mathcal{R}$ (e.g., $\mathcal{R} = \mathbb{R}^3$) we can define Λ_0 instead as the set of functions from $\{1, \ldots, N\}$ to $\mathcal{R}$, and Λ_1 as the set of functions from $\{1, \ldots, M\}$ to $\mathcal{R}$. In this case we identify Λ_0 with the tensor product $\mathbb{R}^N \otimes \mathcal{R}$, and Λ_1 with the tensor product $\mathbb{R}^M \otimes \mathcal{R}$. Furthermore, the incidence matrix $\mathcal{D}$ defines a linear map $\mathcal{D} \otimes I : \Lambda_1 \to \Lambda_0$, where I is the identity map on $\mathcal{R}$. In matrix notation $\mathcal{D} \otimes I$ equals the Kronecker product of the incidence matrix $\mathcal{D}$ and the identity matrix I. See [21, 22] for further details.

3. Resistive electrical networks

Consider an electrical network, where the underlying connected graph, with N nodes and M edges, is specified by its incidence matrix $\mathcal{D}$. Then the relation between the vector of *voltage potentials*

$$\psi = (\psi_1, \ldots, \psi_N)^T \tag{3.1}$$

at the nodes, and the vector of *nodal currents*

$$J = (J_1, \ldots, J_N)^T \tag{3.2}$$

entering the nodes (e.g., the currents taken from generators or delivered to loads) is determined as follows. By Kirchhoff's *voltage laws*, the vector of *voltages* $V = (V_1, \ldots, V_M)^T$ across the edges is determined by the vector of nodal voltage potentials as

$$V = \mathcal{D}^T \psi. \tag{3.3}$$

Dually, the relation between the vector of nodal currents J and the vector $I = (I_1, \ldots, I_N)^T$ of currents through the edges is given by Kirchhoff's *current laws*

$$J = \mathcal{D} I. \tag{3.4}$$

It follows that $I \in \Lambda_1$, $V \in \Lambda^1$, $\psi \in \Lambda^0$, and $J \in \Lambda_0$. In particular, the products $V^T I$ and $\psi^T J$ are intrinsically defined, while

$$V^T I = \psi^T \mathcal{D} I = \psi^T J, \tag{3.5}$$

expressing that the total power flowing through the edges is equal to the total power entering or leaving the nodes.

In the case of a linear resistive network each edge corresponds to a *linear resistor*. That is, for the kth edge the current I_k through the edge, and the voltage V_k across the edge, are related as

$$I_k = g_k V_k, \tag{3.6}$$

with g_k the *conductance* of the resistor, i.e., $g_k = \frac{1}{r_k}$ with $r_k > 0$ the resistance of the kth edge. Hence

$$I = GV \tag{3.7}$$

with G the $M \times M$ diagonal matrix of conductances. Putting all this together, one obtains

$$J = \mathcal{D} G \mathcal{D}^T \psi. \tag{3.8}$$

The matrix $\mathcal{L} = \mathcal{D} G \mathcal{D}^T$ is called the *Laplacian matrix*[3] of the electrical network. Laplacian matrices are fully characterized by the following properties.

Proposition 3.1 ([18]). *The Laplacian matrix $\mathcal{L}$ is a symmetric matrix, with all its diagonal elements positive, and all off-diagonal elements non-positive. Furthermore, the row and column sums of $\mathcal{L}$ are zero, i.e., $\mathcal{L}\mathbb{1} = 0, \mathbb{1}^T \mathcal{L} = 0$.*

'Fully characterized' means that conversely for any matrix $\mathcal{L}$ satisfying the properties listed in Proposition 3.1 there exists an incidence matrix $\mathcal{D}$ and positive diagonal matrix G such that $\mathcal{L} = \mathcal{D} G \mathcal{D}^T$. In particular, any non-zero off-diagonal element $\mathcal{L}_{ij}$ defines an edge ij with conductance equal to $-\mathcal{L}_{ij}$. Furthermore, it is well known (see, e.g., [3]) that $\mathcal{L}$ is independent of the orientation of the graph. Thus if we take another orientation of the graph (corresponding to an incidence matrix $\mathcal{D}$ where some of the columns have been multiplied by -1), then $\mathcal{L}$ remains the same. Moreover, by the assumption of connectedness $\ker \mathcal{L} = \operatorname{span} \mathbb{1}$.

Also note that any Laplacian matrix $\mathcal{L}$ is positive-semidefinite; in fact, the quadratic form associated to $\mathcal{L}$ is equal to

$$\psi^T \mathcal{L} \psi = \sum_{ij} g_{ij} (\psi_i - \psi_j)^2, \tag{3.9}$$

where the summation is over all the M edges ij. Note that $\psi^T \mathcal{L} \psi = \psi^T J$ is equal to the total power at the nodes, which by (3.5) is equal to the total power $V^T I$ through the edges.

[3] Classically, in the context of resistive electrical networks $\mathcal{L}$ is called the *Kirchhoff matrix*; however we will stick to the currently used terminology in general network models.

Remark 3.2. The same equations hold for other classes of physical systems [19]. For example, mechanical damper systems are analogous, with the voltage potentials ψ replaced by the *velocities* of the nodes, the nodal currents J by the nodal *forces*, and the kth edge corresponding to a linear damper $F_k = d_k v_k$, with d_k the damping constant, F_k the damping force, and v_k the difference of the velocities of the tail and head node of this edge.

4. Kirchhoff's problem and its dual

Now suppose the N nodes of the network graph are *split* into N_C internal *connection* nodes (denoted by C), and N_B remaining external *boundary* nodes (denoted by B), with $N = N_B + N_C$. Correspondingly we split (possibly after reordering) the vectors J and ψ of nodal currents and voltage potentials as

$$J = \begin{bmatrix} J_B \\ J_C \end{bmatrix}, \quad \psi = \begin{bmatrix} \psi_B \\ \psi_C \end{bmatrix} \tag{4.1}$$

and the incidence matrix into

$$\mathcal{D} = \begin{bmatrix} \mathcal{D}_B \\ \mathcal{D}_C \end{bmatrix}. \tag{4.2}$$

One then obtains the following equations

$$\begin{bmatrix} J_B \\ J_C \end{bmatrix} = \mathcal{L} \begin{bmatrix} \psi_B \\ \psi_C \end{bmatrix}, \tag{4.3}$$

where

$$\mathcal{L} = \begin{bmatrix} \mathcal{D}_B \\ \mathcal{D}_C \end{bmatrix} G \begin{bmatrix} \mathcal{D}_B^T & \mathcal{D}_C^T \end{bmatrix} =: \begin{bmatrix} \mathcal{L}_{BB} & \mathcal{L}_{BC} \\ \mathcal{L}_{CB} & \mathcal{L}_{CC} \end{bmatrix}. \tag{4.4}$$

4.1. Kirchhoff's problem

The classical *Kirchhoff problem* and its solution, dating back to [15], can be formulated in this notation as follows. Fix the boundary voltage potentials $\psi_B = \psi_B^* \in \mathbb{R}^{N_B}$, and let $0 = J_C \in \mathbb{R}^{N_C}$ (no nodal currents at the connection nodes). Then [3] there exists a *unique* vector of voltage potentials $\psi_C^* \in \mathbb{R}^{N_C}$ at the connection nodes such that

$$\begin{bmatrix} J_B^* \\ 0 \end{bmatrix} = \begin{bmatrix} \mathcal{L}_{BB} & \mathcal{L}_{BC} \\ \mathcal{L}_{CB} & \mathcal{L}_{CC} \end{bmatrix} \begin{bmatrix} \psi_B^* \\ \psi_C^* \end{bmatrix}, \tag{4.5}$$

with the corresponding vector of currents $J_B^* \in \mathbb{R}^{N_B}$ at the boundary nodes given by[4]

$$J_B^* = (\mathcal{L}_{BB} - \mathcal{L}_{BC}\mathcal{L}_{CC}^{-1}\mathcal{L}_{CB})\psi_B^* = \mathcal{L}_S \psi_B^*, \tag{4.6}$$

with $\mathcal{L}_S := (\mathcal{L}_{BB} - \mathcal{L}_{BC}\mathcal{L}_{CC}^{-1}\mathcal{L}_{CB})$ the Schur complement of the Laplacian matrix $\mathcal{L}$ with respect to $\mathcal{L}_{CC}$.

A key observation is that the Schur complement $\mathcal{L}_S$ is again a Laplacian matrix. Explicit statements of this result in the literature before its formulation and proof in [18] seem hard to find; although [8] contains the closely related result

[4]The assumption of connectedness implies that the submatrix $\mathcal{L}_{CC}$ is invertible.

that the Schur complement of any M-matrix is again an M-matrix[5]. See also [17] for the use of Schur complements in reduction of large-scale resistive networks.

Theorem 4.1 **([18]).** *If the graph $\mathcal{G}$ is connected, then all diagonal elements of $\mathcal{L} = \mathcal{D}G\mathcal{D}^T$ are > 0. Furthermore, all Schur complements of $\mathcal{L} = \mathcal{D}G\mathcal{D}^T$ are well defined, symmetric, with diagonal elements > 0, off-diagonal elements ≤ 0, and with zero row and column sums. In particular, all Schur complements of $\mathcal{L} = \mathcal{D}G\mathcal{D}^T$ are Laplacian, and thus can be written as $\bar{\mathcal{D}}\bar{G}\bar{\mathcal{D}}^T$, with $\bar{\mathcal{D}}$ the incidence matrix of some connected graph with the same set of nodes, and $\bar{G}$ a positive definite diagonal matrix with the same dimension as the number of edges of the graph defined by $\bar{\mathcal{D}}$.*

The proof given in [18] is based on two observations. First, it is immediately checked that the Schur complement of a Laplacian matrix $\mathcal{L}$ with respect to a scalar diagonal element is again a Laplacian matrix; see also [26]. Second, any Schur complement can be obtained by the successive application of taking Schur complements with respect to diagonal elements. In fact, this follows from the following quotient formula given in [8]. Denote by M/P the Schur complement of the square matrix M with respect to a leading square submatrix P. Then for any leading submatrix Q of P we have the equality

$$M/P = (M/Q)/(P/Q). \tag{4.7}$$

Thus $\mathcal{L}_S$ as defined in (4.6) is again a Laplacian matrix. Therefore we have obtained a *reduced* resistive network with only boundary nodes, where the map from boundary nodal voltages to boundary nodal currents is given by

$$J_B = \mathcal{L}_S \psi_B. \tag{4.8}$$

This reduced resistive network is *equivalent* (as seen from the boundary nodes) to the original one with the constraint $J_C = 0$. The transformation from electrical network with boundary and internal nodes satisfying $J_C = 0$ to the reduced network *without* internal nodes is called *Kron reduction* [16]; see, e.g., [12] for a review of the literature and applications of Kron reduction.

Remark 4.2. A *special* case is obtained by considering just *two* boundary nodes; the rest being connection nodes. Then the Schur complement $\mathcal{L}_S$ is of the form

$$\mathcal{L}_S = \begin{bmatrix} g & -g \\ -g & g \end{bmatrix}, \tag{4.9}$$

with g the *effective conductance* between the two boundary nodes, and $\frac{1}{g}$ the *effective resistance.*

The uniquely determined vector of internal voltage potentials ψ_C^*, called the *open-circuit* internal voltage potentials, corresponding to the boundary voltage potentials ψ_B^*, is given as

$$\psi_C^* = -\mathcal{L}_{CC}^{-1} \mathcal{L}_{CB} \psi_B^*. \tag{4.10}$$

[5] I thank Nima Monshizadeh, University of Groningen, for pointing out this reference to me.

Since $\mathcal{L}_{CC}$ is an invertible M-matrix it follows that $\mathcal{L}_{CC}^{-1}$ has all non-negative elements, and thus also $-\mathcal{L}_{CC}^{-1}\mathcal{L}_{CB}$ is a matrix with all non-negative elements. Furthermore, it follows from the discrete *Maximum Modulus principle* [3] that for all ψ_B^*

$$\|\psi_C^*\|_{\max} \leq \|\psi_B^*\|_{\max}, \tag{4.11}$$

where $\|\cdot\|_{\max}$ denotes the max-norm $\|\psi\|_{\max} = \max_j |\psi_j|$. Hence the linear map $\psi_B^* \mapsto \psi_C^*$ given by the matrix $-\mathcal{L}_{CC}^{-1}\mathcal{L}_{CB}$ has induced norm ≤ 1. In fact, since $\psi_B^* = \mathbb{1}$ yields $\psi_C^* = \mathbb{1}$, it follows that

$$\| - \mathcal{L}_{CC}^{-1}\mathcal{L}_{CB}\|_{\max} = 1\,. \tag{4.12}$$

The *open-circuit* internal voltage potential ψ_C^* has the following classical *minimization* interpretation, called Thomson's (or also Dirichlet's) principle [3, 18]. Given ψ_B^*, then ψ_C^* is the unique minimizer of

$$\min_{\psi_C} \begin{bmatrix}\psi_B^{*T} & \psi_C^T\end{bmatrix} \mathcal{L} \begin{bmatrix}\psi_B^* \\ \psi_C\end{bmatrix} = \min_{\psi_C} \sum_{(ij)} g_{(ij)}(\psi_i - \psi_j)^2 \tag{4.13}$$

(with $g_{(ij)}$ denoting the conductance of the edge ij, with summation over all the edges ij of the circuit graph). Indeed, the gradient vector of this expression with respect to ψ_C is equal to two times $I_C = \mathcal{L}_{CB}\psi_B^* + \mathcal{L}_{CC}\psi_C$. The expression $\sum g_{(ij)}(\psi_i - \psi_j)^2$ equals the dissipated power in the resistive network. Thus the open-circuit internal voltage potential ψ_C^* corresponds to *minimal dissipated power* in the resistive network.

This is summarized in the following proposition.

Proposition 4.3. *The linear map $\psi_B^* \mapsto \psi_C^*$ given by (4.10) has induced* max*-norm equal to* 1. *Given ψ_B^*, then ψ_C^* is the unique minimizer of the dissipated power $\sum g_{(ij)}(\psi_i - \psi_j)^2$. Any other resistive network inducing by Kron reduction the same linear map $J_B = \mathcal{L}_S\psi_B$ has the same dissipated power.*

4.2. The dual problem

Dually to Kirchhoff's problem there is the (easier) *short-circuit* problem, corresponding to taking the boundary voltage potentials $\psi_B = 0$, and instead to consider a prescribed value of the vector of internal nodal currents $J_C = \bar{J}_C$. Then the corresponding internal voltage potentials $\bar{\psi}_C$ are given as

$$\bar{\psi}_C = \mathcal{L}_{CC}^{-1}\bar{J}_C, \tag{4.14}$$

while the short-circuit boundary nodal currents $\bar{J}_B$ are determined by $\bar{J}_C$ as

$$\bar{J}_B = \mathcal{L}_{BC}\mathcal{L}_{CC}^{-1}\bar{J}_C. \tag{4.15}$$

Note that the matrix $\mathcal{L}_{BC}\mathcal{L}_{CC}^{-1}$ has all non-positive elements, and equals minus the transpose of the matrix $\mathcal{L}_{CC}^{-1}\mathcal{L}_{CB}$ of the Kirchhoff problem. Hence the map $\bar{J}_C \mapsto -\bar{J}_B$ is *dual* to the map $\psi_B^* \mapsto \psi_C^*$. Thus

$$\bar{J}_C^T\psi_C^* = -\bar{J}_B^T\psi_B^* \tag{4.16}$$

for any $\psi_B^*, \bar{J}_C$ and corresponding $\psi_C^*, \bar{J}_B$. In particular the induced 1-norm of $\mathcal{L}_{BC}\mathcal{L}_{CC}^{-1}$ is equal to 1, i.e.,

$$\|\mathcal{L}_{BC}\mathcal{L}_{CC}^{-1}\|_1 = 1\,. \tag{4.17}$$

This follows from considering $\psi_B^* = \mathbb{1}$ and $\psi_C^* = \mathbb{1}$, and computing

$$\|\bar{J}_B\|_1 = |\bar{J}_B^T\mathbb{1}| = |\bar{J}_B^T\psi_B^*| = |\bar{J}_C^T\psi_C^*| = \|\bar{J}_C\|_1 \tag{4.18}$$

for all $\bar{J}_C$ and corresponding $\bar{J}_B$. This is summarized in the following proposition.

Proposition 4.4. *The map $\bar{J}_C \mapsto -\bar{J}_B$ given by* (4.14) *has induced* 1*-norm equal to* 1*. It is dual to the map $\psi_B^* \mapsto \psi_C^*$ given by* (4.10).

4.3. Combining Kirchhoff's problem and its dual

We can *combine* Kirchhoff's problem and its dual. Prescribe ψ_B^* and $\bar{J}_C$. Then

$$(\psi_B^*, \psi_C^* + \bar{\psi}_C, J_B^* + \bar{J}_B, \bar{J}_C) \tag{4.19}$$

with ψ_C^* given by (4.10), and $\bar{J}_B$ given by (4.14), is the unique solution of the resulting flow equation. Furthermore, this solution is the minimizer of

$$\min_{\psi_C} \begin{bmatrix} \psi_B^{*T} & \psi_C^T \end{bmatrix} \mathcal{L} \begin{bmatrix} \psi_B^* \\ \psi_C \end{bmatrix} - 2\psi_C^T\bar{J}_C. \tag{4.20}$$

This last fact follows by noting that the gradient vector (with respect to ψ_C) of this expression is given by $2(J_C - \bar{J}_C)$.

4.4. The prescribed power problem

A third version of the flow problem for linear resistive networks is the following. Next to the voltage potentials ψ_B^* at the boundary nodes prescribe the *power* $P_j = \psi_j J_j$ at each of the connection nodes $j = N_B + 1, \ldots, N_B + N_C = N$. Defining for every vector $z \in \mathbb{R}^n$ the $n \times n$ matrix $[z]$ as the diagonal matrix with diagonal elements $z_1, \ldots, z_n$, the problem is now to find ψ_C satisfying

$$\bar{P}_C = [\psi_C]J_C = [\psi_C](\mathcal{L}_{CC}\psi_C + \mathcal{L}_{CB}\psi_B^*) = [\psi_C]\mathcal{L}_{CC}(\psi_C - \psi_C^*). \tag{4.21}$$

Being quadratic, this equation has generally multiple solutions.

In terms of the deviation $\psi_C - \psi_C^*$ with respect to the open-circuit load voltage potentials ψ_C^* equation (4.21) can be further rewritten as

$$[\psi_C - \psi_C^*]\mathcal{L}_{CC}(\psi_C - \psi_C^*) + [\psi_C^*]\mathcal{L}_{CC}(\psi_C - \psi_C^*) - \bar{P}_C = 0. \tag{4.22}$$

Furthermore, we can *scale*, see also [23], the equation (4.22) by defining the column vector $x \in \mathbb{R}^{N_C}$ with ith element given as

$$x_j := \frac{\psi_j - \psi_j^*}{\psi_j^*}, \quad j = N_B + 1, \ldots, N_B + N_C = N. \tag{4.23}$$

Then equation (4.22) corresponds to

$$\left([x]\mathcal{L}_{CC}[x] - \mathcal{L}_{CC}[x] + [\psi_C^*]^{-1}[\bar{P}_C][\psi_B^*]^{-1}\right)\mathbb{1} = 0 \tag{4.24}$$

in the unknown diagonal matrix $[x]$. This equation bears obvious similarities with a matrix *Riccati equation*, but its properties are not well understood.

The *complex* version of this problem, of much interest for *electrical power networks*, will be indicated in Section 6.

5. The discrete version of Calderón's inverse problem

The map $J_B = \mathcal{L}_S \psi_B$ from boundary nodal voltage potentials to boundary nodal currents can be regarded as the *discrete analog* of the Dirichlet-to-Neumann (DtN) map used, e.g., in tomography. This map arises from considering the partial differential equation

$$\nabla \cdot (\gamma \nabla u) = 0 \ \text{ in } \Omega, \tag{5.1}$$

in the unknown function $u : \Omega \to \mathbb{R}$, with Ω some bounded domain in $\mathbb{R}^n$, together with the Dirichlet boundary conditions

$$u \mid_{\partial\Omega} = v, \tag{5.2}$$

where v is a prescribed voltage potential function on the boundary $\partial\Omega$ of the domain Ω. Here the function $\gamma : \Omega \to \mathbb{R}$ denotes a conductance function. The Neumann boundary variables $j := (\gamma \nabla u \mid_{\partial\Omega}) \cdot n$, with n the normal to the boundary $\partial\Omega$, are equal to the *boundary currents*. Thus the DtN map $v \mapsto j$ maps boundary voltage potentials to boundary currents. The correspondence to the discrete setting is provided by identifying v with ψ_B, j with J_B, the vector $(g_1, \ldots, g_M)$ with γ, and the DtN map with the boundary map $J_B = \mathcal{L}_S \psi_B$.

The inverse problem studied by Calderón [6] concerns the *identifiability* of the conductance function γ from the knowledge of the DtN map $v \mapsto j$. Surprisingly, under rather general conditions, the conductance function γ is indeed uniquely determined by the DtN map; see, e.g., [4] for a review of Calderón's problem.

In the discrete version Calderón's inverse problem amounts to the question when and how the conductances $g := (g_1, \ldots, g_M)$ of the resistors in the resistive network are uniquely determined by the knowledge of the map $\mathcal{L}_S$. Here, similar to the continuous case, it is throughout assumed that the incidence matrix $\mathcal{D}$ of the circuit graph is *known*. The discrete version of Calderón's problem has been studied in a number of papers; see in particular [7, 9–11]. Contrary to the continuous case, the values of the conductances are in general *not* uniquely determined by the boundary map $J_B = \mathcal{L}_S \psi_B$ for *arbitrary* circuit graphs. It *is* true for electrical networks with specific topology, such as the rectangular graphs studied in [9, 10] and the circular graphs studied in [7, 11].

Key notion in proving the identifiability of the conductance parameters g_1, $\ldots$, g_M from the boundary map $J_B = \mathcal{L}_S \psi_B$ is the following approach, which is directly extending the approach in the continuous case taken in the original paper by Calderón [6]; see also [4]. Consider for a given network graph with incidence matrix $\mathcal{D}$ the map T from the vector $g := (g_1, \ldots, g_M)$ of conductances to $\mathcal{L}_S$, or equivalently the map from g to the quadratic form defined by $\mathcal{L}_S$ given by

$$T : g \mapsto Q_g, \quad Q_g(\psi_B) := \psi_B^T \mathcal{L}_S \psi_B \tag{5.3}$$

Reconstructibility of g from the boundary map $J_B = \mathcal{L}_S \psi_B$ is thus equivalent to *injectivity* of the map T. This implies the following *necessary* condition for reconstructibility. Since the $N_B \times N_B$ Laplacian matrix $\mathcal{L}_S$ is symmetric and has column and row sums zero, it follows that the dimension of the set of all Laplacian matrices $\mathcal{L}_S$ is equal to $\frac{N_B(N_B-1)}{2}$. Hence a necessary condition for invertibility of the map $g \mapsto \mathcal{L}_S$, and thus for reconstructibility of g, is that

$$M \leq \frac{N_B(N_B-1)}{2}. \tag{5.4}$$

In order to derive a *sufficient* condition let us recall the equality

$$Q_g(\psi_B) = \begin{bmatrix}\psi_B^T & \psi_C^T\end{bmatrix} \mathcal{L} \begin{bmatrix}\psi_B \\ \psi_C\end{bmatrix}, \text{ with } \psi_C \text{ s.t. } J_C = 0. \tag{5.5}$$

Then the differential of T at g in the direction of a vector $\kappa \in \mathbb{R}^M$ is easily seen to be given by the quadratic form

$$(dT(g)(\kappa))(\psi_B) = \begin{bmatrix}\psi_B^T & \psi_C^T\end{bmatrix} \mathcal{L}^\kappa \begin{bmatrix}\psi_B \\ \psi_C\end{bmatrix}, \text{ with } \psi_C \text{ s.t. } J_C = 0, \tag{5.6}$$

where $\mathcal{L}^\kappa := \mathcal{D}[\kappa]\mathcal{D}^T$. Hence the differential $dT(g)$ is injective whenever

$$\begin{bmatrix}\psi_B^T & \psi_C^T\end{bmatrix} \mathcal{L}^\kappa \begin{bmatrix}\psi_B \\ \psi_C\end{bmatrix} = 0 \text{ for all } \psi_B, \psi_C \text{ s.t. } J_C = 0 \tag{5.7}$$

implies that $\kappa = 0$. Indeed, for rectangular[6] resistive networks as defined in [9, 10], this holds; thus showing reconstructibility of g. All this is summarized in the following proposition.

Proposition 5.1. *Identifiability of he conductance parameters g is equivalent to injectivity of the map T defined in* (5.3). *A necessary condition for this is* (5.4). *A sufficient condition is* (5.7) *implying* $\kappa = 0$.

The problem of characterizing all graph topologies (i.e., incidence matrices $\mathcal{D}$) for which the reconstructibility property holds is an open problem.

Let us finally mention that instead of the discrete version of the DtN map, we may also consider the discrete version of the Neumann-to-Dirichlet map, implicitly given by

$$J_B \mapsto \psi_B, \quad J_B = \mathcal{L}_S \psi_B, \ \mathbb{1}^T J_B = 0. \tag{5.8}$$

(Note that indeed the inverse of the map $\mathcal{L}_S$ is well defined on the subspace of all boundary nodal currents J_B satisfying $\mathbb{1}^T J_B = 0$.) Identifiability of g is the same as before.

[6]Reconstructibility for circular planar graphs was proved in [11] using different methods.

6. RLC electrical networks

The previous framework for linear resistive networks can be directly extended to the *steady-state behavior* of linear RLC electrical networks. Indeed, the steady-state behavior of a linear capacitor given by $C\dot{V} = I$ in the frequency-domain is given by the complex relation

$$I = j\omega CV, \quad I, V \in \mathbb{C}, \tag{6.1}$$

with ω the frequency. Similarly, a linear inductor $L\dot{I} = V$ is described in the frequency-domain as

$$I = \frac{1}{j\omega L}V, \quad I, V \in \mathbb{C} \tag{6.2}$$

Thus the Laplacian matrix $\mathcal{L}_{\mathbb{C}}$ of the steady-state behavior of an RLC-network with incidence matrix $\mathcal{D}$ is given by

$$\mathcal{L}_{\mathbb{C}} = DG_{\mathbb{C}}\mathcal{D}^T, \tag{6.3}$$

where $G_{\mathbb{C}}$ is the complex diagonal matrix with diagonal elements determined by the corresponding edges: real conductances in case of resistors, imaginary numbers $j\omega C$ in case of capacitors, and imaginary numbers $\frac{1}{j\omega L}$ in case of inductors. Conversely, similar to Proposition 3.1, it can be seen that any symmetric complex matrix $\mathcal{L}_{\mathbb{C}}$ with row and column sums zero, and with off-diagonal elements equal to either $-g$, $-j\omega C$, $-\frac{1}{j\omega L}$, where g, C, L are positive constants, corresponds to an RLC circuit. Indeed, if the (i,j)th element of $\mathcal{L}$ is equal to one of these expressions, then the edge between node i and j corresponds to, respectively, a resistor, capacitor, or inductor. On the other hand, taking the Schur complement of a complex Laplacian matrix $\mathcal{L}_{\mathbb{C}}$ in general leads to a matrix of a more general type; see [27] for some illuminating observations.

A particular case, which has received much attention motivated by direct applications in power networks, is the case where all edges correspond to inductors (*inductive transmission lines*); see, e.g., [5, 23–25] and the references quoted therein. In the notation of this paper, the *load flow problem* is to prescribe the vector of *complex powers* $[\psi_C]\bar{J}_C$ (with $\bar{\ }$ denoting complex conjugate) at the connection nodes (now called the *load buses*), as well as the *real part* of the complex powers $[\psi_B]\bar{J}_B$ at the boundary nodes (now called the *generator buses*), together with the *angles* of the complex voltage potentials ψ_B. The real part of a complex power $\psi_k\bar{J}_k$ is commonly called the *active power*, and the imaginary part the *reactive power*.

7. Conclusions

Resistive electrical networks constitute a beautiful example of open, interconnected, large-scale systems, giving rise to an elegant classical mathematical theory, still posing open problems and suggesting important extensions.

References

[1] J.A. Ball, A.J. van der Schaft, "J-inner-outer factorization, J-spectral factorization, and robust control for nonlinear systems", *IEEE Trans. Autom. Contr.*, 41, pp. 379–392, 1996.

[2] J.A. Ball, M.A. Petersen, A.J. van der Schaft, "Inner-outer factorization for nonlinear noninvertible systems", *IEEE Trans. Autom. Contr.*, 49, pp. 483–492, 2004.

[3] B. Bollobas, *Modern Graph Theory*, Graduate Texts in Mathematics 184, Springer, New York, 1998.

[4] L. Borcea, "Electrical impedance tomography", Topical review *Inverse Problems*, 18, R99–R136, 2002.

[5] S. Bolognani, S. Zampieri. On the existence and linear approximation of the power flow solution in power distribution networks. *IEEE Transactions on Power Systems*, 31(1):163–172, 2016.

[6] A.P. Calderón, "On an inverse boundary value problem", *Seminar on Numerical Analysis and its Applications to Continuum Physics*, Soc. Brasileira de Matématica, Rio de Janeiro, pp. 65–73, 1980.

[7] Y. Colin De Verdière, "Réseaux électriques planaires", *Publ. de l'Institut Fourier*, 225, pp. 1– 20, 1992.

[8] D.E. Crabtree, E.V. Haynsworth, "An identity for the Schur complement of a matrix", *Proc. Amer. Math. Soc.*, 22, pp. 364–366, 1969.

[9] E.B. Curtis, J.A. Morrow, "Determining the resistors in a network", *SIAM J. Appl. Math.*, 50(3), pp. 918–930, 1990.

[10] E.B. Curtis, J.A. Morrow, "The Dirichlet to Neumann map for a resistor network", *SIAM J. Appl. Math.*, 51(4), pp. 1021–1029, 1991.

[11] E.B. Curtis, D. Ingerman, J.A. Morrow, "Circular planar graphs and resistor networks", *Linear Algebra and its Applications*, 283, pp. 115–150, 1998.

[12] F. Dörfler, F. Bullo, "Kron reduction of graphs with applications to electrical networks", *IEEE Transactions on Circuits and Systems I: Regular Papers*, 60(1), 2013.

[13] F. Dörfler, J. Simpson-Porco, F. Bullo, "Electrical networks and algebraic graph theory: models, properties, and applications", *Proceedings of the IEEE*, 106(5):977-1005, 2018.

[14] C. Godsil, G. Royle, *Algebraic graph theory*, Graduate Texts in Mathematics 207, Springer, New York, 2004.

[15] G. Kirchhoff, "Über die Auflösung der Gleichungen, auf welche man bei der Untersuchung der Linearen Verteilung galvanischer Ströme geführt wird", *Ann. Phys. Chem.* 72, pp. 497–508, 1847.

[16] G. Kron, *Tensor Analysis of Networks*, John Wiley & Sons, 1939.

[17] J. Rommes, W.H.A. Schilders, "Efficient methods for large resistor networks", *IEEE Transactions on Computer-Aided Design of Integrated Circuits and Systems*, 29(1):28–39, 2009.

[18] A.J. van der Schaft, "Characterization and partial synthesis of the behavior of resistive circuits at their terminals", *Systems & Control Letters*, 59, pp. 423–428, 2010.

[19] A.J. van der Schaft, "Modeling of physical network systems", *Systems & Control Letters*, 101, pp. 21–27, 2017.

[20] A. van der Schaft, D. Jeltsema, "Port-Hamiltonian Systems Theory: An Introductory Overview," *Foundations and Trends in Systems and Control*, 1(2/3), pp. 173–378, 2014.

[21] A.J. van der Schaft, B.M. Maschke, "Port-Hamiltonian systems on graphs", *SIAM J. Control Optim.*, 51(2), 906–937, 2013.

[22] A.J. van der Schaft, B.M. Maschke, "Conservation Laws and Lumped System Dynamics", in *Model-Based Control; Bridging Rigorous Theory and Advanced Technology*, P.M.J. Van den Hof, C. Scherer, P.S.C. Heuberger, eds., Springer, pp. 31–48, 2009.

[23] J.W. Simpson-Porco, "A theory of solvability for lossless power flow equations–Part I: fixed-point power flow", *IEEE Transactions on Control of Network Systems*, to appear.

[24] J.W. Simpson-Porco, "A theory of solvability for lossless power flow equations – Part II: Conditions for radial networks", *IEEE Transactions on Control of Network Systems*, to appear.

[25] J.W. Simpson-Porco, F. Dörfler, F. Bullo, "Voltage collapse in complex power grids", *Nature communications*, 7(34), 2016.

[26] J.C. Willems, E.I. Verriest, "The behavior of resistive circuits", in *Proc. joint 48th IEEE Conference on Decision and Control and 28th Chinese Control Conference*, Shanghai, China, pp. 8124–8129, 2009.

[27] E.I. Verriest, J.C. Willems, "The behavior of linear time invariant RLC circuits", In *Proc. IEEE Conf. on Decision and Control*, pp. 7754–7758, Atlanta, GA, USA, December 2010.

[28] J.C. Willems, "Terminals and ports", *IEEE Circuits and Systems Magazine*, 10(4): 8–16, 2010.

Arjan van der Schaft
Bernoulli Institute for Mathematics, CS and AI
Jan C. Willems Center for Systems and Control
University of Groningen
PO Box 407
9700AK, Groningen, the Netherlands
e-mail: `A.J.van.der.Schaft@rug.nl`

Operator Theory:
Advances and Applications, Vol. 272, 343–353

Control and the Analysis of Cancer Growth Models

Allen Tannenbaum, Tryphon T. Georgiou,
Joseph O. Deasy and Larry Norton

Dedicated to our colleague and friend Joe Ball on the occasion of his 70th birthday

Abstract. We analyze two dynamical models of cancer growth from a system-theoretic point of view. The first model is based upon stochastic controlled versions of the classical Lotka–Volterra equations. Here we investigate from a controls point of view the utility of employing ultrahigh dose flashes in radiotherapy. The second is based on the Norton–Simon–Massagué growth model that takes into account the heterogeneity of a tumor cell population. We indicate an optimal strategy based on linear quadratic control applied to a linear transformed model. The models and analysis are very preliminary and only give an indication of possible therapies in the treatment of cancer.

Mathematics Subject Classification (2010). Primary 37N25; Secondary 37N25.

Keywords. Cancer growth models, Lotka–Voltera equations, linear-quadratic control.

1. Introduction

In this note, we analyze certain models of cancer growth from a control-theoretic point of view. These include the Lotka–Volterra and Norton–Simon–Massagué models [4–9]. We are interested in formulating applicable control techniques, so as to formulate more effective therapeutic strategies. In particular, we are interested in exploring from a controls perspective the idea of using ultrahigh dose flashes in radiotherapy [4]. To this end, we first study stochastic controlled versions of the classical Lotka–Volterra equations. We next turn to the Norton–Simon–Massagué model [6, 8, 9]. This particular approach takes into account the heterogeneity of a tumor cell population following a Gompertzian-type growth curve. A key

---

This work was supported by the Breast Cancer Research Foundation.

implication is that therapy should be given at reduced intervals to maximize the probability of tumor eradication and to minimize the chances of tumor regrowth. We will study this hypothesis more closely via a controlled version of the Norton–Massagué equation [7].

We will only sketch the basic mathematical theory of the Lotka–Volterra and generalized Gompertzian models used in order to justify our use of certain control laws. A good reference for a more complete mathematical treatment may be found in the text [10] that also has an extensive set of references.

2. Lotka–Volterra and radiotherapy

We consider the following controlled stochastic version for a *competitive Lotka–Volterra* system of equations [10], in order to model the competition between healthy tissue (and/or immune system) on the one hand and cancer cells on the other:

$$\begin{aligned} dN_{\text{health}}(t) &= N_{\text{health}}(t)\left[(a_1(1-N_{\text{health}}(t)/k_1)-b_1N_{\text{cancer}}(t)/k_2\right. && \text{(2.1a)}\\ &\quad \left.-\gamma_1 u(t,x,y))dt+\sigma_1 dw_1(t)\right], \\ dN_{\text{cancer}}(t) &= N_{\text{cancer}}(t)\left[(a_2(1-N_{\text{cancer}}(t)/k_2)-b_2N_{\text{health}}(t)/k_1\right. && \text{(2.1b)}\\ &\quad \left.-\gamma_2 u(t,x,y))dt+\sigma_2 dw_2(t)\right]. \end{aligned}$$

Here, $a_i > 0$, $k_i > 0$, $b_i \geq 0$, and $\gamma_i \geq 0$ for $i \in \{1,2\}$, while $u(x,y,t)$ represents intensity of radiation which constitutes the available control variable. In this model, N_{health} represents the number of healthy cells (or the potency of the immune system) and N_{cancer} the number of cancer cells. Further

$$\sigma = \begin{bmatrix} \sigma_1 \\ \sigma_2 \end{bmatrix}$$

is a 2×2 symmetric positive definite matrix (i.e., σ_i are 1×2 row vectors), and

$$w(t) = \begin{bmatrix} w_1 \\ w_2 \end{bmatrix}$$

is a two-dimensional Brownian motion (Wiener process) that models randomness. The parameters γ_i represent the susceptibility of the respective populations to radiation, the parameters a_i, b_i represent the rates of growth and interaction between the two populations respectively, while k_i, for $i \in \{1,2\}$, represents the saturation value for the respective population. Typically $a_2 >> a_1$ and $k_2 >> k_1$. Without loss of generality we may assume that $\gamma_2 = 1$ and therefore redefine $\gamma_1 =: \gamma$ to simplify notation.

We normalize the above populations by setting

$$x(t) = N_{\text{health}}(t)/k_1, \qquad \text{(2.2a)}$$

$$y(t) = N_{\text{cancer}}(t)/k_2, \qquad \text{(2.2b)}$$

and thereby we re-write the system dynamics as follows:

$$dx(t) = x(t)\left[(a_1(1 - x(t)) - b_1 y(t) - \gamma u(t,x,y))dt + \sigma_1 dw_1(t)\right], \tag{2.3a}$$

$$dy(t) = y(t)\left[(a_2(1 - y(t)) - b_2 x(t) - u(t,x,y))dt + \sigma_2 dw_2(t)\right]. \tag{2.3b}$$

The problem we consider is to choose a control strategy (i.e., shape and duration of the radiation intensity $u(t, x, y)$) so as to reduce the tumor from a given starting point

$$x(0) = x_0 \geq 0, \; y(0) = y_0 \geq 0.$$

We will consider both open loop as well as closed loop control. The latter assumes an online estimate of the effectiveness of radiation.

One particular question that we set to answer is the question of whether, given the total amount of radiation, **it is preferable to deliver the treatment over a wide window of time or a short one with high intensity.**

Interestingly, depending on model parameters, there are cases where cancer always wins out unless ameliorated by control. Those cases require periodic treatment in perpetuity. For other model parameters, there are typically two regions for (x, y). In the first region, where concentration of cancer cells is significant, cancer wins out. In the second, the relative proportion of cancer cells is insignificant and the immune system is seen as capable of eliminating all cancerous cells. Accordingly,

$$\text{Case I:} \quad (x_e, y_e) \in \{(0,1)\} \tag{2.4a}$$

$$\text{Case II:} \quad (x_e, y_e) \in \{(1,0), (0,1)\} \tag{2.4b}$$

are the corresponding points of equilibrium. The corresponding values for N_{health}, N_{cancer} are k_1, k_2, respectively.

2.1. Analysis of system

We first consider and analyze the deterministic and autonomous system:

$$\frac{dx(t)}{dt} = x(t)(a_1(1 - x(t)) - b_1 y(t)), \tag{2.5a}$$

$$\frac{dy(t)}{dt} = y(t)(a_2(1 - y(t)) - b_2 x(t)). \tag{2.5b}$$

In general, there are three possible points (x_e, y_e) of equilibrium,

$$P = (1,0)$$
$$Q = (0,1)$$
$$R = \left(\frac{1 - \frac{b_1}{a_1}}{1 - \frac{b_1}{a_1}\frac{b_2}{a_2}}, \frac{1 - \frac{b_2}{a_2}}{1 - \frac{b_1}{a_1}\frac{b_2}{a_2}} \right).$$

Linearization of the dynamics about these points of equilibrium gives

$$\frac{d}{dt}\begin{pmatrix} x(t) - x_e \\ y(t) - y_e \end{pmatrix} = A \begin{pmatrix} x(t) - x_e \\ y(t) - y_e \end{pmatrix}$$

where A is a 2×2 matrix taking values

$$A_P = \begin{pmatrix} -a_1 & -b_1 \\ 0 & a_2 - b_2 \end{pmatrix}$$
$$A_Q = \begin{pmatrix} a_1 - b_1 & 0 \\ -b_2 & -a_2 \end{pmatrix}$$
$$A_R = \begin{pmatrix} a_1 & 0 \\ 0 & a_2 \end{pmatrix},$$

respectively. Thus, P and Q are stable equilibria provided $a_1 < b_1$ and $a_2 < b_2$, respectively, whereas R is always unstable.

Thus, without control, in this deterministic case, cancer cells will always win the competition when $a_1 < b_1 < 0$, and $a_2 > b_2$. In this case, $N_{\text{health}}(t) \to 0$ and $N_{\text{cancer}}(t) \to k_2$ as $t \to \infty$, or equivalently, $x(t) \to 0$ and $y(t) \to 1$.

2.2. Case I: cancer always wins without control

In the simulation of an academic example below, we chose $k_1 = k_2 = 1$, $a_1 = 0.01$, $a_2 = 0.2$, $b_1 = 0.02$, $b_2 = 0.1$, $x_0 = y_0 = .5$ and control input $u = e^{-x^2-y^2/t^2} y/x$. This gives an impulsive nature to the control.

The control contains a burst and then continues on at some level, otherwise cancer will re-appear. See Figures 1 and 2.

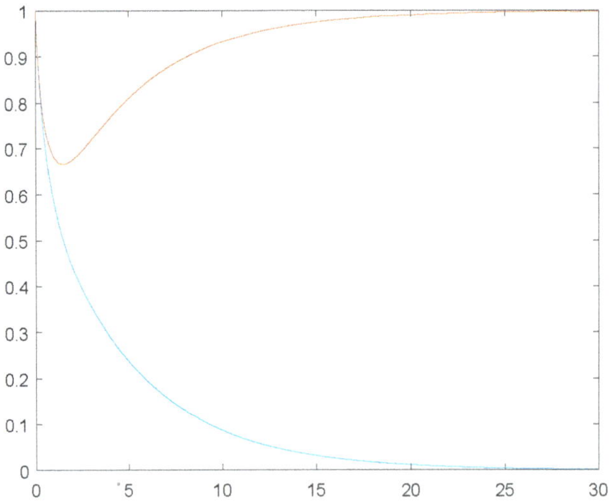

FIGURE 1. Cancer/Normal Cell Concentrations: Time series of Lotka–Volterra system with no control applied. Red represents cancer and blue healthy cell concentrations.

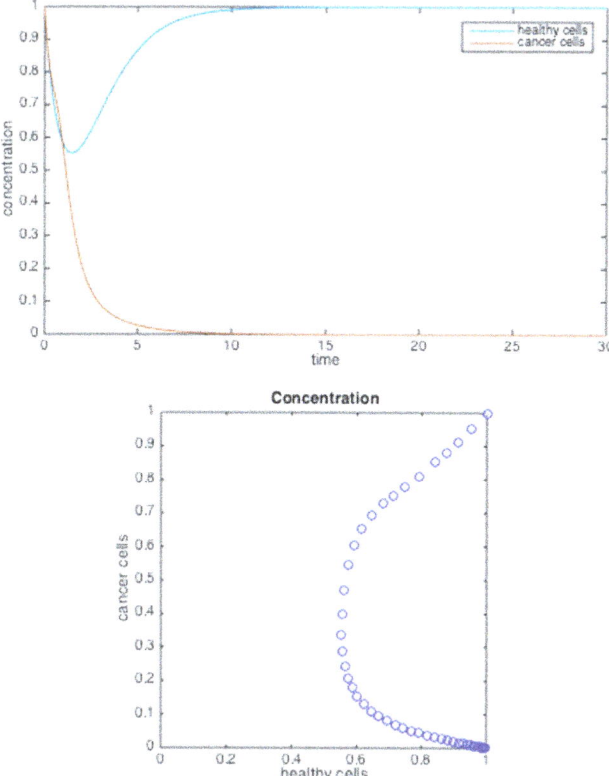

FIGURE 2. Cancer/Normal Cell Concentrations: i) time series, ii) phase plot of Lotka–Volterra system with impulsive control applied. Red represents cancer and blue healthy cell concentrations.

2.3. Case II: if tumor sufficiently shrinks, it is eliminated by the body

In this case, we assume that both P and Q are attractive points of equilibrium and, therefore, proper radiation treatment has a chance to lead the dynamics towards P which represents cure. However, the region in phase space (x, y) where P is an attractive equilibrium and thereby the immune system is capable of eliminating cancer, could be very small. The task then is to steer the system into this region with suitable radiation treatment.

We demonstrate that if a certain amount of radiation is allowable to be delivered (i.e., intensity $\times$ duration $=$ fixed), then it is advantageous to deliver all the radiation in the shortest possible amount of time. Below, we present a numerical example where we assume $a_1 = 1$, $a_2 = 1$, $b_1 = 2$, $b_2 = 1.1$, $\gamma = 0.2$, u constant over an interval T such that $u \times T = 3$. We show phase plots marking the path of the concentrations $(x(t), y(t))$ starting from $(0.7333, 0.3833)$ for a range of values for the intensity u and duration T as follows

duration	15	10	7.5	6	5	4.2857	3.75	3
intensity	0.2	0.3	0.4	0.5	0.6	0.7	0.8	1

We observe that short duration and higher intensity treatment is preferential since it is more effective in steering the state into the region of attraction of the point of equilibrium P representing cure. See Figures 3 and 4.

3. Norton–Massagué dynamics: system identification and control

A key hypothesis in the Norton–Simon–Massagué models [6–9], which is supported by experimental evidence, is that small tumors grow faster, but they are also more susceptible to treatment. The accentuated effect of small-size in growth rate is captured by a fractional power ¡1 in population dynamics, as in

$$\frac{dN(t)}{dt} = kN^{b/3} - hN. \tag{3.1}$$

We will refer to this as the *Norton–Massagué equation.* There are several simple observations to make. First, small b gives a rather high slope near $N = 0$. In this case, N will not go to infinity. When $b > 3$ the system will blow up, but for small N, linear dissipation dominates.

We propose to add a control aspect to the model. For example, the effect of control u (effect of radiation or drugs) on such an evolution can be modeled via

$$\frac{dN(t)}{dt} = k(1 + u(t))N^{b/3} - hN. \tag{3.2}$$

The choice of u specifies the control strategy. When optimizing a quadratic cost, the optimal control is typically in the form of state feedback, i.e., a function of $N(t)$.

We point out that the Norton–Massagué equation may be solved in closed form, namely,

$$N(t) = N(0)\left[\frac{k}{h}N(0)^{b/3-1} + e^{(h(\frac{b}{3}-1)t)}(1 - \frac{k}{h}N(0)^{b/3-1})\right]^{\frac{3}{b-3}},$$

as may be checked by direct differentiation. Indeed, the Norton–Massagué equation is a special case of a Bernoulli equation [2], i.e., an equation of the form

$$\frac{dy}{dt} + p(t)y = q(t)y^n. \tag{3.3}$$

In the Norton–Massagué case, $y(t) = N(t)$, $p(t) = h$, $q(t) = k$, and $n = b/3$. Specifically, via a simple transformation equation (3.3) may be converted into a linear equation. We set $y' = dy/dt$. Now divide both sides of (3.3) by y^n to get

$$y^{-n}y' + py^{1-n} = q.$$

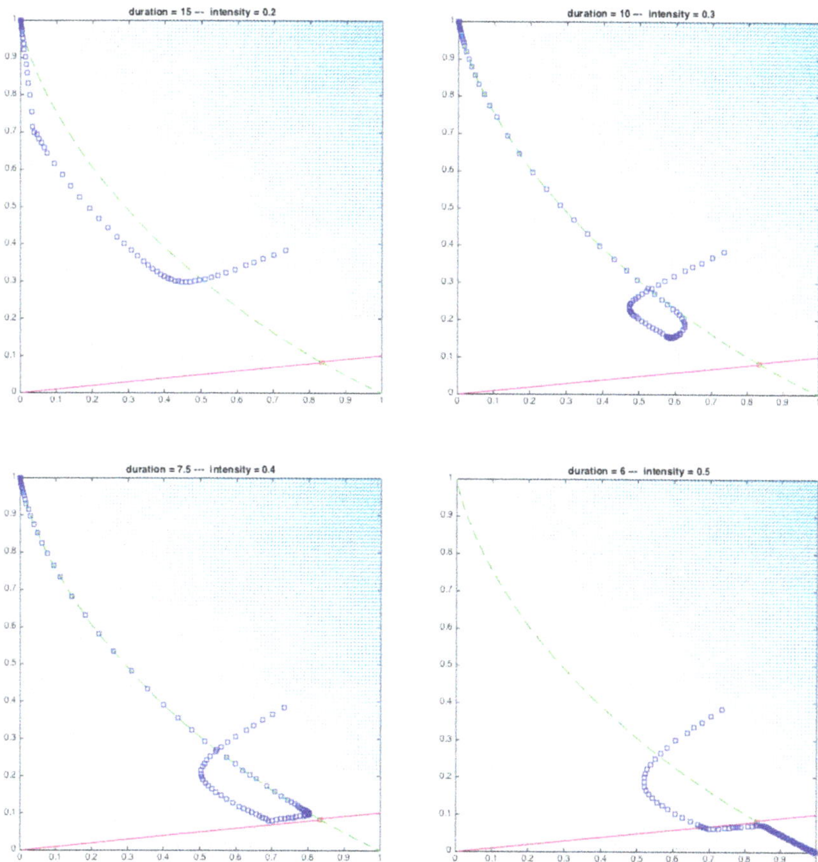

FIGURE 3. Cancel/Normal Cell phase diagram (longer duration/weak intensity): In these simulations, we start near the unstable equilibrium. We plot both the forward-in-time (green dashed) and time-reversed (magenta) trajectories of the Lotka–Volterra system that intersect at the unstable equilibrium. Note that for very small changes in the initial condition, the forward trajectory can either go the cancerous equilibrium point $(0, 1)$ or to the point that represents cure, in which the cancer is eradicated $(1, 0)$. Here we added a control of relatively weak intensity and longer duration. Note that the system does not converge to the healthy point $(1, 0)$ in the first three plots (blue curves). Only when we increase the intensity beyond a certaint threshhold is the system driven to $(1, 0)$. In Figure 4 we continue this analysis, with short duration inputs of higher intensity.

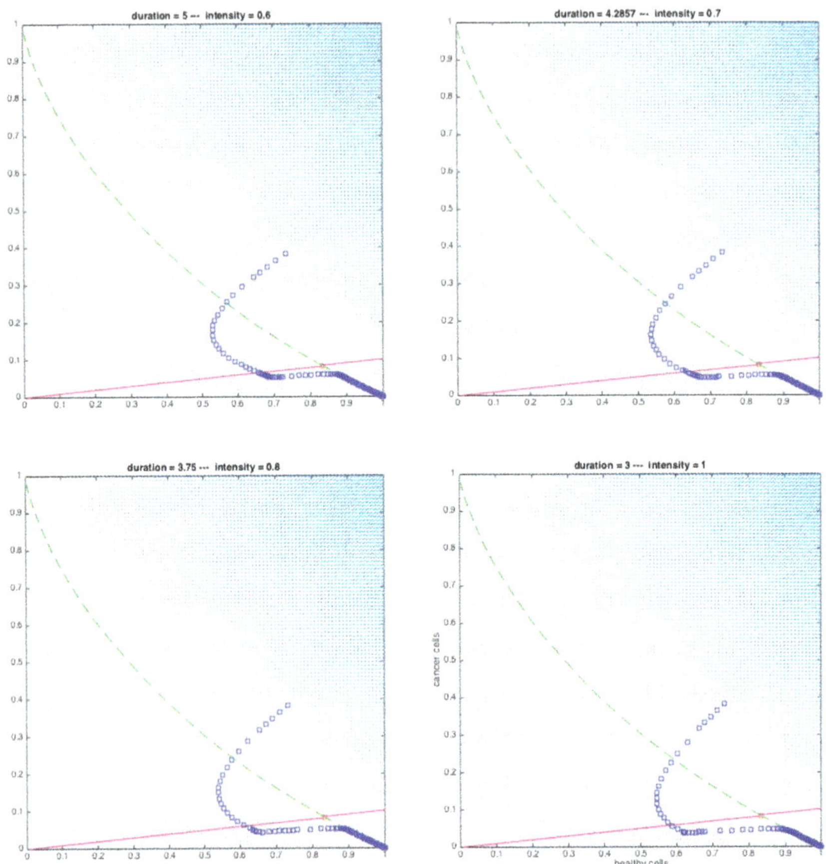

FIGURE 4. Cancer/Normal Cell phase diagram (short duration/high intensity): In these simulations, we start at the unstable equilibrium point. We plot both the forward (green dashed) and time-reversed (magenta) trajectories of the Lotka–Volterra system that intersect at the unstable equilibrium. Note that for very small changes in the initial condition, the forward trajectory can either go the cancerous equilibrium point $(0,1)$ or the "healthy point" in which the cancer is eradicated $(1,0)$. Adding control of a sufficiently high intensity and short duration drives the system to the healthy equilibrium $(1,0)$ (blue curves).

Set $v := y^{1-n}$, so that $v' = (1-n)y^{-n}y'$. Therefore, (3.3) becomes

$$\frac{1}{1-n}v' + p(t)v = q(t), \tag{3.4}$$

which is linear. This linear model motivates our choice of the control strategy for the Norton–Massagué equation explained next.

3.1. Optimal control

We now consider a linear quadratic regulator performance index [1] for the evolution according to the Norton–Massagué equation. Set $n(t) := N(t)^c, c := 1 - b/3$. Then $n(t)$ satisfies the equation

$$n'(t) = -chn(t) + ck.$$

We modify this equation by adding a control $u(t)$

$$n'(t) = -chn(t) + cku(t). \tag{3.5}$$

A suitable performance index to be minimized over choice of control-input in equation (3.5) is:

$$P(u) := fn^2(T) + \int_0^T gn^2(t) + hu^2(t)dt, \tag{3.6}$$

where the time T is the therapeutic treatment horizon, and $f, g, h > 0$ are parameters to be chosen. Thus, we seek to minimize $P(u)$ over all piecewise continuous functions $u : [0, T] \to [0, u_{\max}]$ subject to the dynamical constraint given by equation (3.5). Here g weights the number of cancer cells remaining at the end of treatment, and h scales the running cost that measures the tumor volume. It is well known that the optimal control has the form $u(t) = -\alpha(t)n(t)$ where α may be found via solving a Riccati equation [1]. When transformed back to the original Norton–Massagué setting, this leads to a control law of the form $u(t) = -\alpha(t)N(t)^c$.

3.2. State feedback

For the controlled Norton–Massagué equation (3.2) and for time-varying $u(t)$, we get the following explicit solution:

$$N(t) = e^{-ht}\left[\left(N(0)^{\frac{3-b}{3}} + (\frac{3-b}{3})\int_0^t a(\tau)e^{(\frac{3-b}{3})h\tau}d\tau\right)\right]^{\frac{3}{3-b}},$$

where $a(t) := k(1 + u(t))$.

3.3. Numerical example

We illustrate the use of state feedback in which the state of the system is used to close the feedback loop by displaying responses of the (controlled and uncontrolled) system in Figure 5 below. In a realistic scenario, one would need an estimator as well since the signals would be corrupted by noise. Here we show the following: we start with a small tumor ($N(0) = 1$) and take $b = 2.5$ in the Norton–Massagué equation. As expected, we get an initial sharp increase, and then much slower growth. We then assume that the exponent b becomes 3.1, and as expected without control, one gets rapid exponential growth. In the final plot, we show what what happens if we add feedback when $b = 3.1$. The tumor growth is controlled. Note that we use the approximate steady state of the Norton–Massagué equation for $b = 2.5$ as the input to the controlled and uncontrolled models.

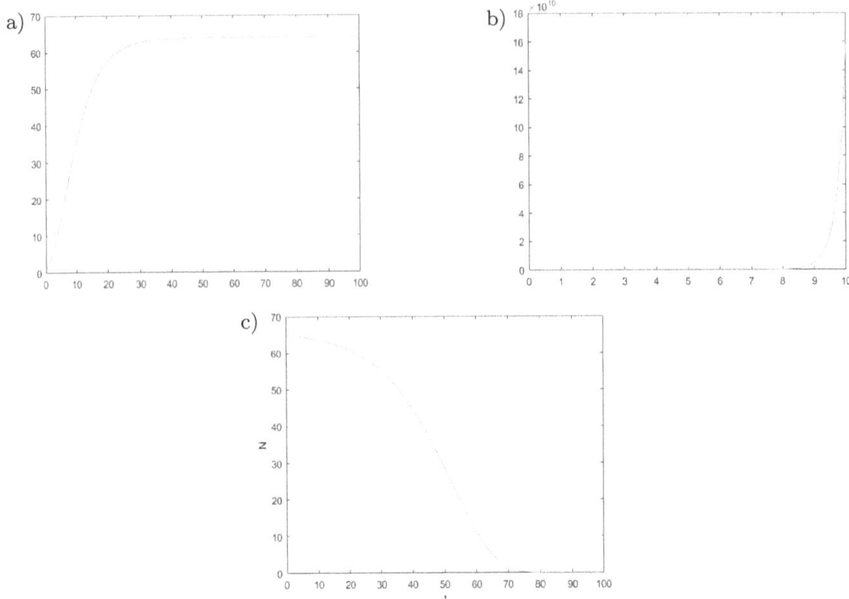

FIGURE 5. Top Row: Growth via the Norton–Massagué equation with $b = 2.5$ and $b = 3.1$ without control; bottom figure: $b = 3.1$ with state feedback.

4. Conclusions

This note sets down some very basic ideas from systems and control that aim at developing more effective cancer therapies. In particular, we anticipate that system identification backed up by laboratory experiments will lead to discovering and tuning more precisely the relevant parameters in dynamical evolution equations of cancer growth model, and of the Norton–Massagué equation in particular. It will be essential to study the effect of noise and uncertainties in such models, and study the behavior of corresponding stochastic equations. Research along this path is underway at Sloan Kettering that will make use of CT lung tumor data before and after radiotherapy. As such, our analysis in the present note is quite preliminary and much deeper work will be necessary in order to have a meaningful impact on cancer treatments.

Finally, it is important to note that, for the Norton–Massagué model, we have only discussed controls that are optimal with respect to a "linear-quadratic" performance index. It is imperative to consider robust control design methodologies as well that would, in addition, ensure small sensitivity to parameter and modeling uncertainly [3] as well as stochastic excitation.

References

[1] B. Anderson and J. Moore, *Optimal Control: Linear Quadratic Methods*, Dover, 2007.

[2] W. Boyce and R. DiPrima, *Elementary Differential Equations and Boundary Value Problems, 9th Edition*, Wiley, NY, 2009.

[3] J. Doyle, B. Francis, and A. Tannenbaum, *Feedback Control Theory*, Macmillan, NY, 1990.

[4] V. Favaudon, L. Caplier, V. Monceau, F. Pouzoulet, M. Sayarath, C. Fouillade, M.-F. Poupon, I. Brito, P. Hupé, J. Bourhis *et al.*, "Ultrahigh dose-rate flash irradiation increases the differential response between normal and tumor tissue in mice," *Science translational medicine*, vol. 6, no. 245, pp. 245ra93–245ra93, 2014.

[5] J.K. Kim and T.L. Jackson, "Mechanisms that enhance sustainability of p. 53 pulses," 2013.

[6] L. Norton, "A Gompertzian model of human breast cancer growth," *Cancer Research* **48** (1988), pp. 7067–7071.

[7] L. Norton and J. Massagué, "Is cancer a disease of self-seeding?," *Nature Medicine* **12(8)** (2006), pp. 875–878.

[8] L. Norton and R. Simon, "Growth curve of an experimental solid tumor following radiotherapy," *J. Natl. Cancer Inst.* **58** (1977), pp. 1735–1741.

[9] L. Norton and R. Simon, "Tumor size, sensitivity to therapy and the design of treatment schedules," *Cancer Treat Rep.* **61** (1977), pp. 1307–1317.

[10] D. Wodarz and N. Komarova, *Dynamics of Cancer: Mathematical Foundations of Oncology*, World Scientific, 2014.

Allen Tannenbaum
Departments of Computer Science
and Applied Mathematics & Statistics
Stony Brook University
Stony Brook, NY 11794, USA
e.mail:
allen.tannenbaum@stonybrook.edu

Tryphon T. Georgiou
Department of Mechanical
& Aerospace Engineering
University of Calfornia
Irvine, CA 92697-3975, USA
e-mail: tryphon@uci.edu

Joseph O. Deasy
Department of Medical Physics
Memorial Sloan Kettering Cancer Center
New York City, NY 10021, USA
e-mail:deasyj@mskcc.edu

Larry Norton
Department of Medicine
Memorial Sloan Kettering Cancer Center
New York City, NY 10021, USA
e-mail: nortonl@mskcc.edu

GPSR Compliance
The European Union's (EU) General Product Safety Regulation (GPSR) is a set of rules that requires consumer products to be safe and our obligations to ensure this.

If you have any concerns about our products, you can contact us on

ProductSafety@springernature.com

In case Publisher is established outside the EU, the EU authorized representative is:

Springer Nature Customer Service Center GmbH
Europaplatz 3
69115 Heidelberg, Germany

www.ingramcontent.com/pod-product-compliance
Ingram Content Group UK Ltd.
Pitfield, Milton Keynes, MK11 3LW, UK
UKHW021919270726
14059UKWH00002B/97
9783030116132